T0145243

# Industrial and Applied Mathematics

The **Industrial and Applied Mathematics** series publishes high-quality research-level monographs, lecture notes, textbooks, contributed volumes, focusing on areas where mathematics is used in a fundamental way, such as industrial mathematics, bio-mathematics, financial mathematics, applied statistics, operations research and computer science.

Harish Garg
Editor

# Advances in Reliability, Failure and Risk Analysis

*Editor*
Harish Garg
School of Mathematics
Thapar Institute of Engineering
and Technology
Patiala, Punjab, India

ISSN 2364-6837 ISSN 2364-6845 (electronic)
Industrial and Applied Mathematics
ISBN 978-981-19-9911-6 ISBN 978-981-19-9909-3 (eBook)
https://doi.org/10.1007/978-981-19-9909-3



This Springer imprint is published by the registered company Springer Nature Singapore Pte Ltd.
The registered company address is: 152 Beach Road, #21-01/04 Gateway East, Singapore 189721, Singapore

# Preface

Reliability is a very crucial concept in a multifaceted term used in several fields from simple product design to complex industrial system such as nuclear power plant and aerospace engineering. The disseminate operation of every subsystem/component of an industrial system is ensured by its Reliability, Availability and Maintainability (RAM) aspects. The ability of equipment to work without failure is the focus of reliability engineering, a sub-discipline of systems engineering. However, with the growing complexity in the systems day-by-day, a failure analysis is an important and challenging aspect of the study of complex systems. A system is defined to be consisting of components, sub-systems, inputs and outputs within system boundaries. The inputs provide physical resources and information to the sub-systems, which are interacting among each other to produce some outputs. All interactions are assumed to take place within the system boundaries. A complex system can be defined as a system structure that is composed of usually a large number of components that have complex interactions. Any failure in performing the required interactions among the system components, or any failure in getting the expected output/result, is considered to be contributing to the system failure. Thus, analysis of a system with its components is a crucial step in determining the difficulties and complexities that the system will experience at any stage.

To increase the performance and life of the system, there is also a need to maintain the schedule of risk analysis at a regular interval of time. Risk and reliability analysis methods can allow for a systematic assessment of these uncertainties, supporting decisions integrating associated consequences in case of unexpected events. Risk is the effect of uncertainty on objectives and the associated likelihood of their occurrence. The development of risk estimates or the determination of risks in a given context is called risk analysis, while risk assessment is the process of evaluating the risks and determining the best course of action. Since, uncertainty is an intrinsic property of risk and is present in all aspects of risk management including risk analysis and risk assessment.

Generally, risk analysis is a systematic tool that facilitates the identification of the weak elements of a complex system and the hazards that mainly contribute to the risk. One of the significant advantages of a coordinated risk analysis approach is that safety, operational, and financial execution are frequently connected.

These days to support the plant design and accidental situations in any industrial plant, a concept of risk tool is considered as a necessity of expertise to tackle any such complicated and multidisciplinary issues in the systems. This entails the acquisition of appropriate reliability modelling and risk analysis tools to complement the basic and specific engineering knowledge for the technological area of application.

The book aims to help the managers and technical specialists with the design and implementation of reliability and risk programs under the following disciplines:

- System safety and risk informed asset management
- Proper strategy to maintain the mechanical components of the systems
- Schedule the proper actions throughout the product life cycle
- To understand the structure and cost of the complex systems
- Plan a proper schedule to improve the reliability and life of the systems
- Identify some unwanted failure and set up their associated preventive and correction actions

The book is useful to not only industry professionals but also for academicians, researchers and scholars. This book publishes the articles dealing with real industrial problems related to uncertainties and vagueness in the expert domain of knowledge. Further, this book also provides the knowledge related to applications of various mathematical and statistical tools in these areas. The results so presented in different chapters of this book will definitely open the gate for researchers and scientists in terms of handling complicated projects in the same domain. This will help in overcoming the various societal problems which the world is facing at presently.

Patiala, Punjab, India Harish Garg

# Contents

# Editor and Contributors

## About the Editor

**Harish Garg** is Associate Professor at the School of Mathematics, Thapar Institute of Engineering and Technology, Patiala, India. He completed his Ph.D. in mathematics from the Indian Institute of Technology Roorkee, India, in 2013. His research interests include computational intelligence, reliability analysis, multi-criteria decision making, evolutionary algorithms, expert systems and decision support systems, computing with words and soft computing. With more than 425 papers published in international journals of repute, he has supervised seven Ph.D. students. He is ranked in the world's top 2% scientists list and rank #1 in India and world rank #229 published by Stanford University in the consecutive three years 2020–2022. He is the recipient of the Obada Prize 2022—Young Distinguished Researchers. He is recipient of the Top-Cited Paper by India-based Author (2015–2019). He also serves as an advisory board member of the Universal Scientific Education and Research Network (USERN). Editor of eight books, Dr. Garg serves as Founding Editor-in-Chief for the *Journal of Computational and Cognitive Engineering*, and Associate Editor for several renowned journals including the *CAAI Transactions on Intelligence Technology*, *Soft Computing*, *Alexandria Engineering Journal*, and *Complex and Intelligent Systems*. His Google citations are over 18820 with H-index 77.

## Contributors

**Fatemeh Afsharnia** Department of Agricultural Machinery and Mechanization Engineering, Agricultural Sciences and Natural Resources University of Khuzestan, Ahvaz, Iran;
Department of Biosystems Engineering, Faculty of Agriculture, Ferdowsi University of Mashhad, Mashhad, Iran;
Department of Mechanical Engineering, Sharif University of Technology, Tehran, Iran

**Irfan Ali** Department of Statistics and Operations Research, Aligarh Muslim University, Aligarh, Uttar Pradesh, India

**Hesam Addin Arghand** Department of Mechanical Engineering, University of Zanjan, Zanjan, Iran

**Mehdi Behzad** Department of Mechanical Engineering, Sharif University of Technology, Tehran, Iran

**Nabaranjan Bhattacharyee** Department of Mathematics, Sidho-Kanho-Birsha University, Purulia, West Bengal, India

**Emre Caliskan** Department of Industrial Engineering, School of Engineering, Gazi University, Ankara, Turkey

**Seyyed Ahmad Edalatpanah** Department of Applied Mathematics, Ayandegan Institute of Higher Education, Tonekabon, Iran

**Ibrahim Ghafir** Department of Computer Science, University of Bradford, Bradford, UK

**Muhammet Gul** School of Transportation and Logistics, Istanbul University, Istanbul, Turkey

**Rakesh Gupta** Department of Statistics, Chaudhary Charan Singh University, Meerut, Uttar Pradesh, India

**Shubham Gupta** Department of Statistics, Chaudhary Charan Singh University, Meerut, Uttar Pradesh, India;
Department of Electronics and Communication Engineering, Manipal University Jaipur, Rajasthan, India

**Lixian Huang** Department of Materials Science and Engineering, B. John Garrick Institute for the Risk Sciences, University of California, Los Angeles (UCLA), Los Angeles, CA, USA

**Mohsen Imeni** Department of Accounting, Ayandegan Institute of Higher Education, Tonekabon, Iran

**Farhana Islam** Department of Education, Bangabandhu Sheikh Mujibur Rahman Digital University, Kaliakair, Gazipur, Bangladesh

**Abdulkareem Lado Ismail** Department of Mathematics, Kano State College of Education, Kano, Nigeria

**Sohag Kabir** Department of Computer Science, University of Bradford, Bradford, UK

**Mehmet Kayra Karacahan** Department of Chemistry and Chemical Processing Technology, Tunceli Vocational School, Munzur University, Tunceli, Turkey

**Komal** Department of Mathematics, School of Physical Sciences, Doon University, Dehradun, Uttarakhand, India

**Mohit Kumar** Department of Mathematics, Institute of Infrastructure Technology Research and Management (IITRAM), Ahmedabad, Gujarat, India

**Sanat Kumar Mahato** Department of Mathematics, Sidho-Kanho-Birsha University, Purulia, West Bengal, India

**Ali Mosleh** B. John Garrick Institute for the Risk Sciences, University of California, Los Angeles (UCLA), Los Angeles, CA, USA

**Hamed Nozari** Department of Industrial Engineering, Iran University of Science and Technology, Tehran, Iran

**Rajesh Paramanik** Department of Mathematics, Sidho-Kanho-Birsha University, Purulia, West Bengal, India

**Tarannom Parhizkar** B. John Garrick Institute for the Risk Sciences, University of California, Los Angeles (UCLA), Los Angeles, CA, USA

**Saeed Ramezani** Department of Industrial Engineering, Faculty of Engineering, Imam Hossein University, Tehran, Iran

**Haiping Ren** Department of Basic Subjects, Jiangxi University of Science and Technology, Nanchang, China

**Abbas Rohani** Department of Biosystems Engineering, Faculty of Agriculture, Ferdowsi University of Mashhad, Mashhad, Iran

**Hamzeh Soltanali** Department of Biosystems Engineering, Ferdowsi University of Mashhad, Mashhad, Iran;
Department of Industrial Engineering, Faculty of Engineering, Imam Hossein University, Tehran, Iran

**Theresa Stewart** Santa Monica, CA, USA

**Ismail Tukur** Department of Art and Humanities, Kano State Polytechnic, Kano, Nigeria

**Nafisatu Muhammad Usman** School of General Studies, Kano State Polytechnic, Kano, Nigeria

**Mohammad Yazdi** Faculty of Engineering and Applied Science, Memorial University of Newfoundland, St. John's, Newfoundland, Canada;
School of Engineering, Macquarie University, Sydney, Australia

**İbrahim Yilmaz** Department of Industrial Engineering, School of Engineering and Natural Sciences, Ankara Yıldırım Beyazıt University, Ankara, Turkey

**Bin Yin** Department of Basic Subjects, Jiangxi University of Science and Technology, Nanchang, China

**Melih Yucesan** Department of Emergency Aid and Disaster Management, Munzur University, Tunceli, Turkey

**Ibrahim Yusuf** Department of Mathematical Sciences, Bayero University, Kano, Nigeria

**Hui Zhou** School of Mathematics and Computer Science, Yichun University, Yichun, China

# Chapter 1
# Degradation and Failure Mechanisms of Complex Systems: Principles

**Tarannom Parhizkar, Theresa Stewart, Lixian Huang, and Ali Mosleh**

**Abstract** A cyber–physical–human complex system failure prevents the accomplishment of the system's intended function. The failure of a complex system could be a breakdown of any system hardware, human-related factors, application software, or the interaction between these components. Having knowledge about all these three components would allow us to better understand the behavior, interactions, and the associated failure mechanisms of the cyber–physical–human systems as a whole. In this study, degradation mechanisms in these three components are classified and discussed. The main categories are hardware-related degradation mechanisms including mechanical, thermal, chemical, electronic and radiation effects degradation mechanisms. In addition to hardware-related degradation mechanisms, human failure modes, software errors, and the failures due to cyber–physical–human interactions are presented and discussed. This chapter covers the main types of failure mechanisms in complex systems and is beneficial for developing conceptual risk and reliability models for complex systems.

**Keywords** Degradation mechanisms · Cyber–physical–human complex system · Material degradation mechanism · Hardware failure · Electronic failure · Human error · Software failure · Complex systems · Risk · Reliability

T. Parhizkar (✉) · A. Mosleh
B. John Garrick Institute for the Risk Sciences, University of California, Los Angeles (UCLA), 404 Westwood Plaza, Los Angeles, CA 90095, USA
e-mail: tparhizkar@g.ucla.edu

A. Mosleh
e-mail: mosleh@ucla.edu

T. Stewart
2911 Washington Ave., Santa Monica, CA 90403, USA
e-mail: theresa@risksciences.ucla.edu

L. Huang
Department of Materials Science and Engineering, B. John Garrick Institute for the Risk Sciences, University of California, Los Angeles (UCLA), 404 Westwood Plaza, Los Angeles, CA 90095, USA
e-mail: lxhuang@g.ucla.edu



H. Garg (ed.), *Advances in Reliability, Failure and Risk Analysis*, Industrial and Applied Mathematics, https://doi.org/10.1007/978-981-19-9909-3_1

## 1.1 Introduction

Complex system components degrade over time due to different degradation mechanisms [1]. Degraded components can negatively affect the system performance, reduce the system lifetime, and even result in a catastrophic failure [2]. In addition, degradation analysis can be used to assess reliability when few or even no failures are expected in a life test.

Degradation mechanisms in a complex system depend on many different factors such as type of the system, components material, system application, operating and environmental conditions [3–5]. In complex systems, we have cyber–physical–human interactions [6, 7]. A cyber–physical–human complex system is a system that is made of interacting components of software, hardware, and human operators. The three different elements in cyber–physical–human systems are hardware, software and humans that have different failure behavior characteristics. It is important to understand the difference between them and their failure modes [8]. However, more importantly, having knowledge about all these three components would also allow us to better understand the behavior, interactions, and the associated failure mechanisms of the cyber–physical–human systems as a whole.

In this chapter, hardware degradation including mechanical, thermal, chemical degradation mechanisms, degradation of electronic devices, radiation effects degradation mechanisms are introduced and discussed.

In this study, degradation mechanisms are categorized into eight main types. The main categories are hardware-related degradation mechanisms including mechanical, thermal, chemical, electronic and radiation effects degradation mechanisms, discussed in Sects. 1.2 to 1.5, respectively. The mechanical, thermal, chemical degradation mechanisms are classified to wear-out and overstress failures. Wear-out mechanisms are degradation mechanisms that happen gradually in the component and result in system aging and performance deterioration. As time passes, the degradation mechanism will cause exceedance of system threshold and system failure. Overstress mechanisms, on the other hand, are sudden degradation mechanisms. In this type of mechanism, operating and environmental conditions are out of nominal range of system operating conditions and results in system failure. This chapter discusses both wear-out and overstress failure mechanisms of mechanical, thermal, and chemical natures. Due to the complexity in electronics, the degradation and failure mechanisms are introduced based on different physics causes.

In addition to hardware-related degradation mechanisms, we could have software and human errors resulting in system failure that are discussed in Sects. 1.4 and 1.5, respectively. Finally, failure mechanisms due to components interaction are introduced. The interactions result in exchange of matter, energy, force, and/or information, and we can have all the combinations between hardware–software, hardware–human, and the human–software, but also the intersection between all three. In Sect. 1.8, different types of interaction failure mechanisms are presented and discussed. It should be noted that this work is not intended to give the reader a thorough understanding of each individual failure mechanism, but rather is intended to introduce the reader to the types of failure or degradation that may be encountered in a wide variety of applications and environments.

For the purpose of reliability analysis of complex systems, proper understanding of cross-filed degradation and failure mechanisms is essential. In this study, different types of material degradation mechanisms in complex systems are classified and presented. In addition to material degradation, human errors, software failure, and cyber–physical–human interaction failure are discussed that is beneficial for developing conceptual risk and reliability models for complex systems.

## 1.2 Mechanical Degradation and Failure Mechanisms

### *1.2.1 Wear-Out Mechanisms*

Mechanical wear could be defined as a process of progressive removal of material from a solid surface while it is in moving contact with another solid, liquid, or gaseous substance (e.g., [9]). The main types of wear-out mechanisms are presented in Fig. 1.1. In the following sections, different types of mechanical wear-out mechanisms are presented and discussed.

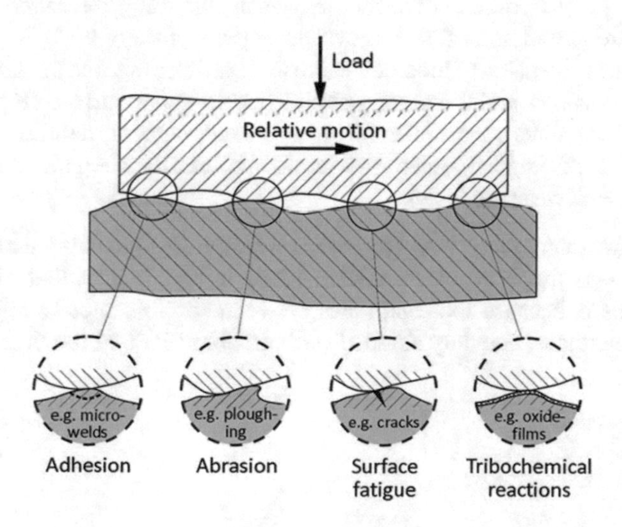


**Fig. 1.1** Four main types of wear-out mechanisms [9]

#### 1.2.1.1 Abrasion

Abrasion wear is one of the most common types of wear and appears across many applications accountings for more than 50% of all wear-related failures in industrial equipment, costing between 1 and 4% of the gross national product of industrialized nations [10]. As a result, abrasion wear resistance is a major consideration in a wide variety of industries and there are at least 17 active ASTM standards directly relating to abrasive wear, and at least 4 relating to erosive wear [11]. The term "abrasion" comprises all groove-forming mechanisms on the surface of a material by microchipping and microplowing.

Abrasive wear can be defined as the removal of material from a surface while it is in relative motion with another contacting surface. This removal can be due to protuberances on the other surface or the presence of hard particles, which may be trapped between the two surfaces or embedded in one of them. In the case where particles are not embedded on either surface, this is called three-body abrasion, while the other cases are referred to as two-body abrasion. Abrasive erosion, which is a special case of erosion where hard particles are present in a moving fluid, is a form of two-body abrasion. Abrasive wear processes are typically classified as:

1. **Two-body abrasion**: This type of abrasion can be further split into two subcategories. In the first, shown in Fig. 1.2a, protuberances or particles attached to body 2 scratch the surface of body 1 as the two bodies move past each other. The second type of two-body abrasion, shown in Fig. 1.2b, describes the scenario where hard particles are free to move along the surface of body 1.
2. **Three-body abrasion**: Three-body abrasive particles are free to slide and/or roll between bodies 1 and 2 and are not attached to either surface (Fig. 1.2c). The origin of the abrasive particles (dust, dirt, sand, aerosol, debris, etc.) is either outside the tribological system (contaminants) or they are generated within the system itself (wear products).

Of these two categories, two-body abrasion results in a much higher rate of material removal—as much as an order of magnitude higher than that of three-body abrasion. This is because loose particles between surfaces mostly roll harmlessly between the surfaces, and only abraded surfaces about 10% of the time [9].

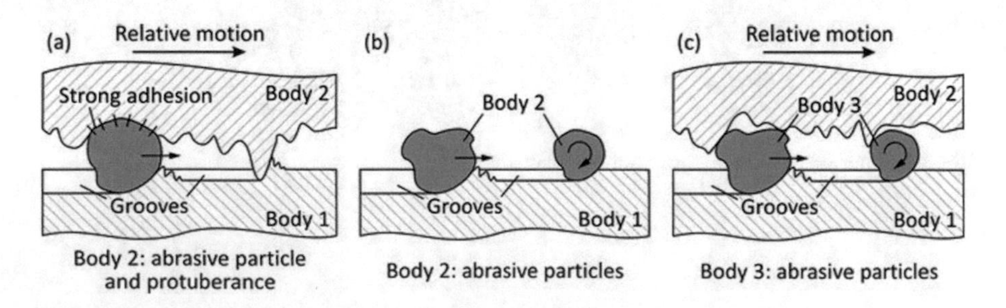


**Fig. 1.2** Two main types of abrasive wear degradation mechanism [9]

#### 1.2.1.2 Adhesion

Adhesive wear is another common type of wear that occurs in systems. It occurs whenever two solid surfaces have rubbing contact and even if all wear mitigation plans have been implemented, this type of wear will remain. As two bodies slide across each other, asperities on each surface may weld together and then either deform or break off as the relative motion continues. Given the exchange of materials, this process is called micro (cold) welding. The real contact area between surfaces is very small. Thus, even with very small loads, there exists pressure between contacted areas that could result in adhesive wear and plastic formation. If two surfaces confront relative sliding motion, the contacted area may break. The break could happen at the original interface or elsewhere, depending on the temperature, material characteristics, and stress distribution. There are different types of adhesive wear including scoring, scuffing, galling, and seizing ordered in terms of increasing severity.

- **Scoring** is the transfer of a small amount of material from one component to another under sliding contact.
- **Scuffing** is a more serious form of scoring and refers to localized surface damage caused by welding of two surfaces.
- **Galling** is cyclic, severe, and large-scale metal transfer between contacting surfaces.
- **Seizing** is a more severe form of galling that becomes so severe that it prevents two surfaces from moving.

Adhesion includes the chemical interaction between the wear materials. Depending on the material properties between two surfaces, a local joining or even welding of both materials may happen. The binding forces may become so high that chips may be pulled out or chipped off from the work material, e.g., the metal debris of the work material may adhere to the ceramic cutting tool. This effect, also known as material transfer, will cause the cutting tool to no longer be in contact with the work material. Figure 1.3 shows several modes to explain the effect of adhesion on materials.

#### 1.2.1.3 Surface Fatigue

This type of wear occurs when two surfaces have rolling contact. Wear here is different from sliding surfaces. As the rolling body rotates, the shear stress ranges from zero to a maximum value, resulting in cyclic stress in the components. This type of wear could result in crack formation and eventual crack growth, leading to fatigue failure. The surface fatigue wear is common in rotatory machines (Fig. 1.4).

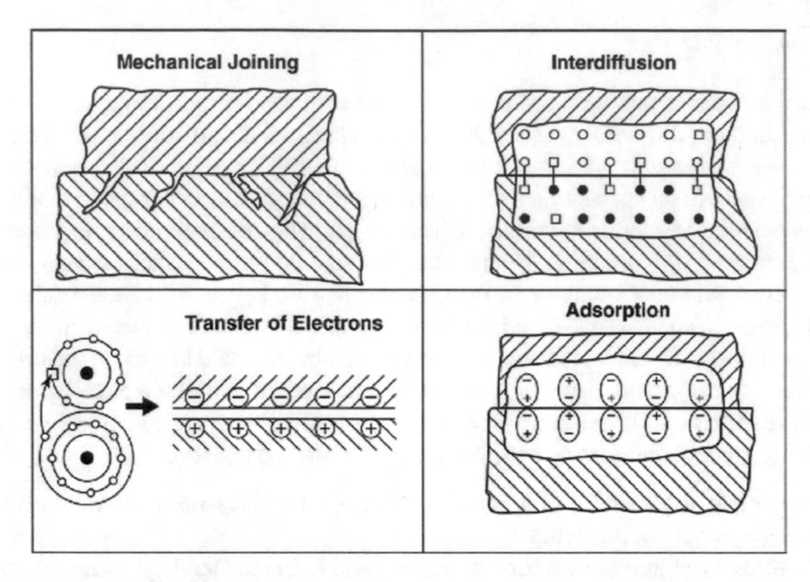


**Fig. 1.3** Effects of adhesion wear-out mechanism [12]

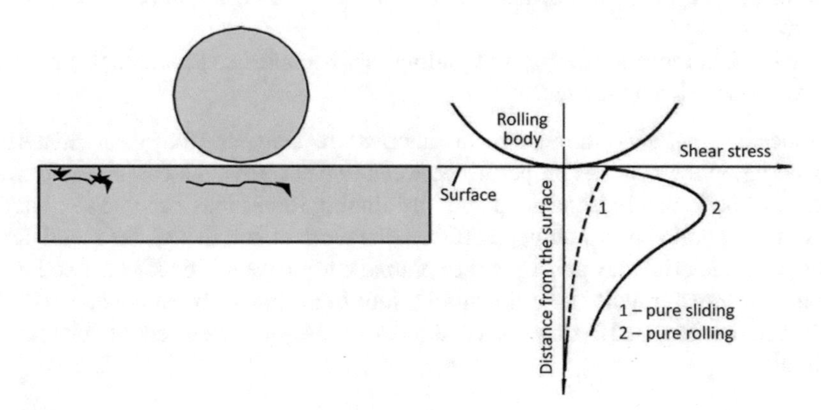


**Fig. 1.4** Surface fatigues wear-out mechanism [13]

#### 1.2.1.4 Erosive Wear

In this type of wear, some particles slide and roll against the surface. The particles could have three different speed levels, as presented in Fig. 1.5. At low speed, we have abrasion wear, and if these particles touch the surface on a cyclic basis, then it will result in fatigue wear out. At medium speed, plastic deformation or erosion by brittle fracture will occur. At higher speed, melting will occur.

As can be seen in Fig. 1.5, each particle removes some small parts of the component. Over time, this phenomenon could result in serious erosion damage to the

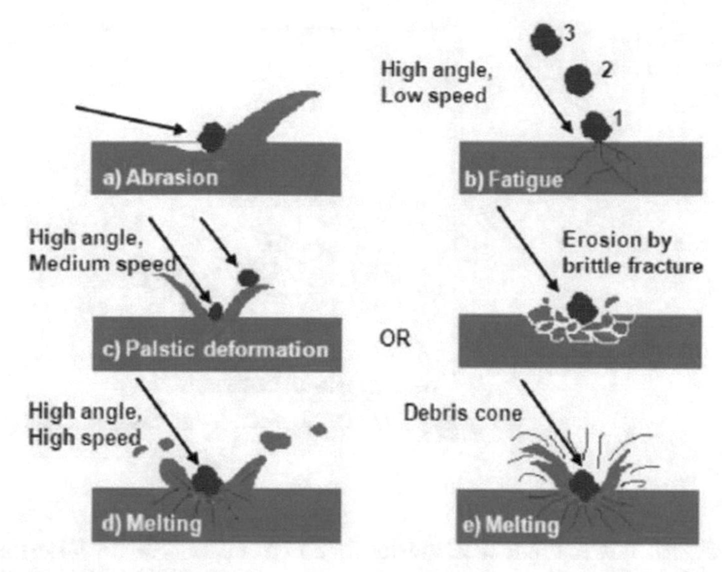


**Fig. 1.5** Different types of erosive wear-out mechanism [14]

component. This type of wear could be used intentionally for some applications. For instance, in paint removal (sanding), the same concept is used to remove the coating over a surface.

#### 1.2.1.5 Fretting Wear

Fretting is a type of wear that occurs in systems with two parts in close contact, vibrating at small amplitudes. During this cyclic contact motion, third body particles (which may be produced by local adhesion or abrasion) become trapped between the contacting parts and accumulate. Harder particles cause plastic deformation, delamination, void formation, and/or cracking damage on the surface of parts which grows over time due to the cyclic loading nature. During system operation, more particles are able to accumulate in the damaged area, and over time pits can form across the surface. Eventually cracks are able to grow deeply enough, or enough material is removed from the surface that the part will fail [15]. This type of wear-out mechanism occurs commonly in components that contain parts with relative motion such as bearings.

#### 1.2.1.6 Creep Deformation

Creep deformation describes the process by which solid materials undergo gradual plastic (permanent) deformation at stresses below the yield stress. Creep is a time-dependent process, and the rate of deformation from creep increases with temperature

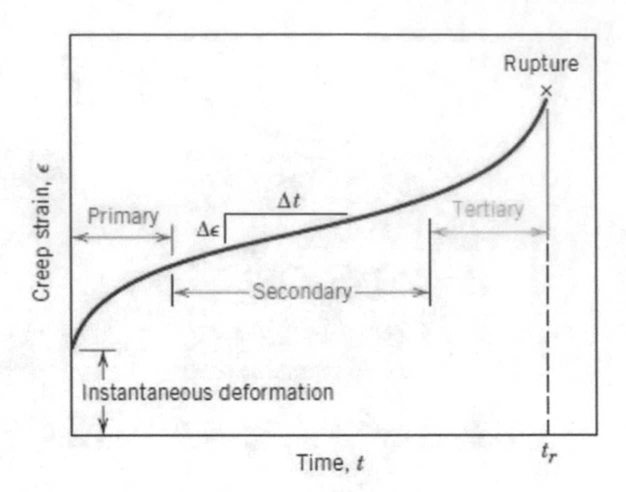


**Fig. 1.6** Three stages of creep deformation process [17]

and stress. There are multiple mechanisms for creep, namely Nabarro–Herring (bulk diffusion), Coble (grain boundary diffusion), dislocation climb, and thermally activated glide, the last of which is exclusive to polymers and viscoelastic materials [16]. Generally, the process of creep deformation can be divided into three-time intervals.

**Stage 1, Primary phase**: This is a short phase that begins when elastic (reversible) deformation occurs on initial loading. Following this, there is an initially high rate of deformation that slows as work hardening occurs [17].

**Stage 2, Secondary phase**: In the secondary phase, the strain rate stabilizes to an approximately linear rate that can be characterized by an equation related to the Arrhenius equation.[1] During the secondary phase, the strain rate is at its lowest, and this phase may last a long time [17].

**Stage 3, Tertiary phase**: In the final phase, the deformation rate increases as damage accumulates in the material structure. During this phase, voids begin to appear in the microstructure and noticeable necking begins to occur leading up to fracture [17] (Fig. 1.6).

#### 1.2.1.7 Fatigue

Fatigue failure occurs as a result of long-term cyclic loading within the design limits of a component. Under these loads, initial defects may become crack initiation sites which grow as the region is loaded and unloaded. Cracks will grow until the stress intensity factor of the crack exceeds the fracture toughness of the material, at which point a complete fracture will occur [16] (Fig. 1.7).

[1] The Arrhenius equation describes processes with a rate that is exponentially dependent on temperature [16].

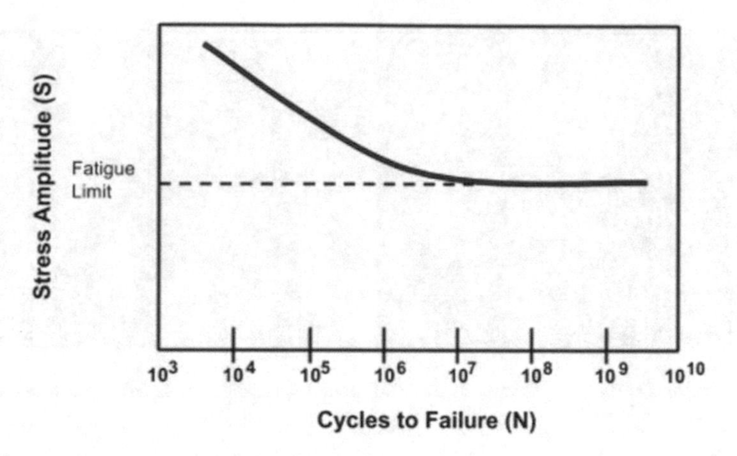


**Fig. 1.7** Stress amplitude versus cycles to failure [18]

Fatigue failure follows three major steps to failure:

1. **Stage 1, Initiation process**: In this step, a crack begins to form at an origin point which may be a surface flaw, geometric feature, grain boundary, or any other form of discontinuity in the material.
2. **Stage 2, Propagation**: In this phase, cracks grow at varying rates depending on the material, environment, stress, and crack shape. A sharper crack will grow faster due to the increased stress concentration at the tip, and crack growth can even be impeded by blunting crack tips.
3. **Stage 3, Failure**: Once many microscopic cracks have been expressed and connected, fracture will occur. When the stress cycle amplitude is low, it generally takes many cycles for the material to fail and this is called high cycle fatigue (HCF). Similarly, if the amplitude is high, failure occurs quickly and is referred to as low cycle fatigue (LCF). Generally, when a material fails under HCF, cracks propagate mostly in an intergranular fashion (along grain boundaries) and allows for significant plastic deformation before failure, but when under LCF will fail in a largely transgranular and brittle fashion (through grains) due to the rapid crack propagation. The differences between the resulting fracture surfaces can be seen in Fig. 1.8.

#### 1.2.1.8 Cavitation Pitting

Cavitation pitting is a type of pitting fatigue as a result of vibration and movement of liquids in contact with solids. It occurs mostly at low-pressure regions where voids can form. Any moving system in liquid will experience cavitation pitting. Generally, corrosion can aggregate the cavitation pitting. In this diagram, the steps of cavitation pitting are simplified by showing the mechanism in slow motion. The

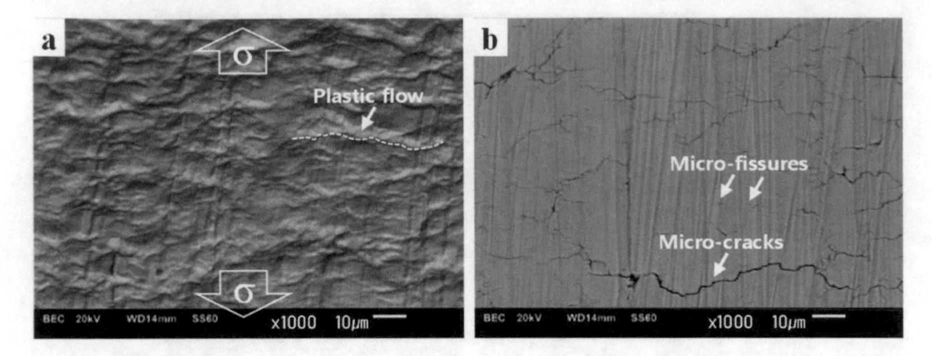


**Fig. 1.8** Fracture surfaces of a material after failing in high cycle fatigue (**a**) and low cycle fatigue (**b**) [19]

actual mechanism happens in milliseconds, depending on vibration and rotation frequency and vapor pressure of the liquid [20] (Fig. 1.9).

- **Step A**: The solid wall moves to the right, building inertia in the water to move right.
- **Step B**: Inertia pushes the water to continue moving to the right.
- **Step C**: The wall starts to move to the left; pressure is reduced and bubbles form. These cavities are known as low-pressure bubbles. At this point, the pressure of bubbles is equal to the vapor pressure of liquid and a phase change from liquid to gas can occur.
- **Step D**: The wall stops moving and pressure starts to increase as the inertia of the water pushes it to move left, compressing the bubbles.

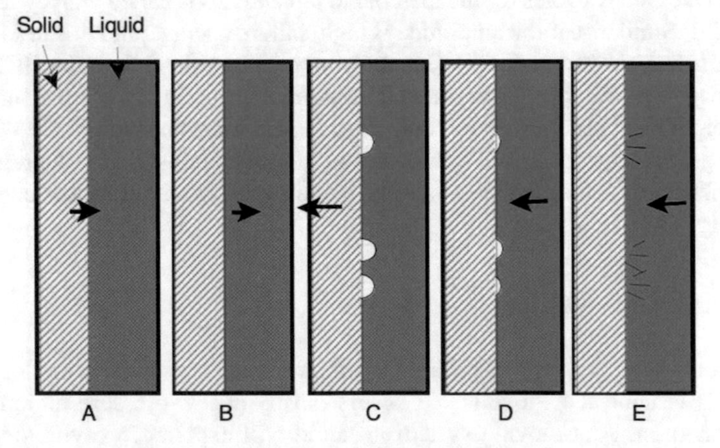


**Fig. 1.9** The process of cavitation pitting degradation mechanism

- **Step E**: Pressure increases until bubbles collapse as vapor changes to liquid. Implosion results in local pressures of many thousand pound-force per square inch (psi).

It should be noted that bubbles nucleate at defect sites, thus no bubbles will form on a perfectly smooth surface.Motion in the fluid caused by vibration and rotation of the system allows for cavities to form in many different places around the part surface, but most will form in low-pressure regions. As time passes, pitting on the surface will result in fatigue failure.

#### 1.2.1.9 Resonance Disaster

When a structure encounters oscillations from wind, earthquakes, or vibrating motions that match the structure's natural frequency of vibration, it is able to pick up motion very easily and may sway violently or fracture. Resonance as a phenomenon is one that is often used in clocks to maintain motion of the timekeeping mechanism, such as a pendulum [21].

### *1.2.2 Overstress Mechanisms*

Overstress failure occurs when a material is subjected to unusually harsh conditions that aren't expected in the design of the part. This may come in the form of ground motions generating high stress near a buried or anchored structural part, damage from accidents, a damaged connecting part creating high stress, or many other potential causes. The behavior of various materials under such conditions has been heavily studied to understand how a material may deform either elastically (reversible) or plastically (irreversible) before fracturing [22].

The response of a material under increasing loads may be characterized using a stress–strain curve. These curves are developed by testing a material with a standardized cross section under increasing stress and recording the deformation in response [23].

Figure 1.10 shows the shape of the stress–strain curves of a typical metal, ceramic, and polymer, respectively. The exact shape and scale of the stress–strain curve varies for each material and loading type, but among similar materials in a standard tensile test, these differences can usually be categorized by the yield and tensile strength.[2]

The stress–strain curve for most solids can be split into at least two regions, with the first region corresponding to elastic (reversible) deformation, and the latter region corresponding to various phases of plastic (permanent) deformation. In linear elastic

[2] Yield strength (or stress) is defined as the minimum stress that will cause permanent (plastic) deformation in a material. Tensile strength reflects the highest point in a stress–strain curve of a material in tension loading. Additional measures can be given for other loading conditions such as compression and bending.

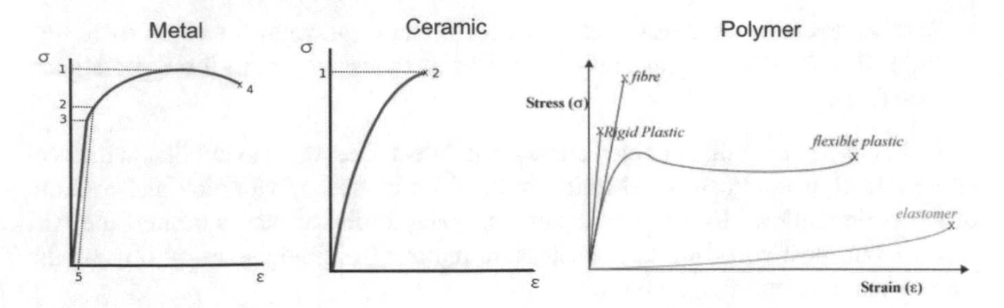


**Fig. 1.10** General form of the stress–strain curves for metal, ceramic, and different types of polymers

materials, such as most ceramics and metals, the elastic region of the curve can be used to determine the Poisson ratio by measuring the strain in the transverse direction relative to the longitudinal strain.

For metals, the stress–strain curve can be divided into four regions [22].

1. First is the linear portion of the curve (proportional region) which starts at no deformation and extends until the yield stress is reached at point 3 on the graph (Fig. 1.10, metal). In this region, the slope corresponds to the Young's modulus, which is a measure of the material's stiffness in tension or compression.
2. The next region is the nonlinear elastic region, which is the final elastic deformation a material can sustain before yielding or fracturing. Point 2 on the graph represents the elasticity limit.
3. The third region is called the strain hardening region and extends from the point where material starts deforming plastically until it reaches its ultimate strength at point 1. As plastic deformation occurs in the metal, deformations accumulate which restrict motion and increase the strength of the material. Eventually this process reaches a limit, at which point necking begins.
4. The final region in the curve is called the necking region. As the material is stretched in one direction, its cross section will thin, and deformation becomes easier. At first, strain hardening counters this effect and the stress needed to cause deformation increases as seen in region 3, but as the strain hardening effect reaches its limit, the cross section starts to rapidly become thinner up until rupture at point 4.

Ceramics usually are brittle, undergoing little to no plastic deformation before fracture. This results in a simpler stress–strain curve characterized by a mostly linear elastic region up to the fracture point, which is nearly equal to the yield point. This is indicated by both points 1 and 2 on the second plot in Fig. 1.10 showing the ultimate strength and yield strength in the same location.

The stress–strain curves of polymers need to be further separated into four major categories of behavior. Flexible plastics have similar mechanical behavior to metals, with various regions of elastic and plastic deformation leading up to fracture. The stress–strain behavior of these polymers can be easily differentiated from that of

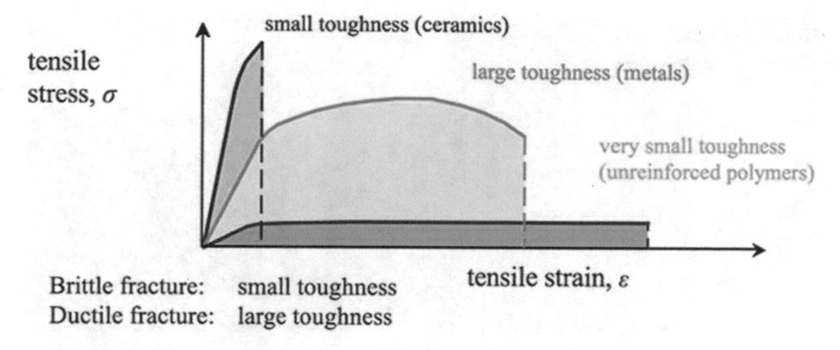


**Fig. 1.11** Stress–strain curve for materials with very small, small, and large toughness [24]

metals by the much greater capacity for plastic deformation before fracture. Elastomers undergo large amounts of viscoelastic (nonlinear) strain and relatively little plastic deformation before failure. Additionally, elastomers tend to have very low ultimate strengths [22].

The behavior of fibers and rigid plastics is almost the same as ceramics. They have a very short elastic region but are very stiff and may sustain large stresses before failure [22].

Toughness is one of the material properties that can be defined based on the stress–strain curve of the material. Toughness defined as a material's ability to absorb energy or shock before fracturing. Toughness as a property is measured in terms of energy per unit volume and can be quantified as the area underneath the stress–strain curve up to the fracture point [22].

Because of the relation of toughness to the stress–strain curve, the differences in toughness between different types of materials can be easily seen in comparisons such as Fig. 1.11. Though the ceramic in the graph presents the highest overall strength and the polymer presents the greatest amount of strain withstood before fracture, both of these materials have a low toughness. The toughest materials are those which have a moderately high strength that does not sacrifice the capacity for plastic deformation such as metals as presented in Fig. 1.11.

#### 1.2.2.1 Plastic Deformation

Whenever a solid material is subjected to a load, it will deform as a result of that load. When the load is relatively small, this deformation, referred to as elastic deformation, is reversible. Elastic deformation involves only the stretching of bonds without any slipping of atoms, and the structure will return to its former shape once the load is removed. At higher stress levels, the change in the shape is non-recoverable. This type is called plastic deformation. When a material is plastically deformed, atoms slide past each other breaking bonds and causing changes in the material's structure (e.g., [26, 27]). Generally speaking, this process cannot occur all at once since a large amount of energy would be needed to do so. Instead, this deformation occurs via

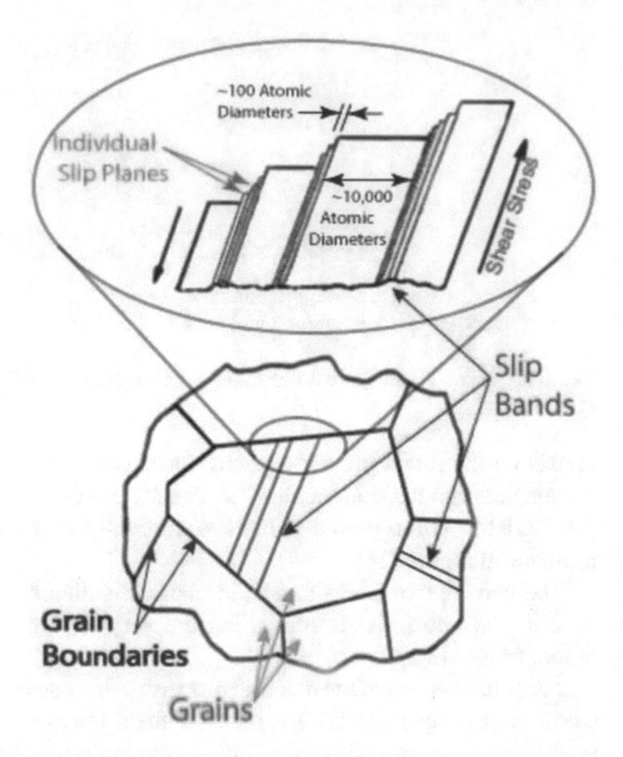


**Fig. 1.12** Slip bands in plastic deformation process

paths of least resistance, where an initial defect in the structure moves across crystal planes (in the case of crystalline materials) or polymer chains slide past each other. In metals, dislocation motion occurs preferentially along the highest density planes. A dislocation can easily move throughout a metal structure until a grain boundary is reached. Using an optical microscope, one can see the paths of these dislocations since they mostly travel along parallel paths. When many parallel paths are close together, they can be seen as slip bands such as in Fig. 1.12 [25].

#### 1.2.2.2 Fracture

Fracture is defined by the formation of new surfaces on a formerly intact piece of material not by growth but by damage from stress. Most often this occurs by the formation of cracks throughout the surface or body of the part, but in extreme cases can result in a part breaking into two or more pieces.

The general definition of cracking refers to those formed perpendicularly to the material surface and can also be called a normal tensile crack. When cracks form tangential to the surface, this is called a shear crack, slip band, or dislocation. Cracks and fracture may occur with or without deformation beforehand. If deformation occurs, this is called a ductile fracture, and a fracture without deformation is likewise called a brittle fracture. A given failed surface may contain a mix of both ductile and

**Fig. 1.13** Sample examples of fracture for steel at low and high temperature, with brittle and ductile behaviors, respectively [26]

brittle fracture regions depending on how the break occurred. On the stress–strain curve, fracture strength is the final point in the curve [16].

In Fig. 1.13, the fracture in a brittle and ductile steels are presented. At lower temperature, steel is more brittle, under stain, the fracture surface is perpendicular to tensile stress. However, as temperature increases, steel is more ductile, and the fracture surface is diagonal.

#### 1.2.2.3 Delamination

Delamination is defined as the separation of layers or plies in a laminate material. These failures are caused by out-of-plane loads, such as loads from bending, impact damage, or tearing.

Composite materials, whether man-made or natural (such as woods), have anisotropic properties and they are more likely to fail in one loading direction over another. These materials are strongest when loaded in the direction parallel to that of the load bearing elements (such as fibers) and are weaker in transverse or out-of-plane loading. These materials are often susceptible to delamination due to their weakness in out-of-plane load conditions. This type of damage may go unseen in a composite structure if it lies beneath the surface. Figure 1.14 shows three different delamination conditions: delamination between two rigid panels,delamination of a thin flexible film or panel on a rigid panel, and delamination between two flexible panels.

#### 1.2.2.4 Degradation Due to Residual Mechanical Stress

When a material undergoes plastic deformation (temperature gradients from thermal cycling, or structural changes from a phase transformation), stress may remain due

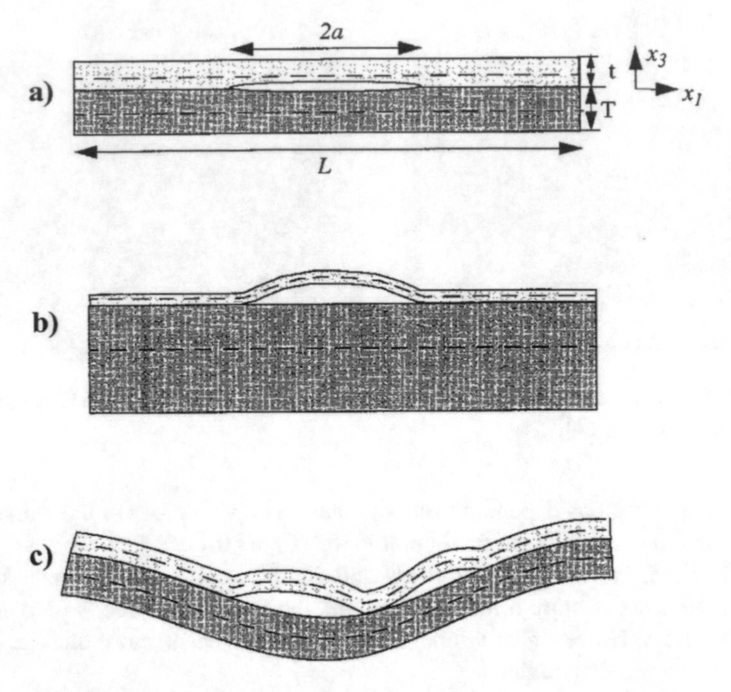


**Fig. 1.14** Three different delamination conditions, two rigid panels (**a**), one flexible panel or film and one rigid panel (**b**), two flexible panels (**c**) [27]

to uneven changes in the dimensions of the part. These stresses are called residual stresses [31].

Residual stresses can cause early failure in a structure. Generally, a large amount of energy is required to propagate a crack through the bulk of a material, but residual stresses may act to reduce the energy needed to do so by adding additional stress. In the case of thermal residual stresses, it is sometimes possible to relieve some of this stress by baking, cryogenic methods, or by introducing stress in the opposite direction to counter the negative stress [25].

Though unwanted residual stress can be detrimental to a material's performance, there are cases in which it is desirable. In many cases, it is beneficial to introduce a residual compressive stress to a surface which acts to close cracks that form under tension. Laser peening (LP) is an example of introducing a desirable residual stress. Laser peening and shot peening are examples of surface processes which introduce compressive stress to the surface to improve its lifetime. The compressive stress acts to close cracks formed in tension, increasing resistance to damage from fatigue and other forms of surface wear. The introduction of these residual stresses can be especially important for brittle materials such as ceramics, which can withstand large amounts of compressive stress but very little tensile stress [25].

## 1.3 Thermal Degradation and Failure Mechanisms

### *1.3.1 Wear-Out Mechanisms*

Long term static or cyclic exposure to thermal loads within normal operating conditions can degrade material properties and cause micro- and macrostructural damage.This section describes thermal degradation via the mechanisms of intermetallic growth and hillocks.

#### 1.3.1.1 Intermetallic Growth

Soldering in electronic components provides a combination of electrical connection and mechanical support, and thus the properties of these solder joints are very important to the performance of the component. During the soldering process of electronic components, intermetallic compound layers will form. During thermal cycling or thermal shock, intermetallic compound (IMC) layers will grow, and the characteristics of the material degrade. If the layers become too thick, this compromises the mechanical strength of the joint. For instance, the precipitation of a new material increases brittleness and cracks can form more easily. Additionally, the formation of IMCs may increase the resistivity of the joint which degrades the electrical performance of the component. This type of failure is more common in electronic devices under extreme temperature environments with large temperature variations, such as in space conditions [28].

#### 1.3.1.2 Hillocks

An important source of mechanical stress in thin films is the thermal mismatch between the film and the substrate material. Depending on the sign of the mismatch and of the temperature change, tensile or compressive stresses can develop in the film. One mechanism of compressive stress relaxation which is specific to thin films is the formation of hillocks. Hillocks form during heating and cooling cycles at layer boundaries with dielectrics [29].

The structure of the hillocks has not been investigated in detail and the exact mechanism of hillock growth remains unclear, but it is known that hillocks preferentially nucleate at a weakly bonded film/substrate interface where delamination occurs due to very high compressive stresses in the films which allow the interface to act as a sort of atom sink. Hillocks tend to be regularly arranged which may be due to the stress relaxation zones around Hillocks which make the surrounding areas less likely to become nucleation sites [30].

Hillocks grow under continued compressive stress which encourages diffusion of atoms from grain boundaries to the hillock sites. Depending on the temperature, the atoms may diffuse along the film/substrate interface or bulk lattice. As atoms diffuse to the hillock site, they may grow in a direction normal to the film surface or grow the grain along the surface plane. Growth of hillocks in the normal direction displaces the original film upward, bending it. Once the bending stresses exceed the strength of the film, it will fracture, and new material is able to penetrate above the film layer and form a spherical cap which reduces overall surface area [31].

#### 1.3.1.3 Fatigue

Similar to mechanical fatigue, thermal fatigue is the weakening of a material caused by cyclic thermal loading resulting in crack growth and structural damage. Thermal fatigue is differentiated from mechanical fatigue in that it may occur by only thermal cycling, without mechanical loads. A key example of this is in turbines, where temperature differentials from the starting and stopping of the turbine produce temperature gradients in the material that leads to thermal fatigue [16].

#### 1.3.1.4 Degradation Due to Residual Thermal Stress

Residual thermal stresses are stresses introduced when a material undergoes heating or cooling and attempts to expand or contract. This can occur in one of two ways [16]:

- If two dissimilar materials are bonded together and heated, dissimilar thermal expansion coefficients create surfaces on both materials as one attempts to expand more than the other.
- A bulk material is heated or cooled unevenly, and one area is at a different temperature or changing temperature at a different rate. Thermal expansion will create stress as one area is expanding or contracting at a different rate than another.

In the following figure, an example of a residual thermal stress process is presented. There is an elastic multilayer structural material (Fig. 1.17a). When we increase the temperature, the thermal strains are generated in the layers (Fig. 1.17b). To satisfy the displacement compatibility condition at the individual interface, thermal stresses should be generated (Fig. 1.17c). Finally, the bending of the whole coating system occurs because of the presence of thermal stresses (Fig. 1.17d).

### *1.3.2 Overstress Mechanisms*

Exposure to temperatures outside of the design limits for a part or system can cause irreverisible damage to the structure and material properties. This section will go over some mechanisms by which this damage can occur.

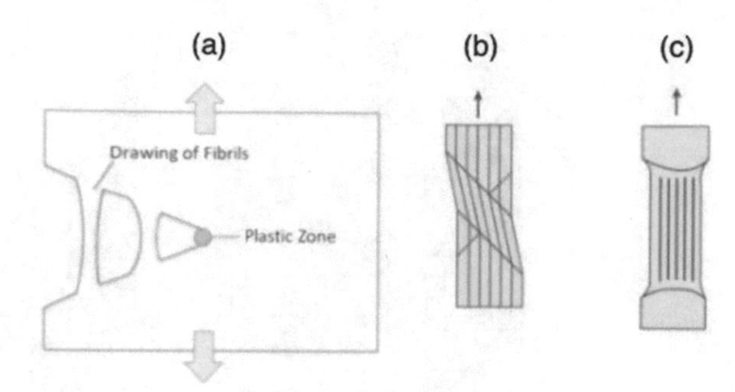


**Fig. 1.15** Crazing [32] (**a**), shear yielding [33] (**b**), yielding [33] (**c**)

#### 1.3.2.1 Ductile to Brittle Transition

A ductile-to-brittle transition (DBT) is the point or range at which a material that is relatively ductile at high temperatures becomes more brittle at low temperatures. It is known to happen to thermoplastic polymers and some metal alloys, most notably ferrous alloys such as steel.

In polymers, the DBT represents a change in the main failure mechanism from yielding as the dominant mechanism at high temperatures to crazing or shear yielding at lower temperatures as shown in Fig. 1.15. Depending on the nature of the main chain and side groups, the DBT temperature will coincide with the α-transition temperature (which is the same as the glass transition temperature), β-transition temperature, or γ-transition temperature which are listed in order of decreasing temperature [32].

In materials with an ordered lattice structure, such as those in metals, the ductility of the structure depends on the energy needed to move a dislocation along slip planes. When the energy needed to move a dislocation is very high, the material is considered "brittle" while also being relatively strong, whereas a structure which requires little energy to move a dislocation is considered "ductile" but is often not as strong as the more brittle structures. A face-centered cubic (FCC) lattice, seen in metals like copper, has many slip planes[3] which allows for easy dislocation movement. On the other hand, a body-centered cubic (BCC) lattice, which is the dominant structure for iron alloys, has fewer slip planes and dislocation motion requires additional energy to occur. Figure 1.16 shows the structure of a BCC (a) and FCC (b) lattice. At high temperatures, energy from heat enables easier motion of dislocations in BCC structures, making the material more ductile, but weaker. At low temperatures, little dislocation motion can occur, and the material becomes harder and more brittle.

[3] A slip plane for a given crystal lattice is the plane with the highest density of atoms (closest packed), and on which slip (motion) may occur with the least energy needed to dislocate the atoms.

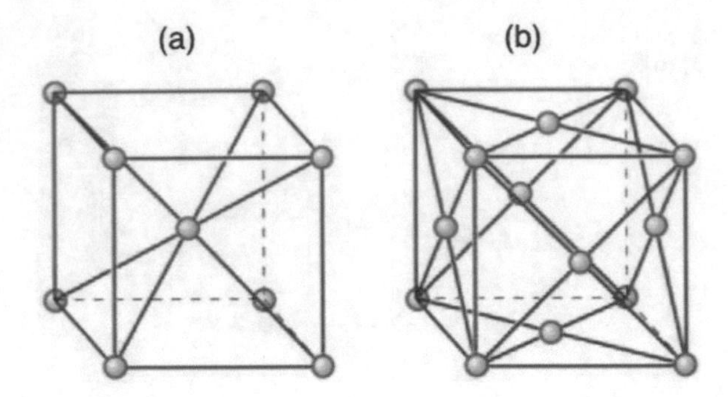


**Fig. 1.16** Body-centered cubic (**a**) and face-centered cubic (**b**) [35]

The presence of alloying elements has a major effect on the DBT in metals, particularly steel. While pure iron is fairly ductile and has a low DBT—about −50 °C [34]—the addition of alloying carbon increases the DBT, and even a small amount of sulfur or phosphorous can greatly decrease the ductility and increase the DBT. The addition of manganese to a steel alloy has a positive effect, however, and can increase ductility as well as reduce DBT [34]. Precipitates, grain boundaries, and other hardening mechanisms all decrease the ductility of the material.

#### 1.3.2.2 Glass Transition Temperature

Glass transition is a second-order transformation which occurs in amorphous materials, namely polymers [32]. This transition occurs over a range of temperatures and marks the point at which the polymer chain is able to move with increasing degrees of freedom without breaking. This change occurs because of the contribution of intermolecular (secondary) bonding between chains. Though these bonds are not as strong as the primary covalent bonds which hold the chains together, at low temperatures it may be difficult for the chains to slide past each other due to these bonds, whereas at high temperatures it is relatively easy to overcome the intermolecular bonds.

The presence of bulky side groups, crosslinking, or polar groups on the chain will all decrease chain mobility and will therefore make glass transition occur at higher temperatures. This can be a problem for polymers that are exposed to temperatures outside of their expected application range and may fracture early or become flimsy.

#### 1.3.2.3 Thermal Shock

Thermal shock is the cracking mechanism as a result of rapid temperature change. It affects material properties such as toughness, thermal conductivity, and thermal expansion. In addition, as different parts expand by different amounts—creating a

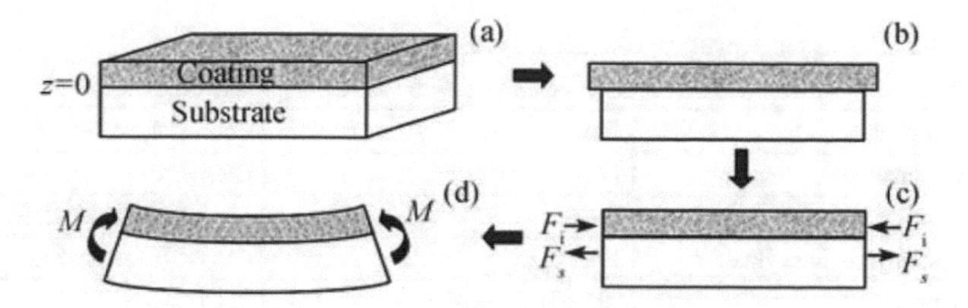


Fig. 1.17 An example of a residual thermal stress process [37]

thermal gradient—it results in a differential strain which creates stress in the material. If this stress grows large enough, it can cause cracks to form [36].

#### 1.3.2.4 Melting

Melting is the phase change of a substance from solid to liquid. It is a first-order transformation that happens at a constant temperature and results in increased molecular vibration. The failure mechanisms that generate heat can result in component melting as a secondary failure. For instance, in electrostatic discharge failure mechanisms heat is generated that can result in melting of electronic devices [28].

## 1.4 Chemical Degradation and Failure Mechanisms

### 1.4.1 *Wear-Out Mechanisms*

In this section, different types of chemical wear-out mechanisms are explained. One of the main materials that suffer from chemical degradation is polymer [38]. Polymers are a wide category of materials which includes synthetic thermoplastic and thermoset polymers, naturally occurring polymers present in organic material, and natural and synthetic rubbers. These materials undergo various forms of degradation depending on the environment and polymer chemistry, but the main forms are [38]:

- Depolymerization
- Thermal-oxidative degradation
- Hydrolytic degradation
- Ozone degradation

In the following section, all these degradation mechanisms are explained briefly. Then, corrosion and outgassing are presented as the other types of chemical wear-out degradation mechanisms.

**Table 1.1** Ceiling temperature of common organic polymers [41]

| Polymer | Ceiling temperature (°C) [3] | Monomer |
|---|---|---|
| Polyethylene | 610 | $CH_2{=}CH_2$ |
| Polyisobutylene | 175 | $CH_2{=}CMe_2$ |
| Polyisoprene (natural rubber) | 466 | $CH_2{=}C(Me)CH{=}CH_2$ |
| Poly(methyl methacrylate) | 198 | $CH_2{=}C(Me)CO_2Me$ |
| Polystyrene | 395 | $PhCH{=}CH_2$ |
| Polytetrafluoroethylene | 1100 | $CF_2{=}CF_2$ |

#### 1.4.1.1 Polymer Degradation-Depolymerization

Depolymerization is a process which occurs at high temperatures and is defined as the decomposition of a polymer into one or more monomers [38, 39]. It is a very common process in nature. For example, in the process of food digestion macromolecules depolymerize. For polymers formed via addition polymerization such as polyethylene, there is no chemical reaction involved and the tendency to depolymerize or polymerize is dependent on the relative size of the entropy gain times the temperature to the enthalpy change from the conversion from a single-bond carbon chain to a double-bond monomer. At low temperatures, polymerization is thermodynamically favored due to the decrease in enthalpy, but at high temperatures depolymerization will become favored [40]. In Table 1.1, ceiling temperatures[4] of different organic polymers are presented.

$$\Delta G = \Delta H - T\Delta S$$

Condensation polymers, on the other hand, do not have the ability to change between monomer and polymer forms in this way. The formation of these polymers occurs via the reaction of two or more constituents, for example, an alcohol reacting with carboxylic acid to form polymers with ester groups. Instead, these polymers depolymerize via reactions with the environment. For example, hydrolysis is the process by which bonds in functional groups are broken in water causing depolymerization [40].

#### 1.4.1.2 Polymer Degradation—Thermal-Oxidative Degradation

Polymers can degrade by chain scission in which a long main chain breaks up into smaller chains of a smaller molecular weight. This type of degradation falls under the general category of thermal-oxidative degradation and is usually caused by exposure

[4] The ceiling temperature of a polymer is the temperature at which polymerization ceases to become thermodynamically favorable. Above this temperature depolymerization will occur.

to heat or UV radiation while in the presence of oxygen. The reaction with oxygen causes the polymer to become brittle, more susceptible to wear, and changes its optical properties such as color and opacity. Thermal-oxidative degradation can be slowed down with the use of UV stabilizers and antioxidants [42].

#### 1.4.1.3 Polymer Degradation—Hydrolytic Degradation

Many polymers are not impermeable to liquids and may either swell or dissolve. The ability for the polymer to take up a liquid depends on the solubility of the polymer and liquid. For polymers, the solubility is typically much higher with organic solvents than water.

As the liquid penetrates the polymer, the smaller molecules are able to fit between the large polymer molecules and gradually increase the distance between polymer chains. This separation of the chains reduces the strength of intermolecular bonds and soften the polymer along with reducing its glass transition temperature. Dissolution is the more extreme form of the same effect, where the polymer is completely soluble [42].

#### 1.4.1.4 Polymer Degradation—Ozone Degradation

Polymers—especially those with doubly bonded carbon atoms in the main chain—are susceptible to chain scission caused or accelerated by reactions with ozone, as well as oxygen, to a lesser extent. If the polymer is unstressed, this reaction will proceed until a protective layer is formed and stop, but tensile stresses form cracks in this layer that expose new material to react. This problem is mainly prevalent in vulcanized rubbers, characterized by the double-bonded carbons [42].

#### 1.4.1.5 Corrosion

Corrosion is the process of metal breakdown by reacting with substances in their environment. One of the well-known examples is rusting, where iron reacts with oxygen and water from the environment to form hydrated iron oxides which are more commonly known as rust.

Corrosion can be divided into several types; the most common of these are general, pitting, crevice, intergranular, stress corrosion cracks, and galvanic corrosion. These types of corrosion are described below, but note that this is not a comprehensive list of all types of corrosion since many environmental, structural, and chemical factors can impact the rate and nature of corrosion in a metal [42].

The simplest form is general corrosion which occurs over the full body of the material but is slow relative to other forms. Pitting and crevice corrosion occur at accelerated rates in small spaces where there is poor circulation, and the local environment is allowed to become highly corrosive as the reaction proceeds. Intergranular corrosion is accelerated corrosion at metal grain boundaries where the material might be more susceptible to a reaction.

Stress corrosion cracking occurs when a material is subjected to mechanical load cycles in a corrosive environment. Stress corrosion cracks greatly accelerate the rate of both corrosion and crack propagation, since the formation of brittle reaction products reduces the strength and ductility of the structure, and the small space created by the crack allows for rapid corrosion propagation similar to pitting corrosion.

Lastly, galvanic corrosion is characterized by the preferential corrosion of one metal when it is in contact with another dissimilar metal when moisture and an electric current are present. Under galvanic corrosion, the two materials form an electrolytic cell, with the more noble metal acting as a cathode and the less noble acting as an anode.

#### 1.4.1.6 Outgassing

Outgassing is the process by which a material loses mass in very low-pressure conditions, most commonly in the vacuum of space [43]. Generally, outgassing describes the process of leaching material out of a part in low-pressure conditions but depending on the makeup of the material/part, the nature and consequentiality of this process vary. Polymers are notably susceptible to undergoing phase changes at these low pressures and may lose mass as material at the surface evaporates. Additionally, even if the bulk of the material is designed to be resistant to this effect, materials trapped in pores or cracks may leach out. This can result in damage due to mass loss, and potential damage to sensors or electronics if some of the leached material attaches itself to these sensitive parts.

### *1.4.2 Overstress Mechanisms*

Overstress chemical degradation mechanisms are chemical degradation mechanisms that result in material breakdown. All degradation mechanisms presented in wear-out chemical degradation mechanisms (Sect. 1.4.1) can occur at varying speeds depending on what elements the material is exposed to. Extremely reactive materials—like strong acids or bases—can quickly break down a material structure as it reacts. These types of degradations are considered as overstress chemical degradation mechanism.

## 1.5 Electronics Degradation and Failure Mechanisms

Electronic components can be divided into three categories: active, passive, and electromechanical. Active components include transistors, diodes, displays, etc. which can be used to amplify signal and power, and whose functions depend on external energy input. Passive components include resistors, capacitors, and inductors, etc. which cannot inject power into a circuit. Electromechanical components include switches, cable assemblies, etc. which perform their electrical functionalities by mechanical movements of their parts. Some of the failure mechanisms in these electrical components have been introduced in former sections of this chapter. For example, coefficient of thermal expansion (CTE) mismatch of packaging materials can cause mechanical stresses inside the components and trigger fatigue. Corrosions due to humidity and chemicals also degrade the packaging materials and PCBs and can further fail the electronics. Outgassing and thermal shocks are also failure mechanisms involved in packaging failures. Glass transitions in the PCB failures can soften the resin matrix and cause contaminant diffusion. In the following sections, more degradation and failure mechanisms in electrical components are introduced.

Semiconductors are widely used in building up electrical systems. For example, semiconductors serve as the dielectric layer in the metal–oxide–semiconductor field-effect transistors (MOSFETs) which are switching and signal amplifying components and as significant members in memory hierarchy. Semiconductors are also used in FLASH memory which serves as solid state memory (SSD). Thus, the following section starts from the degradation mechanisms commonly seen in the MOSFETs, moving on to the crosstalk of the neighboring circuits, and the metallic degradation of interconnections and metallic components.

### *1.5.1 Dielectric Breakdown*

Dielectric breakdown can be divided into five stages as shown in Fig. 1.18. The schematics of how defect clustering affects the leakage current evolution are shown in Fig. 1.19. There is a maximum electric field that the dielectric material can tolerate. When the applied electric field is larger than the maximum tolerable voltage in a dielectric material, the hard breakdown (HBD) happens (as shown in Fig. 1.18 region E), and the gate oxide is no longer insulated. HBD is the most destructive phase of the breakdown. It can also be referred to as thermal runaway. The metal atoms/ions in the gate migrate from the gate to the dielectric at this stage, causing "punch-through". The transitive characteristics are completely lost in the HBD regime.

Under lower electric fields, the gate oxide can also wear out with time-dependent dielectric breakdown (TDDB) process. Stress-induced generation of traps can happen inside the oxide and at the interface. As the dielectric degradation continues, the breakdown initiates (as shown in Fig. 1.18 region B). The degradation process and breakdown position can vary greatly with same size transistors because of the random behavior of the degradation mechanism. Because of the localized defects in the grain boundary (GB) regions, TDDB is observed to propagate faster in GBs [44]. TDDB can be controlled to a certain degree by the external compliance current. Similar to stress-induced leakage current (SILC), TDDB can be partially recovered, and the extent of recovery relies on the electrode material oxygen solubility [45].

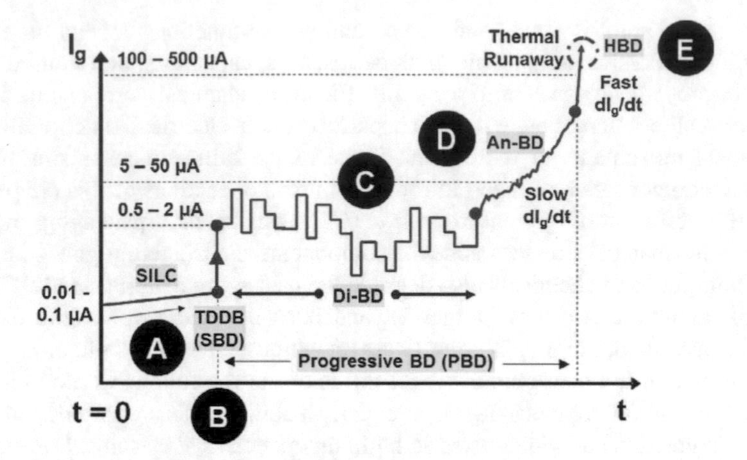


**Fig. 1.18** Leakage current evolution with time for a dielectric stressed at time $t = 0$. Five regimes of dielectric breakdown are shown [48]. A: Stress-induced leakage current (SILC); B: time-dependent dielectric breakdown (TDDB); C: digital breakdown (Di-BD); D: analog breakdown (An-BD); E: hard breakdown (HBD). Different kinetics of degradation are related with different regimes. A and B are device area dependent while the rest are location dependent [48]

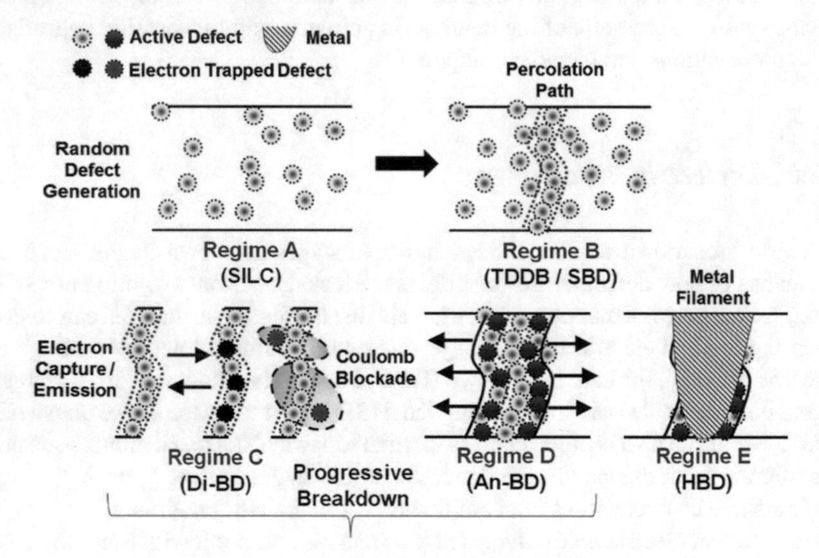


**Fig. 1.19** Correspondence of dielectric defect density and spatial distribution with the five regimes of breakdown [48]. Yellow and blue dots are active defects in the pre-TDDB and post-TDDB stage. Black and red dots are the trapped charges. Regimes A–D involve the role of oxygen vacancy defects, while regime E relates to the metal migration from the gate electrode and/or silicon from the substrate [48]

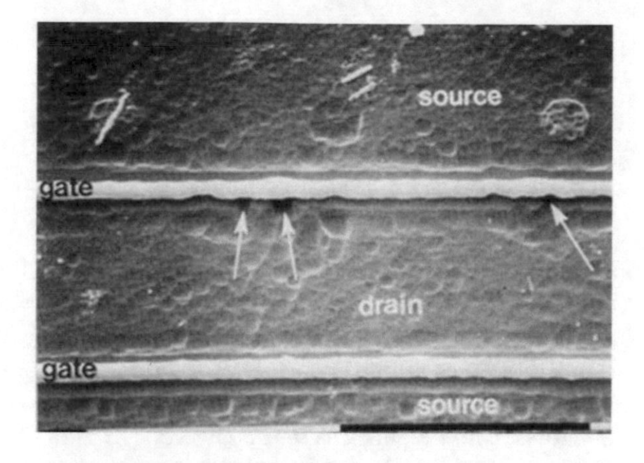


**Fig. 1.20** Multiple breakdown spots at the drain junction of an nMOS transistor [49, 50]

The breakdown (BD) modes depend on the thickness of the gate oxide. HBD marks a complete loss of the oxide dielectric properties. For sub-180 nm complementary metal–oxide–semiconductor (CMOS), soft-BD (SBD) happens more frequently, resulting in a large increase of gate current noise and small increase of gate current [46]. For ultra-thin oxides (< 2.5 nm), progressive-BD (PBD) follows SBD till HBD.

There are other types of breakdown. For instance, in digital breakdown (Di-BD), random telegraph noise (RTN) happens resulting in random jumps in current levels. The electron captures and emission by the traps cause the current fluctuations. Traps from GBs and interface can also contribute to this phenomenon besides traps in the percolation paths.

In analog breakdown (An-BD) (as shown in Fig. 1.18 region D), additional defects are created near the percolation path, enhancing Joule heating, and increasing localized temperature. Percolation paths become wider at this stage [47]. The device is significantly degraded at this An-BD stage and can no longer provide reliable functional operation.

The continuous clustering of defects can create percolation paths that link the gate to the substrate, and the TDDB process happens. Figure 1.20 shows the breakdown spots of an nMOS transistor.

## *1.5.2 Bias Temperature Instability (BTI)*

Bias temperature instability (BTI) is a degradation mechanism of MOSFETs. A threshold voltage shift may happen due to BTI when the MOS gate is applied a bias voltage under elevated temperature. The increase or decrease of threshold voltage changes the sensitivity of the transistor. There are two types of BTI; negative BTI (NBTI) that occurs in pMOS, and positive BTI (PBTI) that occurs in nMOS. At present, there is no consensus about the origins of both types of BTI phenomena. However, many researchers believe that NBTI originates from hole trapping in oxide defects [51–53], and PBTI, on the other hand, originates from electron trapping in oxide traps [54, 55].

### *1.5.3 Hot Carrier Injection (HCI)*

Hot carrier injection is a wear-out mechanism in MOSFETs. Hot carriers refer to the carriers that are accelerated under high electric fields and build up enough kinetic energy to inject into "forbidden" regions like the gate oxide. These carriers can be trapped in those regions or generate interface states, which can lead to threshold voltage shift and output conductance shift. This type of degradation first came to attention in mid-eighties as the supply voltage didn't decrease along with the transistor scaling [56–58]. As the transistors reduced power consumption and lowered the operating voltage, HCI became less severe since mid-nineties. Generally, HCI is more of a problem in nMOS devices [59], but it can also severe the NBTI in pMOS [60].

There are four commonly encountered hot carrier injection mechanisms including [56] (1) channel hot electron injection (CHE); (2) substrate hot electron injection (SHE); (3) drain avalanche hot carrier injection (DAHC); and (4) secondary generated hot electron injection (SGHE). In principle, all these four mechanisms work in a similar manner. Their difference is in the condition that hot carriers are generated. In the following subsections, these mechanisms are explained briefly.

#### 1.5.3.1 Channel Hot Electron (CHE) Injection

It usually happens when some electrons gain sufficient energy to pass the $Si/SiO_2$ barrier at the drain end of the channel of the transistor without energy loss after collisions with channel atoms [61]. CHE is worse in higher voltage because larger electric fields can cause avalanche multiplication and generate hot carriers. Since holes are heavier than electrons, CHE in nMOS is more significant than in pMOS [62]. The schematic of CHE is shown in Fig. 1.21.

#### 1.5.3.2 Substrate Hot Electron (SHE) Injection

This mechanism happens when the transistor bulk is applied with a very large positive or negative bias. The carriers in the substrate are driven to the $Si/SiO_2$ interface, gaining sufficient kinetic energy to by-pass the interface barrier, and injecting to the oxide [61]. Compared with other HCI mechanisms, hot carriers in this type are mostly uniformly distributed along the channel as shown in Fig. 1.22. The phenomenon usually happens in stacked devices with nonzero bulk bias.

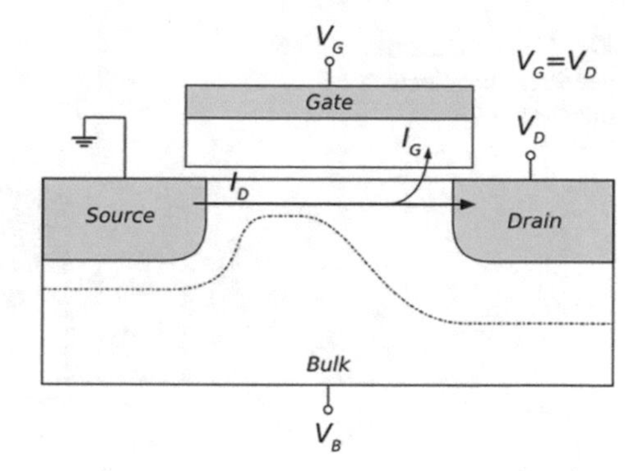


Fig. 1.21 Channel hot electron (CHE) injection process [61]

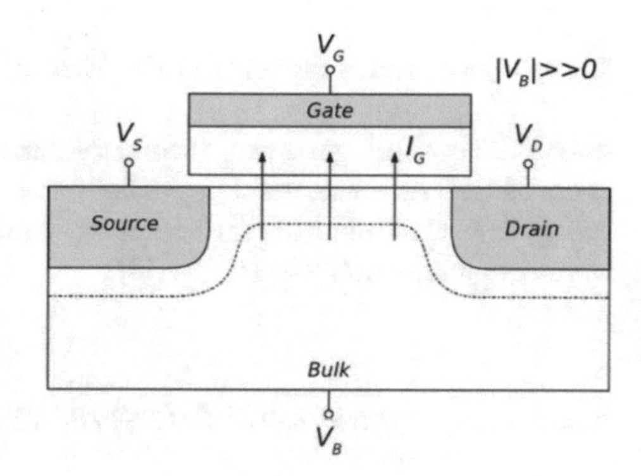


Fig. 1.22 Substrate hot electron injection process [61]

#### 1.5.3.3 Drain Avalanche Hot Carrier Injection (DAHC)

Under high drain voltage and low gate voltage, impact ionization of the channel current near the transistor drain can create electron–hole pairs as shown in Fig. 1.23 [61]. It can also be called as avalanche multiplication which leads to drain avalanche hot carrier generation (DAHC). Bulk current can also be generated in some cases. The gate oxide can heavily degraded because of this avalanche hot carrier injection.

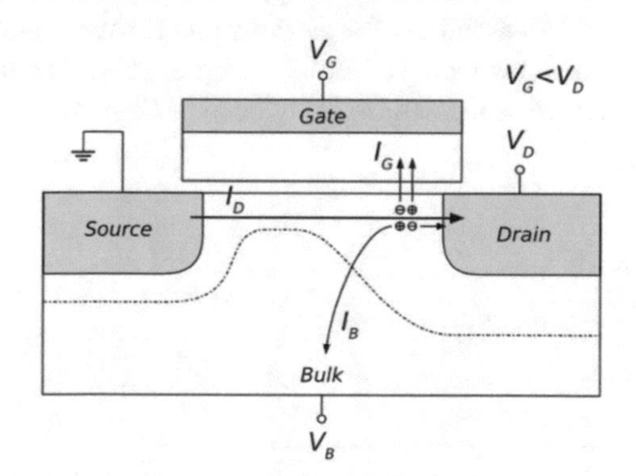


Fig. 1.23 Drain avalanche hot carrier injection process [61]

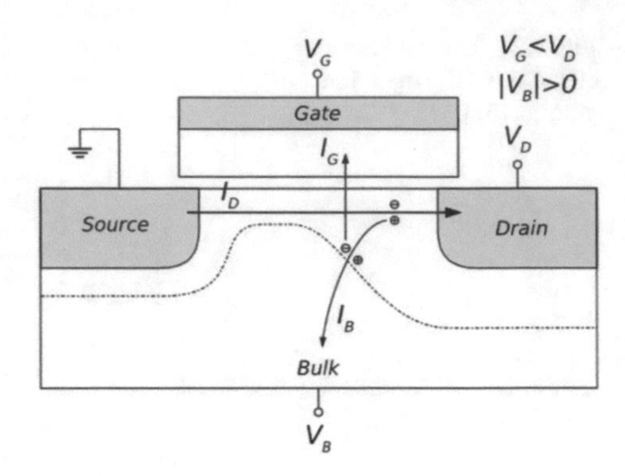


**Fig. 1.24** Secondary generated hot electron injection [61]

#### 1.5.3.4 Secondary Generated Hot Electron Injection (SGHE)

Impact ionization[5] generates secondary carriers. These carriers can be generated under DAHC or bremsstrahlung radiation near the drain, gaining sufficient kinetic energy and overcoming the surface energy barrier, that results in secondary generated hot electron injection [61] (Fig. 1.24).

### *1.5.4 Electromagnetic Interference (EMI)*

Electromagnetic interference (EMI) is the disturbance caused by external sources to the target electrical circuits. External sources can cause disturbance on the target electrical circuits because of EMI. Such disturbance can be conductive, capacitive, magnetic, or radiative between the source and the victim circuits. Types of EMI include on-chip crosstalk, simultaneous switching noise (SSN), energetic particles, and radiated EMI [61].

On-chip crosstalk happens between two circuits or circuit elements (like interconnect wires). The signal waves from one circuit can be influenced by the other. SSN is a case of common impedance crosstalk where subcircuits share the same power distribution bus. Large transient current spikes can be produced with the simultaneous switching of multiple digital gates. The energetic particles and radiation-related interference will be introduced in Sect. 1.6.

[5] Happens when there is a sufficient large, applied voltage.

### 1.5.5 Electrostatic Discharge (ESD)

When two charged objects come close to each other, electrostatic discharge (ESD) can happen and generate electricity flow suddenly. Triboelectric charging, electrostatic induction, and energetic particles can cause ESD. Triboelectric charging happens between two different materials which experience contact electrification in contact and then separate. Electrostatic induction happens when a charged object is close enough to another uncharged object, resulting in charge redistribution in the certain object. Energetic particles can be particles from space, like particles in cosmic rays. Typically, ESD can cause failure in semiconductor chip dialectics or metal interconnects [63]. Severe ESD can cause junction burn-out.

Electrostatic discharge can generate high voltage and cause damage to the electronics. Thus, it must be carefully prevented in manufacturing and quality control processes. Moreover, ESD can be the cause of explosions with just a spark in some other industries.

### 1.5.6 Electrical Overstress (EOS)

Electrical overstress (EOS) can be induced by temperature, electric field, and electromigration. EOS can cause thermal runaway, which results in local thermal conductivity loss and producing more heat locally [64]. Current crowding can be caused by the localized heat, further severe the heat localization and can even burn the device as shown in Figs. 1.25 and 1.26.

### 1.5.7 Electromigration (EM)

Electromigration involves the metal ion migration due to momentum transfers under electrical stresses [61]. Voids and hillocks can be created due to the gradual shift of metals. Images of voids and hillocks are shown in Fig. 1.27. EM is sensitive with the current density in the circuit. As integrated circuits scale down these days, the effect of EM becomes more significant, and some design modifications have been performed. For instance, in advanced semiconductor manufacturing processes aluminum is being replaced by copper. Since copper has much higher melting point so the atomic diffusion is slower than aluminum.

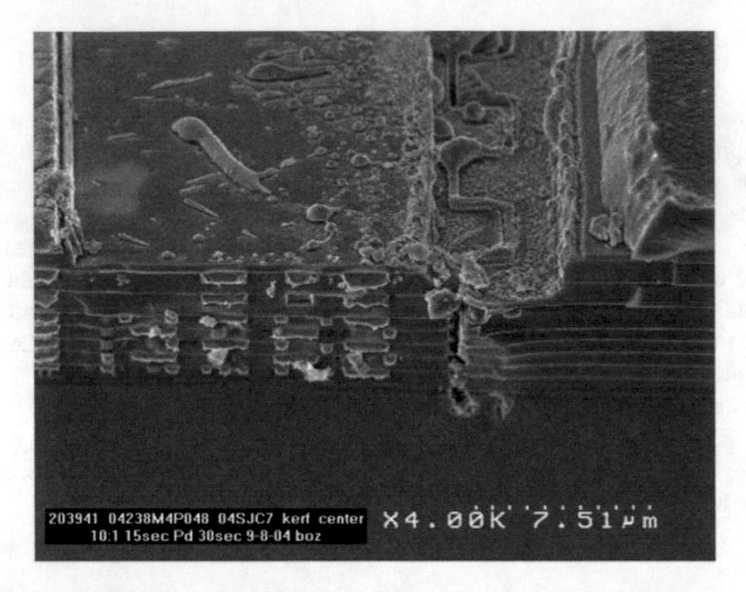


**Fig. 1.25** Plasma arcing in semiconductor manufacturing in 300 mm wafer fabricators causes EOS failure [65]

**Fig. 1.26** EOS failure in packaged semiconductor [66]

### *1.5.8 Tin Whiskers*

Contact resistance in electronics is carefully controlled to ensure the device's functionalities. The contact degradation can be mechanical-induced, thermal-induced, and electrical-induced. For example, similar to packaging failure, thermal expansion mismatch can induce internal mechanical stress and lead to fatigue cracking. Corrosion can generate non-conductive oxides and raise the resistance high up. Electrically, electromigration and brittle intermetallic layer formation can both cause contact

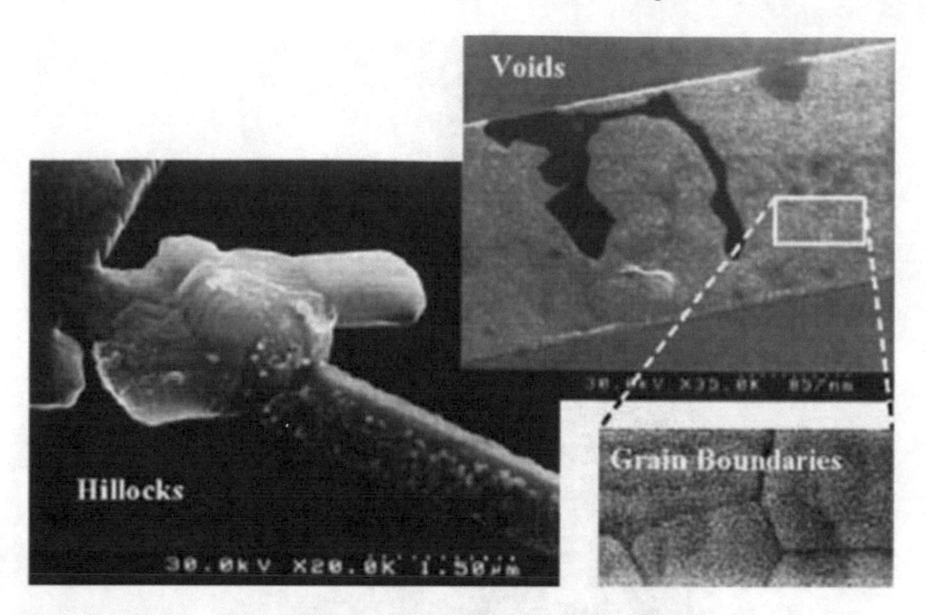


**Fig. 1.27** Hillock and void formations in wires due to electromigration [67]

degradation and failure. Beyond EM discussed in the former section (Sect. 1.5.7), this section will further introduce the tin whisker formation and its possible mechanisms.

Tin whiskers are the whisker-shaped crystalline structures generally grown from pure tin or zinc-finished surfaces. Different from dendrites which grow along a surface (*X*–*Y* plane), tin whiskers grow outward from a surface (*Z* axis). Typically, tin whiskers have length around 1 mm and diameter around 1 μm [68], but the sizes can vary depending on the cases. The actual growth mechanism of tin whiskers has not yet come to a consensus, but this degradation is thought to be related to the diffusion of atoms within the finish surfaces. Several theories have been raised: (1) Tin whiskers can be a response to the stress relief within the tin plating; (2) the growth of the tin whiskers may result from the recrystallization and abnormal grain growth in tin grains [68]. It is shown that neither metal dissolution nor electromagnetic field is required for the formation of tin whiskers. Figure 1.28 shows TEM image of long tin whiskers, and Fig. 1.29 is a schematic of probable tin whisker formation mechanism.

Tin whiskers can cause short circuits, metal vapor arc, and contamination. The whisker can be fused if the current flow exceeds the fusing current of the whisker, and in this case, the short circuit is transient. Otherwise, the short circuit caused by the whisker is permanent. Under high levels of current and voltage, the tin whisker can be vaporized into conductive plasma of metal ions that form destructive metal vapor arcs. The required initiation and sustaining power for metal vapor arcs is smaller under lower air pressure. Thus, in space-based applications, this phenomenon needs to be paid more attention to. Additionally, the presence of whiskers can interfere with sensitive optics and cause contamination [71].

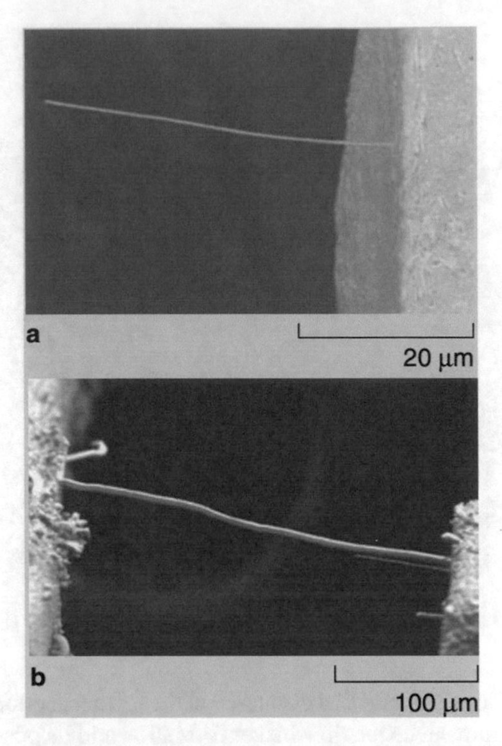


**Fig. 1.28** Examples of long tin whiskers on a 15 μm thick tin film over Cu alloy (ASTM BB42) after exposure to room ambient for 222,981 h [69]

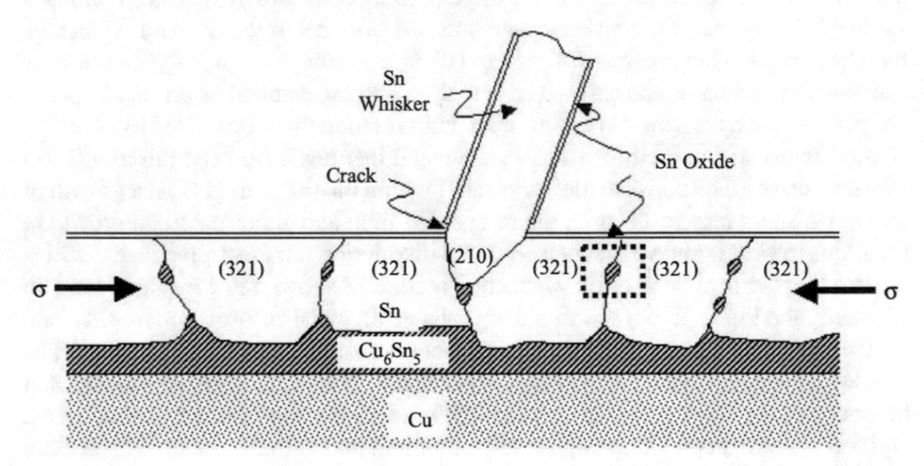


**Fig. 1.29** Diffusion of Cu into the shaded intermetallic compound to react with Sn generates a compressive stress in the volume [70]

### *1.5.9 Self-healing Accumulation*

When overvoltage is applied on non-metallized film capacitors, small localized dielectric breakdown can occur and short the electrodes. However, in metallized film capacitors, since the electrode foils are thin, the foils can vaporize under high energy density in the fault area and avoid breakdown. This is the self-healing phenomenon [72]. Three factors affect self-healing, which are the working voltage, mechanical pressure between the winding layers, and the metallization thickness [73]. A thin layer of Al or Zn coating on the plastic film is used in the metallized film capacitor, and the thickness of this coating determines the energy needed for self-healing process. Typical plastic film dielectrics include polypropylene (PP), polycarbonate, and polyester, whose self-healing capabilities are different. Certain amount of energy is required to initiate the vaporization process. However, if the energy exceeds the tolerance level, weak spots will be created and may further cause avalanche breakdowns.

The capacitance can also gradually decrease because of the accumulation of multiple self-healing events. The metal layer vaporization reduces the electrode area over time. Additionally, self-healing events increase the equivalent series resistance (ESR) of the capacitor, which also reduces the capacitor's lifetime.

### *1.5.10 Self-heating*

The self-heating in the metallized film capacitors happens when body generated power exceeds the surface power dissipation capability. Self-heating raises the temperature of the capacitor, reduces the breakdown voltage, and even melts the capacitor [74]. Several factors can affect the self-heating temperature: ESR, current across the capacitor, and the thermal resistance between the case and the ambient temperature [74].

### *1.5.11 Electrochemical Corrosion*

For metallized film capacitors, the electrochemical corrosion of the thin electrode can be caused by the high ripple current and voltage [75]. The oxygen and/or moisture of the polymer migrates to the polymer and metallization interface during this corrosion process. Factors such as temperature, electrode thickness, stress, and frequency affect the corrosion process. Also, the material selection of the dielectric layer in metallized film capacitors can make a difference in the sensitivity of ripple current, for example, PP is less sensitive to the ripple current than polyethylene terephthalate (PET). Consequently, metallized film capacitors using PP as the dielectric have better resistance to the electrochemical corrosion.

### *1.5.12 Thermal Overstress Induced Electrolyte Evaporation*

Electrolyte evaporation is the primary wear-out mechanism in electrolyte capacitors. This mechanism is caused by high temperatures within the capacitor core. For electrolytic capacitors under overstress, the electrolyte diffuses as vapor through the selling material and the diffusion rate can be accelerated by the increasing vapor pressure due to elevated temperature [75]. The loss of electrolyte decreases the capacitance and increases ESR.

## 1.6 Radiation Degradation and Failure Mechanisms

Radiation is defined as emission or transmission of energy through space or a medium [78]. The word radiation came up from waves radiating that moves outward in every direction from a source [76]. There are two fundamental damage mechanisms caused by radiations including lattice displacement and ionization effects [77].

### *1.6.1 Lattice Displacement*

Threshold displacement energy (TDE) is the most common quantity for describing radiation damage in materials. Thus, displacement energy is defined as the minimum required energy to create a stable defect in a material [81]. At the higher energy than the threshold level, a stable displacement will be generated in the structure. TDE is unique to each direction in a crystal lattice, and a solid crystalline material will typically have a minimum and an average TDE value separately identified. Lattice displacement may occur from proton, neutron, heavy ion, or high energy gamma radiation if the energy of that radiation exceeds the TDE in that crystal direction [78, 79]. The rearrangement of atoms in the crystal lattice caused by this radiation can cause lasting damage and increase the number of recombination centers [79]. Additionally, high doses in a short time span can cause partial annealing of the lattice [80] (Fig. 1.30).

### *1.6.2 Ionization Effects*

Ionization effects result from the emission of energy from a medium through space or a medium that is sufficient to ionize molecules or atoms. The common sources of the ionizing radiation are X-rays, Alpha, Beta, Gamma, and Neutron radiations (e.g., [85, 86]). The ionizing and nonionizing effects of energy particles are hard to define since each atom and molecule need different energy to be ionized. For example,

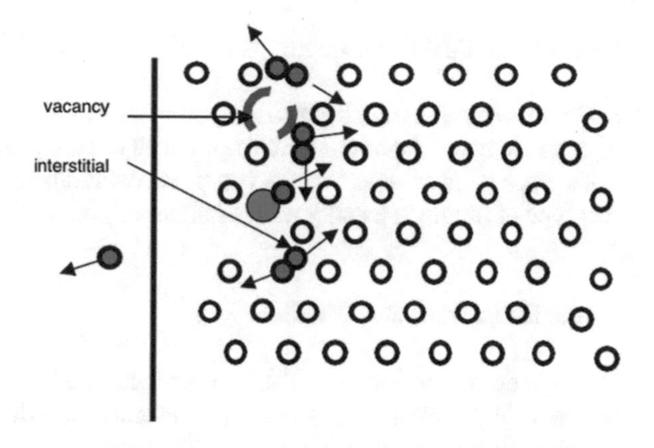


**Fig. 1.30** Lattice displacement

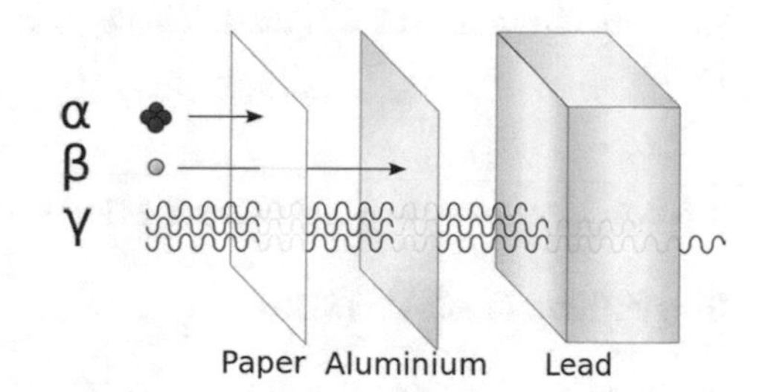


**Fig. 1.31** Alpha, Beta, and Gamma radiation

if an X-ray photon collides an atom, the atom may absorb the energy and knock one electron to a higher orbit level. If the photon has enough energy to remove the electron from the atom, however, ionization will take place [81]. An example of this effect can be seen in an X-ray machine. Bones and soft tissues absorb different levels of energies and X-ray machines utilize this fact to see bones and tissues (Fig. 1.31).

A single event effect (SEE) is an ionization process by one single ionizing particle (ion, electron, or photon) in an electronic device. The ionization process results in change of state in or close to an important node of a logic element. This process is known as a soft error. The type of SEE in electronic devices depends on device type and technology, localization, and amount of radiation. In this section, SEEs are categorized into non-destructive and destructive single event effects [82].

#### 1.6.2.1 Non-destructive Single Event Effects

Single Event Upsets (SEU)

Single event upsets occur in memories and sequential logic circuits. A single event upset causes changes of bits on a memory array, i.e., data loss. It can happen at any time, either at reading, writing, or the idle state of a circuit. A re-write can correct the error. This effect can show up as multiple bit upsets caused by a single ion [83].

Single Event Functional Interrupts (SEFI)

A single event upset that leads to temporary loss of device functionality is considered as a single event functional interrupt (SEFI). This phenomenon occurs in control registers, e.g., processors. SEFIs can be recovered by soft reset or power cycle, and this type of failure is usually more challenging than SEU [84, 85].

Single Event Transients (SET)

In single event transients (SETs), an ion hit caused by cosmic rays leads to a voltage transient in the circuit. It is similar to electrostatic discharge and occurs in logic circuits. As circuit speed increases, SETs happen more often and become indistinguishable from intended signals. This type of failure is recoverable with soft reset [86].

#### 1.6.2.2 Destructive Single Event Effects

Single Event Latch-Up (SEL)

A single event latch-up (SEL) occurs due to a low-impedance path generated by heavy ions or protons between two inner-transistor junctions. This results in a latch-up of the device until it is power-cycled, or the power supply and substrate creates a parasitic structure that could cause current overflow and device damage [87]. SEL is a strongly temperature-dependent event [88].

Single Event Gate Rupture (SEGR)

Under a high electric field between gate silicon, heavy ions passing through the gate oxide will leave a path to discharge the capacitor. This process causes local overheating and microscopic explosion of the gate dielectric region and leads to permanent damage to gate dielectric [89].

Single Event Burn-Out (SEB)

A single event burn-out (SEB) occurs when the substrate under the source region gets forward-biased and the drain-source voltage is higher than the breakdown voltage of the parasitic structures. This process requires a sufficient magnitude of current to overheat the device to destruction. It occurs often in bipolar and MOSFET power transistors and leads to permanent damage [89].

Single Event Hard Errors (SEHE)

A single event hard error (SEHE) occurs when heavy ions strike on memory arrays, leading to a large energy transfer. It causes an unchangeable state, called a stuck bit in memory, and leads to permanent damage [90, 91].

## 1.7 Human Failure Modes

Human failure modes can be classified into human error and violation. These two failures are covered in more detail in the following sections [92].

### *1.7.1 Human Error*

Human error is defined as an unintentional decision or action and can be classified to slips and lapses (skill-based errors), and mistakes. Human errors are possible to happen to even the well-trained and/or experienced person [92].

#### 1.7.1.1 Skill-Based Errors (Slip and Lapse)

Slips and lapses occur in tasks that we can perform with low conscious attention; for instance, driving a vehicle. The tasks with low conscious attention requirements are very vulnerable to slips and lapses when the attention is diverted for a very short time. Slip refers to a failure to perform a task as intended. It includes but not limited to cases such as [92]:

- Performing a task at the wrong time; for instance, braking too late in a car
- Missing one or more steps in a task; for instance, forgetting to turn on the oven before cooking
- Performing an action with too little or too much strength; for instance, breaking a paperclip from too much bending
- Moving an object in the wrong direction; for instance, changing to the wrong gear in a car by moving the joystick in the wrong direction
- Performing a correct action on the wrong object; for instance, reading the wrong value on a thermostat.

Lapse is defined by either forgetting to do a necessary action or losing your place partway through a task [92]. Some examples of lapse include:

- Forgetting to tighten a screw after placing it
- Forgetting to retrieve a jacket that has been set down

- Being interrupted in the middle of an action and missing a step upon returning to it
- Forgetting to protect door and window borders before painting a wall.

#### 1.7.1.2 Mistakes

Mistakes are defined as decision-making failures, i.e., it happens when a person performs a wrong task, believing it to be right. There are two main types of mistakes, including knowledge-based and rule-based mistakes [92].

An example of a knowledge-based mistake is poor judgment during a lane change resulting in an accident or disruption of traffic. The driver used their own assessment of the situation and unintentionally made a poor decision. A rule-based mistake may occur if an operator misinterprets procedure, such as pressing a wrong button in an emergency scenario after misunderstanding the purpose of the controls in an operation manual [92].

### *1.7.2 Violation*

Violations are intentional failures and deliberately performing the wrong task. The violation of health and safety principles and rules is one of the biggest causes of injuries and accidents at workplaces [92]. Some common examples of violations include:

- Not using protective equipment (such as seatbelts or safety goggles)
- Skipping procedures to check a machine before starting it
- Allowing employees to operate machinery they are not trained on
- Beginning work on a new project or site without receiving permission.

## 1.8 Software Errors and Failure Mechanisms

Software errors and failure mechanisms can be classified to function failure mode, value-related and timing failure modes, and interaction failure modes [93]. Figure 1.32 represents the functional decomposition of a software that includes a generic function with inputs and outputs, with outputs being used as input to other functions. The elements of the function can each have different failure modes. The process section is where the functional behavior and computation are executed, turning input(s) to output(s). Types of failure modes associated with this stage are called functional failures. A function in principle has at least one output. This may be a numerical value or a function call to other functions. Input is the output of another function or is given from an external interface. An output of a function can be an

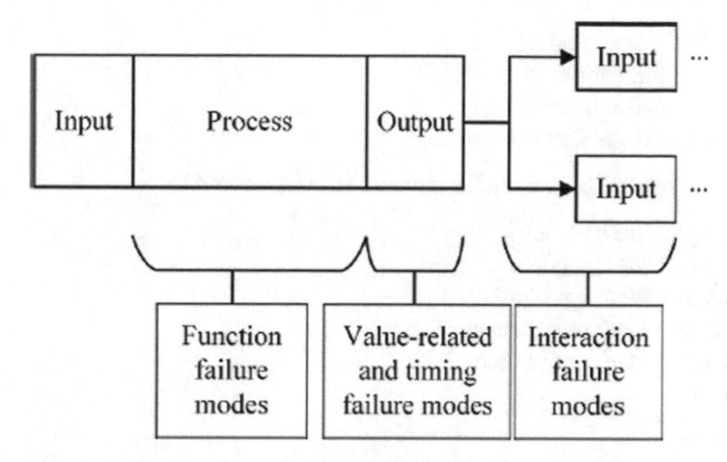


**Fig. 1.32** Functional decomposition of software [94, 95]

input to several other functions. Each function can have several inputs and outputs with a specific data type and range [94, 95].

The input, process and output could have different failure modes. The input could have interaction failure modes, the process can fail because of functional failure modes, and the output could fail due to value and time-related failure modes. In the following, some examples of each type of failure modes are presented [93–95].

The function failure modes include, but are not limited to:

- Omission of a function/missing operation
- Incorrect functionality
- Additional functionality
- No voting
- Incorrect voting
- Failure in failure handling.

The interaction failure modes include, but not limited to:

- Diverted/incorrect functional call
- No call of next function
- No priority for concurrent functions
- Incorrect priority for concurrent functions
- Communication protocol-dependent
- Unexpected interaction with input/output (IO) boards
  - File/database wrong name
  - File/database invalid name/extension
  - File/database does not exist
  - File/database is open
  - Wrong/invalid file format
  - File head contains error

- File ending contains error
- Wrong file length
- File/database is empty
- Wrong file/database contents.

The time-related failure modes include, but not limited to:

- Output provided: too early
- Output provided: too late
- Output provided: spuriously
- Output provided: out of sequence
- Output provided: not in time
- Output rate
  - Too fast
  - Too slow
  - Inconsistent
  - Desynchronized
- Duration
  - Too short
  - Too long
- Recurrent functions scheduled incorrectly.

The value-related failure modes include, but not limited to:

- No value
- Incorrect value
  - Too high
  - Too low
  - Opposite/inverse value
  - Value is zero
- Value out of range
  - Datatype allowable range
  - Application allowable range
- Noisy value/precision error
- Value with wrong data type
- Elements in a data array/structure
  - Too many
  - Too few
  - Data in wrong order
  - Data in reversed order
  - Enumerated value incorrect
- Non-numerical value

  - NaN
  - Infinite

- Redundant/frozen value
- Correct value is validated as incorrect
- Incorrect value is validated as correct
- Data is not validated.

## 1.9 Cyber–Physical–Human (CPH) Systems' Interaction Failure Mechanisms

A cyber–physical–human complex system is a system that is made of interacting components of software, hardware, and human operators (Fig. 1.33). In these types of systems, we have cyber–physical–human interactions [104, 105].

The three different elements in cyber–physical–human systems are hardware, software, and humans that have different failure behavior characteristics. It is important to understand the difference between them and their failure modes, presented in Sects. 1.2 to 1.8. But more importantly, having knowledge about all these three components would also allow us to better understand the behavior, interactions, and the associated failure mechanisms of the cyber–physical–human systems as a whole. Interaction in CPH systems is mutual or reciprocal action or influence in relation to certain function(s). The interactions result in exchange of matter, energy, force, and/or information, and we can have all the combinations as presented as the shaded area between hardware, software, hardware, human, and the human software, but also the intersection between all three.

In this definition, the interaction is connected to functions. In the design of systems, interactions are established between two or more components to accomplish some functions. Essentially, the system functions are realized through interactions among

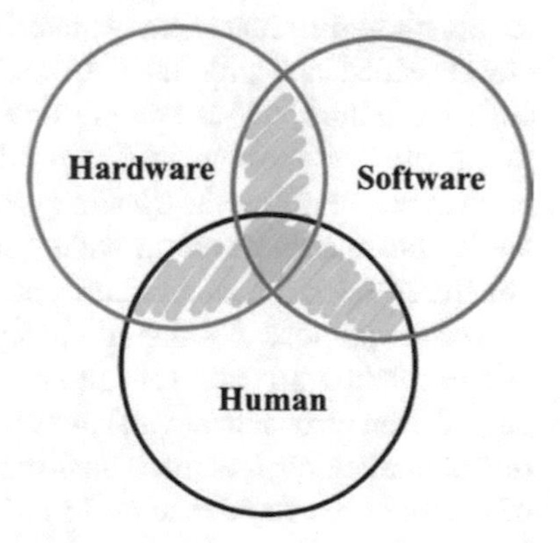


**Fig. 1.33** Cyber–physical–human system

**Table 1.2** Cyber–physical–human system interactions (P: physical, C: cyber, H: human)

| | P–P | P–C | P–H | C–C | C–H | H–H |
|---|---|---|---|---|---|---|
| Form | Physical contact, signal | Signal | Physical contact, signal | Statement | Signal | Physical contact, signal |
| Content | Mass, momentum, energy, information | Information | Momentum, energy, information | Information | Information | Momentum, information |
| Functional dependency | B is required for A to work, B prevents A, unavailability of A activates B, coordination, competition | | | | | |
| Time pattern | Continuous processes, discrete events | | | | | |

the system components. Without interaction, the system functions cannot be implemented even when all components can provide their own functions. However, that does not say that interaction always contributes to the implementation of system functions. Some interactions may actually prevent system functions from being realized. So having ways of representing, identifying, and recognizing, the interactions between these elements are very important. Table 1.2 presents a classification for the cyber–physical–human system interactions. This classification is very important because it helps us identify failure modes and mechanisms. In general, identification of CPH interactions' failure mechanisms is a very challenging task in complex systems, and the provided table presents a structure for analyzing and identifying the potential interaction failure mechanisms in a complex system.

## 1.10 Discussion and Conclusion

In this study, different types of material degradation mechanisms in complex systems are classified and presented. In addition to material degradation, human errors, software failure, and cyber–physical–human interaction failure are presented and discussed. The mechanism-based reliability studies are capable of facilitating the engineers to monitor the system health, help decision makers to weigh among reliability factors, and even provide guidance on the selections of materials, system structures, operation conditions, operating strategies, operators, and software.

For the purpose of reliability analysis of complex systems, proper understanding of cross-field degradation and failure mechanisms is essential. In many cases, various degradation mechanisms can have certain dependencies and consequently, the initiation of one degradation mechanism can positively or negatively affect the propagation of others in the same system. In order to have a more realistic reliability analysis, the interdependencies between failure mechanisms and cascading effects should be considered. In fact, an accurate reliability estimation of a complex system requires engineers not only to investigate the material, human and software reliability factors, but also to understand the interactions and dependencies among them. The current reliability analysis methods can consider certain degradation and failure mechanisms in a complex system. However, the next generation of reliability analysis methods is likely to be techniques that can consider degradation and failure interdependencies in a complex system.

## References


1. Parhizkar, T.: Long-term degradation-based modeling and optimization framework. In: Handbook of Research on Predictive Modeling and Optimization Methods in Science and Engineering, pp. 192–220. IGI Global (2018)
2. Sotoodeh, A.F., Parhizkar, T., Mehrgoo, M., Ghazi, M., Amidpour, M.: Aging based design and operation optimization of organic rankine cycle systems. Energy Convers. Manag. **199**, 111892 (2019)
3. Parhizkar, T., Mosleh, A., Roshandel, R.: Aging based optimal scheduling framework for power plants using equivalent operating hour approach. Appl. Energy **205**, 1345–1363 (2017)
4. Parhizkar, T., Hafeznezami, S.: Degradation based operational optimization model to improve the productivity of energy systems, case study: solid oxide fuel cell stacks. Energy Convers. Manag. **158**, 81–91 (2018)
5. Roshandel, R., Parhizgar, T.: A new approach to optimize the operating conditions of a polymer electrolyte membrane fuel cell based on degradation mechanisms. Energy Syst. **4**(3), 219–237 (2013)
6. Parhizkar, T., Utne, I.B., Vinnem, J.E., Mosleh, A.: Dynamic probabilistic risk assessment of decision-making in emergencies for complex systems, case study: dynamic positioning drilling unit. Ocean Eng. **237**, 109653 (2021)
7. Hu, Y., Parhizkar, T., Mosleh, A.: Guided simulation for dynamic probabilistic risk assessment of complex systems: concept, method, and application. Reliab. Eng. Syst. Saf. 108047 (2021)
8. Parhizkar, T., Vinnem, J.E., Utne, I.B., Mosleh, A.: Supervised dynamic probabilistic risk assessment of complex systems, part 1: general overview. Reliab. Eng. Syst. Saf. 107406 (2020)
9. Kandeva-Ivanova, M., Vencl, A., Karastoyanov, D.: Advanced Tribological Coatings for Heavy-Duty Applications: Case Studies. Bulgarian Academy of Sciences, Institute of Information and Communication Technologies; Prof. Marin Drinov Publishing House (2016)
10. Davis, J.R.: Surface Engineering for Corrosion and Wear Resistance, p. 56. ASM International (2001)
11. Wikimedia Foundation: Abrasion (mechanical). Wikipedia (2021, July 17). https://en.wikipedia.org/wiki/Abrasion_(mechanical) [Online]
12. Doi, T., Uhlmann, E., Marinescu, I.D. (eds.): Handbook of Ceramics Grinding and Polishing. William Andrew (2015)
13. Vencl, A., Gašić, V., Stojanović, B.: Fault tree analysis of most common rolling bearing tribological failures. In: IOP Conference Series: Materials Science and Engineering, vol. 174, no. 1, p. 012048. IOP Publishing (2017, Feb)
14. Abbasi, S.: Characterisation of airborne particles from rail traffic. Doctoral dissertation. KTH Royal Institute of Technology, Chicago (2011)
15. Shaffer, S.J., Glaeser, W.A.: Fretting fatigue. In: Fatigue and Fracture. ASM Handbook, vol. 19, pp. 321–330. ASM International (1996)
16. Callister, W.D., Rethwisch, D.G.: Failure. In: Fundamentals of Materials Science and Engineering: An Integrated Approach, chap. 9, 4th edn. Wiley, New York (2012)

17. Creep and Stress Rupture Properties: Total Materia Article (2010, Dec). www.totalmateria.com
18. Samuel Share, S.: Fatigue failure. Punchlist Zero (2021, Mar 2). https://www.punchlistzero.com/fatigue-failure/
19. Kim, Y., Hwang, W.: High-cycle, low-cycle, extremely low-cycle fatigue and monotonic fracture behaviors of low-carbon steel and its welded joint. Materials **12**(24), 4111 (2019)
20. Fujisawa, N., Fujita, Y., Yanagisawa, K., Fujisawa, K., Yamagata, T.: Simultaneous observation of cavitation collapse and shock wave formation in cavitating jet. Exp. Thermal Fluid Sci. **94**, 159–167 (2018)
21. Hughes, J.: Mechanical Resonance. Department of Physics and Astronomy, State University of New Jersey Rutgers, Piscataway, NJ (2005)
22. Callister, W.D., Rethwisch, D.G.: Mechanical properties. In: Fundamentals of Materials Science and Engineering: An Integrated Approach, chap. 7, 4th edn. Wiley, New York (2012)
23. Roylance, D.: Stress-Strain Curves. Department of Materials Science and Engineering, Massachusetts Institute of Technology, Cambridge, MA (2001, Aug 23)
24. Resiliency and Toughness | MATSE 81: Materials in Today's World (n.d.). https://www.e-education.psu.edu/matse81/node/2105
25. Callister, W.D., Rethwisch, D.G.: Deformation and strengthening mechanisms. In Fundamentals of Materials Science and Engineering: An Integrated Approach, chap. 8, 4th edn. Wiley, New York
26. Carmack, B.: Brittle Fracture (2013, Mar 1). https://failuremechanisms.wordpress.com/2013/03/01/brittle-fracture-2/ [Online]
27. Zeng, L., Olsson, R.: Buckling-induced delamination analysis of composite laminates with soft-inclusion. Swedish Defence Research Agency Technical Report, Feb (2002)
28. Callister, W.D., Rethwisch, D.G.: Phase diagrams. In: Fundamentals of Materials Science and Engineering: An Integrated Approach, chap. 10, 4th edn. Wiley, New York (2012)
29. Holzer, S.: Optimization for enhanced thermal technology CAD purposes. Thesis (2007, Nov 19). Retrieved from https://www.iue.tuwien.ac.at/phd/holzer/diss.html [Online]
30. Mirabella, D.A., Suárez, M.P., Aldao, C.M.: Understanding the hillock-and-valley pattern formation after etching in steady state. Surf. Sci. **602**(8), 1572–1578 (2008)
31. Xiao, C., Jiang, C.S., Liu, J., Norman, A., Moseley, J., Schulte, K., et al.: Carrier-transport study of gallium arsenide hillock defects. Microsc. Microanal. **25**(5), 1160–1166 (2019)
32. Brittle-Ductile Transition Temperature (2020, July 14). https://polymerdatabase.com/polymer%20physics/Crazing2.html
33. https://docplayer.net/56816427-Aae-590-mechanical-behavior-of-materials-deformabon-of-polymers-lecture-23-polymer-deformabon-fracture-and-crazing-4-9-13.html
34. Föll, H.: Ductile to Brittle Transition or Cold Shortness (n.d.). https://www.tf.uni-kiel.de/matwis/amat/iss/kap_9/illustr/s9_1_1.html [Online]
35. https://www.britannica.com/science/face-centred-cubic-structure
36. Askeland, D.R.: 22-4 thermal shock. In: Wright, W.J. (ed.) The Science and Engineering of Materials, 7th edn., pp. 792–793. Boston, MA (Jan 2015). ISBN 978-1-305-07676-1. OCLC 903959750
37. Zhang, X., Wu, Y., Xu, B., Wang, H.: Residual stresses in coating-based systems, part I: mechanisms and analytical modeling. Front. Mech. Eng. China **2**(1), 1–12 (2007)
38. Parhizkar, T., Roshandel, R.: Long term performance degradation analysis and optimization of anode supported solid oxide fuel cell stacks. Energy Convers. Manage. **133**, 20–30 (2017)
39. Roshandel, R., Parhizkar, T.: Degradation based optimization framework for long term applications of energy systems, case study: solid oxide fuel cell stacks. Energy **107**, 172–181 (2016)
40. Callister, W.D., Rethwisch, D.G.: Polymer structures. In: Fundamentals of Materials Science and Engineering: An Integrated Approach, chap. 4, 4th edn. Wiley, New York (2012)
41. Wikimedia Foundation: Depolymerization. Wikipedia (2021, Apr 8). https://en.wikipedia.org/wiki/Depolymerization

42. Callister, W.D., Rethwisch, D.G.: Corrosion and degradation of materials. In: Fundamentals of Materials Science and Engineering: An Integrated Approach, chap. 16, 4th edn. Wiley, New York (2012)
43. Jiao, Z., Jiang, L., Sun, J., Huang, J., Zhu, Y.: Outgassing environment of spacecraft: an overview. In: IOP Conference Series: Materials Science and Engineering, vol. 611, no. 1, p. 012071. IOP Publishing (2019, Oct)
44. Shubhakar, K., Pey, K.L., Kushvaha, S.S., O'Shea, S.J., Raghavan, N., Bosman, M., et al.: Grain boundary assisted degradation and breakdown study in cerium oxide gate dielectric using scanning tunneling microscopy. Appl. Phys. Lett. (2011)
45. Raghavan, N., Pey, K.L., Wu, X., Liu, W., Li, X., Bosman, M., Kauerauf, T.: Oxygen-soluble gate electrodes for prolonged high-$\kappa$ gate-stack reliability. IEEE Electron Device Lett. **32**(3), 252–254 (2011)
46. Gielen, G., De Wit, P., Maricau, E., Loeckx, J., Martin-Martinez, J., Kaczer, B., et al.: Emerging yield and reliability challenges in nanometer CMOS technologies. In: Proceedings of the Conference on Design, Automation and Test in Europe (2008, Mar)
47. Raghavan, N., Padovani, A., Li, X., Wu, X., Lip Lo, V., Bosman, M., et al.: Resilience of ultra-thin oxynitride films to percolative wear-out and reliability implications for high-$\kappa$ stacks at low voltage stress. J. Appl. Phys. (2013)
48. Raghavan, N., Pey, K.L., Shubhakar, K.: High-κ dielectric breakdown in nanoscale logic devices—scientific insight and technology impact. Microelectron. Reliab. **54**(5), 847–860 (2014)
49. Yazdani, S.: Electrostatic discharge (ESD) explained (2011)
50. http://www.electronicspub.com/article/%20%2026/7/ElectroStatic-Discharge-(ESD)-Exp
51. Schroder, D.K., Babcock, J.A.: Negative bias temperature instability: road to cross in deep submicron silicon semiconductor manufacturing. J. Appl. Phys. **94**(1), 1–18 (2003)
52. Kaczer, B., Grasser, T., Roussel, J., Martin-Martinez, J., O'Connor, R., O'Sullivan, B.J., Groeseneken, G.: Ubiquitous relaxation in BTI stressing—new evaluation and insights. In: 2008 IEEE International Reliability Physics Symposium (2008, Apr)
53. Grasser, T., Kaczer, B.: Evidence that two tightly coupled mechanisms are responsible for negative bias temperature instability in oxynitride MOSFETs. IEEE Trans. Electron Devices **56**(5), 1056–1062 (2009)
54. Crupi, F., Pace, C., Cocorullo, G., Groeseneken, G., Aoulaiche, M., Houssa, M.: Positive bias temperature instability in nMOSFETs with ultra-thin Hf-silicate gate dielectrics. Microelectron. Eng. 80, 130–133 (2005)
55. Ioannou, D.P., Mittl, S., La Rosa, G.: Positive bias temperature instability effects in nMOSFETs with $HfO_2$/TiN gate stacks. IEEE Trans. Device Mater. Reliab. **9**(2), 128–134 (2009)
56. Takeda, E., Suzuki, N., Hagiwara, T.: Device performance degradation to hot-carrier injection at energies below the Si-$SiO_2$ energy barrier. In: 1983 International Electron Devices Meeting, pp. 396–399. IEEE (1983, Dec)
57. Tam, S., Ko, P.K., Hu, C.: Lucky-electron model of channel hot-electron injection in MOSFET's. IEEE Trans. Electron. Devices **31**(9), 1116–1125 (1984)
58. Hu, C., Tam, S.C., Hsu, F.C., Ko, P.K., Chan, T.Y., Terrill, K.W.: Hot-electron-induced MOSFET degradation-model, monitor, and improvement. IEEE J. Solid-State Circuits **20**(1), 295–305 (1985)
59. Lunenborg, M.M.: MOSFET hot-carrier degradation: failure mechanisms and models for reliability circuit simulation (1998)
60. Parthasarathy, C.R., Denais, M., Huard, V., Ribes, G., Roy, D., Guérin, C., et al.: Designing in reliability in advanced CMOS technologies. Microelectron. Reliab. **46**(9–11), 1464–1471 (2006)
61. Maricau, E., Gielen, G.: Analog IC Reliability in Nanometer CMOS. Springer Science & Business Media, Berlin (2013)

62. Hu, C., Tam, S.C., Hsu, F.C., Ko, P.K., Chan, T.Y., Terrill, K.W.: Hot-electron-induced MOSFET degradation-model, monitor, and improvement. IEEE J. Solid-State Circuits **20**(1), 295–305 (1985)
63. Stellari, F., Song, P., McManus, M.K., Weger, A.J., Gauthier, R.J., Chatty, K.V., et al.: Study of critical factors determining latchup sensitivity of ICs using emission microscopy. In: International Symposium for Testing and Failure Analysis (2003, Oct)
64. Díaz, C.H., Duvvury, C., Kang, S.M.: Studies of EOS susceptibility in 0.6 μm nMOS ESD I/O protection structures. J. Electrostat. **33**(3), 273–289 (1994)
65. Voldman, S.H.: Electrostatic discharge (ESD) and electrical overstress (EOS)—the state of the art for methods of failure analysis, and testing in components and systems. In: 2016 IEEE 23rd International Symposium on the Physical and Failure Analysis (2016, July)
66. Voldman, S.H.: Evolution, revolution, and technology scaling—the impact on ESD and EOS reliability. Front. Mater. **5**, 33 (2018)
67. Jerke, G., Lienig, J.: Hierarchical current-density verification in arbitrarily shaped metallization patterns of analog circuits. IEEE Trans. Comput. Aided Design Integr. Circuits Syst. **23**(1), 80–90 (2004)
68. https://nepp.nasa.gov/whisker/background/index.htm
69. Osenbach, J.W.: Tin whiskers: an illustrated guide to growth mechanisms and morphologies. JOM **63**(10), 57–60 (2011)
70. Tu, K.N.: Spontaneous tin whisker growth: mechanism and prevention. In: Solder Joint Technology, pp. 153–181. Springer, New York (2007)
71. https://nepp.nasa.gov/whisker/background/index.htm (2021)
72. Gebbia, M.: Introduction to Film Capacitors. Illinois Capacitor, Inc. (2013)
73. Qin, S., Ma, S., Boggs, S.A.: The mechanism of clearing in metalized film capacitors. In: 2012 IEEE International Symposium on Electrical Insulation, pp. 592–595. IEEE (2012, June)
74. Gallay, R.: Metallized film capacitor lifetime evaluation and failure mode analysis (2016). arXiv preprint arXiv:1607.01540
75. Gupta, A., Yadav, O.P., DeVoto, D., Major, J.: A review of degradation behavior and modeling of capacitors. In: International Electronic Packaging Technical Conference and Exhibition, vol. 51920, p. V001T04A004. American Society of Mechanical Engineers (2018, Aug)
76. Rohrlich, F.: The definition of electromagnetic radiation. Il Nuovo Cimento (1955–1965) **21**(5), 811–822 (1961)
77. Slater, J.C.: The effects of radiation on materials. J. Appl. Phys. **22**(3), 237–256 (1951)
78. Varley, J.O.: A mechanism for the displacement of ions in an ionic lattice. Nature **174**(4436), 886–887 (1954)
79. Kikuchi, R., Beldjenna, A.: Continuous displacement of "lattice" atoms. Phys. A Stat. Mech. Appl. **182**(4), 617–634 (1992)
80. Kozlovskiy, A.L., Kenzhina, I.E., Zdorovets, M.V.: FeCo–$Fe_2CoO_4$/$Co_3O_4$ nanocomposites: phase transformations as a result of thermal annealing and practical application in catalysis. Ceram. Int. **46**(8), 10262–10269 (2020)
81. Awad, H., Khamis, M.M., El-Aneed, A.: Mass spectrometry, review of the basics: ionization. Appl. Spectrosc. Rev. **50**(2), 158–175 (2015)
82. Suge, Y., Xiaolin, Z., Yuanfu, Z., Lin, L., Hanning, W.: Modeling and simulation of single-event effect in CMOS circuit. J. Semiconductors **36**(11), 111002 (2015)
83. Hussein, J., Swift, G.: Mitigating single-event upsets (2012)
84. Maqbool, S.: A system-level supervisory approach to mitigate single event functional interrupts in data handling architectures. University of Surrey, UK (2006)
85. Aranda, L.A., Sánchez-Macián, A., Maestro, J.A.: A methodology to analyze the fault tolerance of demosaicking methods against memory single event functional interrupts (SEFIs). Electronics **9**(10), 1619 (2020)
86. Ferlet-Cavrois, V., Massengill, L.W., Gouker, P.: Single event transients in digital CMOS—a review. IEEE Trans. Nuclear Sci. **60**(3), 1767–1790 (2013)
87. Rui, C., Jian-Wei, H., Han-Sheng, Z., Yong-Tao, Y., Shi-Peng, S., Guo-Qiang, F., Ying-Qi, M.: Comparative research on "high currents" induced by single event latch-up and transient-induced latch-up. Chin. Phys. B, **24**(4), 046103 (2015)

88. Karp, J., Hart, M.J., Maillard, P., Hellings, G., Linten, D.: Single-event latch-up: increased sensitivity from planar to FinFET. IEEE Trans. Nuclear Sci. **65**(1), 217–222 (2017)
89. Titus, J.L.: An updated perspective of single event gate rupture and single event burnout in power MOSFETs. IEEE Trans. Nuclear Sci. **60**(3), 1912–1928 (2013)
90. Haran, A., Barak, J., David, D., Keren, E., Refaeli, N., Rapaport, S. Single event hard errors in SRAM under heavy ion irradiation. IEEE Trans. Nuclear Sci. **61**(5), 2702–2710 (2014)
91. G. M. P. D. J. &. J. A. H. (. A. n. c. o. s. e. h. e. [. c. I. t. o. n. s. 4. 2.-2. Swift
92. Wiegmann, D.A., Shappell, S.A.: A Human Error Approach to Aviation Accident Analysis: The Human Factors Analysis and Classification System. Routledge (2017)
93. Feiler, P., Hudak, J., Delange, J., Gluch, D.P.: Architecture fault modeling and analysis with the error model annex, version 2. Carnegie-Mellon University, Pittsburgh, PA, USA (2016)
94. Thieme, C.A., Mosleh, A., Utne, I.B., Hegde, J.: Incorporating software failure in risk analysis—part 1: software functional failure mode classification. Reliab. Eng. Syst. Saf. **197**, 106803 (2020)
95. Thieme, C.A., Mosleh, A., Utne, I.B., Hegde, J.: Incorporating software failure in risk analysis––part 2: risk modeling process and case study. Reliab. Eng. Syst. Saf. **198**, 106804 (2020)
96. Groth, K., Mosleh, A.: Data-driven modeling of dependencies among influencing factors in human-machine interactions. In: Proceedings of ANS PSA 2008 Topical Meeting on CD-ROM. American Nuclear Society, Knoxville (2008)
97. Liu, J.: Autonomous retailing: a frontier for cyber-physical-human systems. In: Principles of Modeling, pp. 336–350. Springer, Cham (2018)

**Dr. Tarannom Parhizkar** is a research scientist in the B. John Garrick Institute for the Risk Sciences, at University of California, Los Angeles (UCLA). She has been technical advisor to several national and international organizations. She conducts research on performance optimization and probabilistic risk analysis of complex systems and has authored over 50 publications including books and technical papers. She has an extensive background and experience both in academia and industry, addressing complex challenges encountered in risk assessment and management practices.

**Theresa Stewart** grew up in the Southern California area and is an ongoing Ph.D. student in Materials Science and Engineering at University of California, Los Angeles. She completed her undergraduate degree in Materials Science at California Polytechnic State University, San Luis Obispo, in 2017. Her primary research area is in the long-term reliability of non-metallic distribution pipelines considering environmental degradation, particularly from diffusion of internal and external fluids and erosion. She has also worked at UCLA's B. John Garrick Institute for the Risk Sciences for several years to study and contribute to a variety of system-level reliability and modeling projects.

**Lixian Huang** is currently a Ph.D. candidate in Materials Science and Engineering department at UCLA. She is a graduate student researcher in B. John Garrick Institute for the Risk Sciences (GIRS) at UCLA. She got her bachelor's degree with honors in Materials Science and Engineering at the University of California, Davis, in 2016. She is an IEEE student member. Her research interests include physic-of-failure (PoF) reliability modeling, kinetic Monte Carlo (KMC) simulation of semiconductors, ASIC statistical yield modeling, resistive random-access memory (RRAM) reliability modeling, Bayesian approach-assisted trade-offs analysis of electronic components, electronic commercial-off-the-shelf (COTS) part reliability estimation, and complex system reliability analysis.

**Ali Mosleh** is Distinguished University Professor and Evelyn Knight Endowed Chair in Engineering. He is also the Director of UCLA's B. John Garrick Institute for the Risk Sciences. Previously he was the Nicole J. Kim Eminent Professor of Engineering and Director of the Center for Risk and Reliability at the University of Maryland. He was elected to the National Academy of Engineering in 2010 and is a Fellow of the Society for Risk Analysis, and the American Nuclear Society, recipient of several scientific achievement awards, including the American Nuclear Society Tommy Thompson Award. He has been technical advisor to numerous national and international organizations. He conducts research on methods for probabilistic risk analysis and reliability of complex systems, holds several patents, and has authored or co-authored over 600 publications including books, guidebooks, and technical papers.

# Chapter 2
# Simplified Approach to Analyse the Fuzzy Reliability of a Repairable System

**Komal**

**Abstract** This chapter presents a simplified approach to analysing the fuzzy reliability of a reparable system by utilising uncertain data collected from different sources. This technique uses fault tree to model the system, triangular fuzzy numbers to quantify uncertain information, and the Lambda-Tau (LT) method to discover functional equations for six distinct system reliability indices, whilst simplified arithmetic operations are using for calculation. The proposed strategy is utilised to estimate the efficiency of a paper producing plant's washing system by determining its fuzzy reliability for various levels of uncertainty. Results are compared with two existing techniques, namely conventional LT and fuzzy Lambda-Tau (FLT). To determine how different operating conditions affect system performance, sensitivity analysis and long-term reliability evaluation are conducted. The significant system components are ranked using the $V$-index. The results indicate that, in comparison with the current FLT technique, the proposed approach is straightforward to implement for assessing the fuzzy reliability of any substantial and intricate repairable industrial system. The presented approach might be very useful to maintenance professionals in designing an effective maintenance strategy for enhancing system performance in a very easy manner.

**Keywords** Fuzzy reliability · Lambda-Tau · Fuzzy Lambda-Tau · Triangular fuzzy numbers · Washing system

## Abbreviation

| | |
|---|---|
| $\tilde{A}$ | Fuzzy set $\tilde{A}$ |
| $\mu_{\tilde{A}}(x)$ | Membership degree of $x$ in $\tilde{A}$ |
| $a_1$, $a_2$ and $a_3$ | Left, modal, and right end values of TFN $\tilde{A} = (a_1, a_2, a_3)$ |
| $\lambda_i$ | System's $i$th component's crisp failure rate |

Komal (✉)
Department of Mathematics, School of Physical Sciences, Doon University, Dehradun, Uttarakhand 248001, India
e-mail: karyadma.iitr@gmail.com



H. Garg (ed.), *Advances in Reliability, Failure and Risk Analysis*, Industrial and Applied Mathematics, https://doi.org/10.1007/978-981-19-9909-3_2

| | |
|---|---|
| $\lambda_s$ | System's crisp failure rate |
| $\mu_s$ | System's crisp repair rate |
| $\tilde{\tau}_i$ | System's $i$th component's fuzzy repair time |
| $\tilde{\tau}_s$ | System's fuzzy repair time |
| $\overline{A}$ | Defuzzified value of fuzzy set $\tilde{A}$ |
| $\tilde{q}_T$ | System's fuzzy failure rate |
| $V(\tilde{q}_T, \tilde{q}_{T_i})$ | $V$-index measures the difference between $\tilde{q}_T$ and $\tilde{q}_{T_i}$ |
| $\text{MTTR}_s$ | Mean time to repair |
| $\text{ENOF}_s(0, t)$ | Expected number of failures |
| $R_s(t)$ | System reliability at time $t$ |
| $x$ | Element of fuzzy set $\tilde{A}$ |
| $X$ | Universal set |
| $+, -, \otimes, /$ | Sum, difference, product, and division between two TFNs |
| $\tau_i$ | System's $i$th component's crisp repair time |
| $\tau_s$ | System's crisp repair time |
| $\tilde{\lambda}_i$ | System's $i$th component's fuzzy failure rate |
| $\tilde{\lambda}_s$ | System's fuzzy failure rate |
| $t$ | Time $t$ |
| $\overline{R}$ | Defuzzified value of any fuzzy reliability index $\tilde{R} = (r_1, r_2, r_3)$ in TFN form |
| $\tilde{q}_{T_i}$ | System's fuzzy failure rate by taking its $i$th component $\tilde{\lambda}_i = \tilde{0}$ |
| $\text{MTTF}_s$ | Mean time to failure |
| $\text{MTBF}_s$ | Mean time between failures |
| $A_s(t)$ | System availability at time $t$. |

## 2.1 Introduction

Nowadays, reliability has become a key index for improving the performance of complex engineering systems such as warships, LNG carrier propulsion systems, underwater dry repair cabins, waste clean-up manipulators, and seabed storage tanks. Experts are expected to develop solutions to issues found during the evaluation process, create maintenance strategies to decrease the frequency of system failures, and even make an effort to lower maintenance costs by identifying vulnerable parts of these critical systems in order to analyse the reliability of complex engineering systems [1]. In general, these systems must be very reliable in order to work safely and without failure. System failure, on the other hand, is a natural occurrence that cannot be totally prevented, but it can be reduced in frequency by planning and implementing suitable maintenance techniques. To appraise the reliability of such complex engineering systems, a variety of qualitative and quantitative techniques are available, such as failure mode and effect analysis (FMEA), root cause analysis (RCA), fault tree analysis (FTA), reliability block diagram (RBD), Petri-nets (PN), Bayesian approach, and Markovian approach [2]. FTA is a well-known, straightforward, and logical strategy that leverages underlying causes of system failure to calculate the likelihood of an undesirable occurrence such as system failure [3]. The qualitative analysis of FTA generates minimal-cut sets, whereas the quantitative analysis uses the system's fundamental event failure probabilities to evaluate the system's top event failure probability.

The literature review found that various studies have used FTA to evaluate the reliability of such complicated repairable systems. Mishra [4] specifically emphasised the Lambda-Tau (LT) approach, which assesses the reliability of complicated repairable systems using FTA, crisp data, probability theory, and fundamental arithmetic operations. Because the LT approach employs crisp data that may include uncertainty, Knezevic and Odoom [5] recognised this limitation and developed the fuzzy Lambda-Tau (FLT) technique. To evaluate the fuzzy reliability of any complicated repairable system, the FLT methodology uses Petri-nets, the LT method, triangular fuzzy numbers (TFNs), and related $\alpha$-cut coupled with interval arithmetic-based sophisticated fuzzy arithmetic operations defined on TFNs [6]. Since its inception, several extensions have been developed by several researchers to handle different kinds of problems. Some researcher designed hybridised techniques by combining FLT and a range of heuristic techniques such as genetic algorithms, artificial bee colonies, particle swarm optimisation, and so on [7–11]. Some researchers replaced TFNs in the FLT technique with triangle intuitionistic fuzzy sets (TIFS) or triangle vague sets (TVS) and employed associated $\alpha$-cut based arithmetic operations [12, 13]. Later, in order to limit the accumulating problem of fuzziness and to adopt other forms of fuzzy environments, some researchers employed $T_\omega$-norm [14] and $\alpha$-cut linked arithmetic operations [6] and adopted either TFNs [15–18], different types of fuzzy membership functions [19, 20], TIFS [21, 22], TVS [23], or different types of vague sets [24] for analysing complicated systems' fuzzy, intuitionistic fuzzy, or vague reliability. From the reviewed literature, it is observed that the main goal of researchers was to quantify uncertainty through different fuzzy environments and the reduction of accumulating phenomenon of fuzziness either using different types of soft computing techniques or $T_\omega$-norm. However, to the best of our knowledge, no study has been carried out to simplify the calculation process in the existing FLT technique by adopting simplified fuzzy arithmetic operations defined on TFN without employing soft computing-based techniques or $T_\omega$-norm.

The objective of this study is to propose a simple and straightforward reliability analysis technique for repairable systems that can be used to analyse the system's current behaviour and, based on its findings, recommend a future course of action to optimise its performance. As a result, the fundamental contribution of the work is a simplified fuzzy Lambda-Tau (SFLT) approach to analyse the fuzzy reliability of any complex engineering systems.

The remaining chapter is organised into five sections. Section 2.2 discusses some fundamental concepts related to the study. A brief summary of the current LT and FLT approaches is given in Sect. 2.3. The section also develops the SFLT technique in a very detailed manner. The efficacy of the proposed technique is demonstrated by evaluating the fuzzy reliability of a critical washing system in a paper manufacturing plant. The brief description of the washing system is provided in Sect. 2.4 along with its fuzzy reliability analysis. In Sect. 2.4, a comparison of the proposed approach has been investigated with traditional LT and existing FLT techniques, whilst to analyse its consistency, a sensitivity analysis has been conducted. The $V$-index approach has been implemented to rate the system's key components based on their criticalities. Finally, Sect. 2.5 concludes the chapter with findings.

## 2.2 Preliminary Concepts

This section provides some preliminary concepts such as definitions of a fuzzy set and triangular fuzzy number (TFN). Section also provides the simplified fuzzy arithmetic operations for TFNs.

### *2.2.1 Fuzzy Set [25]*

A fuzzy set $\tilde{A}$ in the universe of discourse $X$ is a collection of elements denoted by $x$ having varying degrees of membership represented by $\mu_{\tilde{A}}(x) \in [0, 1]$ in the fuzzy set $\tilde{A}$. The mathematical representation of a fuzzy set $\tilde{A}$ is as follows [15, 25]:

$$\tilde{A} = \{(x, \mu_{\tilde{A}}(x)) : x \in X\}. \tag{2.1}$$

### *2.2.2 TFN [26]*

A TFN $\tilde{A} = (a_1, a_2, a_3)$ is a fuzzy set with a triangle membership function defined in Eq. 2.2 and plotted in Fig. 2.1, where $a_1$, $a_2$, and $a_3$ indicate the left, modal, and right end values of $\tilde{A}$ [26].

$$\mu_{\tilde{A}}(x) = \begin{cases} \frac{x-a_1}{a_2-a_1}, & a_1 \leq x \leq a_2 \\ \frac{a_3-x}{a_3-a_2}, & a_2 \leq x \leq a_3 \end{cases} \tag{2.2}$$

### *2.2.3 Simplified Fuzzy Arithmetic Operations for TFNs [26]*

Let $\tilde{A} = (a_1, a_2, a_3)$ and $\tilde{B} = (b_1, b_2, b_3)$ be two TFNs, then simplified fuzzy arithmetic operations defined for TFNs are provided in Table 2.1 [26].

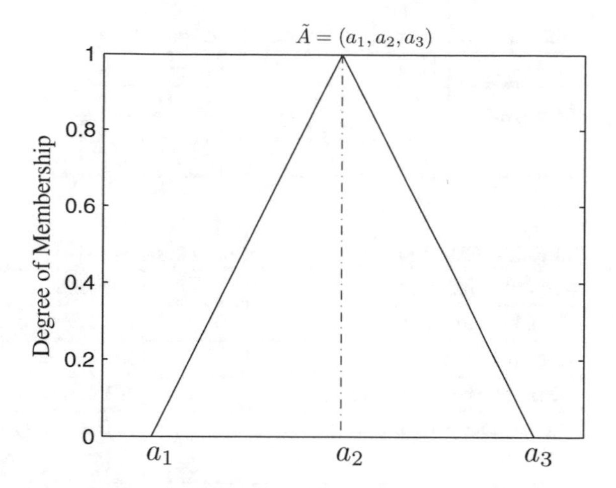


**Fig. 2.1** A TFN
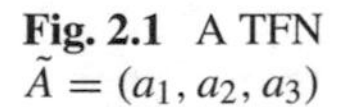
$\tilde{A} = (a_1, a_2, a_3)$

**Table 2.1** Simplified fuzzy arithmetic operations for TFNs [26]

| Operation | Fuzzy expression for TFNs |
|---|---|
| Addition | $\tilde{A} + \tilde{B} = (a_1 + b_1, a_2 + b_2, a_3 + b_3)$ |
| Subtraction | $\tilde{A} - \tilde{B} = (a_1 - b_3, a_2 - b_2, a_3 - b_1)$ |
| Multiplication | $\tilde{A} \times \tilde{B} = (a_1 b_1, a_2 b_2, a_3 b_3)$ |
| Division | $\tilde{A}/\tilde{B} = (a_1/b_3, a_2/b_2, a_3/b_1)$ |
| Compliment | $\tilde{1} - \tilde{A} = (1 - a_3, 1 - a_2, 1 - a_1)$ |

## 2.3 The LT, FLT, and Proposed SFLT Techniques

This section gives a quick review of existing LT and FLT approaches, as well as a full discussion of the proposed SFLT methodology. The conventional LT approach evaluates system reliability using crisp data without measuring the data uncertainty. The limitation of data uncertainty in the LT technique has been dealt with in the FLT technique by assuming data in the form of TFNs. However, the FLT technique relies on complicated $\alpha$-cut and interval arithmetic coupled arithmetic operations for analysing fuzzy reliability of a system. The proposed SFLT technique rectifies the complexity of the FLT technique by applying simple arithmetic operations defined on TFNs. The following assumptions are considered to apply these techniques in this study [4, 5]:

- The failure and repair rates of components are exponentially distributed and statistically independent;
- Each component has a separate maintenance facility;
- When a unit fails, the repair procedure begins immediately; and
- The restored component is deemed as good as the new following repairs.

**Table 2.2** Expressions used in the LT method [4]

| Gate → | $\lambda_{AND}$ | $\tau_{AND}$ | $\lambda_{OR}$ | $\tau_{OR}$ |
|---|---|---|---|---|
| Expressions | $\prod_{j=1}^{n} \lambda_j \left[ \sum_{i=1}^{n} \prod_{\substack{j=1 \\ i \neq j}}^{n} \tau_j \right]$ | $\frac{\prod_{i=1}^{n} \tau_i}{\sum_{j=1}^{n} \left[ \prod_{\substack{i=1 \\ i \neq j}}^{n} \tau_i \right]}$ | $\sum_{i=1}^{n} \lambda_i$ | $\frac{\sum_{i=1}^{n} \lambda_i \tau_i}{\sum_{i=1}^{n} \lambda_i}$ |

**Table 2.3** Reliability indices for repairable systems [5]

| Reliability indices | Expressions |
|---|---|
| Mean time to failure | $\mathrm{MTTF}_s = \frac{1}{\lambda_s}$ |
| Mean time to repair | $\mathrm{MTTR}_s = \frac{1}{\mu_s} = \tau_s$ |
| Mean time between failures | $\mathrm{MTBF}_s = \mathrm{MTTF}_s + \mathrm{MTTR}_s$ |
| Expected number of failures | $\mathrm{ENOF}_s(0, t) = \frac{\lambda_s \mu_s t}{\lambda_s + \mu_s} + \frac{\lambda_s^2}{(\lambda_s + \mu_s)^2} [1 - \mathrm{e}^{-(\lambda_s + \mu_s)t}]$ |
| Availability | $A_s(t) = \frac{\mu_s}{\lambda_s + \mu_s} + \frac{\lambda_s}{\lambda_s + \mu_s} \mathrm{e}^{-(\lambda_s + \mu_s)t}$ |
| Reliability | $R_s(t) = \mathrm{e}^{-\lambda_s t}$ |

### *2.3.1 LT Method [4]*

The LT method is a commonly used approach to estimate the reliability of any repairable system [4]. The technique applies FTA, crisp data, and classical arithmetic operations to analyse system reliability. FTA's qualitative analysis yields all of the minimal-cut sets that reflect probable system failure scenarios, whilst its quantitative analysis yields an estimate of system failure and repair rates [27]. A minimal-cut set represents a group of system components, if they fail simultaneously, then the system fails. Let us suppose that a system contains *n* repairable components with failure rates $\lambda_i$ and repair times $\tau_i$, where $i = 1, 2, \ldots, n$. The mathematical formulae to obtain system failure rate ($\lambda_s$) and system repair time ($\tau_s$) are provided in Table 2.2 by assuming that all the constituting components are arranged either in parallel(AND-gate) or series (OR-gate) configuration. Using the system $\lambda_s$ and $\tau_s$, an expert can evaluate the system's different reliability indices as mentioned in Table 2.3, where $\mu_s$ is the system repair rate.

### *2.3.2 FLT Technique [5]*

To use the FLT technique, uncertain $\lambda_i$ and $\tau_i$ are fuzzified using TFNs with the assistance of system experts, and apprised values are represented by $\tilde{\lambda}_i$ and $\tilde{\tau}_i$, respectively. The FLT technique applies to fault trees, their minimal-cut sets, the LT method, and fuzzified data ($\tilde{\lambda}_i$ and $\tilde{\tau}_i$) in the form of TFNs for estimating different fuzzy reliability indices as provided in Table 2.3 by adopting $\alpha$-cut coupled fuzzy arithmetic operations [5, 6, 28].

### 2.3.3 The Proposed SFLT Technique

The step-wise details of the proposed SFLT technique are provided in Fig. 2.2 and described as follows.

***Step 1***: The failure rate ($\lambda_i$) and repair time ($\tau_i$) data for the main components of the system under consideration are obtained from a variety of sources, such as historical records, logbooks, literature surveys, and so on, and then finalised with the assistance of system specialists.

***Step 2***: With the assistance of system experts, uncertainty in $\lambda_i$ and $\tau_i$ is quantified using TFNs, and their fuzzified values are represented by $\tilde{\lambda}_i$ and $\tilde{\tau}_i$, respectively. However, other kind of fuzzy numbers, such as trapezoidal fuzzy numbers, may also be used in the SFLT technique to quantify data uncertainty as per the needs of the problem and the availability of the data type.

***Step 3***: In this step, evaluation of system minimal-cut sets is performed using the system's fault tree and matrix method [27]. A particular minimal-cut set has a list of the system's components in which if all the listed components fail at the same time, then the system fails.

***Step 4***: Using the minimal-cut sets produced in Step 3 and the results provided in Table 2.2, the mathematical formulae for $\lambda_s$ and $\tau_s$ are evaluated in terms of $\lambda_i$ and $\tau_i$. Assume that the system is comprised of $n$ components with constant failure rates $\lambda_i$ and repair times $tau_i$. The fundamental equations for the system's $\lambda_s$ and $\tau_s$ are given in Eqs. 2.3 and 2.4, respectively.

$$\lambda_s = f(\lambda_1, \lambda_2, \ldots, \lambda_n, \tau_1, \tau_2, \ldots, \tau_n) \tag{2.3}$$

$$\tau_s = g(\lambda_1, \lambda_2, \ldots, \lambda_n, \tau_1, \tau_2, \ldots, \tau_n) \tag{2.4}$$

***Step 5***: In this step, Eqs. 2.3 and 2.4 are fuzzified to get the mathematical expression for fuzzified $\tilde{\lambda}_s$ and $\tilde{\tau}_s$. The fuzzified equations for $\tilde{\lambda}_s$ and $\tilde{\tau}_s$ are given in Eqs. 2.5 and 2.6, respectively.

$$\tilde{\lambda}_s = f(\tilde{\lambda}_1, \tilde{\lambda}_2, \ldots, \tilde{\lambda}_n, \tilde{\tau}_1, \tilde{\tau}_2, \ldots, \tilde{\tau}_n) \tag{2.5}$$

$$\tilde{\tau}_s = g(\tilde{\lambda}_1, \tilde{\lambda}_2, \ldots, \tilde{\lambda}_n, \tilde{\tau}_1, \tilde{\tau}_2, \ldots, \tilde{\tau}_n) \tag{2.6}$$

***Step 6***: Using Eqs. 2.5 and 2.6 and fuzzified data obtained in Step 2, some common fuzzy reliability indices (Table 2.3) can be evaluated by adopting simplified fuzzy arithmetic operations as given in Table 2.1.

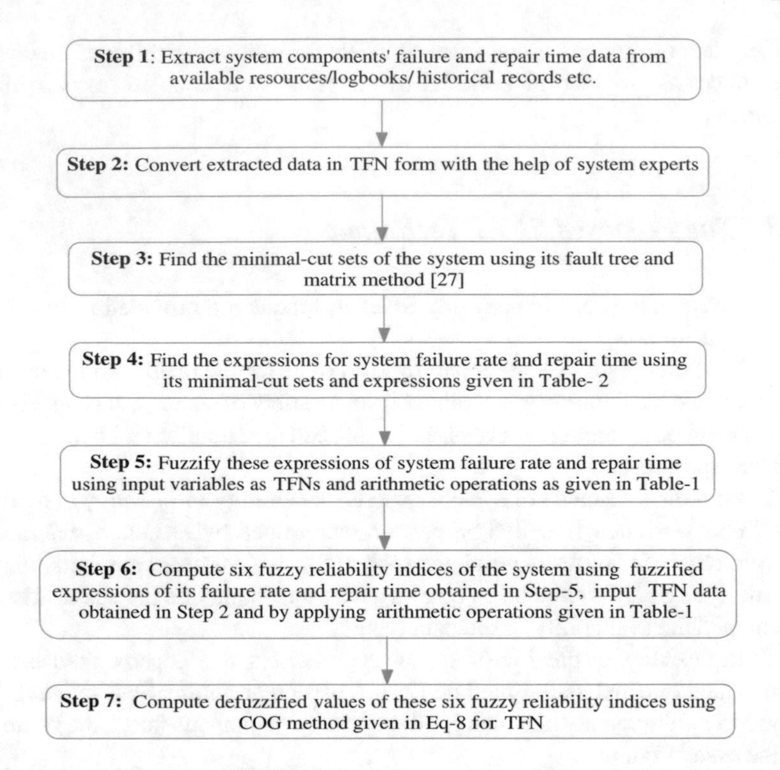


**Fig. 2.2** Steps of the proposed SFLT technique

***Step 7***: In this step, defuzzification is executed for each computed fuzzy reliability index by employing the center of gravity (COG) method [29]. To understand the COG method, let $\tilde{A}$ be a fuzzy set with a membership function $\mu_{\tilde{A}}(x)$, then defuzzified value $\bar{A}$ of $\tilde{A}$ can be evaluated using Eq. 2.7.

$$\bar{A} = \frac{\int x\mu_{\tilde{A}}(x)\mathrm{d}x}{\int \mu_{\tilde{A}}(x)\mathrm{d}x} \tag{2.7}$$

Now, using Eq. 2.7, the defuzzified value $\overline{R}$ of a fuzzy reliability index $\tilde{R} = (r_1, r_2, r_3)$ in the form of a TFN can be computed by applying the following Eq. 2.8.

$$\overline{R} = \frac{\int_{r_1}^{r_2}\left(\frac{x-r_1}{r_2-r_1}\right)x\mathrm{d}x + \int_{r_2}^{r_3}\left(\frac{r_3-x}{r_3-r_2}\right)x\mathrm{d}x}{\int_{r_1}^{r_2}\left(\frac{x-r_1}{r_2-r_1}\right)\mathrm{d}x + \int_{r_2}^{r_3}\left(\frac{r_3-x}{r_3-r_2}\right)\mathrm{d}x} = \frac{r_1+r_2+r_3}{3} \tag{2.8}$$

## 2.4 An Illustration

This section exhibits the application of the presented technique by analysing the fuzzy reliability of an important system known as a washing system in a typical manufacturing plant for different levels of uncertainty [9, 21, 24, 28]. Initially, a brief idea of the main functioning of the washing system and its constituting components is provided along with its schematic and fault tree diagrams. Then, using the presented SFLT technique, a fuzzy reliability analysis of the washing system is conducted for different levels of uncertainty ($\pm 10\%, \pm 20\%, \pm 30\%$). The results are compared with existing LT and FLT techniques. Then, system reliability estimation for the long-term period and sensitivity analyses has also been conducted to examine the efficiency of the proposed technique. A $V$-index-based ranking method has been applied to rank the system components according to their criticality.

### 2.4.1 *A Brief Overview of the System*

In a typical paper manufacturing plant, the washing system is an essential subsystem that is primarily used to remove any blackness present in the pulp by washing, which is done in numerous phases. After washing, pulp with fine fibres is obtained. The system mainly consists of four types of subsystems, namely a filter, cleaners, screeners, and deckers, arranged in a series configuration. Their working descriptions are presented herein [9, 21, 24, 28].

- **Filter (A)**: A filter unit segregates the black liquor from the cooked pulp.
- **Cleaners (B)**: The pulp is cleaned by three cleaner units that are connected in parallel and use rotary action.
- **Screeners (C)**: To filter out oversize, uncooked, and odd-shaped fibres from the pulp, two screeners are interconnected in series.
- **Deckers (D)**: Two decker units, connected in parallel, are used to reduce the blackness of the pulp.

Figures 2.3a and 2.3b show the system's schematic and fault tree diagrams, respectively.

### 2.4.2 *System Fuzzy Reliability Assessment Using SFLT Technique*

The following steps are being used to estimate the fuzzy reliability of the washing system using the suggested SFLT technique.
***Step 1***: The failure and repair data ($\lambda_i$ and $\tau_i$) associated with the main components of the system as shown in Fig. 2.3a are retrieved from different resources and, after

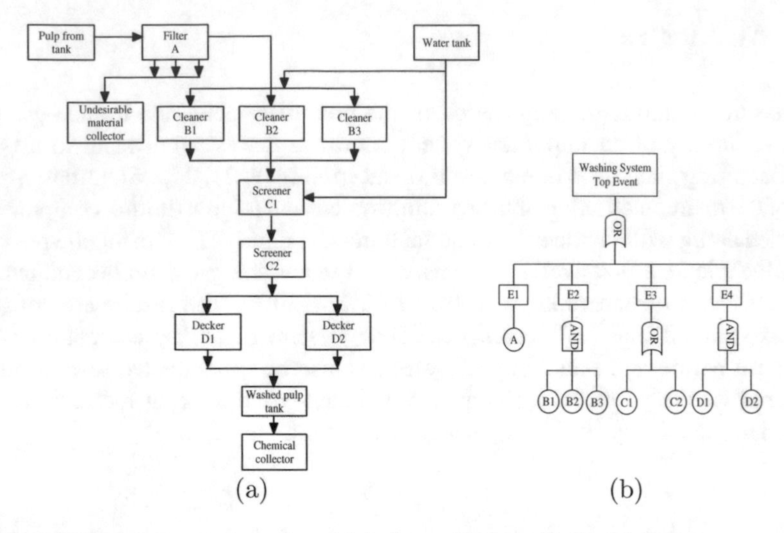


**Fig. 2.3** Washing system **a** schematic and **b** fault tree diagrams [9, 21, 24, 28]

consulting with system experts, are provided in Table 2.4 against the LT data column [9, 21, 24, 28].

***Step 2***: To quantify data uncertainty, fuzzification of the retrieved $\lambda_i$ and $\tau_i$ in terms of TFNs is done with the help of system experts by taking uncertainty levels of $\pm 10\%$, $\pm 20\%$, and $\pm 30\%$. The fuzzified values of $\lambda_i$ and $\tau_i$ in the form of TFNs are now represented by $\tilde{\lambda}_i = (\lambda_{i_1}, \lambda_{i_2}, \lambda_{i_3})$ and $\tilde{\tau}_i = (\tau_{i_1}, \tau_{i_2}, \tau_{i_3})$, respectively.

***Step 3***: The washing system has five minimal-cut sets {{1},{2,3,4},{5},{6},{7,8}} that have been derived using the fault tree (Fig. 2.3b) and matrix method [24, 27].

***Step 4***: The minimal-cut sets and results in Table 2.2 are used to develop the fundamental equations for $\lambda_s$ and $\tau_s$, which are given in Eqs. 2.9 and 2.10, respectively.

$$\lambda_s = \lambda_1 + \lambda_5 + \lambda_6 + \lambda_2\lambda_3\lambda_4(\tau_2\tau_3 + \tau_3\tau_4 + \tau_2\tau_4) + \lambda_7\lambda_8(\tau_7 + \tau_8) \quad (2.9)$$

$$\tau_s = \frac{(\lambda_1\tau_1 + \lambda_5\tau_5 + \lambda_6\tau_6 + \lambda_2\lambda_3\lambda_4\tau_2\tau_3\tau_4 + \lambda_7\lambda_8\tau_7\tau_8)}{\lambda_s} \quad (2.10)$$

***Step 5***: As $\lambda_i$ and $\tau_i$ have uncertainty, so using their fuzzified values $\tilde{\lambda}_i$ and $\tilde{\tau}_i$, the fuzzified version of Eqs. 2.9 and 2.10 is provided in Eqs. 2.11 and 2.12 for $\tilde{\lambda}_s$ and $\tilde{\tau}_s$,respectively.

$$\tilde{\lambda}_s = \tilde{\lambda}_1 + \tilde{\lambda}_5 + \tilde{\lambda}_6 + \tilde{\lambda}_2\tilde{\lambda}_3\tilde{\lambda}_4(\tilde{\tau}_2\tilde{\tau}_3 + \tilde{\tau}_3\tilde{\tau}_4 + \tilde{\tau}_2\tilde{\tau}_4) + \tilde{\lambda}_7\tilde{\lambda}_8(\tilde{\tau}_7 + \tilde{\tau}_8) \quad (2.11)$$

$$\tilde{\tau}_s = \frac{\left(\tilde{\lambda}_1\tilde{\tau}_1 + \tilde{\lambda}_5\tilde{\tau}_5 + \tilde{\lambda}_6\tilde{\tau}_6 + \tilde{\lambda}_2\tilde{\lambda}_3\tilde{\lambda}_4\tilde{\tau}_2\tilde{\tau}_3\tilde{\tau}_4 + \tilde{\lambda}_7\tilde{\lambda}_8\tilde{\tau}_7\tilde{\tau}_8\right)}{\tilde{\lambda}_s} \quad (2.12)$$

***Step 6***: Different fuzzy reliability indices of the system are evaluated for mission time $t = 10$ (h) at different uncertainty levels ($\pm 10\%$, $\pm 20\%$, and $\pm 30\%$) using $\tilde{\lambda}_s$

**Table 2.4** Data for the main components of the system [9, 21, 24, 28]

| Components | Crisp data for LT | | Fuzzified data for FLT and SFLT at different levels of uncertainty | | | | | |
|---|---|---|---|---|---|---|---|---|
| | $\lambda_i$ ($h^{-1}$) | $\tau_i$ (h) | For ± 10% uncertainty | | For ± 20% uncertainty | | For ± 30% uncertainty | |
| | | | $\tilde{\lambda}_i = (\lambda_{i1}, \lambda_{i2}, \lambda_{i3})$ ($h^{-1}$) | $\tilde{\tau}_i = (\tau_{i1}, \tau_{i2}, \tau_{i3})$ (h) | $\tilde{\lambda}_i = (\lambda_{i1}, \lambda_{i2}, \lambda_{i3})$ ($h^{-1}$) | $\tilde{\tau}_i = (\tau_{i1}, \tau_{i2}, \tau_{i3})$ (h) | $\tilde{\lambda}_i = (\lambda_{i1}, \lambda_{i2}, \lambda_{i3})$ ($h^{-1}$) | $\tilde{\tau}_i = (\tau_{i1}, \tau_{i2}, \tau_{i3})$ (h) |
| A: Filter ($i = 1$) | 0.001 | 3 | (0.0009, 0.0010, 0.0011) | (2.70, 3.00, 3.33) | (0.0008, 0.0010, 0.0012) | (2.40, 3.00, 3.60) | (0.0007, 0.0010, 0.0013) | (2.10, 3.00, 3.90) |
| B: Cleaners ($i = 2, 3, 4$) | 0.003 | 2 | (0.0027, 0.0030, 0.0033) | (1.80, 2.00, 2.20) | (0.0024, 0.0030, 0.0036) | (1.60, 2.00, 2.40) | (0.0021, 0.0030, 0.0039) | (1.40, 2.00, 2.60) |
| C: Screeners ($i = 5, 6$) | 0.005 | 3 | (0.0045, 0.0050, 0.0055) | (2.70, 3.00, 3.30) | (0.0040, 0.0050, 0.0060) | (2.40, 3.00, 3.60) | (0.0035, 0.0050, 0.0065) | (2.10, 3.00, 3.90) |
| D: Deckers ($i = 7, 8$) | 0.005 | 3 | (0.0045, 0.0050, 0.0055) | (2.70, 3.00, 3.30) | (0.0040, 0.0050, 0.0060) | (2.40, 3.00, 3.60) | (0.0035, 0.0050, 0.0065) | (2.10, 3.00, 3.90) |

Table 2.5 Comparison between crisp and defuzzified values

| Reliability indices | Crisp | Defuzzified values | | |
|---|---|---|---|---|
| | | ± 10% | ± 20% | ± 30% |
| Failure rate | 0.011150 | FLT: 0.011153<br>SFLT: 0.011153 | 0.011159<br>0.011162 | 0.011171<br>0.011178 |
| Repair time | 2.979753 | 3.043123<br>3.064065 | 3.240947<br>3.328544 | 3.598923<br>3.812036 |
| MTBF | 92.66325 | 93.18673<br>93.36001 | 94.81281<br>95.53839 | 97.72529<br>99.49199 |
| ENOF | 0.108919 | 0.109060<br>0.109131 | 0.109545<br>0.109908 | 0.110603<br>0.111794 |
| Availability | 0.968846 | 0.968264<br>0.968084 | 0.966603<br>0.966033 | 0.964162<br>0.963426 |
| Reliability | 0.894488 | 0.894498<br>0.894501 | 0.894525<br>0.894537 | 0.894571<br>0.894598 |

and $\tilde{\tau}_s$ (Eqs. 2.11 and 2.12), fuzzified reliability indices given in Table 2.3, fuzzified $\tilde{\lambda}_i$ and $\tilde{\tau}_i$ given in Table 2.4, and simplified fuzzy arithmetic operations (Table 2.1). The computed results have been plotted in Fig. 2.4.

***Step 7***: The defuzzified values are evaluated for all the fuzzy reliability indices at different uncertainty levels (± 10%, ± 20%, and ± 30%) by applying the COG method, and the results are tabulated in Table 2.5.

### 2.4.3 Comparative Analysis

The results obtained from the SFLT technique have been compared with results obtained from LT and FLT techniques and are plotted in Fig. 2.4 [24]. Similarly, the crisp results computed by the LT method [4] and the defuzzified values of each fuzzy reliability index obtained from the FLT technique have been plotted in Table 2.5 for different uncertainty levels. Regardless of the uncertainty levels, the results displayed in Fig. 2.4 and recorded in Table 2.5 clearly show that SFLT produces results that are extremely close to those obtained using the FLT technique. Thus, the SFLT technique is simpler than the FLT technique and provides results that are very close to those of the FLT technique. So, the SFLT technique can be considered as an alternative approach to analyse the fuzzy reliability of any complex repairable system.

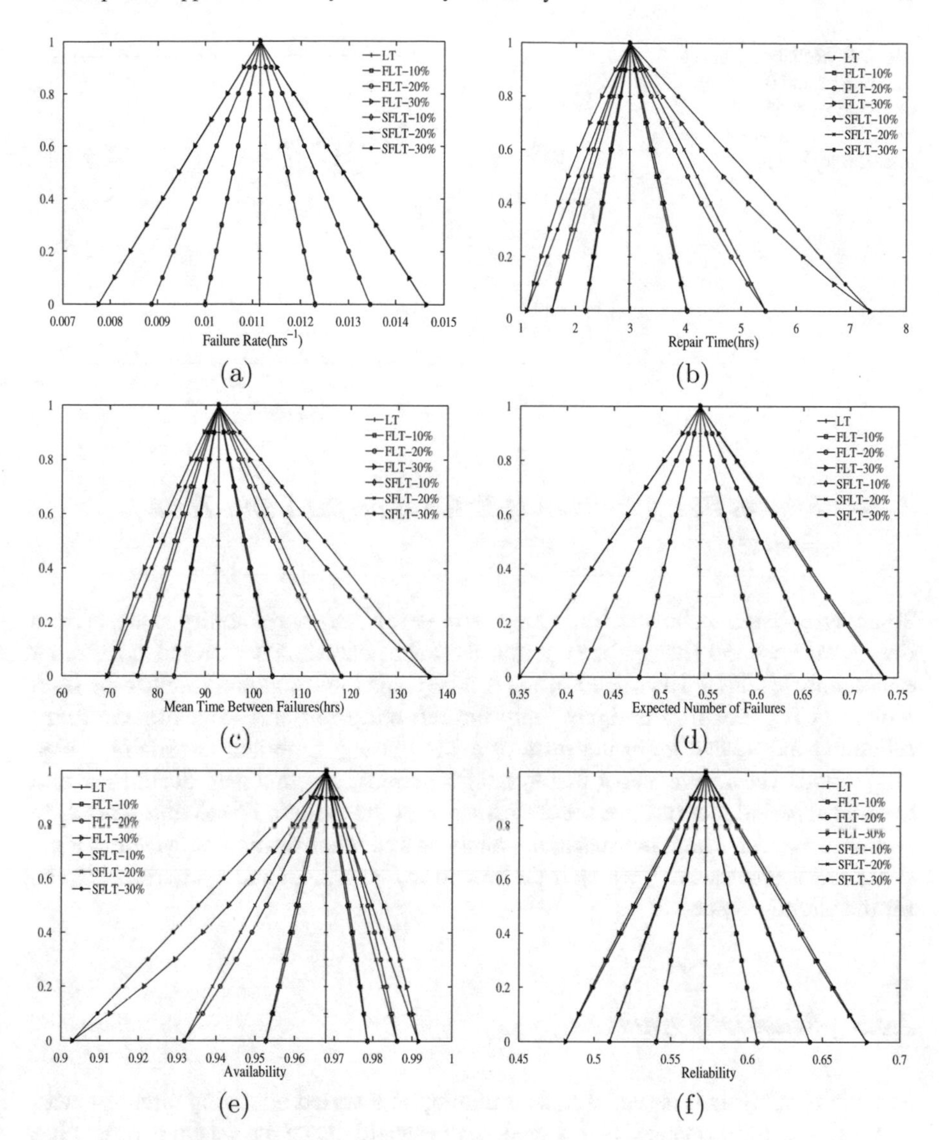


**Fig. 2.4** Fuzzy reliability indices for washing system at $\pm 10\%$, $\pm 20\%$, and $\pm 30\%$ uncertainty levels using LT, FLT, and SFLT techniques

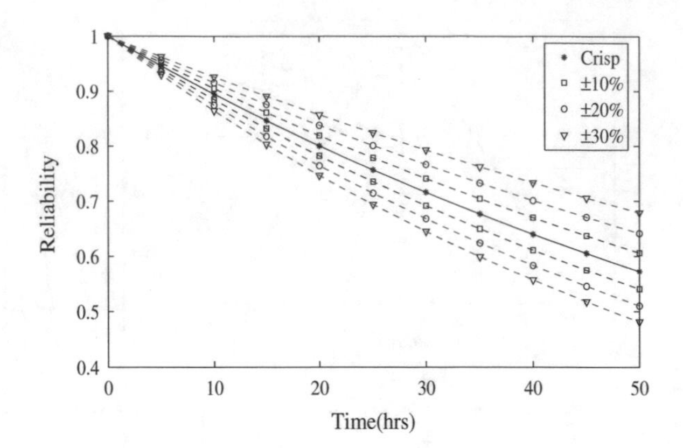


**Fig. 2.5** Reliability curve for time period 0–50 (h) obtained from SFLT technique at different uncertainty levels

### 2.4.4 *System Fuzzy Reliability Estimation for Long-Term Period*

To analyse system behaviour for a long-term period, fuzzy reliability analysis for a time span $t$ = 0–50 (h) has been performed using the SFLT technique at different uncertainty levels ($\pm$ 10%, $\pm$ 20%, and $\pm$ 30%), and the computed results have been plotted in Fig. 2.5. In this figure, only the left and right end values of the fuzzy reliability index, that is, in the form of a TFN at any time and uncertainty level, are plotted. The results show that system reliability continuously declines over a long-term period. To improve the performance of the system, its reliability needs to be improved, for which the sensitivity analysis and identification of system critical components are essential, which is performed in Sects. 2.4.5 and 2.4.6, respectively, for the washing system.

### 2.4.5 *Sensitivity Analysis*

Sensitivity analysis is essential in determining how varied operating circumstances affect system performance. In this section, the sensitivity of the washing system has been analysed by taking nine combinations of reliability, failure rate, and availability, and their impact on system MTBF has been recorded. For all the combinations, the ranges of system repair time and ENOF are fixed and taken from SFLT results, as shown in Figs. 2.4b and d, respectively, for $\pm$ 10% uncertainty level. In these figures, the range of repair time is 2.182966–4.029475, whilst the range of ENOF is 0.095768–0.122705, respectively. The maximum and minimum values of system MTBF for all these nine combinations are recorded in Table 2.6. Consider the first scenario, in which the selected values of reliability, availability, and failure rate are 0.875, 0.992, and $2.3 \times 10^{-4}$, respectively. The computed MTBF range for this combination is 4732–6063 h. Similar effects can be seen in the remaining combinations

**Table 2.6** Impact on washing system MTBF by taking different combinations of its reliability indices

| S.No. | Reliability, failure rate, availability | MTBF (h) |
|---|---|---|
| 1. | [0.875, 2.3 × $10^{-4}$, 0.992] | Min: 4732<br>Max: 6063 |
| 2. | [0.875, 2.8 × $10^{-4}$, 0.992] | 3887<br>4980 |
| 3. | [0.875, 3.3 × $10^{-4}$, 0.992] | 3298<br>4226 |
| 4. | [0.895, 2.3 × $10^{-4}$, 0.995] | 3931<br>5037 |
| 5. | [0.895, 2.8 × $10^{-4}$, 0.995] | 3229<br>4137 |
| 6. | [0.895, 3.3 × $10^{-4}$, 0.995] | 2740<br>3510 |
| 7. | [0.915, 2.3 × $10^{-4}$, 0.998] | 3148<br>4033 |
| 8. | [0.915, 2.8 × $10^{-4}$, 0.998] | 2585<br>3313 |
| 9. | [0.915, 3.3 × $10^{-4}$, 0.998] | 2194<br>2811 |

as well. As a consequence of the analysis, one may conclude that a little change in system failure rate may lead to a significant variation in MTBF, hence altering system performance. The analysis also advises that, to increase system performance, the present operating conditions of equipment and its constituent components be altered to enable the construction of an effective maintenance programme as needed. As a result, rating the system's fundamental components is essential, as indicated in the next section.

### 2.4.6 Ranking of System Critical Components

The ranking of key system components is significant for enhancing system performance through the design and implementation of suitable maintenance action plans. The present study utilises the $V$-index-based ranking approach as suggested by Komal [20] for the ranking of the system's critical components'. The $V$-index determines the disparity in $\tilde{q}_T$ and $\tilde{q}_{T_i}$, where $\tilde{q}_T$ is the system's fuzzy failure rate when its $i$th component's fuzzy failure rate is included in the analysis, whilst $\tilde{q}_{T_i}$ is the recomputed value of the system's fuzzy failure rate when the possibility of the system's $i$th component's failure is neglected by setting $\tilde{\lambda}_i = \tilde{0} = (0, 0, 0)$. The expressions for $\tilde{q}_T$ and $\tilde{q}_{T_i}$ are mentioned in Eqs. (2.13) and (2.14), respectively.

**Table 2.7** Ranking results

| Component | $\tilde{q}_{T_i}$ | $V(\tilde{q}_T, \tilde{q}_{T_i})$ | Rank |
|---|---|---|---|
| A: Filter ($i = 1$) | $(9.109541 \times 10^{-3}, 1.015032 \times 10^{-2}, 1.120017 \times 10^{-2})$ | $3.000000 \times 10^{-3}$ | 2 |
| B: Cleaners ($i = 2, 3, 4$) | $(1.000935 \times 10^{-2}, 1.115000 \times 10^{-2}, 1.229965 \times 10^{-2})$ | $1.037124 \times 10^{-6}$ | 4 |
| C: Screeners ($i = 5, 6$) | $(1.009541 \times 10^{-3}, 1.150324 \times 10^{-3}, 1.300172 \times 10^{-3})$ | $3.000000 \times 10^{-2}$ | 1 |
| D: Deckers ($i = 7, 8$) | $(9.900191 \times 10^{-3}, 1.100032 \times 10^{-2}, 1.210052 \times 10^{-2})$ | $4.590000 \times 10^{-4}$ | 3 |

$$\tilde{q}_T = \tilde{q}_T(\tilde{\lambda}_1, \tilde{\lambda}_2, \ldots \tilde{\lambda}_i \ldots, \tilde{\lambda}_n, \tilde{\tau}_1, \tilde{\tau}_2, \ldots \tilde{\tau}_i \ldots, \tilde{\tau}_n) \equiv (a_T, b_T, c_T) \quad (2.13)$$

$$\tilde{q}_{T_i} = \tilde{q}_T(\tilde{\lambda}_1, \tilde{\lambda}_2, \ldots \tilde{0} \ldots, \tilde{\lambda}_n, \tilde{\tau}_1, \tilde{\tau}_2, \ldots \tilde{\tau}_i \ldots, \tilde{\tau}_n) \equiv (a_{T_i}, b_{T_i}, c_{T_i}) \quad (2.14)$$

The $V$-index calculates the extent of any improvement in system performance when it is presumed that its $i$th component is in a fully working state. Mathematically, $V$-index is explained as follows:

$$V(\tilde{q}_T, \tilde{q}_{T_i}) = (a_T - a_{T_i}) + (b_T - b_{T_i}) + (c_T - c_{T_i}) \quad (2.15)$$

If $V(\tilde{q}_T, \tilde{q}_{T_i}) \geq V(\tilde{q}_T, \tilde{q}_{T_j})$, then the impact of the $i$th component failure is greater than the $j$th component failure on the system performance. The value of the $V$-index has been computed for each component of the system by taking its fuzzy failure rate as $\tilde{q}_T = (0.007752, 0.011150, 0.014631)$, computed by the SFLT technique at an uncertainty level of $\pm$ 10% (see Fig. 2.4a). Table 2.7 provides the ranking order of the system's important components based on their $V$-index values in decreasing order. The screener is obviously the most significant component of the washing system, whereas the cleaner is the least significant component, as shown in Table 2.7.

## 2.5 Conclusion

An SFLT approach has been proposed in the study to analyse the fuzzy reliability of any repairable system. The SFLT method uses a system fault tree to describe the problem, TFNs to quantify data uncertainty, the LT method to formulate mathematical expressions for six reliability indices, and simplified fuzzy arithmetic operations defined on TFS to compute the results. The approach has been illustrated by evaluating the fuzzy reliability of a washing system in a paper plant at different uncertainty levels ($\pm 10\%$, $\pm 20\%$, and $\pm 30\%$). The findings of the SFLT approach have been compared to the results of the conventional LT method and the existing FLT methodology. The findings clearly reveal that the SFLT results are incredibly close to the FLT method outcomes, regardless of the uncertainty level, and require less computational effort. To make sound and effective maintenance strategies for controlling system failure and enhancing its performance, the defuzzified values of reliability indices, reliability estimation for long-term period, sensitivity analysis, and ranking of critical components using $V$-index have been performed, and the results are either plotted or tabulated. The findings generated using the SFLT approach are consistent, according to the study, and may be utilised to plan future actions to improve system performance. The ranking results inferred that the screener and cleaner are the most and least critical components of the system, respectively. Using the results of the system reliability analysis, the maintenance professionals may develop an efficient maintenance plan to improve the system's performance by lowering failure time and increasing MTBF.

**Acknowledgements** The author would like to thank the anonymous referees for their valuable suggestions for improving the readability of the chapter.

## References

1. Zio, E.: Reliability engineering: old problems and new challenges. Reliab. Eng. Syst. Saf. **94**, 125–141 (2009)
2. Ebeling, C.: An Introduction to Reliability and Maintainability Engineering. Tata McGraw-Hill Company Ltd., New York (2001)
3. Mahmood, Y.A., Ahmadi, A., Verma, A.K., Srividya, A., Kumar, U.: Fuzzy fault tree analysis: a review of concept and application. Int. J. Syst. Assur. Eng. Manag. **4**(1), 19–32 (2013)
4. Mishra, K.B.: Reliability Analysis and Prediction: A Methodology Oriented Treatment. Elsevier, Amsterdam (1992)
5. Knezevic, J., Odoom, E.R.: Reliability modeling of repairable systems using petri nets and fuzzy lambda-tau methodology. Reliab. Eng. Syst. Saf. **73**(1), 1–17 (2001)
6. Cheng, C.H., Mon, D.L.: Fuzzy system reliability analysis by interval of confidence. Fuzzy Sets Syst. **56**(1), 29–35 (1993)
7. Komal, Sharma, S.P., Kumar, D.: Stochastic behavior analysis of the press unit in a paper mill using GABLT technique. Int. J. Intell. Comput. Cybern. **2**(3), 574–593 (2009)
8. Sharma, S.P., Kumar, D., Komal: Stochastic behavior analysis of the feeding system in a paper mill Using NGABLT technique. Int. J. Qual. Reliab. Manag. **27**(8), 953–971 (2010)
9. Komal, Sharma, S.P., Kumar, D.: RAM analysis of repairable industrial systems utilizing uncertain data. Appl. Soft Comput. **10**(4), 1208–1221 (2010)
10. Garg, H., Sharma, S.P., Rani, M.: Stochastic behavior analysis of an industrial systems using PSOBLT technique. Int. J. Uncertainty Fuzziness Knowl. Based Syst. **20**(5), 741–761 (2012)
11. Garg, H., Rani, M., Sharma, S.P.: Fuzzy RAM analysis of the screening unit in a paper industry by utilizing uncertain data. Int. J. Qual. Stat. Reliab. 14 pp, Article ID 203842 (2012)

12. Verma, M., Kumar, A., Singh, Y.: Vague reliability assessment of combustion system using Petri nets and vague lambda-tau methodology. Eng. Comput. Int. J. Comput. Aided Eng. Softw. **30**(5), 665–681 (2013)
13. Garg, H.: Reliability analysis of repairable systems using petri nets and vague lambda-tau methodology. ISA Trans. **52**(1), 6–18 (2013)
14. Hong, D.H., Do, H.Y.: Fuzzy system reliability analysis by the use of $T_\omega$ (the weakest t-norm) on fuzzy number arithmetic operations. Fuzzy Sets Syst. **90**(3), 307–316 (1997)
15. Komal, Chang, D., Lee, S.Y.: Fuzzy reliability analysis of dual-fuel steam turbine propulsion system in LNG carriers considering data uncertainty. J. Nat. Gas Sci. Eng. **23**, 148–164 (2015)
16. Panchal, D., Chatterjee, P., Pamucar, D., Yazdani, M.: A novel fuzzy based structured framework for sustainable operation and environmental friendly production in coal fired power industry. Int. J. Intell. Syst. **37**(4), 2706–2738 (2022)
17. Kumar, M., Singh, K.: Fuzzy fault tree analysis of chlorine gas release hazard in Chlor-Alkali industry using $\alpha$-cut interval-based similarity aggregation method. Appl. Soft Comput. **125**, 109199 (2022)
18. Liu, X., Xu, L., Yu, X., Tong, J., Wang, X., Wang X.: Mixed uncertainty analysis for dynamic reliability of mechanical structures considering residual strength. Reliab. Eng. Syst. Saf. **209**, 107472 (2021)
19. Komal: Fuzzy reliability analysis of a phaser measurement unit using generalized fuzzy lambda-tau (GFLT) technique. ISA Trans. **76**, 31–42 (2018)
20. Komal: Fuzzy reliability analysis of DFSMC system in LNG carriers for components with different membership function. Ocean Eng. **155**, 278–294 (2018)
21. Garg, H.: A novel approach for analyzing the behavior of industrial systems using weakest $t$-norm and intuitionistic fuzzy set theory. ISA Trans. **53**(4), 1199–1208 (2014)
22. Garg, H., Rani, M.: An approach for reliability analysis of industrial systems using PSO and IFS technique. ISA Trans. **52**, 701–710 (2013)
23. Verma, M., Kumar, A., Singh, Y., Abraham, A.: Application of the weakest $t$-norm ($T_\omega$) based vague lambda-tau methodology for reliability analysis of gas turbine system. J. Intell. Fuzzy Syst. **25**, 907–918 (2013)
24. Komal: Novel approach to analyse vague reliability of repairable industrial systems. Comput. Ind. Eng. **169**, 108199 (2022)
25. Zadeh, L.A.: Fuzzy sets. Inf. Control **8**(3), 338–353 (1965)
26. Chen, S.M.: Fuzzy system reliability analysis using fuzzy number arithmetic operations. Fuzzy Sets Syst. **64**(1), 31–38 (1994)
27. Liu, T.S., Chiou, S.B.: The application of Petri nets to failure analysis. Reliab. Eng. Syst. Saf. **57**(2), 129–142 (1997)
28. Komal, Sharma, S.P.: Fuzzy reliability analysis of repairable industrial systems using soft-computing based hybridized techniques. Appl. Soft Comput. **24**, 264–276 (2014)
29. Ross, T.J.: Fuzzy Logic with Engineering Applications, 2nd edn. Wiley, New York (2004)

## Author Biography

**Komal** is an Assistant Professor in the Department of Mathematics, School of Physical Sciences, Doon University, Dehradun, Uttarakhand, India. He received his MSc in Applied Mathematics from Indian Institute of Technology Roorkee, Roorkee in 2004. He received his Ph.D. in Reliability Analysis from Indian Institute of Technology Roorkee, Roorkee in 2010. Before joining Doon University, he was working as an Assistant Professor in the Department of Mathematics, H.N.B. Garhwal University (a central university), Srinagar (Garhwal), Uttarakhand. He has published more than 25 research papers in different journals of repute like Applied soft computing, Computers and Industrial Engineering, ISA Transactions, Ocean Engineering, Soft computing,

Mapan, Arabian Journal for Science and Engineering, Journal of Industrial and Management Optimisation, International Journal of Quality & Reliability Management etc. His current research areas include fuzzy reliability analysis, optimisation using soft computing, decision making under uncertain fuzzy environment etc.

# Chapter 3
# Bayesian Reliability Analysis of Topp-Leone Model Under Different Loss Functions

**Haiping Ren, Hui Zhou, and Bin Yin**

**Abstract** The aim of this work is to study Bayesian reliability analysis of Topp-Leone distribution, which is an important lifetime distribution. Based on complete samples, the Bayesian estimation of the parameters of Topp-Leone model based on three loss functions (i.e., squared error, LINEX and entropy) is investigated under the prior distribution of the parameter as uninformative quasi-prior distribution. To compare the performances of Bayesian estimators, risk functions are derived under squared error and LINEX loss, respectively. Based on record values, the Bayesian estimation problem of Topp-Leone distribution is studied under a new class of composite LINEX loss function. Statistical simulation is used to discuss the performances of obtaining Bayesian estimators.

**Keywords** Topp-Leone distribution · Bayesian estimation · LINEX loss function · Quasi-prior distribution

## 3.1 Topp-Leone Distribution Model

Topp-Leone (briefly, T-L) distribution is a lifetime distribution which can well characterize the failure data of a model. Topp and Leone in the literature [1] introduced T-L distribution and stated that it is very useful in the field of characterizing life phenomena and they studied some indicators of T-L distribution such as average remaining life and random order. Al-Zahrani [2] studied the T-L distribution for goodness-of-fit testing. Al-Zahrani and Alshomrani [3] discussed the estimation of

H. Ren (✉) · B. Yin
Department of Basic Subjects, Jiangxi University of Science and Technology, Nanchang 330013, China
e-mail: chinarhp@163.com

B. Yin
e-mail: yinoobin123@163.com

H. Zhou
School of Mathematics and Computer Science, Yichun University, Yichun 336000, China
e-mail: huihui7978@126.com



H. Garg (ed.), *Advances in Reliability, Failure and Risk Analysis*, Industrial and Applied Mathematics, https://doi.org/10.1007/978-981-19-9909-3_3

reliability of a stress-strength model when the samples obey the T-L distribution. Bayoud [4] investigated the estimation of shape parameters of T-L distribution on the basis of timed truncated samples from both classical and Bayesian statistics. El-Sayedet al. [5] obtained the classical and Bayesian estimator of the coefficient of variation of the T-L distribution based on incremental type-II truncated samples. Deng [6] not only discussed the maximum likelihood (ML) estimation but also discussed Bayesian estimation of the parameter of T-L distribution under the squared loss and precautionary loss, respectively. Feroze et al. [7] also discussed Bayesian estimation under progressively type-II censoring test. They derived the approximation of Bayesian estimator with the help of Lindley's approximation algorithm. For more studies and applications of the T-L distribution, one can see the literature [8–14].

Let the random variable $\xi$ obey the T-L distribution, and the corresponding distribution function and probability density function (briefly, pdf) are (Deng [6]):

$$F(\xi;\sigma) = (2\xi - \xi^2)^{\frac{1}{\sigma}}, \quad 0 < \xi < 1, \sigma > 0, \tag{3.1}$$

$$f(\xi;\sigma) = \frac{2}{\sigma}(1-\xi)(2\xi - \xi^2)^{\frac{1}{\sigma}-1}, \quad 0 < \xi < 1, \sigma > 0, \tag{3.2}$$

respectively.

Figure 3.1 gives the graph of the pdf curve of the T-L distribution at taking 0.2, 0.5 and 5.0.

## 3.2 Bayesian Estimation of Parameters of T-L Distribution with Complete Sample

Bayesian estimation and risk function comparison problems of T-L distributed parameters based on three loss functions will be investigated under the prior distribution of parameters as uninformative quasi-prior distribution (briefly, QPD), respectively.

### *3.2.1 Bayesian Estimation Under Quasi-prior Distribution*

In Bayesian statistical analysis, the prior distribution and the loss function occupy a very important position. In the next discussion of this paper, let the prior distribution of the parameter $\theta$ be the uninformative QPD and the corresponding pdf be [15]

$$\pi(\sigma) \propto \sigma^{-d}, \quad \sigma > 0, d > 0. \tag{3.3}$$

The loss functions are introduced as follows, and the details about the three loss functions can also be found in [15].

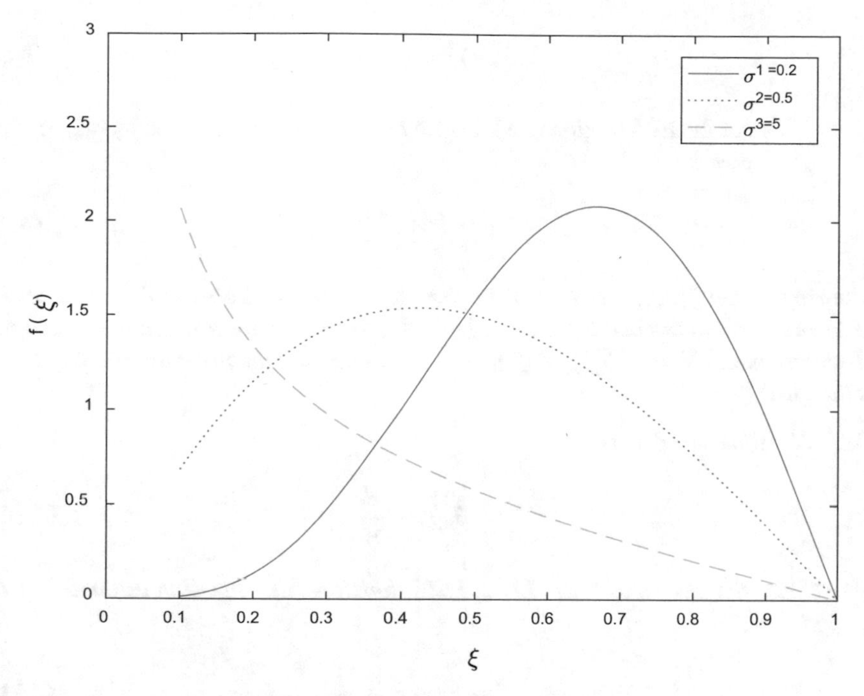


**Fig. 3.1** Graph of pdf of T-L distribution at 0.2, 0.5 and 5.0

(i) Squared error loss (briefly, SEL) function is defined as follows [16]:

$$L(\hat{\sigma}, \sigma) = (\hat{\sigma} - \sigma)^2 \tag{3.4}$$

Suppose that $X$ is the history data information about the parameter $\sigma$. Under the SEL function, the Bayesian estimator of $\sigma$ is

$$\hat{\sigma}_{\text{BS}} = E[\sigma|X] \tag{3.5}$$

(ii) LINEX loss (briefly, LL) function is defined as follows [17–19]:

$$L(\delta) = \mathrm{e}^{c\delta} - c\delta - 1, \quad c \neq 0 \tag{3.6}$$

where $\delta = (\hat{\sigma} - \sigma)/\sigma$ and $c$ are the shape parameter. Under the LL function (3.6), the Bayesian estimator, denoted by $\hat{\sigma}_{\text{BL}}$, is the solution of the following equation

$$E\left[\sigma^{-1}\exp\left(\frac{c\hat{\sigma}_{\text{BL}}}{\sigma}\right)|X\right] = \mathrm{e}^{c}E[\sigma^{-1}|X] \tag{3.7}$$

(iii) Entropy loss (briefly, EL) function is defined as follows [20]:

$$L(\hat{\sigma}, \sigma) = \frac{\hat{\sigma}}{\sigma} - \ln \frac{\hat{\sigma}}{\sigma} - 1 \tag{3.8}$$

Under the EL function (3.8), the Bayesian estimator, denoted by $\hat{\sigma}_{\text{BE}}$, of the parameter $\sigma$ is

$$\hat{\sigma}_{\text{BE}} = [E(\sigma^{-1}|X)]^{-1} \tag{3.9}$$

**Theorem 1** *Let* $\xi_1, \xi_2, \ldots, \xi_m$ *be a random sample drawn from the T-L distribution* (3.1), *and for convenience,* $\xi_1, \xi_2, \ldots, \xi_m$ *is also expressed the corresponding sample observation. Let* $V = -\sum_{i=1}^{m} \ln(2\xi_i - \xi_i^2)$. *Assume that the prior distribution of* $\sigma$ *is the QPD* (3.3), *then*

(i) *ML estimator of* $\sigma$ *is*

$$\hat{\sigma}_{\text{ML}} = \frac{V}{n}, \tag{3.10}$$

(ii) *Under the three loss (SEL, LL and EL) functions, the Bayesian estimator of* $\sigma$ *is*

$$\hat{\sigma}_{\text{BS}} = \frac{V}{n+d-2}, \tag{3.11}$$

$$\hat{\sigma}_{\text{BL}} = \frac{V}{c}\left[1 - \exp\left(-\frac{c}{n+d}\right)\right] \tag{3.12}$$

$$\hat{\sigma}_{\text{BE}} = \frac{V}{n+d-1} \tag{3.13}$$

***Proof*** The likelihood function of the parameter $\sigma$ is

$$\begin{aligned} l(\sigma; \xi_1, \xi_2, \ldots, \xi_m) &= \prod_{i=1}^{m} f(\xi_i; \sigma) \\ &= \sigma^{-m} 2^m \prod_{i=1}^{m} \frac{1 - \xi_i}{2\xi_i - \xi_i^2} \exp\left(-\frac{V}{\sigma}\right) \end{aligned} \tag{3.14}$$

where $V = -\sum_{i=1}^{m} \ln(2\xi_i - \xi_i^2)$.

By solving the log-likelihood equation with respect to $\sigma$ and equal to 0, that is

$$\frac{\mathrm{d} \ln l(\sigma; \xi_1, \xi_2, \ldots, \xi_m)}{\mathrm{d}\sigma} = 0,$$

(i) The ML estimator of $\sigma$ is solved as

$$\hat{\sigma}_{\mathrm{ML}} = \frac{V}{n}.$$

Then from (3.2), (3.14) and Bayesian formula, the posterior pdf of the parameter $\sigma$ is

$$\begin{aligned} h(\sigma|\xi_1, \xi_2, \ldots, \xi_m) &\propto l(\sigma; \xi_1, \xi_2, \ldots, \xi_m) \cdot \pi(\sigma) \\ &\propto \sigma^{-m} \mathrm{e}^{-V/\sigma} \sigma^{-d} \\ &\propto \sigma^{-(m+d)} \mathrm{e}^{-V/\sigma} \end{aligned} \tag{3.15}$$

Thus, the posterior pdf of $\sigma$ is

$$h(\sigma|\xi_1, \xi_2, \ldots, \xi_m) = \frac{V^{m+d-1}}{\Gamma(m+d-1)} \sigma^{-(m+d)} \mathrm{e}^{-V/\sigma} \tag{3.16}$$

that is $\sigma|\xi_1, \xi_2, \ldots, \xi_m \sim I\Gamma(m+d-1, V)$. Then,

(ii) Under the SEL function, the Bayesian estimator of $\sigma$ is

$$\hat{\sigma}_{\mathrm{BS}} = E(\sigma|\xi_1, \xi_2, \ldots, \xi_m) = \frac{V}{m+d-2}.$$

(iii) From (3.16), we have

$$\begin{aligned} E\left[\frac{1}{\sigma} \exp\left(\frac{c\hat{\sigma}_{\mathrm{BL}}}{\sigma}\right)|\xi_1, \xi_2, \ldots, \xi_m\right] &= \int_0^\infty \frac{1}{\sigma} \exp\left(\frac{c\hat{\sigma}_{\mathrm{BL}}}{\sigma}\right) \frac{V^{m+d-1}}{\Gamma(m+d-1)} \sigma^{-(m+d)} \\ &\quad \mathrm{e}^{-\frac{V}{\sigma}} \mathrm{d}\sigma = \frac{(m+d-1)V^{m+d-1}}{(V - c\hat{\sigma}_{\mathrm{BL}})^{m+d}} \end{aligned}$$

and

$$\begin{aligned} E[\sigma^{-1}|\xi_1, \xi_2, \ldots, \xi_m] &= \int_0^\infty \sigma^{-1} \frac{V^{m+d-1}}{\Gamma(m+d-1)} \sigma^{-(m+d)} \mathrm{e}^{-\frac{V}{\sigma}} \mathrm{d}\sigma \\ &= \frac{m+d-1}{V} \end{aligned}$$

Substituting them into Eq. (3.7) solves the Bayesian estimator of the parameter $\sigma$ as

$$\hat{\sigma}_{\mathrm{BL}} = \frac{V}{c}\left[1 - \exp\left(-\frac{c}{m+d}\right)\right].$$

(iv) Under the EL function (3.8), the Bayesian estimator of the parameter $\sigma$ is

$$\begin{aligned}\hat{\sigma}_{\mathrm{BE}} &= [E(\sigma^{-1}|\xi_1, \xi_2, \ldots, \xi_m)]^{-1} \\ &= \left[\int_0^{\infty} \frac{1}{\sigma} \frac{V^{m+d-1}}{\Gamma(m+d-1)} \sigma^{-(m+d)} \mathrm{e}^{-\frac{V}{\sigma}} \mathrm{d}\sigma\right]^{-1} \\ &= \frac{V}{m+d-1}\end{aligned}$$

***Remark 1*** It is easy to prove that $V = -\sum_{i=1}^{m} \ln(2\xi_i - \xi_i^2)$ obeys the gamma distribution $\Gamma(m, \sigma^{-1})$ and the corresponding pdf is

$$\rho(V) = \frac{1}{\Gamma(m)\sigma^m} V^{m-1} \mathrm{e}^{-\frac{V}{\sigma}}, \quad V > 0. \tag{3.17}$$

***Remark 2*** For given time $t$, the reliability $R(t) = P(\xi > t) = 1 - (2t - t^2)^{\frac{1}{\sigma}}$. Let $\hat{\sigma}$ be an estimator of the parameter $\sigma$, then the estimator of $R(t)$ is $\hat{R}(t) = 1 - (2t - t^2)^{\frac{1}{\hat{\sigma}}}$. Thus, according to Theorem 1, we can also get the ML and Bayesian estimators of $R(t)$.

## *3.2.2 Comparative Study of Risk Functions for These Bayesian Estimators*

### 3.2.2.1 Comparative Study Under SEL Functions

Let $\hat{\sigma}$ be an estimator of the parameter $\sigma$, then the risk function of $\hat{\sigma}$ under the SEL function is defined as

$$R(\hat{\sigma}) = \int_0^{\infty} (\hat{\sigma} - \sigma)^2 \rho(V) \mathrm{d}V \tag{3.18}$$

Then, using Eq. (3.17), a simple calculation leads to the following risk functions for each of the three Bayesian estimators.

$$R(\hat{\sigma}_{\mathrm{BS}}) = \sigma^2 \left[\frac{m(m+1)}{(m+d-2)^2} - \frac{2m}{m+d-2} + 1\right], \tag{3.19}$$

$$R(\hat{\sigma}_{\mathrm{BL}}) = \sigma^2 \left[\frac{m(m+1)}{c^2}\left(1 - \mathrm{e}^{-c/(m+d)}\right)^2 - \frac{2m}{c}\left(1 - \mathrm{e}^{-c/(m+d)}\right) + 1\right], \tag{3.20}$$

and

$$R(\hat{\sigma}_{\text{BE}}) = \sigma^2 \left[ \frac{m(m+1)}{(m+d-1)^2} - \frac{2m}{m+d-1} + 1 \right]. \tag{3.21}$$

In order to compare each risk function, each risk function is compared with $\sigma^2$ and we get the following three ratio risk functions:

$$\frac{R(\hat{\sigma}_{\text{BS}})}{\sigma^2} = B_1, \quad \frac{R(\hat{\sigma}_{\text{BL}})}{\sigma^2} = B_2 \text{ and } \frac{R(\hat{\sigma}_{\text{BE}})}{\sigma^2} = B_3.$$

The graphs of the variation of $B_1$, $B_2$, $B_3$ with the a priori parameters are given below and for the SEL function, taking $c = 1$.

As shown in Figs. 3.2, 3.3, 3.4 and 3.5, the images of each risk function differ significantly when $m$ is small, but as $m$ increases, especially when $m > 50$, the various risk functions tend to be consistent. For small $m$, if one estimates the risk functions is smallest among all estimates, then it should be used as alternative estimate of the parameter, and when m is larger, each Bayesian estimate can be used as an alternative estimate of the parameter at this time because each Bayesian estimate is less influenced by the prior parameter $d$.

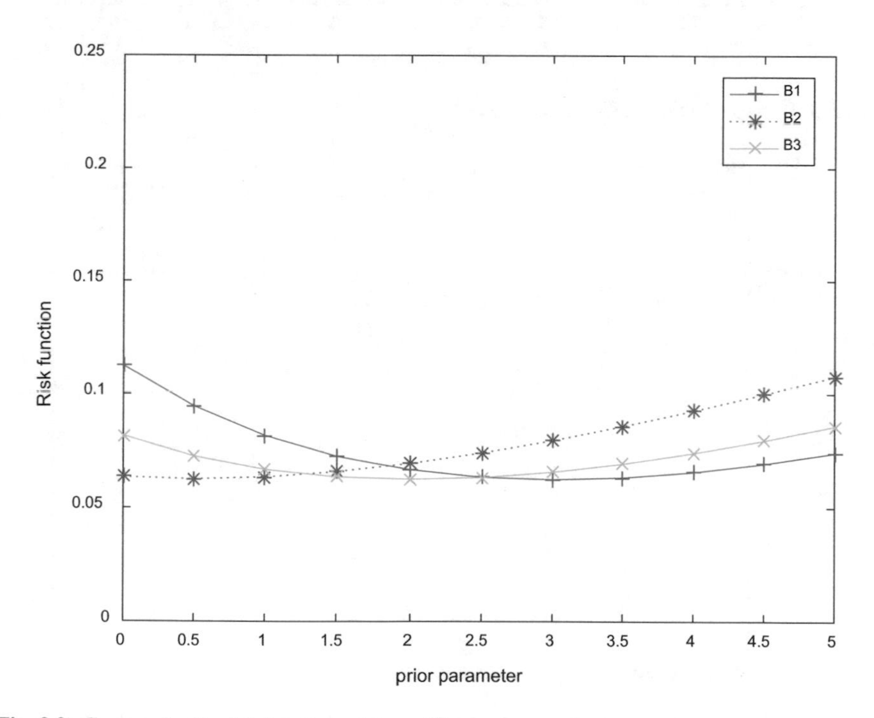


**Fig. 3.2** Curves of ratio risk functions at $m = 15$

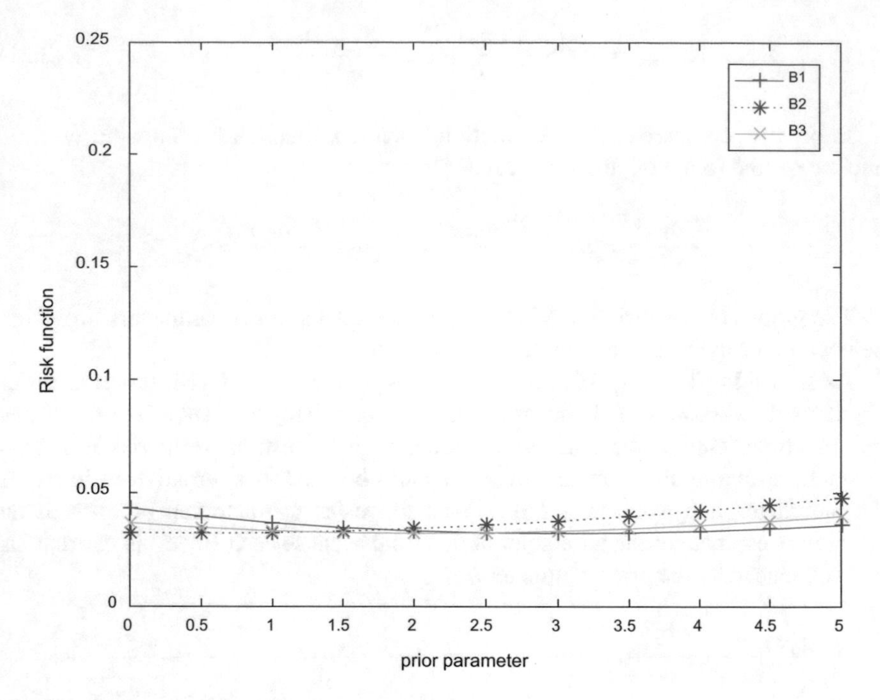


**Fig. 3.3** Curves of ratio risk functions at $m = 30$

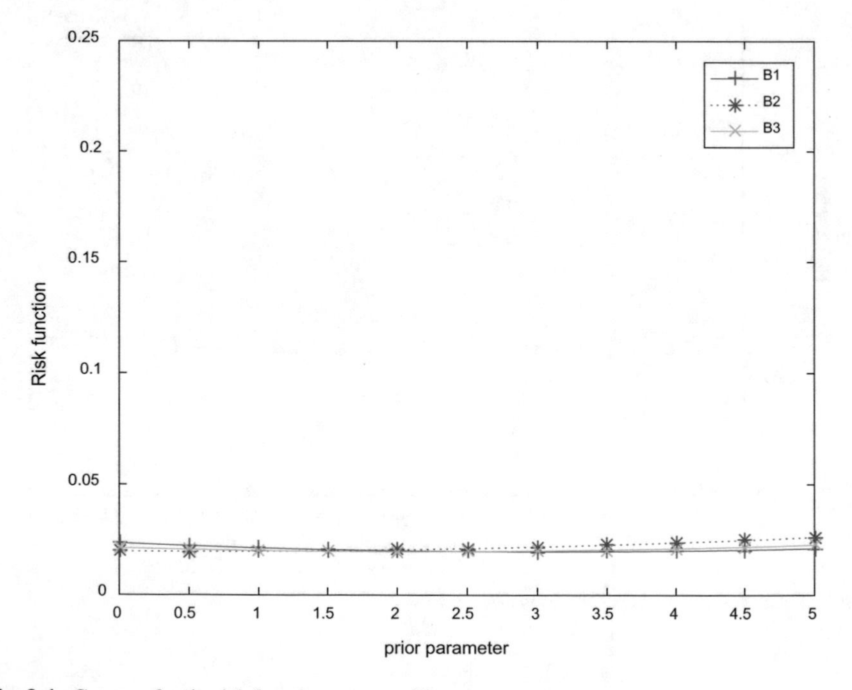


**Fig. 3.4** Curves of ratio risk functions at $m = 50$

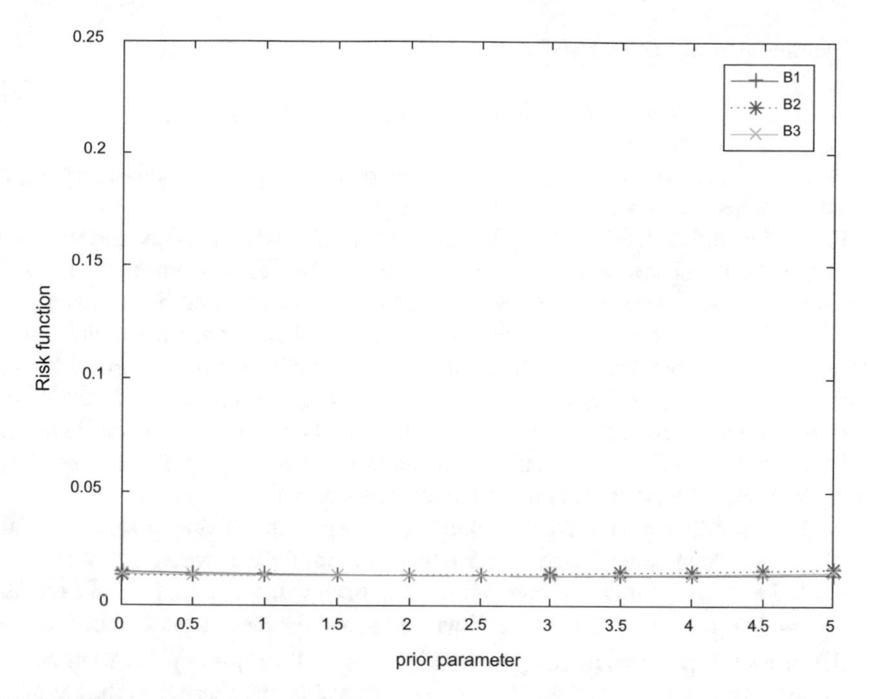


**Fig. 3.5** Curves of ratio risk functions at $m = 75$

#### 3.2.2.2 Comparative Study Under LL Functions

Let $\hat{\sigma}$ be an estimator of the parameter $\sigma$, then the risk function of $\hat{\sigma}$ under the LL function is defined as

$$R(\hat{\sigma}) = \int_{0}^{\infty} \left[ e^{\frac{c(\hat{\sigma}-\sigma)}{\sigma}} - \frac{c(\hat{\sigma}-\sigma)}{\sigma} - 1 \right] \rho(V) \mathrm{d}V \tag{3.22}$$

Then, using Eq. (3.17), a simple calculation leads to the following risk functions for each of the three Bayesian estimators.

$$R_L(\hat{\sigma}_{\mathrm{BS}}) = e^{-c} \left( 1 - \frac{c}{m+d-2} \right)^{-m} - \frac{mc}{m+d-2} + c - 1, \tag{3.23}$$

$$R_L(\hat{\sigma}_{\mathrm{BL}}) = e^{-cd/(m+d)} - m(1 - e^{-c/(m+d)}) + c - 1, \tag{3.24}$$

and

$$R_L(\hat{\sigma}_{\mathrm{BE}}) = e^{-c} \left( 1 - \frac{c}{m+d-1} \right)^{-n} - \frac{cm}{m+d-1} + c - 1. \tag{3.25}$$

In order to compare each risk function, let

$$R_L(\hat{\sigma}_{\mathrm{BS}}) = L_1, \quad R_L(\hat{\sigma}_{\mathrm{BL}}) = L_2 \text{ and } R_L(\hat{\sigma}_{\mathrm{BE}}) = L_3.$$

Plots of $L_1$, $L_2$ and $L_3$ with a priori hyper parameters are given below for LL function, taking $c = 1.5$ and $-1.5$, respectively.

From Figs. 3.6, 3.7, 3.8, 3.9, 3.10, 3.11, 3.12 and 3.13, the LINEX loss function is affected by the shape parameter $c$. Therefore, the risk function under the LINEX loss function and the obtained Bayesian estimates is also affected by it. When $m$ is small, the images of each risk function differ significantly, but as the sample size $m$ increases, especially when $m > 50$, the various risk functions converge. For small $m$, if one estimates the risk functions is smallest among all estimates, then it should be used as alternative estimate of the parameter, and when $m$ is large, each Bayesian estimate can be used as an alternative estimate of the parameter at this time because each Bayesian estimate is less influenced by the prior parameter $d$.

A set of samples from the T-L distribution (3.1) with capacities $m = 15, 30, 45, 60,$ $75, 90$ is generated using Monte Carlo numerical simulation, where the parameter $\sigma = 1.0$. Let $N$ be the times of calculation. The mean value of each type of estimate (i.e., $\hat{\sigma} = \frac{1}{N}\sum_{i=1}^{N}\hat{\sigma}_i$) is used as the estimate of $\sigma$, and the mean square error (briefly, MSE) of each type of estimate (i.e., $\mathrm{ER}(\hat{\sigma}) = \frac{1}{N}\sum_{i=1}^{N}(\hat{\sigma}_i - \sigma)^2$) is used as the criterion to measure the goodness of each type of estimate, where $\hat{\sigma}_i$ is the estimate

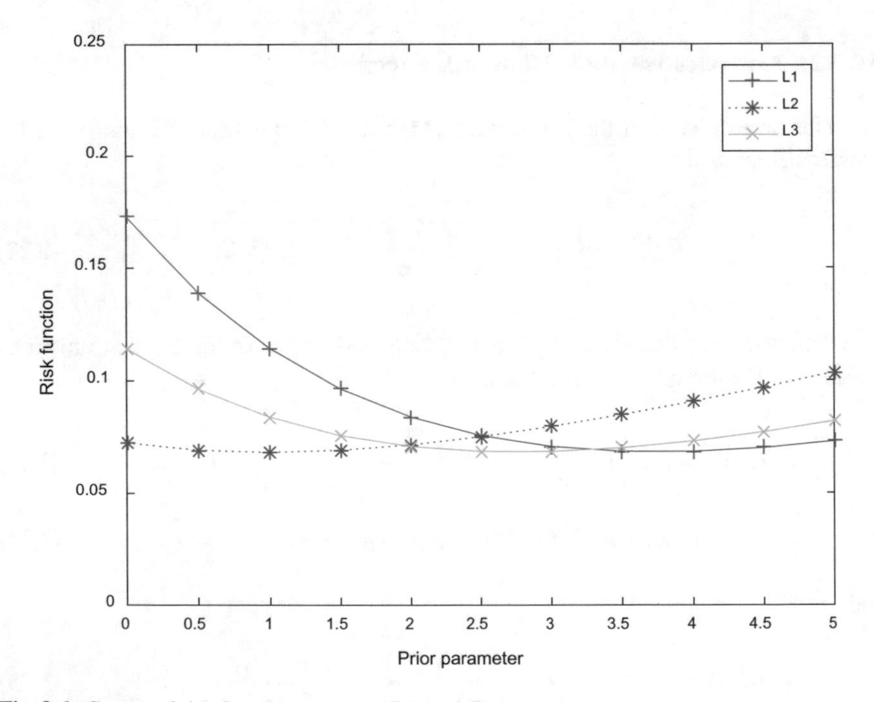


**Fig. 3.6** Curves of risk functions at $m = 15$ ($c = 1.5$)

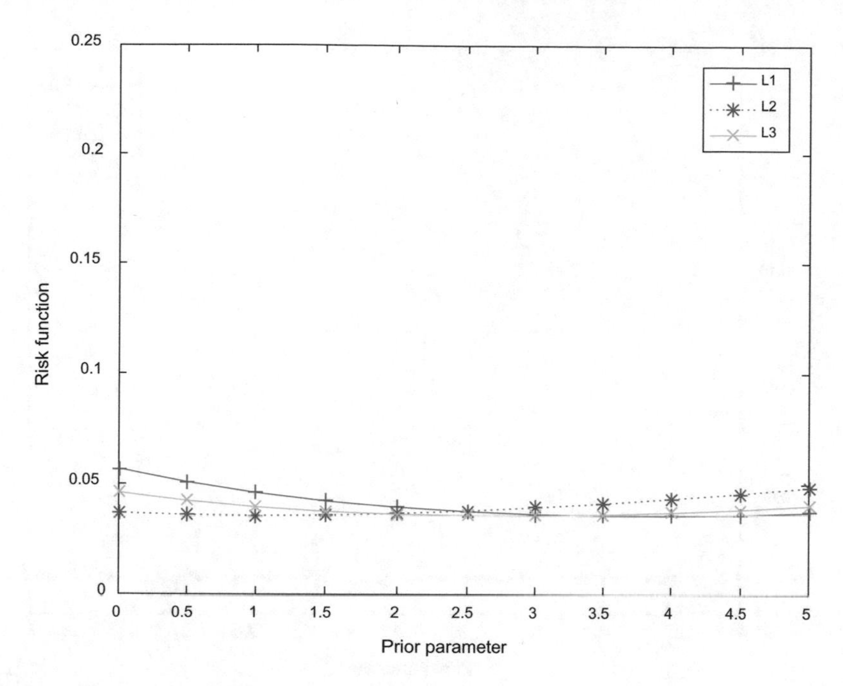


**Fig. 3.7** Curves of risk functions at $m = 30$ ($c = 1.5$)

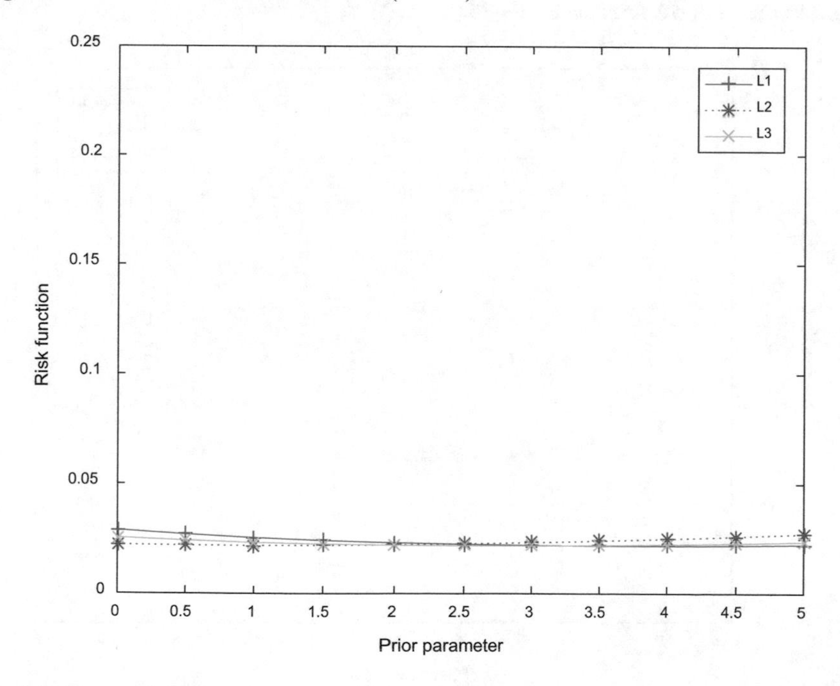


**Fig. 3.8** Curves of risk functions at $m = 50$ ($c = 1.5$)

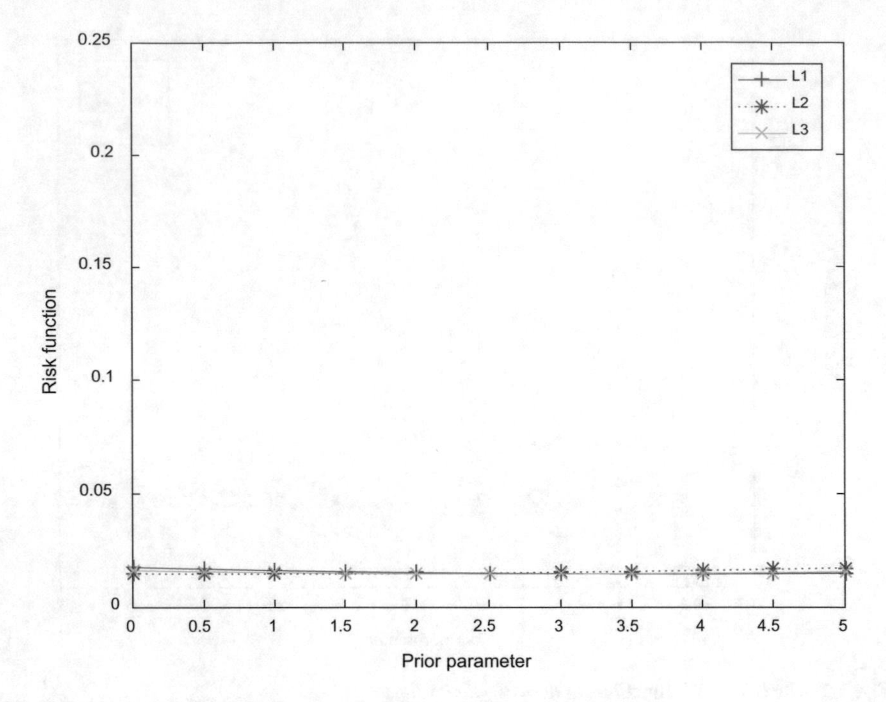


**Fig. 3.9** Curves of risk functions at $m = 75$ ($c = 1.5$)

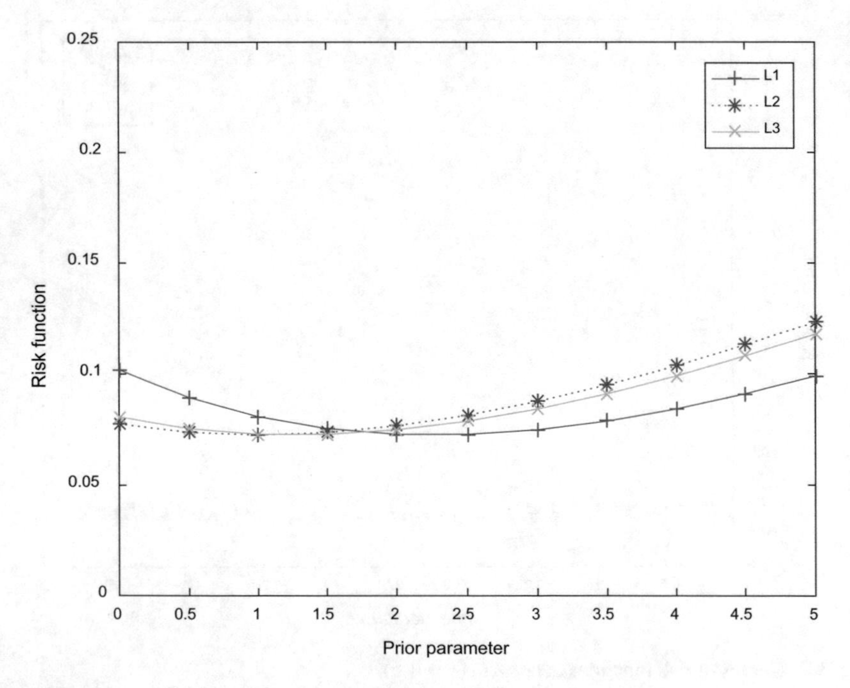


**Fig. 3.10** Curves of risk functions at $m = 15$ ($c = -1.5$)

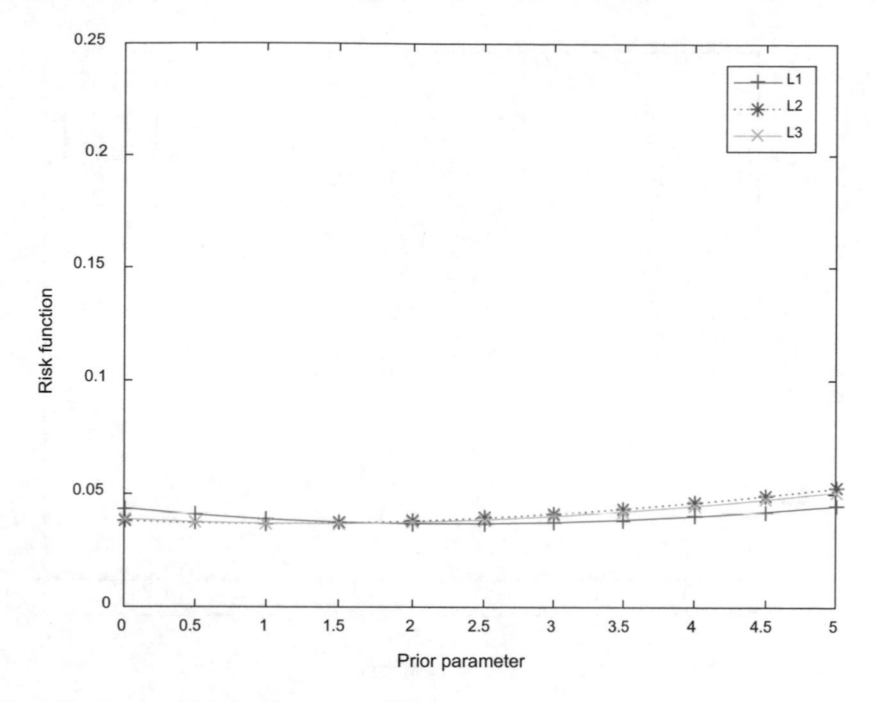


**Fig. 3.11** Curves of risk functions at $m = 30$ ($c = -1.5$)

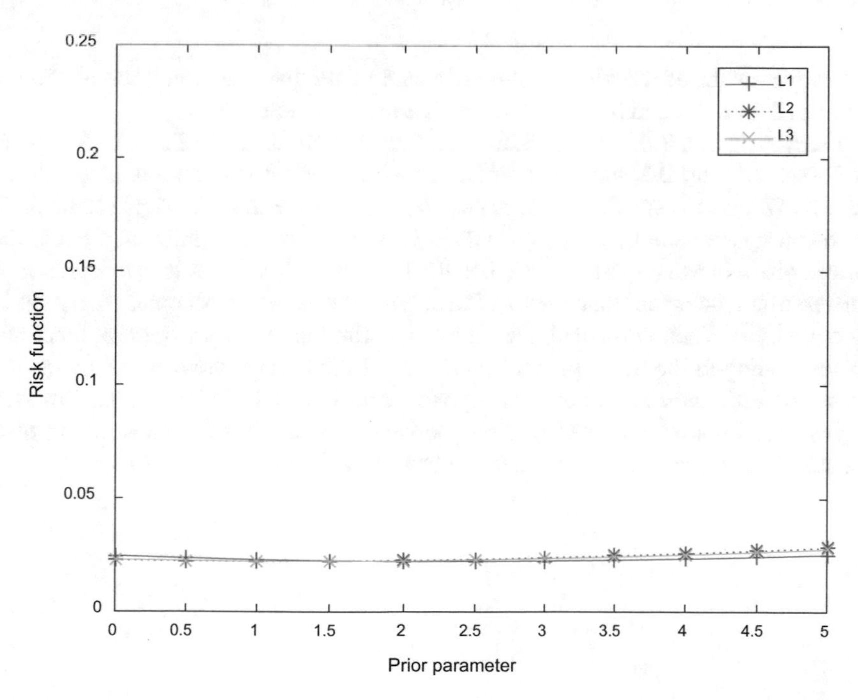


**Fig. 3.12** Curves of risk functions at $m = 50$ ($c = -1.5$)

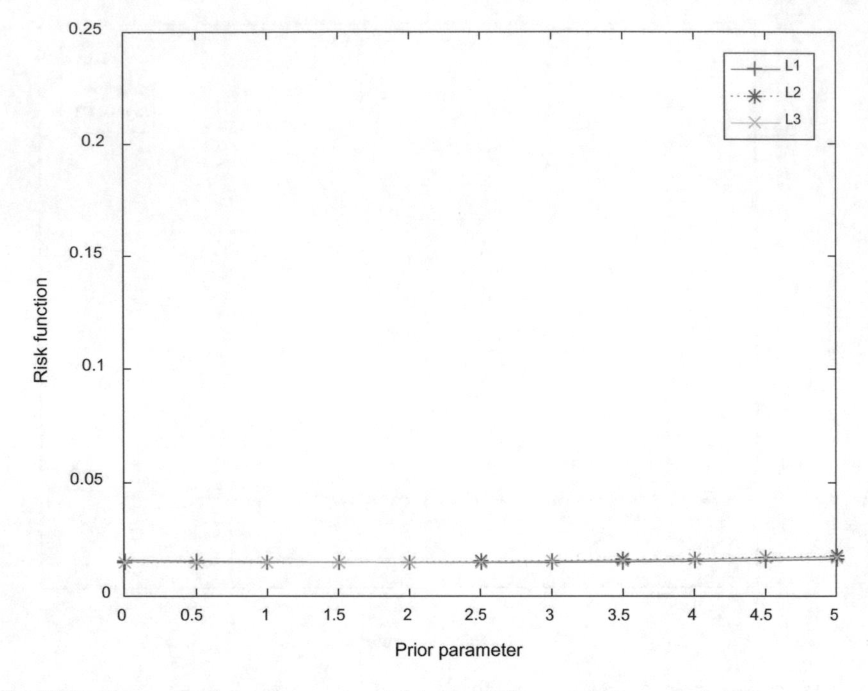


**Fig. 3.13** Curves of risk functions at $m = 75$ ($c = -1.5$)

of the parameter of the $i$th trial. All estimates of the parameter and the MSEs are listed in Tables 3.1 and 3.2, where the MSEs are in parentheses.

From Figs. 3.2, 3.3, 3.4, 3.5, 3.6, 3.7, 3.8, 3.9, 3.10, 3.11, 3.12 and 3.13, as well as Tables 3.1 and 3.2 and a large number of numerical simulations, it is known that: (i) When $m$ is small, the images of each risk function differ significantly, and the mean square error of each type of estimation also differs significantly, but as the sample size $n$ increases, especially when $n$ is larger than 50, the various risk functions converge and the mean square error of each type of estimation becomes smaller as $m$ increases. (ii) When $m$ is small, the Bayesian estimators with smaller risk functions corresponding to the hyper parameters $d$ and $c$ in the image are used as alternative parameter estimates, and when $m$ is large (especially when $n$ is larger than 50), each Bayesian estimate can be used as an alternative estimate of the parameter at this time because each Bayesian estimate is less affected by the prior parameter $d$.

**Table 3.1** Estimates and MSEs for different sample sizes ($d = 0.5$)

| $n$ | $\hat{\sigma}_{ML}$ | $\hat{\sigma}_{BS}$ | $\hat{\sigma}_{BL}$ | | | | $\hat{\sigma}_{BE}$ |
|---|---|---|---|---|---|---|---|
| | | | $c = -1$ | $c = -0.5$ | $c = 0.5$ | $c = 1$ | |
| 15 | 0.9997 | 1.1535 | 1.0338 | 1.0165 | 0.9832 | 0.9671 | 1.0711 |
| | (0.0720) | (0.1195) | (0.0782) | (0.0748) | (0.0700) | (0.0685) | (0.0877) |
| 30 | 0.9999 | 1.0713 | 1.0167 | 1.0082 | 0.9916 | 0.9834 | 1.0343 |
| | (0.0364) | (0.0468) | (0.0379) | (0.0371) | (0.0358) | (0.0355) | (0.0401) |
| 45 | 0.9973 | 1.0437 | 1.0084 | 1.0028 | 0.0205 | 0.9863 | 1.0199 |
| | (0.0206) | (0.0245) | (0.0212) | (0.0209) | (0.9918) | (0.0204) | (0.0220) |
| 60 | 0.9971 | 1.0315 | 1.0055 | 1.0013 | 0.9930 | 0.9888 | 1.0140 |
| | (0.0156) | (0.0177) | (0.0159) | (0.0158) | (0.0155) | (0.0155) | (0.0164) |
| 75 | 0.9964 | 1.0237 | 1.0030 | 0.9997 | 0.9930 | 0.9897 | 1.0098 |
| | (0.0126) | (0.0139) | (0.0128) | (0.0127) | (0.0126) | (0.0126) | (0.0131) |
| 90 | 0.9994 | 1.0221 | 1.0050 | 1.0022 | 0.9966 | 0.9939 | 1.0106 |
| | (0.0101) | (0.0111) | (0.0103) | (0.0102) | (0.0101) | (0.0100) | (0.0105) |

**Table 3.2** Estimates and MSEs for different sample sizes ($d = 1.0$)

| $n$ | $\hat{\sigma}_{ML}$ | $\hat{\sigma}_{BS}$ | $\hat{\sigma}_{BL}$ | | | | $\hat{\sigma}_{BE}$ |
|---|---|---|---|---|---|---|---|
| | | | $c = -1$ | $c = -0.5$ | $c = 0.5$ | $c = 1$ | |
| 15 | 0.9997 | 1.0711 | 0.9671 | 0.9520 | 0.9227 | 0.9085 | 0.9997 |
| | (0.0720) | (0.0877) | (0.0685) | (0.0676) | (0.0673) | (0.0679) | (0.0720) |
| 30 | 0.9999 | 1.0343 | 0.9834 | 0.9755 | 0.9599 | 0.9522 | 0.9999 |
| | (0.0364) | (0.0401) | (0.0355) | (0.0352) | (0.0351) | (0.0353) | (0.0364) |
| 45 | 0.9973 | 1.0199 | 0.9863 | 0.9809 | 0.9703 | 0.9651 | 0.9973 |
| | (0.0206) | (0.0220) | (0.0204) | (0.0203) | (0.0204) | (0.0205) | (0.0206) |
| 60 | 0.9971 | 1.0140 | 0.9888 | 0.9848 | 0.9768 | 0.9728 | 0.9971 |
| | (0.0156) | (0.0164) | (0.0155) | (0.0155) | (0.0155) | (0.0156) | (0.0156) |
| 75 | 0.9964 | 1.0098 | 0.9897 | 0.9865 | 0.9800 | 0.9768 | 0.9964 |
| | (0.0126) | (0.0131) | (0.0126) | (0.0126) | (0.0126) | (0.0127) | (0.0126) |
| 90 | 0.9994 | 1.0106 | 0.9939 | 0.9911 | 0.9857 | 0.9830 | 0.9994 |
| | (0.0101) | (0.0105) | (0.0100) | (0.0100) | (0.0100) | (0.0101) | (0.0101) |

## 3.3 Bayesian Reliability Analysis of T-L Distribution Based on Record Values

### 3.3.1 Record Values and Compound LINEX Loss Function

The record value (briefly, RV), first proposed by Chandler in 1952, is an important value that delineates the trend of a sequence of random variables, and it was extended by Dziubdziela and Kopocinski in 1976 to define the k-RV.

RVs have been used in many fields such as climatology, hydrology, earthquakes, genetics, actuarial insurance, mechanical engineering and sports events. It is important to study the trend of RVs and the theory of statistical inference for the development of national economy.

Zhang [21] introduced the compound LINEX symmetric loss function and pointed out some excellent properties of it and the paper also studied the problem of Bayesian estimation of normally and exponentially distributed parameters. The literature [22–24] studied the Bayesian estimation problem for the parameters of Poisson, Pareto and Burr XII distributions under the Zhang's [21] loss function, respectively. In the literature [25], another compound LINEX symmetric loss function (briefly, CLL) was proposed and the Bayesian estimation problem for exponential distributions was studied.

El-Sayed et al. [26] studied the Bayesian estimation problem of T-L distribution based on record-valued samples under SEL function. Ali et al. [27] studied the Bayesian estimation, prediction and correlation properties of Gumbel model parameters based on record-valued; Jaheen [28] studied the Bayesian estimation problem of Gompertz model parameters based on record-valued samples Ahmadi and Doostparast [29] studied Bayesian estimation and prediction of several common types of distribution models based on record-valued samples. Asgharzadeh [30] studied Bayesian estimation of parameters of exponential distribution based on RVs under SE function and discussed issues such as estimation tolerability estimation.

Now, we will study the Bayesian estimation of T-L model based on RVs under CLL function, whose mathematical expression is [25]

$$L(\delta) = L_c(\delta) + L_{-c}(\delta) = \mathrm{e}^{c\delta} + \mathrm{e}^{-c\delta} - 2, \tag{3.26}$$

where $\delta = (\hat{\sigma} - \sigma)/\sigma$ and $c$ are the shape parameter of this loss function.

Suppose that $X$ is the history data information about the parameter $\sigma$. Let $\eta$ be an estimator of $\sigma$. Under the CLL function (3.26), the Bayesian estimator of $\sigma$ is the solution of the following equation [25]:

$$\mathrm{e}^{-c} E(\sigma^{-1}\mathrm{e}^{c/\eta}|X) = \mathrm{e}^{c} E(\sigma^{-1}\mathrm{e}^{-c/\eta}|X) \tag{3.27}$$

Figure 3.14 is the image of CLL function under different values of $c$.

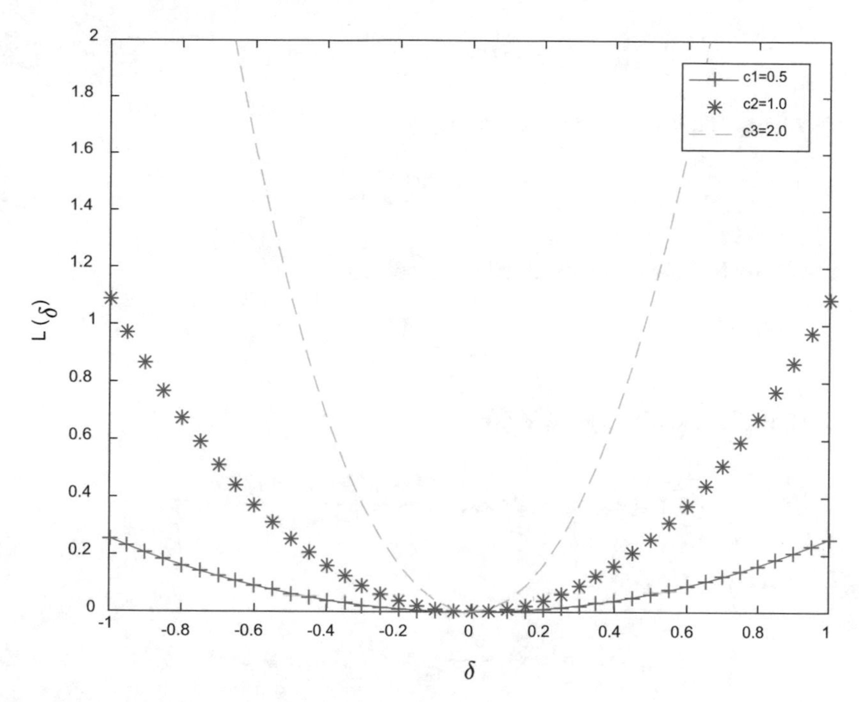


**Fig. 3.14** Graph of CLL function with different $c$

## 3.3.2 ML and Minimum Variance Unbiased Estimation

**Definition 1 [26]** Let $\xi_1, \xi_2, \ldots$ be a sequence of independent identically distributed (i.i.d.) random variables. Let

$$L(1) = 1, L(m+1) = \min\{k : k > L(m), \xi_k < \xi_{L(m)}\},$$

Then, $\xi_{L(m)}$,$m = 1, 2, \ldots$ is called a sequence of lower RVs.

Let $\xi_{L(i)} = \gamma_i, i = 1, 2, \ldots, m$ be a sample of lower RVs from the T-L distribution (3.1) and $\vartheta = -\ln(2\gamma_m - \gamma_m^2)$ be a sample observation of the statistic $V = -\ln[2\xi_{L(m)} - \xi_{L(m)}^2]$. Then, given the sample observation $\gamma = (\gamma_1, \gamma_2, \ldots, \gamma_m)$, the likelihood function of the parameters is [26]

$$\begin{aligned} l(\sigma; \gamma) &= f(\gamma_m; \sigma) \prod_{i=1}^{m-1} \frac{f(\gamma_i; \sigma)}{F(\gamma_i; \sigma)} \\ &= \frac{2}{\sigma}(1-\gamma_m)(2\gamma_m - \gamma_m^2)^{\frac{1}{\sigma}-1} \cdot \prod_{i=1}^{m-1} \frac{\frac{2}{\sigma}(1-\gamma_i)(2\gamma_i - \gamma_i^2)^{\frac{1}{\sigma}-1}}{(2\gamma_i - \gamma_i^2)^{\frac{1}{\sigma}}} \end{aligned}$$

Then, we have

$$l(\sigma;\gamma) \propto \sigma^{-m} \mathrm{e}^{-\frac{-\ln(2\gamma_m-\gamma_m^2)}{\sigma}} = \sigma^{-m}\mathrm{e}^{-\frac{\vartheta}{\sigma}} \tag{3.28}$$

By solving the log-likelihood equation

$$\frac{\mathrm{d}\ln l(\sigma;\gamma)}{\mathrm{d}\sigma} = 0. \tag{3.29}$$

The ML estimator of the parameter $\sigma$ is solved as

$$\hat{\sigma}_{\mathrm{ML}} = \frac{V}{m}.$$

The pdf of the statistic $\gamma_{L(m)}$ is [26]

$$f_m(\gamma_m;\sigma) = f(\gamma_m;\sigma)\frac{[-\ln(F(\gamma_m;\sigma)]^{m-1}}{(m-1)!}.$$

Then, it is easy to derive

$$f_m(\gamma_m;\sigma) = \frac{\sigma^{-m}}{(m-1)!}\vartheta^{m-1}\mathrm{e}^{-\frac{\vartheta}{\sigma}}, \quad \vartheta > 0 \tag{3.30}$$

According to Eq. (3.30), the pdf of the statistic $T$ is

$$f_V(\vartheta) = \frac{\sigma^{-m}}{\Gamma(m)}\vartheta^{m-1}\mathrm{e}^{-\frac{\vartheta}{\sigma}}, \quad \vartheta > 0 \tag{3.31}$$

This implies that $V$ obeys a gamma distribution $\Gamma(m, \sigma^{-1})$, such that there is

$$EV = m\sigma.$$

Further, there are

$$E(\hat{\sigma}_{\mathrm{MLE}}) = E\left(\frac{V}{m}\right) = \sigma.$$

Thus, $\hat{\sigma}_{\mathrm{MLE}}$ is an unbiased estimate of $\sigma$. According to Eq. (3.28), $V$ is a sufficient statistic. Thus, $\hat{\sigma}_{\mathrm{MLE}}$ is also a minimum variance unbiased estimator of $\sigma$.

### *3.3.3 Bayesian Estimation Under CLL Function*

**Theorem 2** *Let* $\xi_{L(1)} = \gamma_1, \xi_{L(2)} = \gamma_2, \ldots, \xi_{L(n)} = \gamma_n$ *be a RVs simple from the T-L distribution* (3.1), *where* $\vartheta = -\ln(2\gamma_m - \gamma_m^2)$ *is a sample observation of*

$V = -\ln[2\xi_{L(m)} - \xi_{L(m)}^2]$. *Assume that the prior distribution of* $\sigma$ *is QPD* (3.3), *then under the CLL function* (3.6), *the Bayesian estimator of* $\sigma$ *is*

$$\hat{\sigma}_{\mathrm{BC}} = \frac{2}{c}\left[\frac{1}{1+\exp(-2c/(m+d)} - \frac{1}{2}\right] \cdot T \tag{3.32}$$

***Proof*** By (3.28) and Bayes' theorem, the posterior pdf of $\sigma$ is

$$\begin{aligned} h(\sigma|\gamma) &\propto l(\sigma;\gamma)\cdot\pi(\sigma) \propto \sigma^{-m}\mathrm{e}^{-\vartheta/\sigma}\cdot\sigma^{-d} \\ &\propto \sigma^{-(m+d)}\mathrm{e}^{-\vartheta/\sigma} \end{aligned}$$

Thus, the posterior pdf of $\sigma$ is

$$h(\sigma|\gamma) = \frac{\vartheta^{m+d-1}}{\Gamma(m+d-1)}\sigma^{-(m+d)}\mathrm{e}^{-\vartheta/\sigma} \tag{3.33}$$

Then,

$$\begin{aligned} E\left[\frac{1}{\sigma}\exp\left(\frac{c\eta}{\sigma}\right)|\gamma\right] &= \int_0^\infty \frac{1}{\sigma}\exp\left(\frac{c\eta}{\sigma}\right)\frac{V^{m+d-1}}{\Gamma(m+d-1)}\sigma^{-(m+d)}\mathrm{e}^{-\frac{V}{\sigma}}\mathrm{d}\sigma \\ &= \frac{m+d-1}{(T-c\eta)^{m+d}}V^{m+d-1}, \\ E\left[\frac{1}{\sigma}\exp\left(\frac{-c\eta}{\sigma}\right)|\gamma\right] &= \int_0^\infty \frac{1}{\sigma}\exp\left(\frac{-c\eta}{\sigma}\right)\frac{V^{m+d-1}}{\Gamma(m+d-1)}\sigma^{-(m+d)}\mathrm{e}^{-\frac{V}{\sigma}}\mathrm{d}\sigma \\ &= \frac{m+d-1}{(T+c\eta)^{m+d}}V^{m+d-1}. \end{aligned}$$

Substituting them into Eq. (3.27) yields the Bayesian estimator of the parameter $\sigma$ as

$$\hat{\sigma}_{\mathrm{BC}} = \frac{2}{c}\left[\frac{1}{1+\exp(-2c/(m+d)} - \frac{1}{2}\right] \cdot V.$$

The risk function of each estimator under the SEL function is derived below, and the results of Monte Carlo simulation are compared. Let $\hat{\sigma}$ be an estimator of the parameter $\sigma$, then the risk function of $\hat{\sigma}_{\mathrm{BC}}$ under the SEL function is

$$R(\hat{\sigma}_{\mathrm{BC}}) = \int_0^\infty (\hat{\sigma}_{\mathrm{BC}} - \sigma)^2 f_V(\vartheta)\mathrm{d}\vartheta$$

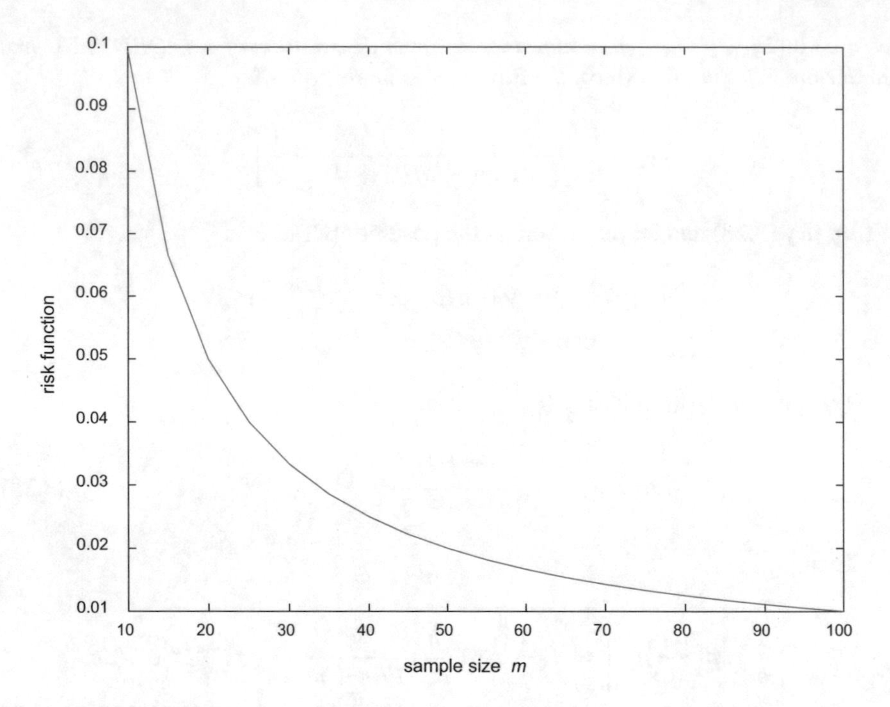


**Fig. 3.15** Risk function's curve of $\hat{\sigma}_{\mathrm{BC}}$ with different sample sizes ($c = 2.5, \sigma = 1$)

$$
\begin{aligned}
&= \int_0^{\infty} (\hat{\sigma}_{\mathrm{BC}} - \sigma)^2 \frac{\sigma^{-m}}{\Gamma(m)} \vartheta^{m-1} \mathrm{e}^{-\frac{\vartheta}{\sigma}} \mathrm{d}\vartheta \\
&= \int_0^{\infty} (\hat{\sigma}_{\mathrm{BC}}^2 - 2\sigma\hat{\sigma}_{\mathrm{BC}} + \sigma^2) \frac{\sigma^{-m}}{\Gamma(m)} \vartheta^{m-1} \mathrm{e}^{-\frac{\vartheta}{\sigma}} \mathrm{d}\vartheta \\
&= (A^2 \cdot EV^2 - 2A \cdot \sigma EV) + \sigma^2 \\
&= [(m + m^2)A^2 - 2mA + 1]\sigma^2,
\end{aligned}
$$

where $A = \frac{2}{c}\left[\frac{1}{1+\exp(-2c/(m+d)} - \frac{1}{2}\right]$ (Fig. 3.15).

As can be seen from the figure, risk function is a decreasing function of sample size $m$, and the effect of $c$ becomes smaller when the sample size increases.

## 3.4 Conclusions

This chapter investigates the problem of Bayesian statistical inference for the parameters of the T-L model. The main works and innovations in this chapter are:

(1) The problem of Bayesian estimation and comparison of risk functions for parameters of T-L distribution based on SEL, LL and EL functions is studied for a complete sample with uninformative QPD of parameters, respectively.
(2) The Bayesian estimation is also studied under a new loss function named CLL function when samples are the RVs.
(3) The performances of different Bayesian estimators are also discussed by comparing their risk functions under SEL and LL functions, respectively.

In our future study, the Bayesian estimation and hypothesis of T-L distribution under other loss functions and censored samples will be considered.

**Acknowledgements** The author Hui Zhou thanks to the support of Foundation of Jiangxi Educational Committee (No. GJJ211604).

## References

1. Ghitany, M.E., Kotz, S., Xie, M.: On some reliability measures and their stochastic orderings for the Topp-Leone distribution. J. Appl. Stat. **32**(7), 715–722 (2005)
2. Al-Zahrani, B.: Goodness-of-fit for the Topp-Leone distribution with unknown parameters. Appl. Math. Sci. **125**, 6355–6363 (2012)
3. Al-Zahrani, B., Alshomrani, A.: Inference on stress-strength reliability from Topp-Leone distributions. J. King Abdulaziz Univ. Sci. **24**(1), 73–88 (2012)
4. Bayoud, H.A.: Estimating the shape parameter of the Topp-Leone distribution based on type I censored samples. Appl. Math. **42**(2), 219–230 (2015)
5. El-Sayed, M.A., Abd-Elmougod, G.A., Abdel Rahman, E.O.: Estimation for coefficient of variation of Topp-Leone distribution under adaptive type-II progressive censoring scheme: Bayesian and non-Bayesian approaches. J. Comput. Theor. Nanosci. **12**(11), 4028–4035 (2015)
6. Deng, Y.L.: The Bayesian estimation and prediction for the Topp-Leone distribution under type-II doubly censored sample. J. Jiangxi Normal Univ. (Nat. Sci. Edn.) **45**(3), 272–277 (2021)
7. Feroze, N., Aslam, M., Khan, I.H., et al.: Bayesian reliability estimation for the Topp-Leone distribution under progressively type-II censored samples. Soft Comput. **25**, 2131–2152 (2021)
8. Zghoul, A.: Efficient plug-in estimators of the Topp-Leone distribution shape parameter. Pak. J. Stat. **36**(4), 305–320 (2020)
9. Li, L.P.: Bayes estimation of Topp-Leone distribution under symmetric entropy loss function based on lower record values. Sci. J. Appl. Math. Stat. **4**(6), 284–288 (2016)
10. Kumar, S.V.: Topp-Leone normal distribution with application to increasing failure rate data. J. Stat. Comput. Simul. **88**(7–9), 1782–1803 (2018)
11. Aldahlan, M.A.: Classical and Bayesian estimation for Topp Leone inverse Rayleigh distribution. Pure Math. Sci. **8**(1), 1–10 (2019)
12. Khan, N., Khan, A.A.: Bayesian analysis of Topp-Leone generalized exponential distribution. Austrian Journal of Statistics **47**(4), 1–15 (2018)
13. Aryuyuen, S., Bodhisuwan, W.: The type II Topp Leone-power Lomax distribution with analysis in lifetime data. J. Stat. Theory Pract. **14**, 31 (2020). https://doi.org/10.1007/s42519-020-00091-x
14. Zeineldin, R.A., Jamal, F., Chesneau, C., et al.: Type II Topp-Leone inverted Kumaraswamy distribution with statistical inference and applications. Symmetry **11**(12), 1–21 (2019)
15. Ren, H.P.: Research on Bayesian Inference of Parameters of Reliability Models. Beijing, China (2020)

16. Alenezi, F.N., Tsokos, C.P.: The Effectiveness of the squared error and Higgins-Tsokos loss functions on the Bayesian reliability analysis of software failure times under the power law process. Engineering **11**(5), 272–299 (2019)
17. Hwang, L.C.: A robust two-stage procedure for the Poisson process under the linear exponential loss function. Stat. Probab. Lett. **163**, 108773 (2020)
18. Zeghdoudi, H., Metiri, F., Remita, M.R.: On Bayes estimates of Lindley distribution under Linex loss function: informative and non informative priors. Glob. J. Pure Appl. Math. **12**(1), 391–400 (2016)
19. Hwang, L.C.: Second order optimal approximation in a particular exponential family under asymmetric LINEX loss. Stat. Probab. Lett. **137**, 283–291 (2018)
20. Rasheed, H.A., Naji, L.F.: Estimate the two parameters of Gamma distribution under entropy loss function. Iraqi J. Sci. **60**(1), 127–134 (2019)
21. Zhang, R.: The Parameter Estimation under the Compound LINEX Symmetric Loss Function. Dalian University of Technology, Dalian (2010)
22. Wei, C.D., Wei, S., Su, H.: Bayes estimation and application of Poisson distribution parameter under compound LINEX symmetric loss. Stat. Decis. **7**, 156–157 (2010)
23. Wei, C.D., Wei, S., Su, H.: E-Bayes estimation of shape parameter of Pareto distribution under compound LINEX symmetric loss and its application. Stat. Decis. **17**, 7–9 (2009)
24. Wei, S., Li, Z.N.: Bayes estimation of Burr XII distribution parameter in the composite LINEX loss of symmetry. Appl. Math. A J. Chin. Univ. **32**(1), 49–54 (2017)
25. Ren, H.P., Chao, S.G.: Bayesian Reliability analysis of exponential distribution model under a new loss function. Int. J. Performability Eng. **14**(8), 1815–1823 (2018)
26. El-Sayed, M.A., Abd-Elmougod, G.A., Abdel-Khalek, S., et al.: Bayesian and non-Bayesian estimation of Topp-Leone distribution based lower record values. Far East J. Theor. Stat. **45**(2), 133–145 (2013)
27. Ali, M.M., Jaheen, Z.F., Ahmad, A.A.: Bayesian estimation, prediction and characterization for the Gumbel model based on records. Statistics **36**, 65–74 (2002)
28. Jaheen, Z.F.: A Bayesian analysis of record statistics from the Gompertz model. Appl. Math. Comput. **145**, 307–320 (2003)
29. Ahmadi, J., Doostparast, M.: Bayesian estimation and prediction for some life distributions based on record values. Stat. Pap. **47**, 373–392 (2006)
30. Asgharzadeh, A.: On Bayesian estimation from exponential distribution based on records. Korean Stat. Soc. **38**(2), 125–130 (2009)

**Haiping Ren** was born in 1979. Now, he is an associate professor of Jiangxi University of Science and Technology. He received the bachelor's degree in applied mathematics from Changsha University of Science and Technology, China, in 2003, and the master's degree in probability and statistics from Central South University, China, in 2005, and the Ph.D. degree in the management science from Jiangxi University of Finance and Economics, China, in 2015, respectively. He has authored or coauthored more than 50 journal papers and 5 books. His research interests include Bayesian statistics and fuzzy decision analysis.

**Hui Zhou** was born in 1979. Now she is an associate professor of Yichun University. In 2005 she graduated from China Central South University for her master degree. She has published about more than 20 research papers (including about 8 SCI/EI-indexed papers) and 2 books. She current major research fields include Bayesian statistics and fuzzy decision-making.

**Bin Yin** is a lecturer at Jiangxi University of Science and Technology. He graduated from Chengdu University of Technology with a master's degree in 2012. He received his Ph.D. degree from China University of Geosciences in 2017. His current research interests include intelligent optimization algorithms and Bayesian inversion theory.

# Chapter 4
# Reliability Metrics of Textile Confection Plant Using Copula Linguistic

**Abdulkareem Lado Ismail and Ibrahim Yusuf**

**Abstract** The goal behind the current chapter is to examine the performance of a textile confection plant through measures such as availability, sensitivity of MTTF, reliability, mean time to failure (MTTF), and expected profit. Traditional performance analysis for textile confection plant frequently overestimates the fundamental dependability of their components, ignoring the importance of the multi-station textile confection plant. This quandary will be addressed by employing a copula approach for analyzing the performance of textile confection plant. The plant is made up of five subsystems: two identical weavers and three identical dry cleaners that operate under the $k$-out-of-$n$: G-policy, a printer, a cross-cut, and a cleaner in serial arrangement. Failure of units can be lighter or heavy, are distributed exponentially, and are repaired through the distributions of general and copula. The repair by general distribution restores the unit to the state prior to their failure, while the repair by copula distribution restores the subsystem to the initial state after installation of the plant. To derive the general analytical solution of the textile confection plant, Laplace transforms and technique of supplementary variable to establish the partial differential equations related to transition diagram are essential to this chapter. The expressions for the factory's reliability metrics of availability, reliability, MTTF, sensitivity, and cost function are obtained using a mathematical tool. Finally, numerical examples are provided to investigate the effects of the parameters used in the analysis. Tables and figures revealed that the results obtained indicated that the repairs used are very good, which would lead to higher system performance and also motivate factory users. This chapter may help textile industries and their maintenance by alleviating some of the difficulties encountered during the factory's maintenance process and increasing income mobilization.

**Keywords** Reliability · Availability · Confection plant · Copula · General repair

A. L. Ismail
Department of Mathematics, Kano State College of Education, Kano, Nigeria

I. Yusuf (✉)
Department of Mathematical Sciences, Bayero University, Kano, Nigeria
e-mail: iyusuf.mth@buk.edu.ng



H. Garg (ed.), *Advances in Reliability, Failure and Risk Analysis*, Industrial and Applied Mathematics, https://doi.org/10.1007/978-981-19-9909-3_4

## 4.1 Introduction

Today's manufacturing systems are extremely advanced and consist of a collection of interlinked machines. These interconnected machines are susceptible to failure, affecting the system's reliability and availability, as well as the revenue generated. Nowadays, consumers expect complete assurance that the goods produced by manufacturing systems are of high quality and will continue to function. The stages of design and manufacturing are critical in ensuring the dependability of the goods produced. Among all manufacturing sectors, the garment industry, textile, agricultural and chemical fibers, apparel, and waste management, has the most complicated production process.

Textile manufacturing is one of the manufacturing industries that has grown tremendously as a result of global population growth, cumulative improvement in living standards, and the sudden increase of fast fashion. Textile confection plant is one of the major industries molding fiber into fabric through yarn to other materials.

Reliability is the tendency of a system to perform satisfactorily under specified working scenario with passage of time. One of the most vital metrics in deriving the system's capability is reliability. Where reliability is low, system is weak and hence cannot deliver all manufacturing tasks or meet production targets.

Several authors have discussed the comprehensive reliability and performance of manufacturing systems, including Ye et al. [1], who examined the reliability of a repairable machine while it was subjected to shocks and degradation from low-quality feedstocks. Ye et al. [2] develop a new model for competing failure to investigate the relationships among the inspection process, product quality, and machine failures. Chang et al. [3] investigated the accuracy with which the reliability of a manufacturing network with finite buffer quantity is assessed in terms of the modification term. Centered on operations and maintenance accurate information, Chen et al. [4] suggested mission reliability for multi-state manufacturing technologies. Zhang et al. [5] capture the development and evaluation of performance of manufacturing systems with serial arrangements that account for rework and product polymorphism. Pundir et al. [6] investigated the reliability metrics of two non-identical cold standby unit systems that used different types of priors for unknown parameters. Kumar et al. [7, 8] introduced the gray wolf optimization technique for life support system's reliability and cost management. Mella and Zio [9] presented an improved nest cuckoo optimization algorithm for the reliability–redundancy optimization framework. Okafor et al. [10] examine the reliability of a multi-state parallel system using an Archimedean copula-based technique. Sharifi et al. [11] proposed the universal generating function technique for assessing and estimating the reliability of a system with weighted-$k$-out-of-$n$ subsystems. Jia et al. [12] proposed multi-state warm standby and multi-state performance communicating mechanisms for power system reliability modeling and assessment. Lin et al. [13] suggested Bayesian copula-based models for predicting component reliance and interactions. Jia et al. [14] developed a power system model that includes warm standby and energy storage in reliability prediction of the system.

Copula can also be used to perform nonparametric analysis on pairs of random variables. In a variety of real-world scenarios, multiple repairs between nearby transition states are possible in order to quickly restore a failing system. Copula is used to repair the system when this occurs. The copula repair method is a powerful tool for explaining variable dependency that has piqued the interest of researchers across a wide range of disciplines. Nelson [15] dealt with analysis and applications of copula. Yusuf et al. [16] suggested performance and reliability models for assessing and measuring the resilience of network systems with serially connected devices. Chopra and Ram [17] use copula to explore the performance indicators of reliability and availability of systems in a parallel network with two distinct units. Gahlot et al. [18] explore the performance of a serial system with two subsystems operating in sequence under the 2-out-of-3: G and 1-out-of-2: G-policies. Lado and Singh [19] investigated the cost of a serial system manned by a human operator. Sha [20] used Clayton copula functions to investigate working unit dependency, and Farlie–Gumbel–Morgenstern established models for parallel–series and series–parallel. Sanusi and Yusuf [21] investigated the resilience of a three-component dispersed data center network topology. Tyagi et al. [22] studied the behavior of stochastic parallel system reliability models with major, unit, and human failure types under copula-coverage, copula, and coverage factor conditions. Yusuf et al. [16] demonstrated the efficacy of a multi-computer system composed of three subsystems linked in series using the copula repair technique. Singh et al. [23] presented an improved model for detecting flaws in previously published models and evaluating performance for different types of failure and repair, claiming that the system outperforms previously evaluated systems.

The aforementioned researcher developed outstanding work on reliability of repairable systems via copula technique, asserting that their operations enhanced the repairable systems' reliability and performance. Nonetheless, a new model with a substantiated and sufficient assessment is obliged. With the aforementioned facts in mind, this chapter dealt with the reliability analysis of a textile confection plant, which consisted of five subsystems in series, namely weaver, dry cleaner, printer, cross-cut, and cleaner, as shown in Fig. 4.1. To the authors' little knowledge, no reliability modeling and performance study has focused on estimating the reliability optimizing of a textile confectionery plant via copula. As a consequence, the current study was intended to fill a research gap. The copula repair approach was used in this chapter to analyze the reliability and performance optimization of the textile confection plant.

The goals of the chapter are: First is to develop the models of availability, sensitivity, cost function, MTTF, and reliability. Second is to numerically determine the behavior of availability, reliability, MTTF, sensitivity, and cost function with passage of time. Third is to predict system performance optimization by employing two repair strategies. In estimating the reliability and optimizing the performance of a textile confection plant, the study suggested two different types of repair techniques. The system is prone to light and heavy of failure. In the event of partial or light failure, the system is repaired using general repair, whereas copula repair is used to recover

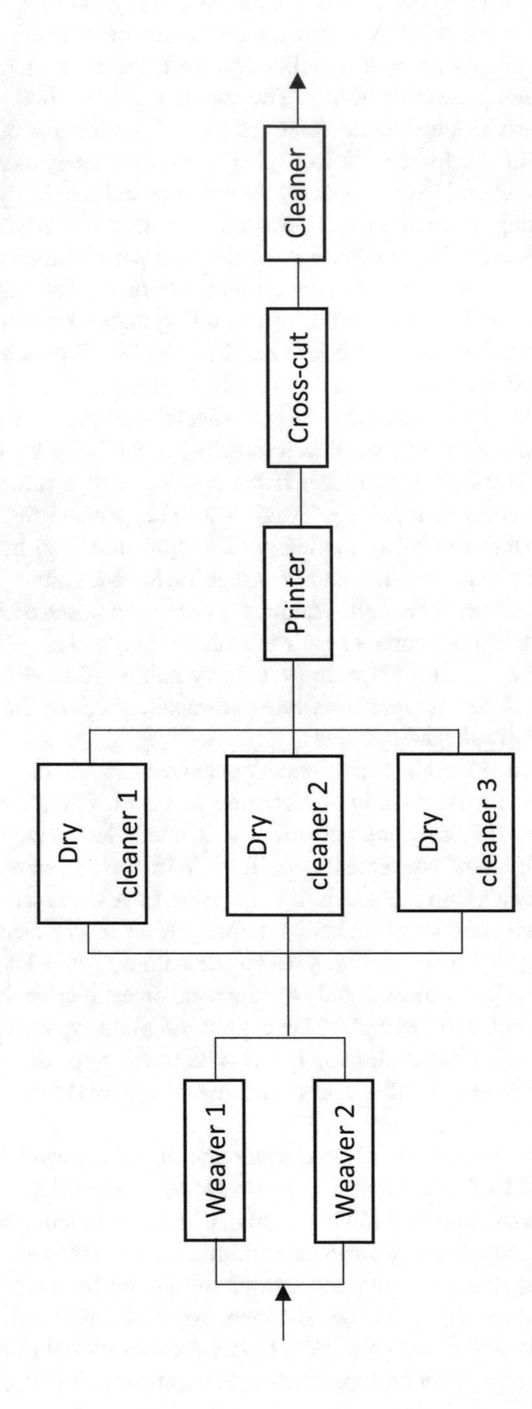


**Fig. 4.1** Factory's block diagram

completely or heavy failure. The interactive study, in our opinion, should focus on increasing the system's reliability, availability, and revenue generation.

## 4.2 Abbreviations, Description, and State of the Confection Plant

### 4.2.1 Abbreviations

| | |
|---|---|
| $q$ | Variable of time |
| $l_k$ | Weavers, dry cleaners, printer, cross-cut, and cleaner failure rate $k = 1, 2, 3, 4, 5$ |
| $f(y_1)/f(y_2)$ | Weaver and dry cleaner repair rate |
| $b_0(y_1)/b_0(y_2)/b_0(y_5)/$ $b_0(y_4)/b_0(y_3)$ | Repair rates for; weavers, dry cleaners, printer, cross-cut, and cleaner that are completely failed |
| $D_i(q)$ | For $i = 0, 1, \ldots, 11$, $S_i$ defined the possible states of the textile confection plant |
| $\overline{D}(\vartheta)$ | Laplace conversion of $D(q)$ |
| $D_i(y_1, q)$ | For $i = 1, 6, 7$, defined the states probability with repair and repair time |
| $D_i(y_2, q)$ | For $i = 2, 3, 4, 5, 8$, defined the states probability with repair and repair time |
| $D_i(y_5, q)$ | For $i = 9$, defined the states probability with repair and repair time |
| $D_i(y_4, q)$ | For $i = 10$, defined the states probability with repair and repair time |
| $D_i(y_3, q)$ | For $i = 11$, defined the states probability with repair and repair time |
| $E_p(q)$ | Profit/revenue expected in the interval [0, $q$] |
| $A_1, A_2$ | The revenue and service costs |
| $\vartheta_f(q)$ | Function distribution $\vartheta_f(q) = f(y_1)\mathrm{e}^{-\int_o^x f(y_1)\mathrm{d}y_1}$ notation |
| $\overline{\vartheta}_f(\vartheta)$ | Laplace transforms of $\vartheta_f(q)$ i.e., $\overline{\vartheta}_f(\vartheta) = \int_0^\infty \mathrm{e}^{-\vartheta y_1} f(y_1)\mathrm{e}^{-\int_0^x f(y_1)\mathrm{d}y_1}\mathrm{d}y_1$ |
| $b_0(x) = C_\theta(b_1(x)b_2(x))$ | Gumbel–Hougaard family copula repair distribution |

defined as:

$$c_\theta(b_1(y_1), b_2(y_1)) = \exp\left(y_1^\theta + \left\{\log\phi(y_1)^\theta\right\}^{\frac{1}{\theta}}\right),$$
$$1 \leq \theta \leq \infty.$$

where $b_1 = f(y_1)$ and $b_2 = e^{y_1}$.

### 4.2.2 The Description of the Confection Plant

The textile confection plant comprises of weavers, dry cleaners, printer, cross-cut, and cleaner, which are all arranged in a series–parallel configuration. The $k$-out-of-$n$ G-policy applies to subsystem 1 and subsystem 2. Originally, one unit from subsystems works, while others are on standby mode. Immediately, the system was in a partial operational state due to the failure of first unit from subsystem 1 and subsystem 2, immediately the standby units switch automatically to the operational mode, the system continues working, and failed units are assigned for repair. The system would experience complete failure if second unit from subsystem 1, third unit from subsystem 2 have failed or at the failure of subsystem 3, subsystem 4, and subsystem 5. General repair is used to fix partially failed states, while Gumbel–Hougaard family copula repair is used to repair completely failed states.

### 4.2.3 Description of the State

- $S_0$ Initial state. The subsystems and the plant are in perfect state. The plant is up and running
- $S_1$ One unit has failed in subsystem 1; the plant is up and running
- $S_2$ One unit has failed in subsystem 2; the plant is up and running
- $S_3$ One unit has failed in subsystem 2; the second and third units are working. The plant is up and running
- $S_4$ One unit has failed in subsystem 1 previously; failure of one in subsystem 2 followed. The plant is up and running
- $S_5$ One unit each has failed in subsystems 1 and 2 previously; failure of second in subsystem 2 followed. The plant is up and running
- $S_6$ One unit has failed in subsystem 2 previously; failure of one in subsystem 1 followed. The plant is up and running
- $S_7$ One unit each has failed in subsystems 1 and 2 previously; failure of second in subsystem 1 followed. The plant is down
- $S_8$ The plant is down due to last operational unit failure in subsystem 2
- $S_9$ The plant is down due to printer failure
- $S_{10}$ The plant is down due to cross-cut failure

$S_{11}$ The plant is down due to cleaner failure.

## 4.3 Formulation of Textile Confection Plant Mathematical Model

The equations associated with Fig. 4.2 of the textile confection plant are

$$\left(\frac{\partial}{\partial q}+2l_1+3l_2+l_3+l_4+l_5\right)D_0(q)=\int_0^\infty f(y_1)D_1(y_1,q)\mathrm{d}y_1+\int_0^\infty f(y_2)D_2(y_2,q)\mathrm{d}y_2+\int_0^\infty b_0(y_1)D_7(y_1,q)\mathrm{d}y_1+\int_0^\infty b_0(y_2)D_8(y_2,q)\mathrm{d}y_2+\int_0^\infty b_0(y_5)D_9(y_5,q)\mathrm{d}y_5+\int_0^\infty b_0(y_4)D_{10}(y_4,q)\mathrm{d}y_4+\int_0^\infty b_0(y_3)D_{11}(y_3,q)\mathrm{d}y_3 \tag{4.1}$$

$$\left(\frac{\partial}{\partial q}+\frac{\partial}{\partial y_1}+l_1+3l_2+l_3+l_4+l_5+f(y_1)\right)D_1(y_1,q)=0 \tag{4.2}$$

$$\left(\frac{\partial}{\partial q}+\frac{\partial}{\partial y_2}+2l_1+2l_2+l_3+l_4+l_5+f(y_2)\right)D_2(y_2,q)=0 \tag{4.3}$$

$$\left(\frac{\partial}{\partial q}+\frac{\partial}{\partial y_2}+l_2+f(y_2)\right)D_3(y_2,q)=0 \tag{4.4}$$

$$\left(\frac{\partial}{\partial q}+\frac{\partial}{\partial y_2}+2l_2+f(y_2)\right)D_4(y_2,q)=0 \tag{4.5}$$

$$\left(\frac{\partial}{\partial q}+\frac{\partial}{\partial y_2}+l_2+f(y_2)\right)D_5(y_2,q)=0 \tag{4.6}$$

$$\left(\frac{\partial}{\partial q}+\frac{\partial}{\partial y_1}+l_1+f(y_1)\right)D_6(y_1,q)=0 \tag{4.7}$$

$$\left(\frac{\partial}{\partial q}+\frac{\partial}{\partial y_1}+b_0(y_1)\right)D_7(y_1,q)=0 \tag{4.8}$$

$$\left(\frac{\partial}{\partial q}+\frac{\partial}{\partial y_2}+b_0(y_2)\right)D_8(y_2,q)=0 \tag{4.9}$$

$$\left(\frac{\partial}{\partial q}+\frac{\partial}{\partial y_4}+b_0(y_4)\right)D_9(y_4,q)=0 \tag{4.10}$$

$$\left(\frac{\partial}{\partial q}+\frac{\partial}{\partial y_3}+b_0(y_3)\right)D_{10}(y_3,q)=0 \tag{4.11}$$

$$\left(\frac{\partial}{\partial q}+\frac{\partial}{\partial y_3}+b_0(y_3)\right)D_{11}(y_3,q)=0 \tag{4.12}$$

**Boundary condition**

$$D_1(0,q)=2l_1D_0(q) \tag{4.13}$$

$$D_2(0,q)=3l_2D_0(q) \tag{4.14}$$

$$D_3(0,q)=2l_2D_2(0,q) \tag{4.15}$$

$$D_4(0,q)=3l_2D_1(0,q) \tag{4.16}$$

$$D_5(0,q)=2l_2D_4(0,q) \tag{4.17}$$

$$D_6(0,q)=2l_1D_2(0,q) \tag{4.18}$$

$$D_7(0,q)=l_1(D_1(0,q)+D_6(0,q)) \tag{4.19}$$

$$D_8(0,q)=l_2(D_3(0,q)+D_5(0,q)) \tag{4.20}$$

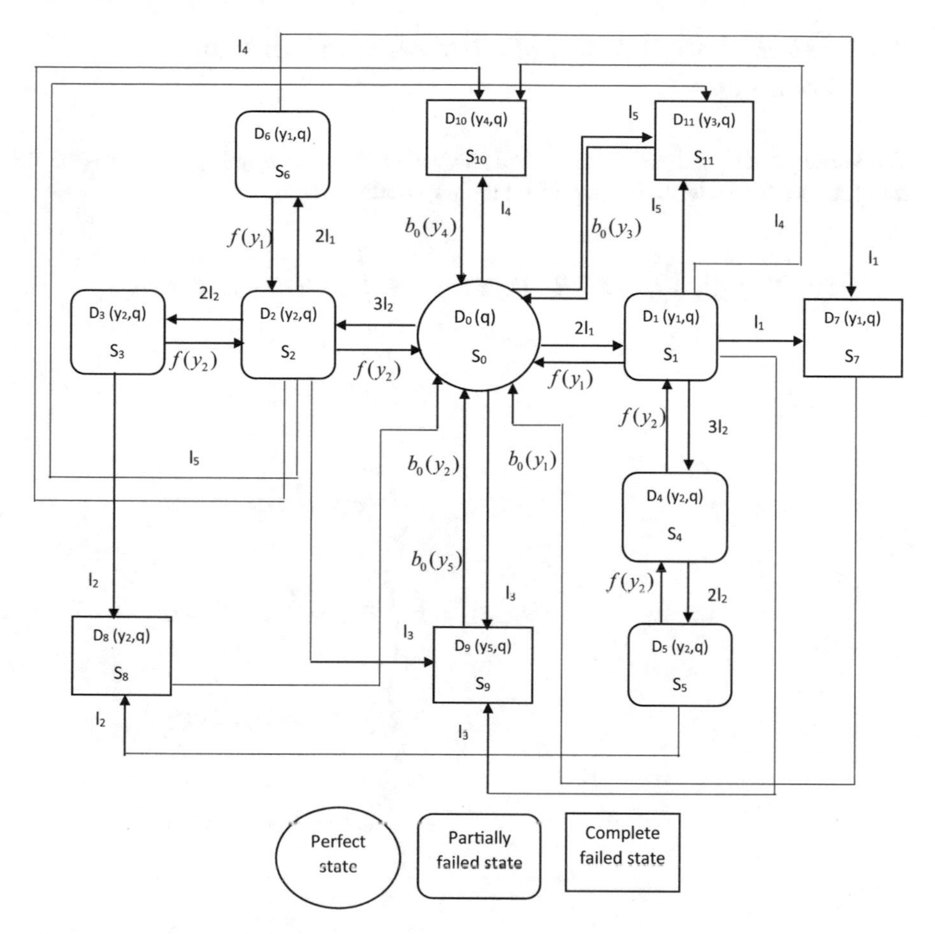


**Fig. 4.2** Factory's plan diagram

$$D_9(0, q) = l_3(D_0(q) + D_1(0, q) + D_2(0, q)) \tag{4.21}$$

$$D_{10}(0, q) = l_4(D_0(q) + D_1(0, q) + D_2(0, q)) \tag{4.22}$$

$$D_{11}(0, q) = l_5(D_0(q) + D_1(0, q) + D_2(0, q)) \tag{4.23}$$

### 4.3.1 Mathematical Model of Textile Confection Plant Solution

The subsequent equations are produced as a result of converting Eqs. (4.1) to (4.23) into Laplace forms with the help of boundary conditions.

$$
\begin{aligned}
(\vartheta + 2l_1 + 3l_2 + l_3 + l_4 + l_5)\overline{D}_0(\vartheta) = 1 &+ \int_0^\infty f(y_1)\overline{D}_1(y_1, \vartheta)\mathrm{d}y_1 \\
&+ \int_0^\infty f(y_2)\overline{D}_1(y_2, \vartheta)\mathrm{d}y_2 \\
&+ \int_0^\infty b_0(y_1)\overline{D}_7(y_1, \vartheta)\mathrm{d}y_1 \\
&+ \int_0^\infty b_0(y_2)\overline{D}_8(y_2, \vartheta)\mathrm{d}y_2 \\
&+ \int_0^\infty b_0(y_5)\overline{D}_9(y_5, \vartheta)\mathrm{d}y_5 \\
&+ \int_0^\infty b_0(y_4)\overline{D}_{10}(y_4, \vartheta)\mathrm{d}y_4 \\
&+ \int_0^\infty b_0(y_3)\overline{D}_{11}(y_3, \vartheta)\mathrm{d}y_3
\end{aligned}
\tag{4.24}
$$

$$\left(\vartheta + \frac{\partial}{\partial y_1} + l_1 + 3l_2 + l_3 + l_4 + l_5 + f(y_1)\right)\overline{D}_1(y_1, \vartheta) = 0 \tag{4.25}$$

$$\left(\vartheta + \frac{\partial}{\partial y_2} + 2l_1 + 2l_2 + l_3 + l_4 + l_5 + f(y_2)\right)\overline{D}_2(y_2, \vartheta) = 0 \tag{4.26}$$

$$\left(\vartheta + \frac{\partial}{\partial y_2} + l_2 + f(y_2)\right)\overline{D}_3(y_2, \vartheta) = 0 \tag{4.27}$$

$$\left(\vartheta + \frac{\partial}{\partial y_2} + 2l_2 + f(y_2)\right)\overline{D}_4(y_2, \vartheta) = 0 \tag{4.28}$$

$$\left(\vartheta + \frac{\partial}{\partial y_2} + l_2 + f(y_2)\right)\overline{D}_5(y_2, \vartheta) = 0 \tag{4.29}$$

$$\left(\vartheta + \frac{\partial}{\partial y_1} + l_1 + f(y_1)\right)\overline{D}_6(y_1, \vartheta) = 0 \tag{4.30}$$

$$\left(\vartheta + \frac{\partial}{\partial y_1} + b_0(y_1)\right)\overline{D}_7(y_1, \vartheta) = 0 \tag{4.31}$$

$$\left(\vartheta + \frac{\partial}{\partial y_2} + b_0(y_2)\right)\overline{D}_8(y_2, \vartheta) = 0 \tag{4.32}$$

$$\left(\vartheta + \frac{\partial}{\partial y_5} + b_0(y_5)\right)\overline{D}_9(y_5, \vartheta) = 0 \tag{4.33}$$

$$\left(\vartheta + \frac{\partial}{\partial y_4} + b_0(y_4)\right)\overline{D}_{10}(y_4, \vartheta) = 0 \tag{4.34}$$

$$\left(\vartheta + \frac{\partial}{\partial y_3} + b_0(y_3)\right)\overline{D}_{11}(y_3, \vartheta) = 0 \tag{4.35}$$

**Boundary conditions**

$$\overline{D}_1(0, \vartheta) = 2l_1\overline{D}_0(\vartheta) \tag{4.36}$$

$$\overline{D}_2(0, \vartheta) = 3l_2\overline{D}_0(\vartheta) \tag{4.37}$$

$$\overline{D}_3(0, \vartheta) = 2l_2\overline{D}_2(0, \vartheta) \tag{4.38}$$

$$\overline{D}_4(0, \vartheta) = 3l_2\overline{D}_1(0, \vartheta) \tag{4.39}$$

$$\overline{D}_5(0, \vartheta) = 2l_2\overline{D}_4(0, \vartheta) \tag{4.40}$$

$$\overline{D}_6(0, \vartheta) = 2l_1\overline{D}_2(0, \vartheta) \tag{4.41}$$

$$\overline{D}_7(0, \vartheta) = l_1\left(\overline{D}_1(0, \vartheta) + \overline{D}_6(0, \vartheta)\right) \tag{4.42}$$

$$\overline{D}_8(0, \vartheta) = l_2\left(\overline{D}_3(0, \vartheta) + \overline{D}_5(0, \vartheta)\right) \tag{4.43}$$

$$\overline{D}_9(0, \vartheta) = l_3\left(\overline{D}_0(\vartheta) + \overline{D}_1(0, \vartheta) + \overline{D}_2(0, \vartheta)\right) \tag{4.44}$$

$$\overline{D}_{10}(0, \vartheta) = l_4\left(\overline{D}_0(\vartheta) + \overline{D}_1(0, \vartheta) + \overline{D}_2(0, \vartheta)\right) \tag{4.45}$$

$$\overline{D}_{11}(0,\vartheta) = l_5\big(\overline{D}_0(\vartheta) + \overline{D}_1(0,\vartheta) + \overline{D}_2(0,\vartheta)\big) \tag{4.46}$$

**Initial condition**

$$D_k(0) = \begin{cases} 1, & k = 0 \\ 0, & k \neq 0 \end{cases} \tag{4.47}$$

The subsequent equations are attained by solving Eqs. (4.25) to (4.33) with the assistance of boundary conditions.

$$\overline{D}_0(\vartheta) = \frac{1}{K(\vartheta)} \tag{4.48}$$

$$\overline{D}_1(\vartheta) = \frac{2l_1}{K(\vartheta)}\left\{\frac{1-\overline{\vartheta}_f(\vartheta + l_1 + 3l_2 + l_3 + l_4 + l_5)}{\vartheta + l_1 + 3l_2 + l_3 + l_4 + l_5}\right\} \tag{4.49}$$

$$\overline{D}_2(\vartheta) = \frac{3l_2}{K(\vartheta)}\left\{\frac{1-\overline{\vartheta}_f(\vartheta + 2l_1 + 2l_2 + l_3 + l_4 + l_5)}{\vartheta + 2l_1 + 2l_2 + l_3 + l_4 + l_5}\right\} \tag{4.50}$$

$$\overline{D}_3(\vartheta) = \frac{6l_2^2}{K(\vartheta)}\left\{\frac{1-\overline{\vartheta}_f(\vartheta + l_2)}{\vartheta + l_2}\right\} \tag{4.51}$$

$$\overline{D}_4(\vartheta) = \frac{6l_1l_2}{K(\vartheta)}\left\{\frac{1-\overline{\vartheta}_f(\vartheta + 2l_2)}{\vartheta + 2l_2}\right\} \tag{4.52}$$

$$\overline{D}_5(\vartheta) = \frac{12l_1l_2^2}{K(\vartheta)}\left\{\frac{1-\overline{\vartheta}_f(\vartheta + l_2)}{\vartheta + l_2}\right\} \tag{4.53}$$

$$\overline{D}_6(\vartheta) = \frac{6l_1l_2}{K(\vartheta)}\left\{\frac{1-\overline{\vartheta}_f(\vartheta + l_1)}{\vartheta + l_1}\right\} \tag{4.54}$$

$$\overline{D}_7(\vartheta) = \frac{\big(2l_1^2 + 6l_1^2l_2\big)}{K(\vartheta)}\left\{\frac{1-\overline{\vartheta}_{b_0}(\vartheta)}{\vartheta}\right\} \tag{4.55}$$

$$\overline{D}_8(\vartheta) = \frac{\big(6l_2^3 + 12l_1l_2^3\big)}{K(\vartheta)}\left\{\frac{1-\overline{\vartheta}_{b_0}(\vartheta)}{\vartheta}\right\} \tag{4.56}$$

$$\overline{D}_9(\vartheta) = \frac{(l_3 + 2l_1l_3 + 3l_2l_3)}{K(\vartheta)}\left\{\frac{1-\overline{\vartheta}_{b_0}(\vartheta)}{\vartheta}\right\} \tag{4.57}$$

$$\overline{D}_{10}(\vartheta) = \frac{(l_4 + 2l_1l_4 + 3l_2l_4)}{K(\vartheta)}\left\{\frac{1-\overline{\vartheta}_{b_0}(\vartheta)}{\vartheta}\right\} \tag{4.58}$$

$$\overline{D}_{11}(s) = \frac{(l_5 + 2l_1 l_5 + 3l_2 l_5)}{K(\vartheta)} \left\{ \frac{1 - \overline{\vartheta}_{b_0}(\vartheta)}{\vartheta} \right\} \tag{4.59}$$

where $K(s)$ is defined as

$$K(\vartheta) = \left\{ \vartheta + 2l_1 + 3l_2 + l_3 + l_4 + l_5 - \begin{pmatrix} 2l_1 \overline{\vartheta}_f(\vartheta + l_1 + 3l_2 + l_3 + l_4 + l_5) \\ +3l_2 \overline{\vartheta}_f(\vartheta + 2l_1 + 2l_2 + l_3 + l_4 + l_5) \\ + \begin{bmatrix} (2l_1^2 + 6l_1^2 l_2) + (6l_2^3 + 12l_2^3 l_1) \\ +(l_3 + 2l_1 l_3 + 3l_2 l_3) + (l_4 + 2l_1 l_4 + 3l_2 l_4) \\ +(l_5 + 2l_1 l_5 + 3l_2 l_5) \end{bmatrix} \overline{\vartheta}_{b_0}(\vartheta) \end{pmatrix} \right\}. \tag{4.60}$$

The chance that the system is up and running is

$$\overline{D}_{\text{up}}(\vartheta) = \left[ \overline{D}_0(\vartheta) + \overline{D}_1(\vartheta) + \overline{D}_2(\vartheta) + \overline{D}_3(\vartheta) + \overline{D}_4(\vartheta) + \overline{D}_5(\vartheta) + \overline{D}_6(\vartheta) \right] \tag{4.61}$$

$$\overline{D}_{\text{up}}(\vartheta) = \frac{1}{K(\vartheta)} \left\{ \begin{array}{l} 1 + 2l_1 \left\{ \frac{1 - \overline{\vartheta}_f(\vartheta + l_1 + 3l_2 + l_3 + l_4 + l_5)}{\vartheta + l_1 + 3l_2 + l_3 + l_4 + l_5} \right\} \\ +3l_2 \left\{ \frac{1 - \overline{\vartheta}_f(\vartheta + 2l_1 + 2l_2 + l_3 + l_4 + l_5)}{\vartheta + 2l_1 + 2l_2 + l_3 + l_4 + l_5} \right\} \\ +6l_2^2 \left\{ \frac{1 - \overline{\vartheta}_f(\vartheta + l_2)}{\vartheta + l_2} \right\} + 6l_1 l_2 \left\{ \frac{1 - \overline{\vartheta}_f(\vartheta + 2l_2)}{\vartheta + 2l_2} \right\} \\ +12l_1 l_2^2 \left\{ \frac{1 - \overline{\vartheta}_f(\vartheta + l_2)}{\vartheta + l_2} \right\} + 6l_1 l_2 \left\{ \frac{1 - \overline{\vartheta}_f(\vartheta + 2l_1)}{\vartheta + 2l_1} \right\} \end{array} \right\} \tag{4.62}$$

$$\overline{D}_{\text{down}}(\vartheta) = 1 - \overline{D}_{\text{up}}(\vartheta) \tag{4.63}$$

## 4.4 Investigation of Textile Confection Plant Model for Numerous Occurrences

In this section, we present numerical simulations using Maple package with respect to availability, reliability, MTTF, sensitivity, and cost function for the established models. The following set of parameter values is fixed for consistency in the simulations: $l_1 = 0.011, l_2 = 0.022, l_3 = 0.033, l_4 = 0.044, l_5 = 0.055$ with $f(y_1) = f(y_2) = 1$.

### 4.4.1 Analysis of Availability

Considering $l_1 = 0.011, l_2 = 0.022, l_3 = 0.033, l_4 = 0.044, l_5 = 0.055$ and the repair rate as $f(y_1) = f(y_2) = 1$, all in Eq. (4.62) and inverting the transformation to have the following model of availability:

$$\overline{D}_{\text{up}}(q) = \begin{bmatrix} 0.051704\text{e}^{-2.86900q} - 0.016253\text{e}^{-1.26790q} \\ -0.000082\text{e}^{-1.20590q} + 0.968452\text{e}^{-0.00248q} \\ -0.001945\text{e}^{-1.02200q} - 0.000896\text{e}^{-0.04400q} \\ 0.000978\text{e}^{-1.01100q} \end{bmatrix} \tag{4.64}$$

Letting $t = 0, 1, \ldots, 10$, in Eq. (4.64), the calculated availability is shown in Table 4.1.

### 4.4.2 Analysis of Reliability

In the process of investigating the reliability of the factory, repairs are assumed to zero and considering $l_1 = 0.011, l_2 = 0.022, l_3 = 0.033, l_4 = 0.044, l_5 = 0.055$ in (4.62), and then inverting the transformed, the reliability expression is obtained as:

$$R(q) = \begin{bmatrix} 3\text{e}^{-0.19800q} + 2\text{e}^{-0.20900q} + 0.014989\text{e}^{-0.02200q} + 0.008250\text{e}^{-0.04400q} \\ -4.030186\text{e}^{-0.22000q} + 0.006947\text{e}^{-0.01100q} \end{bmatrix} \tag{4.65}$$

Letting $t = 0, 1, \ldots, 10$, in Eq. (4.65), the computed reliability is presented in the subsequent table.

### 4.4.3 Analysis of MTTF

All repairs are assumed to zero and $s$ approaches zero, all in Eq. (4.62), the expression for MTTF is derived as:

$$\text{MTTF} = \lim_{\vartheta \to 0} \overline{D}_{\text{up}}(\vartheta) = \frac{1}{2l_1 + 3l_2 + l_3 + l_4 + l_5} \begin{Bmatrix} 1 + \frac{2l_1}{l_1+3l_2+l_3+l_4+l_5} \\ + \frac{3l_2}{2l_1+2l_2+l_3+l_4+l_5} \\ +3l_1 + 12l_1l_2 + 12l_2 \end{Bmatrix} \tag{4.66}$$

Letting $l_1 = 0.011, l_2 = 0.022, l_3 = 0.033, l_4 = 0.044, l_5 = 0.055$, MTTF of the required failure rate is varied as 0.001, 0.002, ..., 0.009, and other failure rates are fixed, all in Eq. (4.66). The results are shown in the subsequent table.

### 4.4.4 *Analysis of Sensitivity*

MTTF expression is differentiated partially to obtain sensitivity expression. Sensitivity of the required failure rate is varied as 0.001, 0.002, ..., 0.009, and other failure rates are fixed to give

### 4.4.5 *Analysis of Cost*

$$E_p(q) = A_1 \int_0^t D_{\text{up}}(q)\mathrm{d}q - A_2 q \tag{4.67}$$

The factory's analysis of cost is purposely done to find out financial implications, in terms of revenue and service cost, over a specific time. As a result, the availability of the factory is integrated with respect to time, at a specific interval, and other cost factors are fixed, and expected profit expression is obtained as:

$$E_p(q) = A_1 \left\{ \begin{array}{l} -0.018021\mathrm{e}^{-2.86900q} + 0.012819\mathrm{e}^{-1.26790q} + 0.000068\mathrm{e}^{-1.20590q} \\ -389.247206\mathrm{e}^{-0.00248q} + 0.001903\mathrm{e}^{-1.02200q} + 0.000858\mathrm{e}^{-1.04400q} \\ +0.000967\mathrm{e}^{-1.01100q} + 389.2486099 \end{array} \right\} - A_2(q) \tag{4.68}$$

Setting $A_2 \in [0.1, 0.5]$ in Eq. (4.68). The profit of the factory is calculated as shown in Table 4.5 when $q \in [0, 10]$.

Setting $A_1 \in \{2, 4, 6, 8, 10\}$ in Eq. (4.68). The profit of the factory is calculated as shown in Table 4.6 when $q \in [0, 10]$.

## 4.5 Discussion and Concluding Remark

On the basis of the obtained data, the behavior of several tables and their accompanying figures will be addressed at the conclusion. Starting with those time-dependent dependability criteria, such as availability, reliability, and cost. Figure 4.3 and Table 4.1 show how the textile confection plant's availability changes over time. The system's availability declines with time, eventually stabilizing at zero with passage of time. The findings also reveal that the system's availability is excellent, constant, and long-lasting. The impact, on the other hand, would stimulate the factory's management. Because the reliability of a textile confectionery plant is affected by time, it is critical to look at how reliability has changed over time. Table 4.2 and its companion Fig. 4.4 depict information about the system's dependability. Although the system's reliability decreases with time, it does so steadily, indicating that it is quite reliable. Having such a result, though, is intriguing since it would inspire management. The textile confection plant's mean time to failure (MTTF) falls as the failure rate rises; the result obtained reveals that each failure rate correlates to MTTF. The system's MTTF is depicted in Fig. 4.5 and Table 4.3. The importance of MTTF analysis is to determine how much the system's MTTF changes with respect to each failure rate, in the event that a failure could be catastrophic to the system, so that failure of that subsystem/unit can be avoided. Table 4.4 reveals that the sensitivity model's performance was dictated by the strength of failure rates; this conclusion is also shown in Fig. 4.6. Where the rate of failure of subsystem/unit rises, sensitivity of the plant rises also, and the magnitude of each failure entailed the magnitude of sensitivity. The impact of sensitivity on the textile confection plant's performance is that if a failure is serious enough, it will lower the plant's performance capacity. Table 4.5 depicts the effects of repair and failure rate, which are also shown in Fig. 4.7, and close examination of the results reveals that Gumbel–Hougaard family copula repair is very successful for the system's performance. However, as the tables and figures show, this would almost certainly result in the management making the most profit possible. It is interesting to observe that profit margins are better when services are provided at a low cost.

Table 4.6 and Fig. 4.8 depict the effect of repair on failure rate, and close examination of the results reveals that Gumbel–Hougaard family copula repair is quite successful for the system's performance. However, as sales rises, profit rises in lockstep, demonstrating that revenue and profit are precisely proportionate. Furthermore, this corresponds to real-life situations, as industrial users are expected to meet the needs of society as a whole.

This chapter predicts the effect of repair approach on reliability characteristics. The adoption of copula repair as a maintenance strategy improves the performance models of textile confection plants, resulting in maximum availability, reliability,

**Fig. 4.3** Availability against $q$

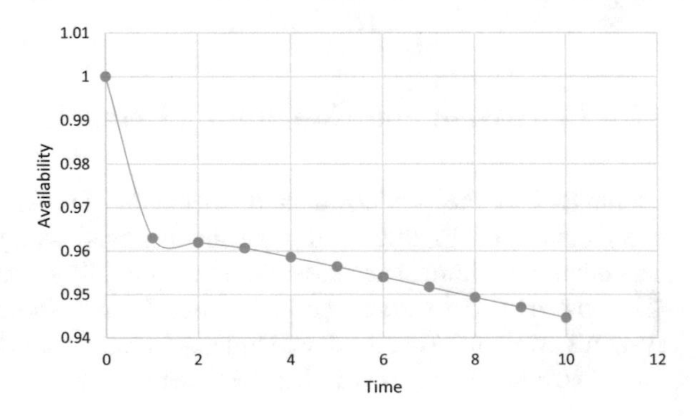

**Table 4.1** Availability with passage time

| $q$ | 0 | 1 | 2 | 3 | 4 | 5 | 6 | 7 | 8 | 9 | 10 |
|---|---|---|---|---|---|---|---|---|---|---|---|
| Availability | 1.0000 | 0.9630 | 0.9620 | 0.9607 | 0.9586 | 0.9564 | 0.9540 | 0.9517 | 0.9493 | 0.9470 | 0.9446 |

**Table 4.2** Reliability

| $q$ | 0 | 1 | 2 | 3 | 4 | 5 | 6 | 7 | 8 | 9 | 10 |
|---|---|---|---|---|---|---|---|---|---|---|---|
| Reliability | 1.0000 | 0.8790 | 0.7688 | 0.6697 | 0.5813 | 0.5032 | 0.4345 | 0.3746 | 0.3225 | 0.2774 | 0.2385 |

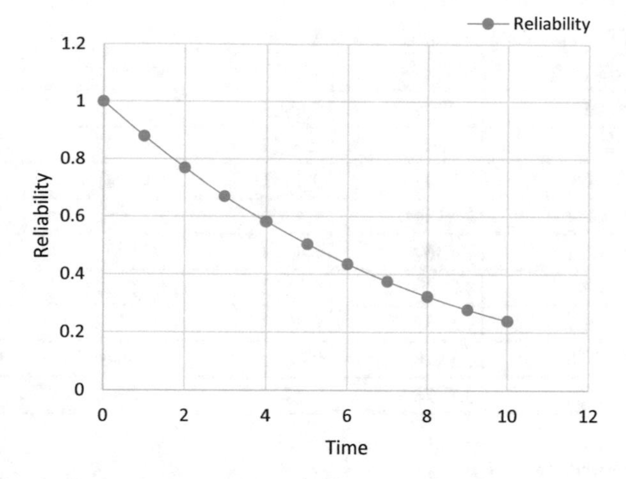


Fig. 4.4 Reliability

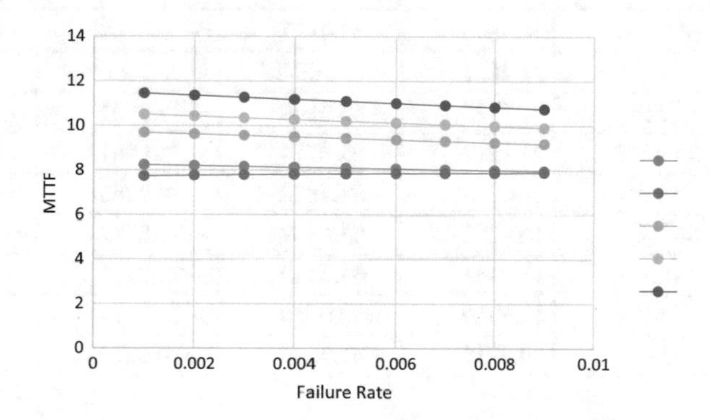


Fig. 4.5 MTTF

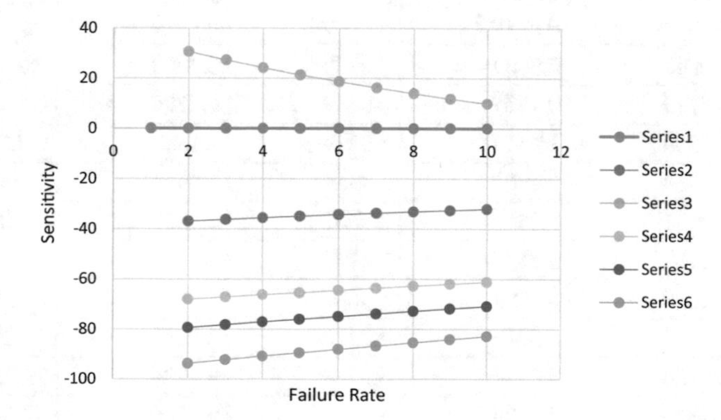


Fig. 4.6 Sensitivity with respect to failure

**Table 4.3** MTTF variation with respect to failure rate

| $l_k$ | MTTF $l_1$ | MTTF $l_2$ | MTTF $l_3$ | MTTF $l_4$ | MTTF $l_5$ |
|---|---|---|---|---|---|
| 0.001 | 8.2405 | 7.7391 | 9.6903 | 10.4985 | 11.4468 |
| 0.002 | 8.2039 | 7.7680 | 9.6227 | 10.4197 | 11.3539 |
| 0.003 | 8.1680 | 7.7937 | 9.5561 | 10.3421 | 11.2624 |
| 0.004 | 8.1327 | 7.8164 | 9.4903 | 10.2655 | 11.1723 |
| 0.005 | 8.0981 | 7.8364 | 9.4254 | 10.1901 | 11.0836 |
| 0.006 | 8.0641 | 7.8539 | 9.3613 | 10.1157 | 10.9962 |
| 0.007 | 8.0307 | 7.8689 | 9.2981 | 10.0423 | 10.9101 |
| 0.008 | 7.9978 | 7.8818 | 9.2357 | 9.9699 | 10.8253 |
| 0.009 | 7.9654 | 7.8926 | 9.1741 | 9.8986 | 10.7418 |

**Table 4.4** Sensitivity against $l_k$

| $l_k$ | $\frac{\partial(\text{MTTF})}{l_1}$ | $\frac{\partial(\text{MTTF})}{l_2}$ | $\frac{\partial(\text{MTTF})}{l_3}$ | $\frac{\partial(\text{MTTF})}{l_4}$ | $\frac{\partial(\text{MTTF})}{l_5}$ |
|---|---|---|---|---|---|
| 0.001 | – 36.9170 | 30.5890 | – 68.0197 | – 79.3448 | – 93.6470 |
| 0.002 | – 36.2274 | 27.2601 | – 67.1094 | – 78.2058 | – 92.1978 |
| 0.003 | – 35.5655 | 24.1775 | – 66.2168 | – 77.0904 | – 90.7810 |
| 0.004 | – 34.9297 | 21.3217 | – 65.4314 | – 75.9981 | – 89.3957 |
| 0.005 | – 34.3185 | 18.6748 | – 64.4828 | – 74.9282 | – 88.0410 |
| 0.006 | – 33.7307 | 16.2206 | – 63.6407 | – 73.8802 | – 86.7161 |
| 0.007 | – 33.1648 | 13.9441 | – 62.8145 | – 72.8534 | – 85.4200 |
| 0.008 | – 32.6198 | 11.8319 | – 62.0040 | – 71.8474 | – 84.1519 |
| 0.009 | – 32.0946 | 9.8715 | – 61.2086 | – 70.8615 | – 82.9111 |

**Table 4.5** Expected revenue with passage of time for different $A_2$

| $q$ | $E_p\ (t)$ $A_2 = 0.1$ | $E_p\ (t)$ $A_2 = 0.2$ | $E_p\ (t)$ $A_2 = 0.3$ | $E_p\ (t)$ $A_2 = 0.4$ | $E_p\ (t)$ $A_2 = 0.5$ |
|---|---|---|---|---|---|
| 0 | 0.0000 | 0.0000 | 0.0000 | 0.0000 | 0.0000 |
| 1 | 0.8725 | 0.7725 | 0.6725 | 0.5725 | 0.4725 |
| 2 | 1.7349 | 1.5349 | 1.3349 | 1.1349 | 0.9349 |
| 3 | 2.5964 | 2.2964 | 1.9964 | 1.6964 | 1.3964 |
| 4 | 3.4561 | 3.0561 | 2.6561 | 2.2561 | 1.8561 |
| 5 | 4.3137 | 3.8137 | 3.3137 | 2.8137 | 2.3137 |
| 6 | 5.1689 | 4.5689 | 3.9689 | 3.3689 | 2.7689 |
| 7 | 6.0218 | 5.3218 | 4.6218 | 3.9218 | 3.2218 |
| 8 | 6.8724 | 6.0724 | 5.2724 | 4.4724 | 3.6724 |
| 9 | 7.7206 | 6.8206 | 5.9206 | 5.0206 | 4.1206 |
| 10 | 8.5664 | 7.5664 | 6.5664 | 5.5664 | 4.5664 |

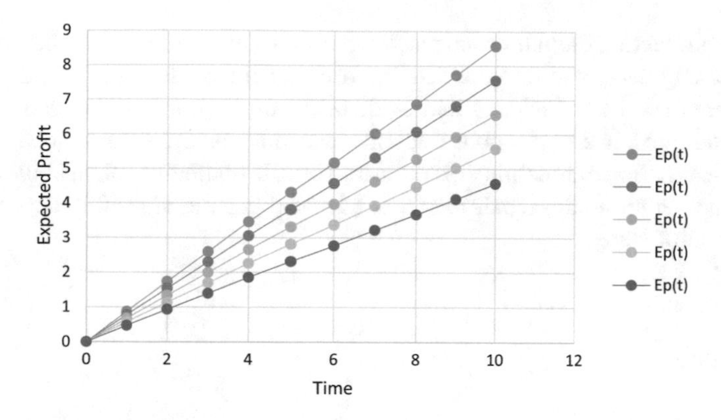


**Fig. 4.7** Revenue with time for different values of $A_2$

**Table 4.6** Expected revenue with passage of time for different $A_1$

| $q$ | $E_p$ $(q)$ $A_1 = 2$ | $E_p$ $(q)$ $A_1 = 4$ | $E_p$ $(q)$ $A_1 = 6$ | $E_p$ $(q)$ $A_1 = 8$ | $E_p$ $(q)$ $A_1 = 10$ |
|---|---|---|---|---|---|
| 0 | 0.0000 | 0.0000 | 0.0000 | 0.0000 | 0.0000 |
| 1 | 0.9451 | 2.8903 | 4.8355 | 6.7807 | 8.7259 |
| 2 | 1.8698 | 5.7397 | 9.6096 | 13.4795 | 17.3494 |
| 3 | 2.7928 | 8.5856 | 14.3784 | 20.1712 | 25.9640 |
| 4 | 3.7122 | 11.4245 | 19.1368 | 26.8491 | 34.5614 |
| 5 | 4.6274 | 14.2548 | 23.8823 | 33.5097 | 43.1371 |
| 6 | 5.5379 | 17.0759 | 28.6138 | 40.1518 | 51.6897 |
| 7 | 6.4437 | 19.8875 | 33.3313 | 46.7750 | 60.2188 |
| 8 | 7.3448 | 22.6897 | 38.0345 | 53.3774 | 68.7242 |
| 9 | 8.2412 | 25.4824 | 42.7236 | 59.9649 | 77.2061 |
| 10 | 9.1328 | 28.2657 | 47.3986 | 66.5315 | 85.6644 |

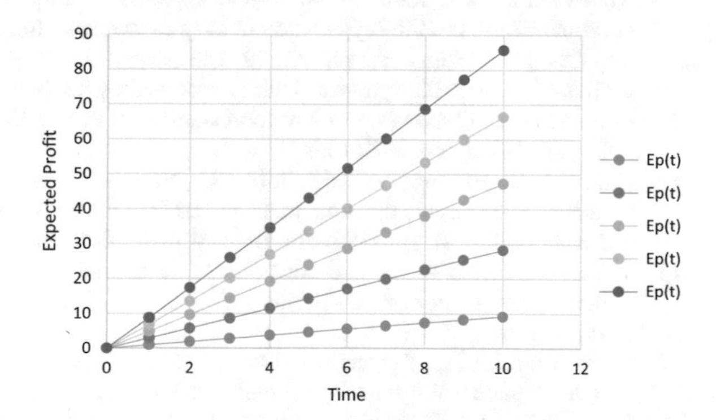


**Fig. 4.8** Revenue with time for different values of $A_1$

and high production output, which results in maximum revenue. The chapter may be useful for engineers and designers in determining the plant reliability in achieving the required level of efficiency and availability under unavoidable economic and repair conditions. In order to more efficiently estimate reliability measures, additional accurate estimation of reliability of multi-unit textile confection plant in the presence of human operation-cum repair man using hybrid genetic algorithm approach may be a subject of study.

## References

1. Ye, Z., Cai, Z., Si, S., Zhang, S., Yang, H.: Competing failure modeling for performance analysis of automated manufacturing systems with serial structures and imperfect quality inspection. IEEE Trans. Ind. Inf. **16**(10), 6476–6486 (2020). https://doi.org/10.1109/TII.2020.2967030
2. Ye, Z., Cai, Z., Yang, H.: Reliability analysis of manufacturing machine with degradation and low-quality feedstocks. In: 2020 Asia-Pacific International Symposium on Advanced Reliability and Maintenance Modeling (APARM), pp. 1–5 (2020). https://doi.org/10.1109/APARM49247.2020.9209452
3. Chang, P.C., Lin, Y.K., Chen, J.C.: System reliability for a multi-state manufacturing network with joint buffer stations. J. Manuf. Syst. **42**, 170–178 (2017)
4. Chen, Z., He, Y., Zhao, Y., Han, X., He, Z., Xu, Y., Zhang, A.: Mission reliability evaluation based on operational quality data for multistate manufacturing systems. Int. J. Prod. Res. **57**(6), 1840–1856 (2019). https://doi.org/10.1080/00207543.2018.1508906
5. Zhang, H., Li., A.P., Liu, X.M.: Modeling and performance evaluation of a multistage serial manufacturing system considering rework and product polymorphism. Chin. J. Mech. Eng. **53**, 191–201 (2017)
6. Pundir, P.S., Patawa, R., Gupta, P.K.: Analysis of two non-identical unit cold standby system in presence of prior information. Am. J. Math. Manag. Sci. **40**(4), 320–335 (2021). https://doi.org/10.1080/01966324.2020.1860840
7. Kumar, A., Pant, S., Ram, M.: Gray wolf optimizer approach to the reliability-cost optimization of residual heat removal system of a nuclear power plant safety system. Qual. Reliab. Eng. Int. **35**(7), 2228–2239 (2019). https://doi.org/10.1002/qre.2499
8. Kumar, A., Pant, S., Ram, M.: Multi-objective grey wolf optimizer approach to the reliability-cost optimization of life support system in space capsule. Int. J. Syst. Assur. Eng. Manag. **10**, 276–284 (2019). https://doi.org/10.1007/s13198-019-00781-1
9. Mellal, M.A., Zio, E.: System reliability-redundancy optimization with cold-standby strategy by an enhanced nest cuckoo optimization algorithm, Reliab. Eng. Syst. Saf. **201** (2020). https://doi.org/10.1016/j.ress.2020.106973
10. Okafor, E.G., Ezugwu, E.O., Jemitola, P.O., Sun, Y.-C., Lu, Z.: Multistate system reliability modeling using copula function. In: 2018 3rd International Conference on System Reliability and Safety (ICSRS), pp. 135–141 (2018). https://doi.org/10.1109/ICSRS.2018.8688860
11. Sharifi, M., Moghaddam, T.A., Shahriar, M.: Multi-objective redundancy allocation problem with weighted-k-out-of-n subsystems. Heliyon **5**, 1–8 (2019). https://doi.org/10.1016/j.heliyon.2019.e02346
12. Jia, H., Liua, D., Lia, Y., Dingb, Y., Liua, M., Peng, R.: Reliability evaluation of power systems with multi-state warm standby and multi-state performance sharing mechanism. Reliab. Eng. Syst. Saf. **204** (2020). https://doi.org/10.1016/j.ress.2020.107139
13. Lin, S.-W., Matanhire, T.B., Liu, Y.-T.: Copula-based Bayesian reliability analysis of a product of a probability and a frequency model for parallel systems when components are dependent. Appl. Sci. **11**, 1697 (2021). https://doi.org/10.3390/app11041697

14. Jia, H., Peng, R., Yang, L., Y. Li, Y.: Reliability assessment of power systems with warm standby and energy storage. In: 2021 Global Reliability and Prognostics and Health Management (PHM-Nanjing), pp. 1–5 (2021). https://doi.org/10.1109/PHM-Nanjing52125.2021.9613081
15. Nelson, R.B.: An Introduction to Copulas, 2nd edn. Springer Publisher, New York (2006)
16. Yusuf, I., Ismail, A.L., Singh, V.V., Ali, U.A., Sufi, N.A.: Performance analysis of multi-computer system consisting of three subsystems in series configuration using copula repair policy. SN Comput. Sci. **1**(5), 1–11 (2020)
17. Chopra, G., Ram, M.: Reliability measures of two dissimilar units' parallel system using Gumbel-Hougaard family copula. Int. J. Math. Eng. Manag. Sci. **4**(1), 116–130 (2019). https://doi.org/10.33889/IJMEMS.2019.4.1-011
18. Gahlot, G., Singh, V.V., Ayagi, H.I., Abdullahi, I.: Stochastic analysis of a two units' complex repairable system with switch and human failure using copula approach. Life Cycle Reliab. Saf. Eng. **9**(1), 1–11 (2020)
19. Lado, A., Singh, V.V.: Cost assessment of complex repairable systems in series configuration using Gumbel Hougaard family copula. Int. J. Qual. Reliab. Manag. **36**(10), 1683–1698 (2019)
20. Sha, N.: A copula approach of reliability analysis for hybrid systems. Reliab. Theory Appl. **16**(1), 231–242
21. Sanusi, A., Yusuf, I.: Reliability assessment and profit analysis of distributed data center network topology. Life Cycle Reliab. Saf. Eng. (2022). https://doi.org/10.1007/s41872-022-00186-3
22. Tyagi, V., Arora, R., Ram, M., Triantafyllou, I.S.: Copula based measures of repairable parallel system with fault coverage. Int. J. Math. Eng. Manag. Sci. (2021). https://doi.org/10.33889/IJMEMS.2021.6.1.021
23. Singh, V.V., Poonia, P.K., Abdullahi, A.H.: Performance analysis of a complex repairable system with two subsystems in a series configuration with an imperfect switch. J. Math. Comput. Sci. **10**(2), 359–383 (2020). https://doi.org/10.28919/jmcs/4399

**Abdulkareem Lado Ismail** is Lecturer at the Department of Mathematics, Kano State College of Education, Kano, Nigeria. His research interest includes system reliability maintenance and replacement modeling.

**Ibrahim Yusuf** is Lecturer at the Department of Mathematical Sciences, Bayero University, Kano, Nigeria. He received his BSc in Mathematics in 1996, M.Sc. in Mathematics in 2007, and Ph.D. in Mathematics in 2014 from Bayero University, Kano, Nigeria. He is currently Associate Professor at the Department of Mathematical Sciences, Bayero University, Kano, Nigeria. He has reviewed papers from *IJSA, JCCE, Life Cycle Reliability and Safety Engineering, JRSSIJRRS,* Inderscience journals*IJQRM, and Operation Research and Decision.* His research includes system reliability theory, maintenance and replacement, and operation research.

# Chapter 5
# An Application of Soft Computing in Oil Condition Monitoring

**Fatemeh Afsharnia, Mehdi Behzad, and Hesam Addin Arghand**

**Abstract** Preventive maintenance strategy can reduce the exorbitant costs of purchasing spare parts, repairs, and consequently downtime, as well as increase efficiency and income by reducing downtime. Oil monitoring is one of the most important policies for preventive maintenance of equipment. This chapter aimed to develop a fuzzy program based on engine oil analysis to investigate the erosive behavior of the engine as well as identify the engine condition. Once 1500 engine oil samples were analyzed, the wear debris was measured in oil including iron, copper, aluminum, lead, tin, silicon, PQ, water content, viscosity, and alkalinity of oil, and the suitable information for analysis was obtained. The findings of this chapter indicate a specific pattern appropriate to the wear debris of oil that can be found by applying fuzzy logic and creating a series of fuzzy rules. Then, using fuzzy logic, it diagnosed and predicted the defects, failures, and conditions of the sugarcane harvester engine.

**Keywords** Engine · Sugarcane harvester · Oil analysis · Fuzzy logic

## Abbreviations

| | |
|---|---|
| OCM | Oil condition monitoring |
| CBM | Condition-based maintenance |
| UOA | Used oil analysis |
| CM | Corrective maintenance |
| PM | Preventive maintenance |

F. Afsharnia (✉)
Department of Agricultural Machinery and Mechanization Engineering, Agricultural Sciences and Natural Resources University of Khuzestan, Ahvaz, Iran
e-mail: phd.afsharnia@asnrukh.ac.ir; afsharniaf@yahoo.com

F. Afsharnia · M. Behzad
Department of Mechanical Engineering, Sharif University of Technology, Tehran, Iran
e-mail: m_behzad@sharif.ir

H. A. Arghand
Department of Mechanical Engineering, University of Zanjan, Zanjan, Iran



H. Garg (ed.), *Advances in Reliability, Failure and Risk Analysis*, Industrial and Applied Mathematics, https://doi.org/10.1007/978-981-19-9909-3_5

Fe Iron
AI Aluminum
Cu Copper
Pb Lead
Sn Tin
Si Silicon
PQ Particle quantifier index is a relative measurement of the total ferrous (iron) metal content of oil irrespective of the debris size or shape by means of detection by a magnetic field
TBN Total base number is a measurement of basicity that is expressed in terms of the number of milligrams of potassium hydroxide per gram of oil sample (mg KOH/g)
Vis40 Measuring viscosity at 40° centigrade
DSS Decision Support System

## 5.1 Introduction

Predictive maintenance programs such as oil condition monitoring (OCM) and used oil analysis (UOA) can track changes in the lubricant quality of machinery, preventing costly equipment failures. This system provides an important "early warning" of impending problems and ensures that the machinery operates as it should. Lubricant oil is often an indicator of the condition of machines, engines, and other components. By regularly monitoring oil conditions or testing used oil, a machine's efficiency can be ensured before costly problems develop later [11, 13].

Despite their vital role in lubrication, lubricants are exposed to a variety of dangers, including water contamination, corrosion, fuel contamination, and air ingested particles. The presence of high amounts of wear particles provides early warning of possible machinery malfunctions, enabling early remedial action. If analysis indicates that no undue wear is taking place, the operator may extend the interval between services or oil changes. By scheduling periodic lubricant testing and consulting experts, failures and unscheduled maintenance can be avoided.

Today, intelligent fault diagnosis models are applied for the implementation of CBM strategy. Soft computing such as fuzzy and neuro-fuzzy systems is the most widely used methods in this field. Lou and Loparo [10] proposed a wavelet transform and fuzzy inference for bearing fault diagnosis. Alizadeh and Ahmadi [2] and Ramezani and Memariani [15] used a fuzzy rule-based system for fault diagnosis of diesel engines. Moreover, Khan et al. [8] presented a combination of fuzzy logic and ANFIS models to eliminate the limitations of conventional methods of transformer fault diagnosis using dissolved gas analysis (DGA). Also, Deng et al. [5] used integrating empirical wavelet transform and fuzzy entropy for motor bearing fault diagnosis. Recently, Pan et al. [14] presented a multi-class fuzzy support matrix machine for classification in fault diagnosis of roller bearing.

In this chapter, after reviewing and studying the available sources, we collected data through oil analysis, and then evaluated the data by comparing the obtained values with the base values, problematic cases were identified, and in addition, they are divided into two categories according to the difference with the base values and type of difference: (1) initial warning and (2) warning with suspension. In the first type, only the initial evaluation is done to obtain assurance. In the second type, a more complete overhaul is carried out, and the defective part of the machine is repaired if it was necessary. As a result, with the help of oil analysis, in addition to obtaining data about oil conditions, we find the presence of erosion particles, which is a sign of wear in the parts. So, by having suitable information sources, an intelligent fuzzy model can be presented for the engine's failure prediction. Also, the condition of the effective debris in oil analysis with the final condition of the engine is examined to discover the effective patterns in the engine wear process, and the influencing factors are recognized based on the state of the engine, and the erosion rules are extracted.

## 5.2 Importance of Oil Condition Monitoring in Agro-industries

Effective maintenance is one of the most important parts of increasing the productivity and effectiveness of the mechanized fleet of agricultural production. A proper maintenance strategy should be able to ensure timely agricultural operations. Sugarcane harvesting as a strategic operation requires the use of mechanization management at the regional level. If the crop is harvested too early or too late, the yield may decrease due to decreasing in the sugar extracted percentage. Depending on the weather conditions, the duration of sugarcane harvest is between 4 and 6 months in Khuzestan Province. Due to the coincidence of the harvest season (autumn and winter) with rainfall and improper field moisture conditions, the harvesting operations are not often carried out on time and are delayed. So, the degree of purity and the sugar content of the product decrease. A delay in sugarcane harvest until May can reduce sugar extraction by around 20–30% [1]. For this reason, reliability, availability, reduction of downtime, and more repairs of sugarcane harvesting equipment are of great importance [6]. Failure of devices and systems disrupts various levels of production and support and can be considered a serious threat to increase production costs [18]. Maintenance is a set of activities that are performed specifically and usually in a planned manner to prevent the sudden breakdown of machinery, equipment, and facilities to increase their reliability and availability [16]. Two categories of maintenance policies include preventive maintenance (PM) and corrective maintenance (CM). In corrective maintenance, repairs are made after breakage or apparent failure. In PM policy, maintenance is implemented to prevent failures and breakdowns. PM policy applies time-based maintenance programs for replacement parts and maintenance activities [3], which includes lubrication schedule, inspection, and daily adjustments [19].

For each type of maintenance programming, it should be noted that the breakdown of machines is not an issue that can be prevented but can be improved by using planning techniques, reliability, and usability of machines [17]. The subsystems of the sugarcane harvesting machine are repairable. Therefore, the reliability of the system will decrease and the failure rate will increase by using the system as well as decreasing its life. As a result, it is essential to make the necessary investments in maintenance and repairs using advanced techniques and methods for maximizing the utilization of the funds spent on the purchase of machines.

An important and noteworthy issue is that every year the engine of some sugarcane harvesting machines breaks down due to neglect of forecasting and prevention of its defects and must be replaced, which will lead to high costs for agro-industrial units. In general, one of the most important issues in agricultural machines and mechanical equipment is the entry of contaminants such as water, fuel, silica, debris, soot particles into the engine. All mechanical equipment use oil for lubrication or transmission; they are always affected by the quality of the oil, and because this fluid is in direct contact with all the mechanical components of the equipment, it must be considered the most important factor, and its contaminants must be cleaned and separated. Coarse and hard particles in the oil are responsible for most of the erosion that leads to component failure. Therefore, in this chapter, the erosive behavior of sugarcane harvester engines was investigated by performing engine oil analysis experiments.

## 5.3 Implementation of Oil Condition Monitoring

Correct and accurate condition monitoring and determining the severity of failures in the engine is one of the serious and challenging issues. By this method, the failures of the engines are identified in time, and their development process can be tracked. It will enable us to take action to fix the failure before the engines are damaged and major repairs are caused. In this process, it is often difficult to find a relationship between the components of the engine, the type of failure, and the causes of the failure. So, the failures usually remain unknown over time. In this chapter, fuzzy logic has been used to prevent such problems and monitor the condition of the engine more precisely.

For instance, an experiment was performed on Austaft 7000 sugarcane harvesting machine with 14 years of operation from the Sugarcane and By-Products Development Company of Khuzestan Province in Iran during the sugarcane harvesting operations of two growing seasons. The main raw materials used in the present study include engine oil samples taken from a sugarcane harvester. For every oil analysis, a sample is required that truly represents the entire system. Although sampling is the simplest step of the oil analysis program, it is a very important step. Since, if the sampling is not correct, the results of oil tests will be invalid. The main items of oil sampling are: tool selection, determining the sampling frequency for different components, identifying oil sampling locations in different components, and the procedure of oil sampling. For sampling, the used accessories include: (1) sampling pump (manual suction pump), (2) polyethylene tube (hoses with an outer diameter

**Fig. 5.1** Oil sampling tool

of 1/4″ or 5/16″), and (3) sample carrying bag (Fig. 5.1). All coded samples were sent to the company's laboratory after coding and numbering, along with a sample of new unused oil as same as the oil that was used in the engine. Erosive elements were tested for 1500 samples of engine oil including iron (Fe), aluminum (Al), copper (Cu), lead (Pb), tin (Sn), silicon (Si), water content, PQ index, TBN, and Vis40. The results of oil analysis include the status of the machine, the status code, status change code, theory code, recommendation and status of each erosive element, and oil wear debris and oil status code. It should be noted that the opinions of experts have been based solely on experience and years of work in oil analysis laboratories. Metallic and non-metallic elements in oil are detected by spectroscopic testing. The amount of water in the oil was measured by the Karl Fischer experiment [7]. The test result reports the exact amount of water in ppm. This test, known as the Crackle test, is performed to approximate the amount of water contamination in the oil. In this method, a few drops of oil are poured on a hot plate (approximately 150 °C). If there is water in the oil, it forms a bubble and comes out of the oil. Most mechanical equipment manufacturers recommend that the amount of water in the oil should not be more than 0.1% because too much water can cause oil spoilage and change the erosion status of the device. Values less than 0.1% are usually not a problem unless the manufacturer has specified lower values for the device.

### 5.3.1 *Fuzzy Systems*

This system is made up of a set of "if → then" conditional rules to create a mapping from the input set $U \subset R^h$ to the output set $V \subset R$. If $U$ is a reference set whose each member is represented by $x$, the fuzzy set $A$ in $U$ is expressed by the ordered pair $A = \{(x, \mu_A(x)|x \epsilon U)\}$, in which $x$ is a linguistic variable and $\mu_A(x)$ is a membership function that indicates the degree to which $x$ belongs to set $A$ [4, 9, 12, 20].

#### 5.3.1.1 Fuzzy Ranking Scale and Fuzzy Number

To define fuzzy sets, we must have a basic knowledge of the scope of the definition related to each variable. According to the experience and knowledge of analysis and the effect of these factors on the engine condition of sugarcane harvesters, iron, aluminum, copper, lead, tin, silicon, 40 Vis viscosity, water content, and PQ were considered as the inputs of the fuzzy system. The output also indicates the status of the engine. According to the information in the oil analysis sheets and available resources and also after consulting with sugarcane harvesting oil experts, four classes including normal (0), boundary (1), boundary unacceptable (2), and critical (3) were coded. For the output variable, language committees including normal, boundary, emergency, and critical processing were considered. MATLAB software fuzzy toolbox was used for calculations [2].

## 5.4 Analysis of Oil Condition Monitoring

According to 1500 laboratory samples of sugarcane harvester engine oil during two harvest seasons, the values of baseline indices including upper limit, middle limit, and lower limit of all states of wear debris related to sugarcane harvester engine oil are given in Table 5.1. As can be seen from Table 5.1, the range of iron wear debris was 0–15 ppm in normal conditions, 15–30 ppm in the boundary condition, 30–60 ppm in boundary unacceptable condition, and 60–100 ppm in critical condition. Also, the range of iron wear debris was 0–8 ppm in normal conditions, 9–15 ppm in the boundary condition, 16–28 ppm in boundary unacceptable conditions, and 30–85 ppm in critical conditions.

Using these limits, the attribution functions for each element and output were plotted (Figs. 5.2 and 5.3). For the inputs, the shape of these functions was selected as Z shape and S shape at the beginning and end, respectively, and as a triangular in the rest.

A fuzzy triangular function is determined by three parameters a lower limit *a*, an upper limit *c*, and a value *b*, where $a \leq b \leq c$. The precise appearance of the function is determined by the choice of the parameters *a*, *b*, and *c* which in turn form a triangle. In this, *a* and *c* locate the feet of the triangle and the parameter *b* locates the peak. A fuzzy triangular membership function of *A* can be calculated by Eq. (5.1):

$$\mu_A(x) = \begin{cases} 0; & x \leq a \\ \left(\frac{x-a}{b-a}\right) - \epsilon; & a < x \leq b \\ \left(\frac{c-x}{c-b}\right) - \epsilon; & b \leq x < c \\ 0; & x \geq c \end{cases} \tag{5.1}$$

**Table 5.1** Baseline index values for oil wear debris

| Fe (ppm) | | | | |
|---|---|---|---|---|
| State | 0 | 1 | 2 | 3 |
| Upper limit | 15 | 30 | 60 | 100 |
| Mean limit | 8 | 22 | 45 | 75 |
| Lower limit | 0 | 15 | 30 | 60 |
| Cu (ppm) | | | | |
| State | 0 | 1 | 2 | 3 |
| Upper limit | 7 | 16 | 30 | 84 |
| Mean limit | 3 | 10 | 23 | 52 |
| Lower limit | 0 | 8 | 20 | 30 |
| Sn (ppm) | | | | |
| State | 0 | 1 | 2 | 3 |
| Upper limit | 7 | 12 | 24 | 65 |
| Mean limit | 4 | 10 | 20 | 45 |
| Lower limit | 0 | 8 | 12 | 24 |
| PQ (ppm) | | | | |
| State | 0 | 1 | 2 | 3 |
| Upper limit | 8 | 18 | 35 | 60 |
| Mean limit | 5 | 15 | 27 | 45 |
| Lower limit | 0 | 10 | 20 | 35 |
| TBN | | | | |
| State | 0 | 1 | 2 | 3 |
| Upper limit | 14 | 7 | 4 | 2 |
| Mean limit | 8 | 5 | 3 | 1 |
| Lower limit | 7 | 4 | 2 | 0 |
| Water content | | | | |
| State | 0 | 1 | 2 | 3 |
| Upper limit | 0.001 | 0.002 | 0.005 | 0.01 |
| Mean limit | 0.0005 | 0.0015 | 0.0035 | 0.008 |
| Lower limit | 0 | 0.001 | 0.002 | 0.006 |
| Al (ppm) | | | | |
| State | 0 | 1 | 2 | 3 |
| Upper limit | 8 | 15 | 28 | 85 |
| Mean limit | 5 | 11 | 20 | 52 |
| Lower limit | 0 | 9 | 16 | 30 |

(continued)

**Table 5.1** (continued)

| Pb (ppm) | | | | |
|---|---|---|---|---|
| State | 0 | 1 | 2 | 3 |
| Upper limit | 10 | 17 | 28 | 90 |
| Mean limit | 5 | 15 | 25 | 58 |
| Lower limit | 0 | 11 | 19 | 30 |
| Si (ppm) | | | | |
| State | 0 | 1 | 2 | 3 |
| Upper limit | 15 | 22 | 45 | 120 |
| Mean limit | 5 | 19 | 30 | 60 |
| Lower limit | 0 | 16 | 22 | 46 |
| Vis40 (CST) | | | | |
| State | 0 | 1 | 2 | 3 |
| Upper limit | 170 | 180 | 200 | 250 |
| Mean limit | 140 | 120 | 145 | 120 |
| Lower limit | 100 | 85 | 65 | 20 |

The precise appearance of the fuzzy S-shaped function is determined by the choice of parameters *a*, *b*, and the parameters locate the extremes of the sloped portion of the curve. A fuzzy S-shaped membership function is defined as Eq. (5.2):

$$\mu_A(x) = \begin{cases} 0; & x \le a \\ 2\left(\frac{x-a}{b-a}\right)^2 - \epsilon; & a < x \le \frac{a+b}{2} \\ 1 - 2\left(\frac{x-b}{b-a}\right)^2 - \epsilon; & \frac{a+b}{2} \le x < b \\ 1 - \epsilon; & x \ge b \end{cases} \tag{5.2}$$

The fuzzy Z-shaped function is given by two parameters, a and b which locate the extremes of the sloped portion of the curve. A fuzzy Z-shaped membership function is defined as Eq. (5.3):

$$\mu_A(x) = \begin{cases} 1 - \epsilon; & x \le a \\ 1 - 2\left(\frac{x-a}{b-a}\right)^2 - \epsilon; & a < x \le \frac{a+b}{2} \\ 2\left(\frac{x-b}{b-a}\right)^2 - \epsilon; & \frac{a+b}{2} \le x < b \\ 0; & x \ge b \end{cases} \tag{5.3}$$

For the output that indicates the condition of the sugarcane harvester engine, the fuzzy triangular function was used (Fig. 5.3). The engine condition is then estimated by considering the fuzzy laws according to the condition of the erosive elements and oil wear debris (Fig. 5.4) as follows (11, 12):

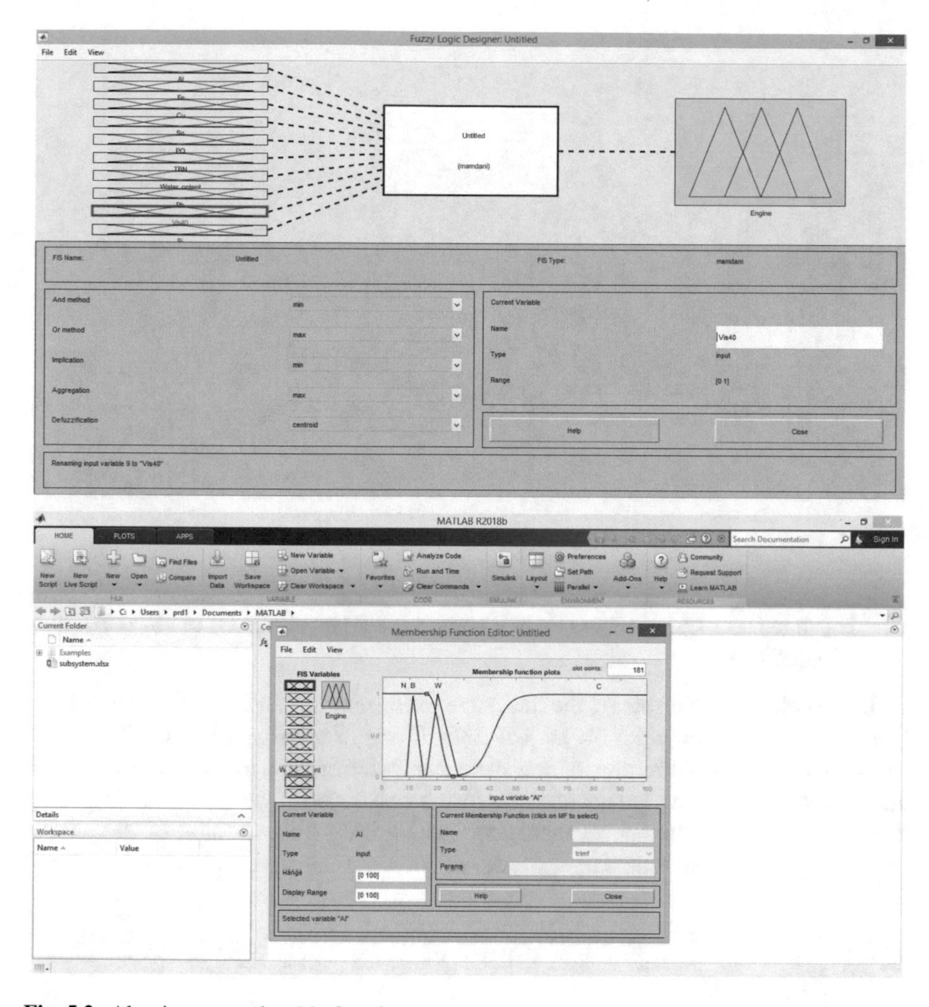


Fig. 5.2 Aluminum membership function

1. If (Si is C) and (Cu is N) and (Vis40 is W low), then (Engine is E)
2. If (Si is B) and (Cu is B) and (Vis40 is W low), then (Engine is E)
3. If (Vis40 is W low) and (Si is C), then (Engine is C)
4. If (Pb is C) and (Fe is C) and (Vis40 is N), then (Engine is C)
5. If (Pb is N) and (Fe is C) and (PQ is C) and (Vis40 is N), then (Engine is E)
6. If (Pb is N) and (Fe is N) and (Al is N) and (Cu is N) and (PQ is N) and (Si is N) and (Vis40 is N) and (TBN is N) and (Cr is N) and (Water is N), then (Engine is N)
7. If (Al is W) or (Fe is W) or (PQ is W) or (Si is W), then (Engine is B)
8. If (Water is B) or (TBN is W), then (Engine is B)
9. If (Pb is B) and (PQ is C) and (Fe is C) and (Vis40 is N), then (Engine is E)

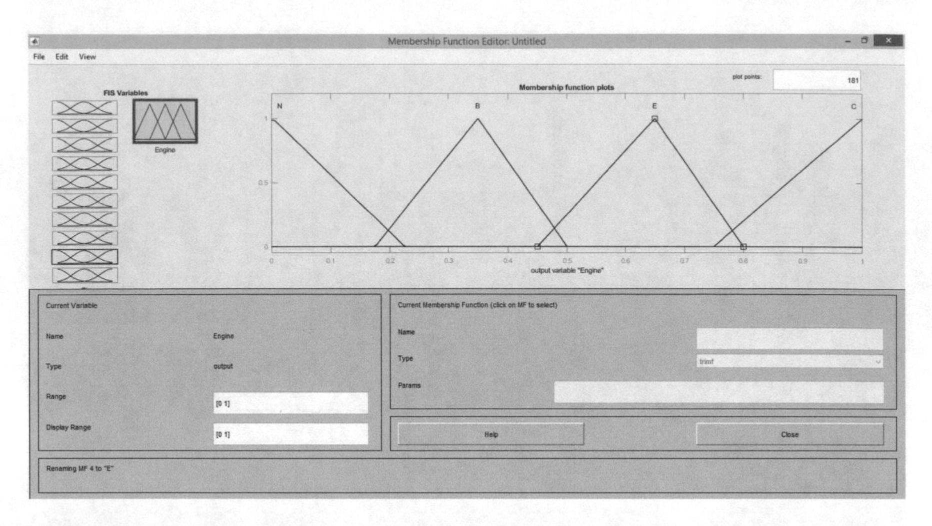


**Fig. 5.3** Engine output membership function

10. If (Pb is B) or (Al is B) or (Fe is B) or (PQ is B) or (Cu is B) or (Si is B), then (Engine is N).

In the defuzzification stage, the final state of the engine can be obtained by giving values to the inputs (Fig. 5.5). In this case, the agro-industrial units will be able to predict the final condition of the cane harvester engine by performing an oil analysis test and giving values as input to the developed fuzzy program. By using this program, the maintenance managers can indicate the condition of the engine based on oil analysis data. The oil analysis data are the input of fuzzy logic, and

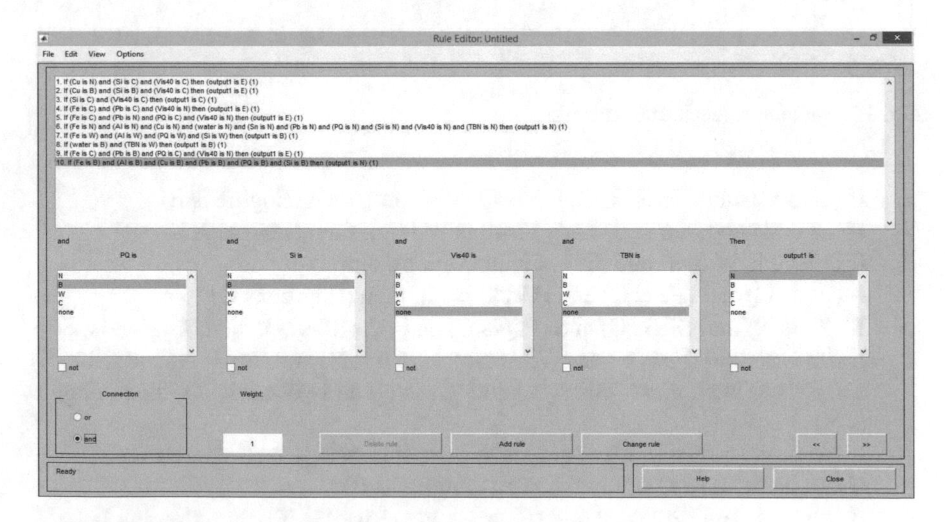


**Fig. 5.4** Fuzzy rules for estimating engine condition

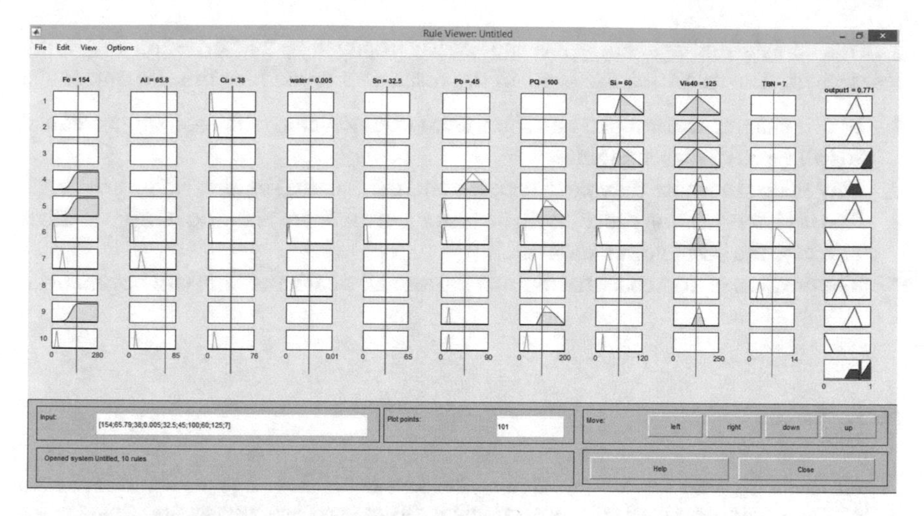


**Fig. 5.5** Engine status concerning fuzzy outputs

the output of the fuzzy program is the condition of the engine (normal, boundary, emergency, and critical). For example, by giving the fuzzy logic input 154, 65.8, 38, 0.005, 32.5, 45, 100, 60, 125, and 7 for iron, aluminum, copper, water content, tin, lead, PQ, silicon, viscosity, and alkali number of oil, respectively; engine condition or the output will be 0.771, which indicates the boundary condition (*B*) for an engine of the sugarcane harvesting machine.

## 5.5 Conclusion

This chapter has established a decision support system (DSS) based on oil analysis results and fuzzy logic to predict the final status to monitor the real condition of the engine. The results of this research, as a valuable basis and criteria, can be used by sugarcane harvester machine operators to determine each oil element, detect the final status of the harvester engine, and create a database for fault diagnosis and wear behaviors as well. As a result of using this non-traditional modeling approach, the following benefits can be expected:

- Physics-based modeling and data-driven modeling can be integrated with an approach based on rule-based knowledge representation.
- Domain experts may contribute directly to model development because rule-based models are consistent with human heuristic reasoning.
- The user can see through the rule-based model. The process of making a choice may be communicated, allowing the system to quickly garner user trust. This is especially crucial in safety–critical applications involving human life.

This study's findings can serve as a useful foundation and criterion for future researchers. The following are some of the research's broad benefits and outcomes:

1. In this chapter, determined baselines were used to determine each oil ingredient for all types of diesel engines.
2. Any linked database may be diagnosed and the eventual engine status predicted.
3. The decision-making process will be faster and more precise, allowing specialists to locate the best answer sooner.
4. The creation of a database for fault scenarios and wear behaviors will be practical.

## References


1. Afsharnia, F., Marzban, A.: Investigating the long-term effect of preventive maintenance strategy on the operational efficiency and failure rate of sugarcane harvester using time series. J. Agric. Mach. **10**(2), 347–359 (2020)
2. Alizadeh, D., Ahmadi, H.: Condition monitoring of diesel engine via oil analysis using fuzzy logic. J. Engine Res. **19**(19), 9–18 (2010)
3. Amari, S.V., McLaughlin, L., Pham, H.: Cost effective condition-based maintenance using Markov decision processes, pp. 1–6. The Institute of Electrical and Electronics Engineers (IEEE) (2006)
4. Dawood, T., Elwakil, E., Novoa, H.M., Delgado, J.F.G.: Watermain's failure index modeling via Monte Carlo simulation and fuzzy inference system. Eng. Fail. Anal. **134**, 106100 (2022)
5. Deng, W., Zhang, S., Zhao, H. and Yang, X.: A novel fault diagnosis method based on integrating empirical wavelet transform and fuzzy entropy for motor bearing. IEEE Access, **6**, 35042–35056 (2018)
6. Ebrahimi, E., Bavandpour, M., Astan, N.: Condition monitoring of clutch mechanism retainer of MF-285 tractor with vibration analysis and ANFIS. J. Biosyst. Eng. **1**, 52–65 (2014)
7. ISO 760: Determination of water—Karl Fischer method (general method) (1978)
8. Khan, S.A., Equbal, M.D., Islam, T.: A comprehensive comparative study of DGA based transformer fault diagnosis using fuzzy logic and ANFIS models. IEEE Trans. Dielectr. Electr. Insul. **22**(1), 590–596 (2015)
9. Köken, E., Başpınar Tuncay, E.: Assessment of rock aggregate quality through fuzzy inference system. Geotech. Geol. Eng. 1–9 (2022)
10. Lou, X., Loparo, K.A.: Bearing fault diagnosis based on wavelet transform and fuzzy inference. Mech. Syst. Signal Process. **18**(5), 1077–1095 (2004)
11. McCoy, B.: The future of oil analysis: converting to electric power and sensors. Tribol. Lubr. Technol. **78**(4), 42–47 (2022)
12. Mydukuri, R.V., Kallam, S., Patan, R., Al-Turjman, F., Ramachandran, M.: Deming least square regressed feature selection and Gaussian neuro-fuzzy multi-layered data classifier for early COVID prediction. Expert. Syst. **39**(4), e12694 (2022)
13. Nazari, N.M., Muhammad, M., Mokhtar, A.A.: Prediction of lubrication oil parameter degradation to extend the oil change interval based on Gaussian process regression (GPR). Tribol. Online **17**(3), 135–143 (2022)
14. Pan, H., Xu, H., Zheng, J., Su, J., Tong, J.: Multi-class fuzzy support matrix machine for classification in roller bearing fault diagnosis. Adv. Eng. Inform. **51**, 101445 (2022)
15. Ramezani, S., Memariani, A.: A fuzzy rule based system for fault diagnosis, using oil analysis results. Int. J. Ind. Eng. Prod. Manag. **22**(2), 91–98 (2011)
16. Rasekhi, R., Shamsi, M.: Determining and evaluating the mathematical model for predicting the maintenance costs of MF 285 tractors: case study. In: 6th Iranian National Congress of Biosystems Engineering and Mechanization, 24–25 Sept, Tehran, Iran (2010)

17. Shirmohammadi, A.: Repair and Maintenance Planning, 315p. Arkane Danesh, Iran (2002)
18. Vafaee, M.R., Mashhadi Meighani, H., Almasi, M., Minaee, S.: Choosing the best method of estimating reliability parameter for grain harvesting machines in Markazi province, Iran. Agroecol. J. **5**(15), 144–151 (2009)
19. Wireman, T.: Developing Performance Indicators for Managing Maintenance. Industrial Press, Inc., New York, NY (2005)
20. Zhou, Q., Thai, V.V.: Fuzzy and grey theories in failure mode and effect analysis for tanker equipment failure prediction. Saf. Sci. **83**, 74–79 (2016)

**Fatemeh Afsharnia** received her BS, MS, and Ph.D. degrees in agricultural mechanization engineering form the Agricultural Sciences and Natural Resources University of Khuzestan, Ahvaz, Iran. Dr. Fatemeh Afsharnia is a member of the Iranian Maintenance Association (IRMA) and has published over 30 publications in several journals. Moreover, she has been working as a reviewer in peer-review journals such as Measurement (Elsevier), International Journal of Quality and Reliability Management (Emerald), Journal of the Brazilian Society of Mechanical Sciences and Engineering (Springer). Her main research interests are Reliability Engineering and Analysis, Maintenance Management, Maintenance Planning, Regression Modeling, Reliability Theory, Risk Assessment and Analysis, Safety Management and Engineering, Probabilistic Risk Analysis, Maintenance Optimization, Condition-Based Maintenance, etc.

**Mehdi Behzad** received the Ph.D. degree in mechanical engineering from the University of New South Wales (UNSW), Australia, in 1995. He is a full professor of mechanical engineering with Sharif University of Technology (SUT), Tehran, Iran. He is a board member of the Iranian Maintenance Association (IRMA) and the founding board member of the Iranian acoustic and vibration society (ISAV). Professor Behzad is the founder and the dean of the condition monitoring and fault diagnostics (CMFD) center at SUT since 2000. He is also the chairman of the CMFD annual conference since 2007. He was selected as one of the top distinguished scientists of Iran for his joint academic-industrial activities in 2020. His research interest includes vibration condition monitoring, diagnostics and prognostics, reliability analysis, rotor dynamics, vibration analysis, and intelligent maintenance systems.

**Hesam Addin Arghand** received his BS and MS degrees in mechanical engineering from the University of Tabriz and his Ph.D. degree in mechanical engineering from Sharif University of Technology (SUT), Iran, in 2019. He was also a visiting scholar at the reliability laboratory at the University of Alberta (UoA) for six months in 2016. He is currently an assistant professor with the department of mechanical engineering at the University of Zanjan, Iran. He is also a senior researcher at the condition monitoring and fault diagnostics (CMFD) center at SUT. His research interest is vibration analysis, diagnostics and prognostics of rotating machines, AI application in vibration condition monitoring, and reliability analysis. Since 2019, he has been the scientific committee member of the annual CMFD conference. He is a member of Iranian society of acoustic and vibration (ISAV) since 2012 and Iranian maintenance association (IRMA) since 2020.

# Chapter 6
# A Multi-parameter Occupational Safety Risk Assessment Model for Chemicals in the University Laboratories by an MCDM Sorting Method

**Muhammet Gul, Melih Yucesan, and Mehmet Kayra Karacahan**

**Abstract** University laboratories are high-risk working environments where many chemicals coexist to conduct teaching and scientific research. The frequent occurrence of such chemical-related occupational accidents in recent times highlights the importance of ensuring these units' safety in universities for researchers, students, workplaces, and institutions. In occupational safety risk assessment, very few studies have grouped risks. Many of them also ignored some important risk parameters. Overcoming these disadvantages, this study developed a multi-parameter and multi-criteria decision-making (MCDM) sorting-based methodology for the occupational safety risk assessment of chemicals in a university laboratory. First, six different risk parameters (probability, severity, exposure, detectability, worsening factor, sensitivity to non-usage of personal protective equipment) were determined. The weight value was calculated using the best–worst method (BWM). Then, a risk priority value and classification for each chemical with TOPSIS-Sort, the MCDM sorting method. Finally, some control measures are recommended to reduce the safety risk of the laboratory. The applicability of the proposed methodology has been tested with a real case study. The methodology is intended to adapt to university laboratories' risk assessment and become a primary reference for university safety analysts.

**Keywords** Risk assessment · University laboratory · Chemical · MCDM · BWM · TOPSIS-sort

---

M. Gul (✉)
School of Transportation and Logistics, Istanbul University, Avcılar, 34320 Istanbul, Turkey
e-mail: muhammetgul@istanbul.edu.tr

M. Yucesan
Department of Emergency Aid and Disaster Management, Munzur University, 62000 Tunceli, Turkey
e-mail: melihyucesan@munzur.edu.tr

M. K. Karacahan
Department of Chemistry and Chemical Processing Technology, Tunceli Vocational School, Munzur University, 62000 Tunceli, Turkey
e-mail: mtanaydin@munzur.edu.tr



H. Garg (ed.), *Advances in Reliability, Failure and Risk Analysis*, Industrial and Applied Mathematics, https://doi.org/10.1007/978-981-19-9909-3_6

## 6.1 Introduction

Laboratories in universities are physical spaces where students and researchers conduct their scientific experiments and make sacrifices for the advancement of science. Depending on the work done in chemistry, biology, environment, construction, mechanical and other engineering, and basic science laboratories, there may be many hazards and associated risks. Robust and accurate risk analyses must be made to combat these risks encountered when working with chemicals and performing operations on mechanical devices. The destructiveness of the accidents when adequate precautions are not taken and not prepared can be very large. Although there are no robust and comprehensive statistical figures about university laboratory accidents specifically, the information obtained from the literature and visual media is that there are very destructive accidents [24]. Risk assessment is a multi-stage process that includes identifying, numerical analysis, classifying hazards with a proactive point of view before such accidents occur, and finally, planning control measures. Most studies on laboratory safety in the literature aim to determine the priority of emerging risk types. For this purpose, fuzzy logic, probability theory, and stochastic processes are used to determine the importance weights of the risk parameters and calculate each risk's priority score due to the difficulties in obtaining data. In addition, the existence of multiple risk parameters and the grading of each risk type according to these parameters show that the structure of the problem is under the concept of "*multi-criteria decision-making—MCDM*".

MCDM-based risk analysis determines the priority order of hazards, assigns risks to certain classes, and selects appropriate control measures. The laboratory risk analysis proposed in this study has three important focal points. First, a chemical risk analysis was performed with five experts. In other words, a priority score calculated depending on the determined risk parameters for each chemical was obtained. Thus, each chemical was evaluated in terms of the identified risk factors. Here, chemicals are considered as alternatives in the MCDM perspective. Second, the proposed approach is a multi-parameters approach. In most risk analysis studies from an occupational health and safety perspective, multiple risk parameters are studied double, triple, or rarely. In this approach, six different parameters are taken into account. These are probability, severity, exposure, detectability, worsening factor, and sensitivity to non-usage of personal protective equipment. Third, it aims to determine the chemical hazard classes by MCDM sorting. Thus, chemicals will be divided into different clusters as high-risk clusters and relatively less risky clusters. Thus, it will be possible to reduce the total risk by focusing on the high-risk chemicals group.

In this study, the proposed BWM-TOPSIS-Sort integrated method is used. The reason for using this combination is briefly explained, and BWM is a very effective weighting method, using a small number of evaluations and allowing the calculation of consistency because it contains two vectors (best to others, others to worst). Classification is needed in the chemistry laboratory to select hazardous chemicals or to provide necessary precautions before applications are made. For this reason, the TOPSIS-Sort method, used successfully in many occupational health and safety risk assessment problems, has been chosen. Applying the BWM-TOPSIS-Sort integrated method aims to prioritize the hazards that may arise and consider the results obtained for hazardous chemicals and related control measures.

## 6.2 Literature Review

For this research paper, we review the previous literature in two sections. While the first presents the summary of studies on university laboratory safety, the second includes an overview of MCDM sorting regarding occupational safety risk assessment studies.

### *6.2.1 Past Studies Carried Out for University Laboratory Safety*

Many approaches have been proposed to identify the hazards that arise in university laboratories to analyze and classify the risks associated with these hazards and plan control measures [3, 26]. For example, Li et al. [20] proposed a semi-quantitative approach for a university chemical laboratory risk assessment. The approach consists of two decision-making methods: matter-element extension theory (MEET) and combination ordered weighted averaging (C-OWA) operator. C-OWA operator is applied to compute the weight of assessment indices, and MEET is applied to determine the correlation degree of assessment indices. After applying both methods, the comprehensive risk of university chemical laboratories is calculated, and some safety measures are suggested. Another study by Ma et al. [24] integrated three methods of human factors analysis and classification system (HFACS), fuzzy set theory (FST), and Bayesian network (BN) to identify the most critical and highly contributing human factors exposed in the laboratory fire and explosion accidents. Via this study, 39 laboratory fire and explosion accidents between 2008 and 2020 in China and the USA were tested. The sensitivity analysis was also conducted to events associated with human factors, foremost in laboratory fire and explosion accidents. Li et al. [21] suggested a new approach to assess the risk of unsafe behavior in university laboratories using the human factor analysis and classification system (HFACS-UL) fuzzy BN. Ouédraogo et al. performed two iterative studies for risk analysis in the research environment. The first study [27] modeled the laboratory criticality index by an improved risk priority number. The second one [28] focused on the laboratory criticality index by analytic hierarchy process (AHP). Both methodologies proposed a laboratory assessment and risk analysis (LORA). Shariff and Norazahar [32] developed a program known as the laboratory at-risk behavior and improvement system (Lab-ARBAIS) to monitor and control students' at-risk behaviors. Leggett [18, 19] developed a hazard identification and risk analysis for the chemical research laboratory (Lab-HIRA) model in two parts. While the first presented a preliminary hazard evaluation [18], the second focused on risk analysis of laboratory operations [19]. Ozdemir et al. [29] assessed occupational hazards and associated risks for a university chemical laboratory using 5S methodology, failure modes and effects analysis (FMEA), interval type-two fuzzy sets (IT2FSs), AHP, and the VlseKriterijumska Optimizacija I Kompromisno Resenje (VIKOR) methods.

It can be seen from this short literature review that the risks arising in these environments are analyzed in the context of laboratory safety. However, there is no study on the risk analysis of the chemicals used in particular. Therefore, the study will remedy an important gap, and in light of determined and weighted criteria for each chemical, it will be determined which hazard class.

### *6.2.2 Past Studies Carried Out on Occupational Safety Risk Assessment via MCDM Sorting*

The second subsection of the literature review section of the article is the compilation of occupational risk assessment studies using MCDM sorting algorithms. Sorting algorithms of a number of MCDM methods have been developed [7, 12–14, 16, 17, 30, 35]. A new literature review on Alvarez et al. [1] can be seen for detailed information. The results obtained from this review study have confirmed that the area where MCDM sorting algorithms are applied the most is risk assessment. Brito et al. [5] integrated Utility Theory and the ELECTRE TRI method to assess risk in natural gas pipelines and classify pipeline sections into risk categories. Gul [10] developed a quantitative occupational risk assessment methodology based on TOPSIS-Sort and applied it to an aluminum extrusion process. Yu et al. [37] performed safety risk grading of coal mines based on an AHPSort II method under fuzzy environment. Qin et al. [30] conducted an ecological risk assessment study using a context-dependent DEASort based on BWM. Lolli et al. [23] applied a FlowSort group decision support system to the failure mode and effect analysis (FMEA). They implemented the approach in a blow molding process. Valipour et al. [34] proposed a two-step approach to prioritize and identify critical health, safety, and environment risks. The critical cluster for risks has been determined. Thus, organizational efforts will not be wasted on non-critical risks. Jahangoshai Rezaee et al. [15] proposed a unique risk priority number (RPN) calculation to avoid traditional RPN limitations. They prioritized HSE risks using the fuzzy inference system (FIS) and DEA integrated model.

Having reviewed the studies regarding the occupational safety risk assessment by MCDM sorting algorithms, it is considered that very few of them focused on occupational safety risk assessment and classified the emerged risks into some clusters [5, 10, 23, 34]. However, these limited papers also have some deficiencies regarding the lack of consideration of all essential risk parameters. To overcome these, the current paper aims to eliminate this drawback and consider six risk parameters (probability, severity, exposure, detectability, worsening factor, sensitivity to non-usage of personal protective equipment). In the first phase of the approach, the risk parameters were determined, and a weight value for each was calculated using the BWM algorithm. Then, a risk priority value and risk classification were determined for each chemical in the university laboratory via the TOPSIS-Sort algorithm. Finally, a control measure suggestion phase is carried out for the laboratory.

## 6.3 The Proposed Methodology

This section discusses the framework of the proposed multi-parameter risk assessment approach. This study consists of four steps. In the first step, six risk parameters were determined to be considered in assessing the risks in the chemistry laboratory. These parameters are probability, severity, exposure, detectability, worsening factor, and sensitivity to not using personal protective equipment. It is assumed that these risk factors are not affected by each other. To evaluate these parameters, 7-point linguistic terms were selected. The second step involves weighting with BWM. In this section, decision-makers identify the risk factor they consider the most important and the risk factors they consider the least important. They then compared these risk factors with other risk factors. The third step is the TOPSIS-Sort stage. The 60 most frequently used chemicals in the chemistry laboratory were determined and evaluated within the scope of the risk parameters determined in step 1. Chemicals were divided into five clusters with the solution: very high risk, high risk, substantial risk, possible risk, and risk. The fourth step presented preventive measures, starting with the very high-risk cluster.

### *6.3.1 Establishing the Multi-parameter Occupational Safety Risk Assessment*

As in all occupational risk assessment studies, some parameters should be considered to prioritize emerging hazards and associated risks. The literature has mostly double, triple, and, rarely, multi-parameter risk assessments. In this study, it considered six risk parameters. (1) Probability: the frequency of occurrence of the hazard, (2) Severity: the degree of danger that the risk will pose to personnel, machinery equipment, environment and continuity of production, (3) Exposure: it is considered as the duration of exposure of a hazardous activity on the respective workers, (4) Detectability: detectability of the risk with the eye or any digital device, (5) Worsening factor: it is related to hazardous circumstances that could amplify the dire outcomes of an accident and focuses on the hazardous situations leading up to an accident [25], and (6) Sensitivity to not using personal protective equipment: to what extent the use of personal protective equipment affects the severity of the risk. The 7-point scale for each parameter mentioned here is given in Table 6.1. The numerical values determined according to this scale are used in the evaluations made in the TOPSIS-Sort phase.

**Table 6.1** 7-point scale for each parameter used in the assessment of hazards and associated risks

| Numerical scale | Linguistic term | Probability (*P*) | Severity (*S*) | Exposure (*E*) |
|---|---|---|---|---|
| 1 | Very low | Close to impossible | Equals to first aid | Exceptionally rare |
| 2 | Low | Practically impossible | Equals minor injury | Rare |
| 3 | Medium low | Practicable but very unlikely | Leads to temporary disability | Seldom |
| 4 | Medium | Only remotely possible | Equals serious injury | Occasional |
| 5 | Medium high | Rare but possible | Culminates in permanent disability | Often |
| 6 | High | Quite possible | Equals fatality | Frequent |
| 7 | Very high | Mostly observed | Leads to many fatalities | Prolonged |
| Numerical scale | Linguistic term | Detectability (*D*) | Worsening factor (WF) | Sensitivity to non-usage of PPE (PPE) |
| 1 | Very low | Extremely easy | Does not amplify it | Risk can be avoided without using PPE |
| 2 | Low | Highly possible | Can negligibly amplify it | The use of PPE can slightly reduce the risk |
| 3 | Medium low | Slightly possible | Can slightly amplify it | The use of PPE can reasonably reduce the risk |
| 4 | Medium | Can sometimes be possible | Can moderately amplify it | The use of PPE reduces the risk moderately |
| 5 | Medium high | To a large extent difficult to be noted | Can to a large extent amplify it | The use of PPE reduces the risk greatly |
| 6 | High | Highly difficult to be noted | Can highly amplify it | It is necessary to use PPE to reduce the risk |
| 7 | Very high | Extremely difficult to be noted | Can extremely amplify it | PPE must be used |

### 6.3.2 Determining the Weight of Risk Parameters via BWM

There are many methods in the literature on elicitation of the weights of criteria (called risk parameters) in the field of occupational safety risk assessment, directly (e.g., AHP; BWM) and indirectly (e.g., robust ordinal regressions) [9, 22]. In this study, the weights of the risk parameters were determined using the BWM. Initially, experts familiar with laboratory research and experienced in risk assessment of chemicals were invited to participate in the study. In this context, communication was established with experts (two chemical engineers, two food engineers, and one chemist, whose academic and research activities continue), and their evaluations were received. The received evaluations were converted into mathematical models and solved using the BWM algorithm. The following BWM algorithm is given step by step [31].

**Step 1**. The criteria to be considered are determined. The criteria to be used in decision-making are shown with $(c_1, c_2, \ldots, c_n)$.

**Step 2**. Among the evaluated criteria, the best and worst criteria are determined.

**Step 3**. It uses numbers one-nine for pairwise comparisons. If two criteria are equally important, one is used, and if there is a huge difference in importance between the two criteria, nine is used. The Best to other vector is created as: $A_B = (a_{B1}, a_{B2}, \ldots, a_{Bn})$ where $a_{Bj}$ shows the predilection of the best criterion $B$ over criterion $j$ Comparison of the criteria with themselves $(a_{BB} = 1)$.

**Step 4**. Similar to step 3, the Other to worst vector is created. Others-to-Worst vector is created as: $A_B = (a_{1W}, a_{2W}, \ldots, a_{nW})$ where $a_{jW}$ shows the predilection of criterion $j$ over the worst criterion $w$.

**Step 5**. Determination of weight $(w_1^*, w_2^*, \ldots, w_n^*)$. Following the steps in Rezaei [31]

$$
\begin{aligned}
&\min \xi \\
&\left|\frac{w_B}{w_j} - a_{Bj}\right| \le \xi \text{ for all } j \\
&\left|\frac{w_j}{w_W} - a_{jW}\right| \le \xi \text{ for all } j \\
&\Sigma w_j = 1, w_j \ge 0, \text{ for all } j
\end{aligned}
$$

By solving the mathematical model, the optimum weights $(w_1^*, w_2^*, \ldots, w_n^*)$ and $\xi^*$ are calculated.

### 6.3.3 Calculating the Risk Priority Classes of Chemicals in the Lab via TOPSIS-Sort

TOPSIS is an MCDM method first proposed by Hwang and Yoon [11]. It is used to select the best alternative from a set of homogeneous alternatives under number of decision criteria. TOPSIS considers the shortest distance of the best alternative to the positive ideal solution and the longest distance to the negative ideal solution. Due to the diversity and prevalence of the application area, many extensions were proposed in the first proposed TOPSIS version and successfully hybridized with other MCDM methods [4]. Since classification is one of the main uses of MCDM problems, a new method, TOPSIS-Sort, has been developed for classification. TOPSIS-Sort was first developed by Faraji Sabokbar et al. [8] suggested. Later, it was used by some scientists and applied to various problems [6, 10, 36]. In this paper, TOPSIS-Sort determines the final priority value, ranking order, and class of each chemical in the laboratory. The steps of the TOPSIS-Sort are summarized below.

**Step 1**. Determine the decision matrix. The decision matrix shows the value obtained by scoring each chemical according to six risk parameters weighted in the BWM phase. These scores are assigned by decision-making experts using a 7-point scale provided in Table 6.1. In this step, the weights determined via BWM should be ready.

**Step 2**. Define the limit profile set. Limit profiles are determined based on the predicted number of classes to which chemicals will be assigned. This paper determines five classes as "very dangerous, dangerous, medium, less dangerous, and non-hazardous".

**Step 3**. Creating the aggregated decision matrix. Each evaluation made by the expert team is brought together and combined with the help of the weighted average aggregation operator.

**Step 4**. Normalization process and creation of the normalized decision matrix.

**Step 5**. Generating the weighted normalized decision matrix. Here, the weight matrix obtained by BWM is multiplied with the normalized decision matrix obtained in step 4.

**Step 6**. Determine the positive and negative ideal solutions.

**Step 7**. Calculate the distances between the positive and negative ideal solutions.

**Step 8**. Calculate the TOPSIS closeness coefficient ($CC_i$) value for each chemical. The chemical with a high closeness coefficient value means the highest priority. According to this value, a priority order is made for all chemicals in the chemical list.

**Step 9**. Obtaining the deviation of the upper and lower limit profiles of each chemical from the ideal solution. In this step, all chemicals are assigned to the appropriate classes.

### 6.3.4 Suggesting Control Measures

Control measures constitute one of the basic components and the most important risk management stage. The assessed risk class of the chemical indicates the importance of the risk resulting from exposure to the hazard. Based on the risk assessment results, the risk score and class found help determine which chemical should be controlled primarily in laboratory risk management. Some useful control measures can then be developed to eliminate or reduce the risks to an acceptable level, improving the safety of university laboratories.

## 6.4 Case Study

### 6.4.1 System Environment of the University Chemical Laboratory and Chemical List

Teaching and academic laboratories are where experimental studies are carried out, which are calculated theoretically and thought to be applied by making preliminary preparations. Experimental work helps us understand the concepts we visualize in our minds, develop problem-solving skills, and improve researchers' manual dexterity and observation skills. Teaching and research laboratories are classified according to their functions: comprehensive laboratories and basic laboratories such as general chemistry laboratory, analytical chemistry laboratory, physical chemistry laboratory, chemical engineering laboratories, and research laboratories. Undergraduate and graduate students and researchers in these laboratories represent a comprehensive laboratory, laboratories used to conduct multidisciplinary experimental research. The basic laboratory serves for regular teaching experiments for undergraduates. In these laboratories, computer-controlled heat exchanger service unit, chemical reactors apparatus, continuous distillation unit, absorption unit, continuous liquid–liquid extraction unit, tray dryer, head loss in piping and fittings, process control, tubular flow reactor, atomic absorption spectrophotometer, gas chromatography-mass spectrometry, ultraviolet–visible spectroscopy, high-performance liquid chromatography, magnetic and mechanical stirrers, autoclave, fume hoods, constant temperature circulators, ultra-pure water systems, orbital shakers, etc. devices are available. Some of the chemicals used in these laboratories are listed in Table 6.2.

**Table 6.2** Chemical list

| # | Chemicals | # | Chemicals | # | Chemicals | # | Chemicals |
|---|---|---|---|---|---|---|---|
| 1 | Acrylonitrile | 16 | Diethyl ether | 31 | Hydrogen peroxide | 46 | Morpholine |
| 2 | Ammonia | 17 | Dimethyl sulfoxide | 32 | Hydrochloric acid | 47 | Naphthalene |
| 3 | Aniline | 18 | Ethyl acetate | 33 | Iodine | 48 | Nitric acid |
| 4 | Acetaldehyde | 19 | Ethylene dichloride | 34 | Calcium hypochlorite | 49 | Perchloric acid |
| 5 | Acetic acid | 20 | Ethylene glycol | 35 | Carbon disulfide | 50 | Potassium hydroxide |
| 6 | Aseton | 21 | Arsenic pentafluoride | 36 | Carbon tetrachloride | 51 | Propylene glycol |
| 7 | Benzaldehyde | 22 | Phenol | 37 | Chloroacetone | 52 | Cyclohexane |
| 8 | Benzene | 23 | Arsenic trichloride | 38 | Chloroform | 53 | Sodium hydroxide |
| 9 | Benzyl chloride | 24 | Formaldehyde | 39 | Hydrogen sulfide | 54 | Sodium hypochlorite |
| 10 | Bromine | 25 | Formic acid | 40 | Chromic acid | 55 | Sulfuric acid |
| 11 | Hydrogen cyanide | 26 | Phosphoric acid | 41 | Methyl ethyl ketone | 56 | Toluen |
| 12 | Butyraldehyde | 27 | Glycerol | 42 | Propionic acid | 57 | Triethanolamine |
| 13 | Dibenzyl ether | 28 | Hexane | 43 | Methylamine | 58 | Trichloroethylene |
| 14 | Dibutyl phthalate | 29 | Hydrobromic acid | 44 | Methylene chloride | 59 | Triphenyl phosphate |
| 15 | Diethanolamine | 30 | Hydrofluoric acid | 45 | Monoethanolamine | 60 | Pentaborane |

### 6.4.2 *The Exploitation of BWM in the Determination of Risk Parameter Weights*

The weight of six risk parameters is calculated using the BWM. By determining the best and worst risk parameter preferences of five experts, best others and others worst matrices were created. After that, each created matrix was solved with BWM Solver, and each expert determined the optimal weights of the risk parameters. Average optimal weight values were obtained by taking the average of the weights obtained with the evaluations of these five experts. Figure 6.1 shows these optimal weights. At the same time, the consistency ratio was calculated for each assessment made by five experts. These consistency values were obtained as 0.022, 0.016, 0.016, 0.044, and 0.027, respectively.

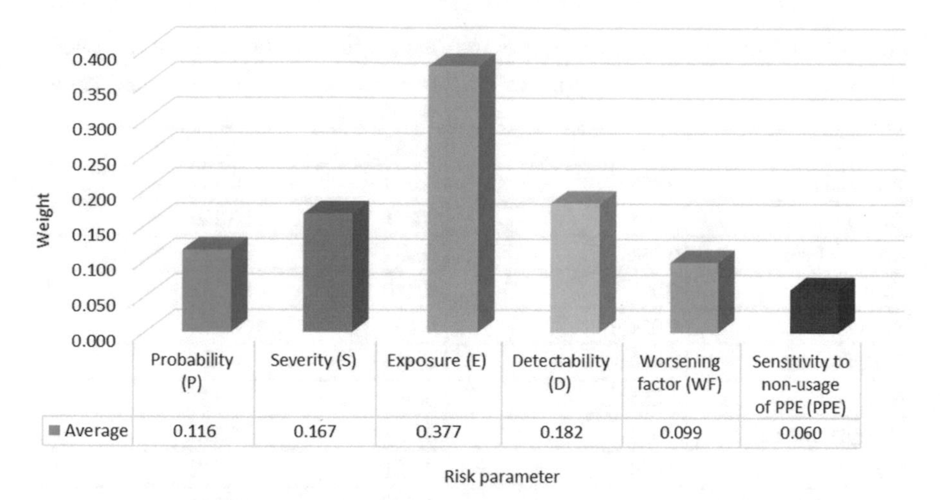


**Fig. 6.1** Averaged optimal weight values of risk parameters

### 6.4.3 *The Exploitation of TOPSIS-Sort in Risk Classification of Chemicals*

In this section, TOPSIS-Sort was applied, and the risk level and risk class were determined for each chemical in the laboratory where the application was carried out. Here, an evaluation was made with the consensus of the expert team who evaluated the BWM part using the numerical scale in Table 6.1. The values obtained from the evaluation were aggregated and inserted into the TOPSIS-Sort algorithm. By applying the steps from 1 to 7, the distance from the ideal and negative ideal solutions and final closeness coefficient values are calculated as in Table 6.3. Finally, the risk class of each chemical was determined by applying the last step of the TOPSIS-Sort algorithm. Accordingly, if the $CC_i$ value is between 0 and 0.192, risk class is defined as "risk", between 0.193 and 0.575 as "possible risk", between 0.576 and 0.763 as "substantial risk", between 0.764 and 0.922 as "high risk", and between 0.923 and 1 as "very high risk". The risk classification according to the calculations obtained is given in Fig. 6.2.

### 6.4.4 *Risk Management of the Laboratory*

Chemical hazards can be classified as corrosivity, explosivity, flammability, toxicity, and reactivity [2]. The chemicals in Table 6.3 are dangerous in only one of these ways, and some are dangerous in more than one way ($CC_i$ value is between 0.923 and 1, pentaborane and hydrogen cyanide, and risk class is defined as "very high risk"). Many chemicals used in chemistry laboratories are dangerous in at least one

**Table 6.3** TOPSIS-Sort outputs: the distances from the ideal and negative ideal and $CC_i$ values

| Chemicals | Distance from the ideal | Distance from the negative ideal | Closeness coefficient |
|---|---|---|---|
| Acrylonitrile | 0.035 | 0.055 | 0.607 |
| Ammonia | 0.055 | 0.034 | 0.384 |
| Aniline | 0.068 | 0.021 | 0.240 |
| Acetaldehyde | 0.058 | 0.032 | 0.358 |
| Acetic acid | 0.061 | 0.030 | 0.331 |
| Asheton | 0.054 | 0.035 | 0.395 |
| Benzaldehyde | 0.069 | 0.020 | 0.227 |
| Benzene | 0.055 | 0.034 | 0.377 |
| Benzyl chloride | 0.036 | 0.054 | 0.596 |
| Bromine | 0.060 | 0.031 | 0.342 |
| Hydrogen cyanide | 0.006 | 0.086 | 0.932 |
| Butyraldehyde | 0.053 | 0.037 | 0.411 |
| Dibenzyl ether | 0.069 | 0.021 | 0.231 |
| Dibutyl phthalate | 0.071 | 0.018 | 0.202 |
| Diethanolamine | 0.055 | 0.034 | 0.384 |
| Diethyl ether | 0.061 | 0.030 | 0.328 |
| Dimethyl sulfoxide | 0.071 | 0.018 | 0.202 |
| Ethyl acetate | 0.068 | 0.021 | 0.237 |
| Ethylene dichloride | 0.070 | 0.019 | 0.216 |
| Ethylene glycol | 0.071 | 0.018 | 0.204 |
| Arsenic pentafluoride | 0.051 | 0.038 | 0.425 |
| Phenol | 0.040 | 0.050 | 0.553 |
| Arsenic trichloride | 0.051 | 0.038 | 0.423 |
| Formaldehyde | 0.050 | 0.041 | 0.446 |
| Formic acid | 0.058 | 0.032 | 0.358 |
| Phosphoric acid | 0.050 | 0.040 | 0.444 |
| Glycerol | 0.055 | 0.034 | 0.384 |
| Hexane | 0.068 | 0.021 | 0.237 |
| Hydrobromic acid | 0.059 | 0.032 | 0.352 |
| Hydrofluoric acid | 0.025 | 0.067 | 0.732 |
| Hydrogen peroxide | 0.051 | 0.039 | 0.430 |
| Hydrochloric acid | 0.055 | 0.034 | 0.384 |
| Iodine | 0.055 | 0.034 | 0.384 |
| Calcium hypochlorite | 0.042 | 0.049 | 0.534 |
| Carbon disulfide | 0.037 | 0.051 | 0.579 |

(continued)

**Table 6.3** (continued)

| Chemicals | Distance from the ideal | Distance from the negative ideal | Closeness coefficient |
|---|---|---|---|
| Carbon tetrachloride | 0.054 | 0.035 | 0.391 |
| Chloroacetone | 0.037 | 0.052 | 0.584 |
| Chloroform | 0.054 | 0.034 | 0.385 |
| Hydrogen sulfide | 0.034 | 0.055 | 0.623 |
| Chromic acid | 0.038 | 0.051 | 0.577 |
| Methyl ethyl ketone | 0.060 | 0.031 | 0.339 |
| Propionic acid | 0.034 | 0.054 | 0.615 |
| Methylamine | 0.021 | 0.068 | 0.763 |
| Methylene chloride | 0.051 | 0.037 | 0.421 |
| Monoethanolamine | 0.055 | 0.034 | 0.384 |
| Morpholine | 0.017 | 0.071 | 0.807 |
| Naphthalene | 0.038 | 0.051 | 0.575 |
| Nitric acid | 0.020 | 0.069 | 0.773 |
| Perchloric acid | 0.051 | 0.038 | 0.423 |
| Potassium hydroxide | 0.049 | 0.040 | 0.446 |
| Propylene glycol | 0.089 | 0.000 | 0.000 |
| Cyclohexane | 0.071 | 0.018 | 0.202 |
| Sodium hydroxide | 0.051 | 0.038 | 0.423 |
| Sodium hypochlorite | 0.055 | 0.035 | 0.393 |
| Sulfuric acid | 0.034 | 0.055 | 0.616 |
| Toluen | 0.035 | 0.054 | 0.603 |
| Triethanolamine | 0.042 | 0.048 | 0.531 |
| Trichloroethylene | 0.067 | 0.022 | 0.249 |
| Triphenyl phosphate | 0.089 | 0.000 | 0.000 |
| Pentaborane | 0.000 | 0.089 | 1.000 |

of these ways. However, the degree of danger can vary from major to minor or in between ($CC_i$ value is between 0.193 and 0.575 as "possible risk", between 0.576 and 0.763 as "substantial risk", between 0.764 and 0.922 as "high risk"). Some chemicals are pretty-safe ($CC_i$ value is between 0 and 0.192, propylene glycol and triphenyl phosphate, and risk class is defined as "risk").

A total of 60 chemical risk factors were estimated, and "risk", "possible risk", "substantial risk", "high risk", and "very high risk" were accepted as risk classes. The number of chemicals at the "very high" risk level accounted for 3.33% of the risk classes. In addition, chemicals with a "high" risk level (5%) contributed. "Substantial risk" was determined by 16.67% of the chemicals identified. In addition, 71.67% of the evaluated chemicals have "possible risk" and 3.33% "risk" ratios.

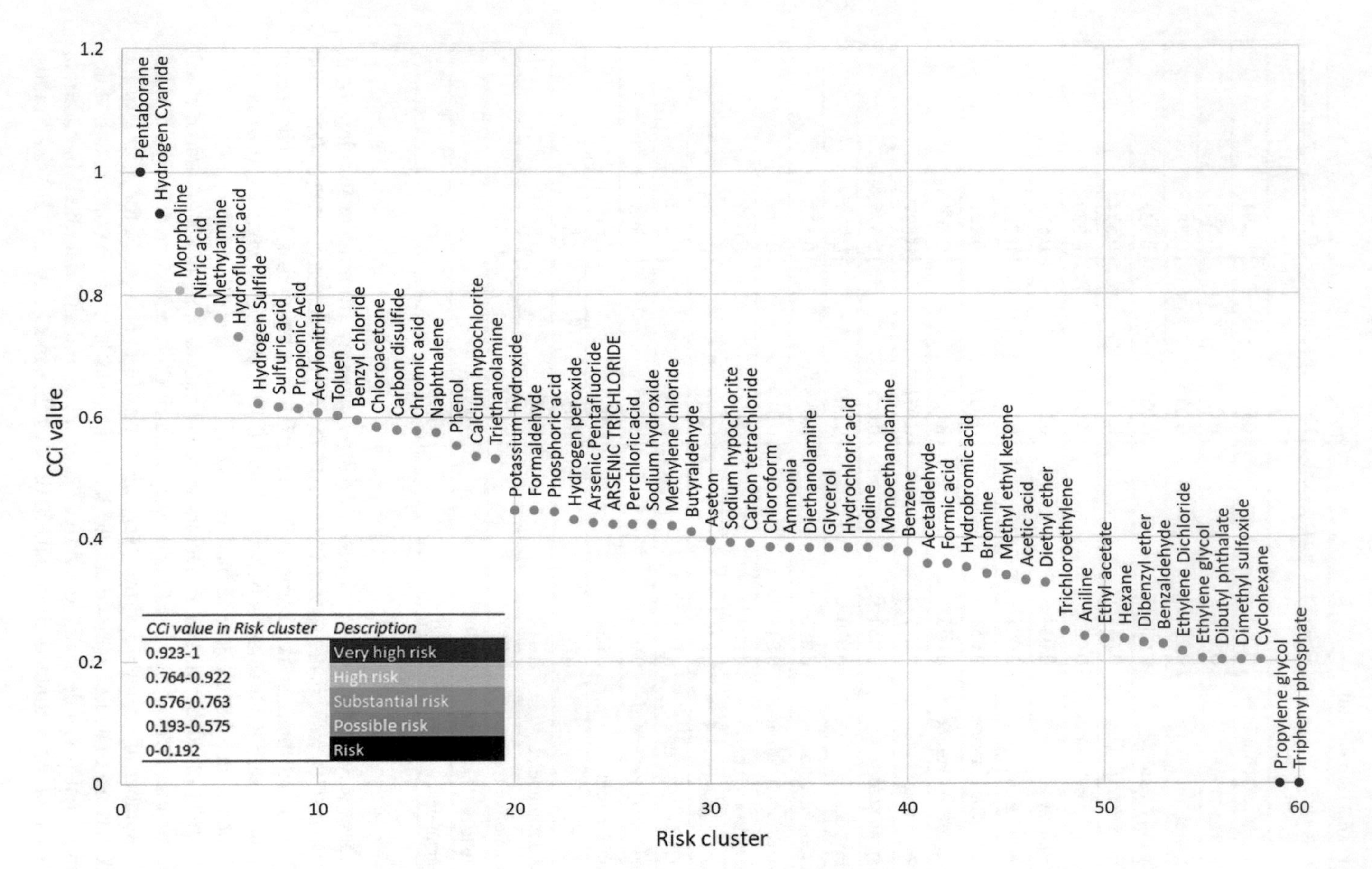


**Fig. 6.2** Risk classes

Using chemicals in laboratory processes generally produces a wide range of risk classes.

If the hazards posed by chemicals are not recognized, unexpected events can result in personal injury or death. Chemical safety in academic laboratories requires laboratory safety documentation, fume hood maintenance, proper chemical storage, correct use of the fume hood for chemical processing, and laboratory safety labeling. Laboratory safety provisions for students and researchers working in laboratories should include safety training, chemical exposure control, medical consultation, suitability of personal protective equipment, control measures, and specific guidelines for hazardous chemicals. Chemicals in the "very high risk" group are quite dangerous. For example, pentaborane is one of the substances harmful to health because it is flammable and reactive. Therefore, brief exposure to small quantities can cause death or permanent injury. The legal exposure limit allowed in the air is a very small value of 0.005 ppm on average in an 8-h work shift, and even this small value can cause enormous damage to human health [33].

Here are some ways to reduce exposure to "high risk and very high-risk chemicals": (1) Hazard and warning information should be posted in the work area, and, as part of an ongoing education and training effort, all information on these chemicals health and safety hazards should be communicated to potentially exposed persons. (2) Protective work clothes and equipment should be worn. (3) Appropriate eye protection should be used to avoid eye contact. (4) Any air-supplied respirator operated in a continuous flow model should be used when working with this type of "very high-risk chemical". (5) If possible, shut off processes and electricity and provide local exhaust ventilation where chemical releases occur. (6) Breathing masks must be worn if local exhaust ventilation or containment is not used. (7) Wash thoroughly immediately after exposure to chemicals and at the end of the work shift. (8) Appropriate personal protective clothing should be worn to prevent contact with the skin. Appropriate eye protection should be used to prevent contact with the eyes. The skin should be washed immediately when contaminated with the skin. It should be removed immediately due to the risk of ignition when it comes into contact with work clothes.

The results of the study determine the risky chemicals to be considered and the associated control measures to be taken. It contributes in terms of knowledge, both methodologically, and in terms of practice, in light of current literature. Injecting the methodology allows easy identification of hazards or risks in laboratories. BWM allows to suggest a method and prioritize it. Evaluation of hazards in a university chemistry laboratory environment reveals that precautions should be taken for some hazardous chemicals. It shows that special working environments should be created, and precautions should be taken to avoid exposure to human health and environmental effects while working with hazardous chemicals. It reveals that experimental studies with hazardous chemicals should not be carried out before these precautions are taken.

## 6.5 Conclusion

This chapter presents a multi-parameter occupational safety risk assessment model for chemicals in university laboratories by an MCDM sorting method. Three parameters, which include probability, severity, exposure, detectability, worsening factor, and sensitivity to non-usage of personal protective equipment issues related to chemicals, are included in the model. Five experts participated in the rating process of relative importance weight of these parameters via the BWM. Then, the final priority value, rank, and class of each chemical were determined via the TOPSIS-Sort algorithm. Based on expert opinion, pentaborane and hydrogen cyanide proved to be the most dangerous chemicals, followed by morpholine, nitric acid, and methylamine.

University chemical laboratories are used extensively in educational and research and development activities. Chemical-based risk analysis has become mandatory due to the large number of experiments and the impossibility of predetermining the experiments to be conducted for future research and development activities. Thanks to the critical chemical list, technicians, researchers, and executives will be able to take risk-preventing activities. In addition, an approach can be made regarding the risk of the research activity to be carried out based on the materials used in the experimental design. The performed study plays a crucial role in deploying occupational risk management of chemicals based on chemical ranking. It also provides a solid framework for other researchers, laboratory personnel, and chemical stakeholders to apply the proposed method to their laboratory safety processes.

Although this risk analysis enriches the literature by addressing the 60 most commonly used chemicals, including six risk factors, the study has some limitations. The fact study that was conducted only for the laboratory of a university reduces the generalizability of the results. Since a process-based evaluation was not made, the situation of the same chemicals creating different risks in different processes could not be evaluated. In addition, the study was carried out with information obtained from analytical chemistry, physical chemistry, instrumental analysis, and R&D laboratories. No privatization has been made on a laboratory basis. The results were interpreted according to 5 expert evaluations. Specific evaluations can be made in the future studies in specialized laboratories and research areas for specific subjects. Considering all these issues in the future studies will inevitably increase the study's effectiveness, consistency, and applicability.

## References


1. Alvarez, P.A., Ishizaka, A., Martínez, L.: Multiple-criteria decision-making sorting methods: a survey. Expert Syst. Appl. **183**, 115368 (2021)
2. American Chemical Society, Committee on Chemical Safety: In: Finster, D.C. (ed) Safety in Academic Chemistry Laboratories, 8th edn. American Chemical Society, Washington, DC, Mar 2017. ISBN 978-0-8412-3732-2

3. Bai, M., Liu, Y., Qi, M., Roy, N., Shu, C.M., Khan, F., Zhao, D.: Current status, challenges, and future directions of university laboratory safety in China. J. Loss Prev. Process Ind. **74**, 104671 (2022)
4. Behzadian, M., Otaghsara, S.K., Yazdani, M., Ignatius, J.: A state-of the-art survey of TOPSIS applications. Expert Syst. Appl. **39**(17), 13051–13069 (2012)
5. Brito, A.J., de Almeida, A.T., Mota, C.M.: A multicriteria model for risk sorting of natural gas pipelines based on ELECTRE TRI integrating utility theory. Eur. J. Oper. Res. **200**(3), 812–821 (2010)
6. de Lima Silva, D.F., de Almeida Filho, A.T. Sorting with TOPSIS through boundary and characteristic profiles. Comput. Ind. Eng. **141**, 106328 (2020)
7. Demir, L., Akpınar, M.E., Araz, C., Ilgın, M.A.: A green supplier evaluation system based on a new multi-criteria sorting method: VIKORSORT. Expert Syst. Appl. **114**, 479–487 (2018)
8. Faraji Sabokbar, H., Hosseini, A., Banaitis, A., Banaitiene, N.: A novel sorting method TOPSIS-SORT: an applicaiton for Tehran environmental quality evaluation. Bus. Adm. Manag. **2**(19), 87–104 (2016)
9. Gul, M.: A review of occupational health and safety risk assessment approaches based on multi-criteria decision-making methods and their fuzzy versions. Hum. Ecol. Risk Assess. Int. J. **24**(7), 1723–1760 (2018)
10. Gul, M.: A quantitative occupational risk assessment methodology based on TOPSIS-Sort with its application in aluminum extrusion industry. Int. J. Pure Appl. Sci. **7**(1), 163–172 (2021)
11. Hwang, C.L., Yoon, K., Hwang, C.L., Yoon, K.: Methods for multiple attribute decision making. Multiple Attribute Decis. Mak.: Methods and Appl a State-of-the-art Surv. 58–191 (1981)
12. Ishizaka, A., Lolli, F., Balugani, E., Cavallieri, R., Gamberini, R.: DEASort: assigning items with data envelopment analysis in ABC classes. Int. J. Prod. Econ. **199**, 7–15 (2018)
13. Ishizaka, A., López, C.: Cost-benefit AHPSort for performance analysis of offshore providers. Int. J. Prod. Res. **57**(13), 4261–4277 (2019)
14. Ishizaka, A., Pearman, C., Nemery, P.: AHPSort: an AHP-based method for sorting problems. Int. J. Prod. Res. **50**(17), 4767–4784 (2012)
15. Jahangoshai Rezaee, M., Yousefi, S., Eshkevari, M., Valipour, M., Saberi, M.: Risk analysis of health, safety and environment in chemical industry integrating linguistic FMEA, fuzzy inference system and fuzzy DEA. Stoch. Env. Res. Risk Assess. **34**(1), 201–218 (2020)
16. Krejčí, J., Ishizaka, A.: FAHPSort: a fuzzy extension of the AHPSort method. Int. J. Inf. Technol. Decis. Mak. **17**(04), 1119–1145 (2018)
17. Labella, Á., Ishizaka, A., Martínez, L.: Consensual group-AHPSort: applying consensus to GAHPSort in sustainable development and industrial engineering. Comput. Ind. Eng. **152**, 107013 (2021)
18. Leggett, D.J.: Lab-HIRA: hazard identification and risk analysis for the chemical research laboratory: Part 1. Preliminary hazard evaluation. J. Chem. Health Saf. **19**(5), 9–24 (2012)
19. Leggett, D.J.: Lab-HIRA: hazard identification and risk analysis for the chemical research laboratory. Part 2. Risk analysis of laboratory operations. J. Chem. Health Saf. **19**(5), 25–36 (2012)
20. Li, X., Zhang, L., Zhang, R., Yang, M., Li, H.: A semi-quantitative methodology for risk assessment of university chemical laboratory. J. Loss Prev. Process Ind. **72**, 104553 (2021)
21. Li, Z., Wang, X., Gong, S., Sun, N., Tong, R.: Risk assessment of unsafe behavior in university laboratories using the HFACS-UL and a fuzzy Bayesian network. J. Saf. Res. (2022)
22. Lolli, F., Coruzzolo, A.M., Balugani, E.: The indoor environmental quality: a TOPSIS-based approach with indirect elicitation of criteria weights. Saf. Sci. **148**, 105652 (2022)
23. Lolli, F., Ishizaka, A., Gamberini, R., Rimini, B., Messori, M.: FlowSort-GDSS—a novel group multi-criteria decision support system for sorting problems with application to FMEA. Expert Syst. Appl. **42**(17–18), 6342–6349 (2015)
24. Ma, L., Ma, X., Xing, P., Yu, F.: A hybrid approach based on the HFACS-FBN for identifying and analysing human factors for fire and explosion accidents in the laboratory. J. Loss Prev. Process Ind. **75**, 104675 (2022)

25. Mohandes, S.R., Durdyev, S., Sadeghi, H., Mahdiyar, A., Hosseini, M.R., Banihashemi, S., Martek, I.: Towards enhancement in reliability and safety of construction projects: developing a hybrid multi-dimensional fuzzy-based approach. Eng. Constr. Archit. Manag. (2022)
26. Olewski, T., Snakard, M.: Challenges in applying process safety management at university laboratories. J. Loss Prev. Process Ind. **49**, 209–214 (2017)
27. Ouédraogo, A., Groso, A., Meyer, T.: Risk analysis in research environment—Part I: Modeling lab criticality index using improved risk priority number. Saf. Sci. **49**(6), 778–784 (2011)
28. Ouédraogo, A., Groso, A., Meyer, T.: Risk analysis in research environment—Part II: Weighting lab criticality index using the analytic hierarchy process. Saf. Sci. **49**(6), 785–793 (2011)
29. Ozdemir, Y., Gul, M., Celik, E.: Assessment of occupational hazards and associated risks in fuzzy environment: a case study of a university chemical laboratory. Hum. Ecol. Risk Assess. Int. J. **23**(4), 895–924 (2017)
30. Qin, J., Zeng, Y., Zhou, Y.: Context-dependent DEASort: a multiple criteria sorting method for ecological risk assessment problems. Inf. Sci. **572**, 88–108 (2021)
31. Rezaei, J.: Best-worst multi-criteria decision-making method. Omega. **53**, 49–57 (2015)
32. Shariff, A.M., Norazahar, N.: At-risk behaviour analysis and improvement study in an academic laboratory. Saf. Sci. **50**(1), 29–38 (2012)
33. URL1. https://www.ilo.org/dyn/icsc/showcard.display?p_lang=en&p_card_id=0819&p_version=. Access date: 29 May 2022
34. Valipour, M., Yousefi, S., Jahangoshai Rezaee, M., Saberi, M.: A clustering-based approach for prioritizing health, safety and environment risks integrating fuzzy C-means and hybrid decision-making methods. Stoch. Env. Res. Risk Assess. **36**(3), 919–938 (2022)
35. Xu, Z., Qin, J., Liu, J., Martínez, L.: Sustainable supplier selection based on AHPSort II in interval type-2 fuzzy environment. Inf. Sci. **483**, 273–293 (2019)
36. Yamagishi, K., Ocampo, L. Utilizing TOPSIS-Sort for sorting tourist sites for perceived COVID-19 exposure. Curr Issues in Tourism. *25*(2), 168–178 (2022)
37. Yu, Q., Liu, J., Tu, Y., Nie, L., Shen, W.: Safety risk grading of coal mine based on AHPsort II method under fuzzy environment. In: International Conference on Management Science and Engineering Management, pp. 289–301. Springer, Cham, Aug 2021

**Muhammet Gul** has been working as an associate professor at the Department of Transportation and Logistics, Istanbul University, Istanbul, Turkey. He has received his MSc and Ph.D. in Industrial Engineering from Yildiz Technical University. His research interests are in simulation modeling, healthcare system management, occupational safety and risk assessment, multi-criteria decision- making, and fuzzy sets. He is the editor-in-chief of two books released in 2020*Computational Intelligence and Soft Computing Applications in Healthcare Management Science* (IGI-Global) and *Fine–Kinney-Based Fuzzy Multi-Criteria Occupational Risk Assessment: Approaches, Case Studies and Python Applications* (Springer). He is also on the Editorial Board of *Complex & Intelligent SystemsMathematical Problems in Engineering*, and *Journal of Healthcare Engineering*.

**Melih Yucesan** completed his Ph.D. and has been working as an associate professor at the Department of Emergency Aid and Disaster Management, Munzur University, Tunceli, Turkey. He received his Ph.D. in Econometrics from Karadeniz Technical University. His research interests are in planning, forecasting, multi-criteria decision-making, and fuzzy sets. His papers appeared in international journals such as International Journal of Disaster Risk Reduction, Applied Soft Computing, Energy Policy, Soft Computing, and International Journal of Healthcare Management.

**Mehmet Kayra Karacahan** completed his doctorate at Munzur University, Department of Chemistry and Chemical Processes, Tunceli Vocational School, Munzur University, and Rare Earth Application and Research Center, Munzur University, Tunceli, Turkey, and works as an assistant professor. He received his BS, MS, and Ph.D. degrees from Inonu University in chemical engineering. His research interests are hydrometallurgy, leaching, solvent extraction, cementation, and electrowinning. His papers appeared in international journals such as Separation Science and Technology, Desalination and Water Treatment, Brazilian Journal of Chemical Engineering, Russian Journal of Non-Ferrous Metals Construction and Building Materials, Journal of Sustainable Metallurgy, Mineral Processing and Extractive Metallurgy Review.

# Chapter 7
# Smart Failure Mode and Effects Analysis (FMEA) for Safety–Critical Systems in the Context of Industry 4.0

**Hamzeh Soltanali and Saeed Ramezani**

**Abstract** In digitalized environments, advanced fault diagnosis and prognosis approaches are widely used for system safety and reliability assessments. As a proactive diagnosis approach, Failure Mode and Effects Analysis (FMEA) plays a critical role in identifying system bottlenecks and mitigating the adverse consequences within high-risk industries. Therefore, this chapter deals with the different types of FMEAs, FMEA in safety–critical systems, current drawbacks, and limitations of classical-FMEA theories, as well as supporting the classical form by introducing hybrid-FMEA models that performs the uncertainty quantification and machine learning techniques, MCDM methods, and other complementary failure analysis approaches. Finally, it discusses about smart-FMEA platform in modern industries and its improvements in the context of Industry 4.0.

## Abbreviations

| **Notation** | **Main acronyms** |
|---|---|
| FMEA | Failure mode and effects analysis |
| MCDM | Multiple-criteria decision-making |
| FTA | Fault tree analysis |
| HACCP | Hazard analysis, critical control points |
| RCA | Root cause analysis deployment |
| QSR | Quality system requirements |
| IATF | International automotive task force |
| AIAG | Automotive industry action group |
| QMS | International quality management system |

H. Soltanali (✉)
Department of Biosystems Engineering, Ferdowsi University of Mashhad, 9177948974 Mashhad, Iran
e-mail: ha.soltanali@mail.um.ac.ir

H. Soltanali · S. Ramezani
Department of Industrial Engineering, Faculty of Engineering, Imam Hossein University, 1698715461 Tehran, Iran



H. Garg (ed.), *Advances in Reliability, Failure and Risk Analysis*, Industrial and Applied Mathematics, https://doi.org/10.1007/978-981-19-9909-3_7

| | |
|---|---|
| ETA | Event tree analysis |
| RCM | Reliability centered maintenance |
| BWM | Best-worst method |
| RAMS | Reliability, availability, maintainability, and safety |
| RPN | Risk priority number |
| S | Severity |
| O | Occurrence |
| D | Detectability |
| QFD | Quality function deployment |
| IoT | Internet of Things |
| FM | Failure mode |
| DEA | Data envelopment analysis |
| HAZOP | Hazard and operability analysis |
| QRA | Quantitative risk assessment |
| PSA | Probabilistic safety assessment |
| ANP | Analytic network process |
| BOFM | Brake oil filling machine |

## 7.1 Introduction

Ensuring system safety and reliability is increasingly becoming an essential dilemma in the digital transformation paradigm, also known as Industry 4.0, with the introduction of new technologies and a growth in system complexity [1–3]. Indeed, concerns about reliability and safety are developing across a range of industries that play a significant role in satisfying demand and enhancing productivity and availability at the lowest possible cost and with the fewest possible unexpected failures [4–6]. In order to identify and reduce process bottlenecks, proactive approaches for analyzing the reliability and safety within high-risk sectors are critical. To achieve this, advanced fault diagnosis and prognostic methods are extensively employed for safety management activities, with hardware and software solutions being provided [7–9].

In general, such advanced methods are divided into two categories: knowledge-based and data-driven approaches to risk and reliability analysis and prediction in a variety of settings [10–12]. Fault Tree Analysis (FTA), Hazard Analysis, Critical Control Points (HACCP), Root Cause Analysis (RCA), and other knowledge-based methodologies can be used for reliability and risk analysis [13–16]. The Failure Mode and Effects Analysis (FMEA) approach is one of them, and it is extensively used in a variety of sectors to analyze and prevent the effects of unexpected events/failures [17–19]. FMEA technique was introduced in 1949 by the U.S. Armed Forces (Military Procedures document MIL-P-1629) to analyze the failures according to their impact on mission success and equipment safety. From the Apollo space program in the 1960s through the semiconductor industry, foodservice, software, and the automobile sector, the application of the FMEA has risen dramatically since then (1980s) [20].

FMEA is one of the most fundamental methods for evaluating the level of risk as a prelude to risk reduction, according to the Quality System Requirements (QSR)-9000.[1] This approach tries to avoid defects rather than discover them, and companies should complete FMEA assessment and approval prior to production stages. For the IATF[2] 16949:2016 standard, industrial businesses must record methods for managing product safety-related products and manufacturing processes, including FMEA. Given the importance of effective product testing and manufacturing process controls in product development, FMEA is also used to enhance test plans and process controls [20]. Furthermore, FMEA has been a well-established process for improving production quality and minimizing the severity and occurrence of failure through the use of corrective actions [21]. In theory, FMEA is a bottom-up risk analysis technique dominated by expert knowledge, with the following steps: identifying failure modes, evaluating their causes and consequences, assessing the risk of failure modes, and lastly prescribing maintenance tasks for high-risk failures [22]. A Risk Priority Number (RPN) is widely used in an FMEA to assess a process's risk level and rank failures and prioritize maintenance activities [23, 24]. The RPN value is computed by multiplying three parameters, namely Occurrence (O), Severity (S), and Detection (D). On a discrete ordinal scale, they are rated from 1 to 10. Finally, the most significant failures may be found by sorting the RPNs in ascending order [25]. As a proactive diagnosis approach, FMEA plays a critical role in identifying system bottlenecks and mitigating the adverse consequences within high-risk industries. With the growth of digitalization and automation, the major aspects of FMEAs, particularly for safety–critical systems, have received less attention in previous research. Hence, the following are the current chapter's main objectives:

- Defining the primary concept and types of FMEAs
- Investigating the FMEAs in safety–critical systems
- Introducing hybrid-FMEA models to overcome current uncertain issues using machine learning techniques, Multi-criteria decision-making (MCDM) methods, etc.
- Proposing a smart-FMEA platform for the needs of Industry 4.0 digital transformation.

---

[1] International quality management system (QMS) standard for the automotive industry originally developed by the American auto industry (Daimler Chrysler Corporation, Ford Motor Company, and General Motors Corporation).

[2] IATF 16949 is a global Quality Management System Standard for the Automotive industry. It was developed by the International Automotive Task Force (IATF) with support from the Automotive Industry Action Group (AIAG).

### 7.1.1 Types of FMEA

FMEA can also be used to establish numerous options (e.g., system, design, process, and service), provide opportunities for fundamental diversity, improve the company's image and competitiveness, and increase customer satisfaction [26]. According to the most basic and widely used handbooks, FMEA is divided into three categories: system-FMEA, design-FMEA, and process-FMEA [27–29]. As indicated in Fig. 7.1, several sorts of FMEAs are utilized to aid in the product development process [26, 30–32]. To analyze a collection of subsystems, system-FMEAs are used. They are used to identify system flaws such as integration, interactions, and interfaces between subsystems; interactions with the immediate or adjacent environment; interactions with workers; and system safety considerations. System functions are in charge of them. A system is a collection of parts or subsystems that work together to perform one or more functions.

Besides, design-FMEA, which is typically managed by product/design engineers, aims to identify and demonstrate engineering solutions that are compliant with system-FMEA requirements and customer specifications. It is used to improve the design of a product in order to ensure its reliability. Another goal of a design-FMEA is to find potential product design failures that could result in product malfunctions, shortened product life, or safety hazards while using the product. Design-FMEAs should be used throughout the design process, from the initial concept to the final product. Furthermore, process-FMEA is concerned with manufacturing processes. The goal is to define how manufacturing and assembly processes can be developed to ensure that products or technologies are built to design specifications while also maximizing the quality, reliability, productivity, and efficiency of the various processes

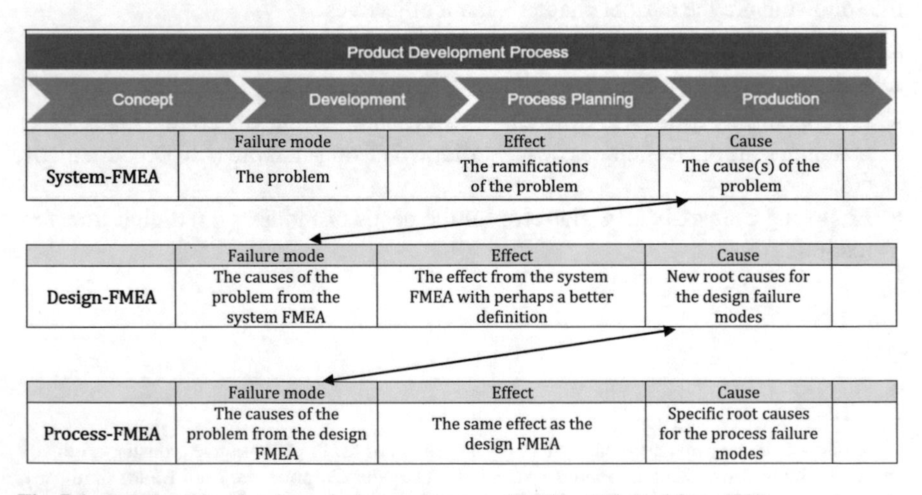


**Fig. 7.1** Relationship of system, design, and process FMEAs (Adapted from [20])

[20]. Process-FMEAs reveal potential failures that may have an impact on product quality, reduce process reliability, cause customer dissatisfaction, pose a safety or environmental risk, and so on. Process-FMEAs should ideally be performed prior to the start-up of a new process, but they can also be performed on existing processes.

## 7.2 FMEA Methodology

### *7.2.1 Classical-FMEA*

The FMEA methodology is based on presenting data in a systematic configuration. The results of the analysis are represented in Fig. 7.2. Three main steps should be considered when implementing the FMEA methodology, which is based on some well-known industrial handbooks [26–28]: (1) functions, potential failures, and effects analysis; (2) cause and detection analysis; and (3) improvement actions.

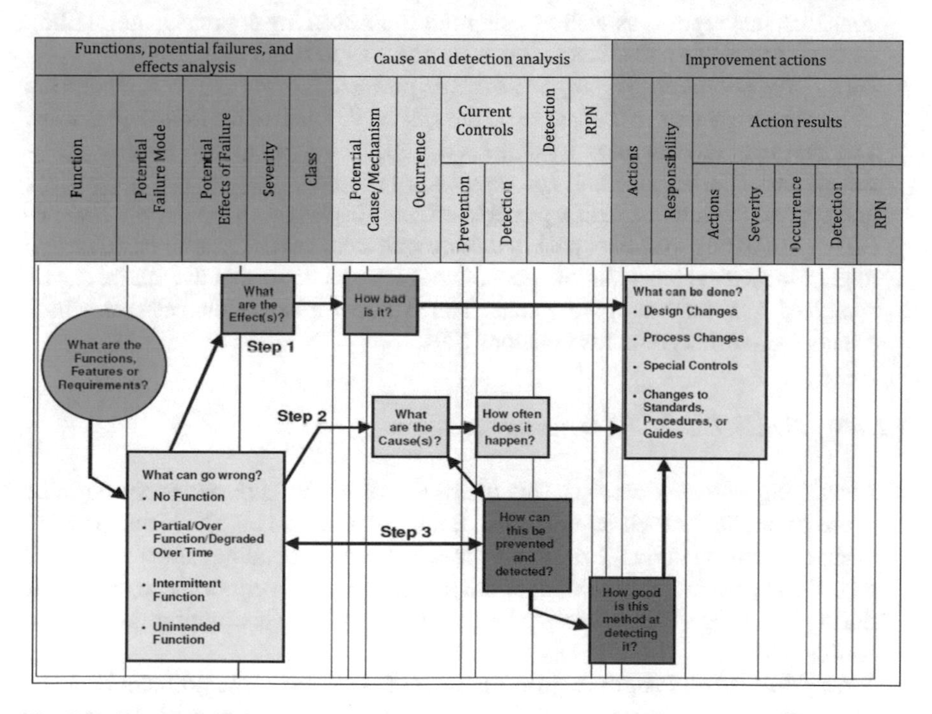


Fig. 7.2 Generic FMEA worksheet [20]

#### 7.2.1.1 Step 1: Functions, Potential Failures, and Effects Analysis

- ***Identifying functions***: The purpose of this activity is to identify, clarify, and understand the functions, requirements, and specifications that are relevant to the specified scope. A functional block diagram for the system- and design-FMEA, as well as a process flowchart, are advised in this case. The task that the system, design, or process must fulfill is referred to as a function. When describing a function, an active verb is generally employed [20].
- ***Identifying potential failure modes***: The goal of this step is to create a list of every possible failure mode connected with the specific function. Failure is assumed to be a possibility but not a requirement. There are four possible failure models: (1) no function (system is completely nonfunctional); (2) partial/over function/degraded over time (degraded performance); (3) intermittent function (complies but loses some functionality or becomes inoperative frequently due to external factors); and (4) unintended function (system is completely nonfunctional) (interaction of several elements whose independent performance is correct adversely affects the product or process). Conducting a review of previous things that have gone wrong, concerns, and reports, as well as using the brainstorming approach, storytelling method, and cause-and-effect diagram, is one way to begin [20].
- ***Identifying potential effects of failure***: The problem is to list and characterize the effects/consequences of the failure on the system for each of the failure scenarios. The investigation of the severity of the consequences is part of determining potential effects. The outcome and consequence of the failure on the system, design, and process are referred to as a possible effect. This is what happens when something goes wrong. Failure's potential consequences must be examined from two angles: local and global ramifications. Local effects denote that the failure can be separated from the rest of the system. The failure can have global effects, which means it can influence other functions [20].

#### 7.2.1.2 Step 2: Cause and Detection Analysis

- ***Identifying potential causes***: This phase's goal is to figure out every possible cause of failure for each failure mode. Each failure mode may have one or more causes, and by definition, if a cause happens, the corresponding failure mode will as well. The occurrence ranking, or the likelihood that a certain cause will occur during the design life, is one factor to consider when determining prospective causes [20].
- ***Identifying current controls (prevention and detection)***: The problem is determining the design or process controls for each cause. The operations that prevent or detect the cause of probable failures are referred to as design or process controls. Controls for prevention define how a cause, failure mode, or effect is avoided based on present or planned actions. The goal is to lessen the likelihood of the

problem occurring. Detection controls define how a failure mode or cause is identified before the product design is put into production. The goal is to maximize the possibility of detecting an issue before it reaches the end-user [20].

#### 7.2.1.3 Step 3: Improvement Actions

The goal of improvement actions is to provide engineering evaluations that will lower overall risk and the possibility of a failure mode occurring. This can be accomplished by estimating the Risk Priority Number (RPN) values based on three parameters: severity (S), occurrence (O), and detectability (D). These factors are combined to calculate the RPN, as in the following expression [33]:

$$\mathrm{RPN} = \mathrm{D} * \mathrm{O} * \mathrm{S} * \tag{7.1}$$

where,

- The possibility of a failure mode occurring is known as occurrence, and it is closely tied to the equipment's failure rate. It can take integer values in the range [1; 10], with 10 being the most likely failure mode. The details of these scenarios are provided in Table 7.1 [34].
- Severity of a failure's influence on the system is measured in terms of its impact. It can take integer values in the range [1; 10], with 10 representing the worst-case scenario. The details of these scenarios are given in Table 7.2 [34].
- The possibility of identifying the failure mode before its effects show in the system is indicated by detection. It can take integer values in the range [1; 10], with 10 being the least diagnosable event. The details of these scenarios are illustrated in Table 7.3 [34].

**Table 7.1** Traditional ratings for occurrence of a failure mode

| Rating | Probability of failure | Possible failure rate |
|---|---|---|
| 10 | Extremely high: failure almost inevitable | ≥1 in 2 |
| 9 | Extremely high: failure almost inevitable | 1 in 3 |
| 8 | Repeated failures | 1 in 8 |
| 7 | High | 1 in 20 |
| 6 | Moderately high | 1 in 80 |
| 5 | Moderate | 1 in 400 |
| 4 | Relatively low | 1 in 2000 |
| 3 | Low | 1 in 15,000 |
| 2 | Remote | 1 in 15,000 |
| 1 | Nearly impossible | ≤1 in 1,500,000 |

**Table 7.2** Traditional ratings for severity of a failure mode

| Rating | Effect | Severity of effect |
|---|---|---|
| 10 | Hazardous without warning | Highest severity ranking of a failure mode, occurring without warning, and consequence is hazardous |
| 9 | Hazardous with warning | Higher severity ranking of a failure mode, occurring with warning, and consequence is hazardous |
| 8 | Very high | Operation of system or product is broken down without compromising safe |
| 7 | High | Operation of system or product may be continued, but performance of system or product is affected |
| 6 | Moderate | Operation of system or product is continued, and performance of system or product is degraded |
| 5 | Low | Performance of system or product is affected seriously, and the maintenance is needed |
| 4 | Very low | Performance of system or product is less affected, and the maintenance may not be needed |
| 3 | Minor | System performance and satisfaction with minor effect |
| 2 | Very minor | System performance and satisfaction with slight effect |
| 1 | None | No effect |

**Table 7.3** Traditional ratings for detection of a failure mode

| Rating | Detection | Criteria |
|---|---|---|
| 10 | Absolutely impossible | Design control does not detect a potential cause of failure or subsequent failure mode, or there is no design control |
| 9 | Very remote | Very remote chance the design control will detect a potential cause of failure or subsequent failure mode |
| 8 | Remote | Remote chance the design control will detect a potential cause of failure or subsequent failure mode |
| 7 | Very low | Very low chance the design control will detect a potential cause of failure or subsequent failure mode |
| 6 | Low | Low chance the design control will detect a potential cause of failure or subsequent failure mode |
| 5 | Moderate | Moderate chance the design control will detect a potential cause of failure or subsequent failure mode |
| 4 | Moderately high | Moderately high chance the design control will detect a potential cause of failure or subsequent failure mode |
| 3 | High | High chance the design control will detect a potential cause of failure or subsequent failure mode |
| 2 | Very high | Very high chance the design control will detect a potential cause of failure or subsequent failure mode |
| 1 | Almost certain | Design control will almost certainly detect a potential cause of failure or subsequent failure mode |

Following that, it is necessary to address corrective activities to decrease or eliminate probable failure modes, as well as detective actions to aid in the identification of a weakness, based on the greatest value of RPN for each failure mode. The first two steps of an FMEA process (prospective failures and effects analysis (identification of potential failures and effects) and cause and detection analysis (identification of potential causes and controls) are critical [20].

### 7.2.2 *Hybrid-FMEA Model*

Despite the widespread use of classical-FMEAs (Sect. 7.2.1) in numerous fields, they are still subject to a variety of uncertainties and variabilities in real-world situations, limiting their ability to be used in a reliable and accurate manner, particularly in risk (safety) and assessment applications. According to the literature [19, 20, 34–36], the following are the key shortcomings and limitations of classical-FMEAs:

(1) The assumption that three failure variables contribute equally to an event's risk factor (RPN). In practice, this is unlikely to be the case, at least in the majority of cases. Because the Severity (S) failure factor is often more critical than other failure factors, practitioners will often examine the occurrence (O) and severity (S) columns of the FMEA separately from the overall RPN. Furthermore, the study does not take into consideration the participants' experience and competence; they are all presumed to have the same level of experience and skills.
(2) The RPN values produced by different combinations of O, S, and D rankings may be identical. This could lead to a false conclusion, claiming that these hazards have the same priority when, in fact, they may have very different priorities. If two events have O, S, and D values of 5, 1, and 10 and 5, 10, and 1, respectively, they will both have an RPN of 50. This suggests that, despite their differences, both hazards require the same level of attention to be mitigated. This may result in inefficient use of limited resources and/or the omission of a high-risk failure mode.
(3) On a discrete ordinal scale, the three risk variables O, S, and D are rated. On the ordinal scale, however, the multiplication is meaningless. As a result, the resulting results are not only meaningless, but also misleading.
(4) It is controversial whether the RPN is a product of O, S, and D. Some scholars dispute why the RPN is calculated by multiplying the numerical numbers of the failure factors.
(5) The rating transitions for the three failure mode components are distinct. The probability table for O and O has a nonlinear relationship, whereas the probability table for D(S) and D(S) has a linear relationship.
(6) It can be difficult to precisely determine the three risk factors. In the absence of data for a comprehensive quantitative analysis, or when the number of failure modes is such that a quantitative analysis is impossible, the procedure relies on

the subjective judgment of the team members. There is no systematic technique to deal such subjectivity within the analysis at the moment.

(7) In the absence of quantitative data, the existing measure of utilizing numerical rankings to grade failure O, S, and D might be erroneous and difficult to award. Natural language usage may be desirable for practitioners and operatives, particularly in poor nations where field operating employees are unlikely to be numerate and would struggle to connect an arbitrary number to the state of a piece of equipment's probable failure O, S, or D.

Such major fluctuations in the real world may have an impact not only on the accuracy of predicted risk and reliability values, but also on the suggested maintenance and safety functions. These are the primary reasons why hybrid-FMEA has attracted the most attention from scientists in recent years. To put it another way, a modified FMEA approach that overcomes some of the limits is required to adapt, regulate, and reduce the existing uncertainty and variability issues of the process and ensure that the classical-FMEA remains appropriate for future applications. According to Fig. 7.3, the hybrid-FMEAs have been applied in four different ways to supplement the classical models in risk (safety), reliability, and maintenance decisions:

(A) ***Combination with failure/event analysis approaches***: Using classical-FMEA in conjunction with other related failure and event analysis techniques (i.e., Root Cause Analysis (RCA), Fault Tree Analysis (FTA), Event Tree Analysis (ETA), Quantitative Risk Assessment (QRA), Probabilistic Safety Assessment (PSA), Probabilistic Risk Analysis (PRA), Brainstorming, Hazard Analysis Critical Control Point (HACCP), Hazard and Operability Analysis (HAZOP), Reliability Centered Maintenance (RCM), etc.) would help present the connections and relationships between various failures more effectively.

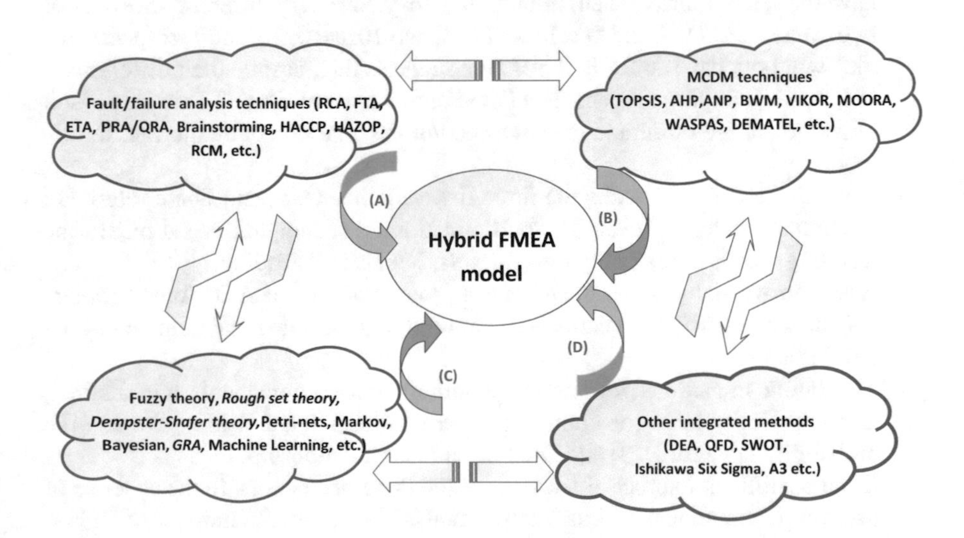


**Fig. 7.3** Type of hybrid-FMEA models under various uncertainty and variability issues

(B) ***Multi-criteria decision-making (MCDM) techniques***: The use of MCDM techniques such as TOPSIS, AHP, ANP, BWM, VIKOR, MOORA, WASPAS, and DEMATEL in integrated FMEA models is extremely advantageous for overcoming the uncertain concerns connected to weighting problems of three elements (S, O, D) for risk and reliability analysis, as well as their discrete ordinal scale issue, which results in meaningless and misleading results.

(C) ***Artificial/computational intelligence techniques***: The use of uncertainty quantification models (fuzzy theory, rough set theory, Shafer theory, Petri-nets, Markov, Bayesian) and other machine learning models, among others, is another option for mitigating the uncertainties of classical-FMEA, particularly determining accurately the risk parameters due to different types of assessment information from the same risk factor, time constraints, inexperience, and insufficient data.

(D) ***Other integrated models***: To support classical-FMEAs in production and service areas and improve their efficiency while estimating RPN that only consider safety and ignore other important factors such as quality and cost, their combination with other systematic approaches such as DEA, QFD, SWOT, Ishikawa Six Sigma, A3, and so on is recommended.

## 7.3 FMEA for Safety–Critical Systems

### *7.3.1 Basic Concept and Definition*

A safety–critical system, often known as a life-critical system, is one whose failure or malfunction might result in one (or more) deaths or major injuries to people, loss or severe damage to equipment/property, economic loss, or environmental harm [33, 37]. Some failures may have immediate negative repercussions, while others may increase the risk of damage. The potential consequences of a system's failure determine whether it is considered safety–critical. The system is said to be safety–critical if a malfunction can result in consequences that are deemed unacceptable [37, 38]. A safety-related system consists of hardware, software, and human components that work together to perform one or more safety functions, and the failure of which would result in a significant increase in the risk of harm to people or the environment. However, safety-related systems are those that do not have complete control over risks such as loss of life, serious injury, or severe environmental damage. A malfunction of a safety-related system would be dangerous only when combined with the failure of other systems or human error [38, 39].

## 7.3.2 Functional Safety Standards

### 7.3.2.1 The Generic IEC 61508 Standards

Significant material and financial assets are lost, people are wounded and killed, and the environment is poisoned as a result of failures of safety–critical systems and a lack of functional safety. Functional safety is often defined as a situation in which the risk has been decreased to, and is maintained at, a level as low as reasonably practical, and the residual risk is widely accepted. The phrase "functional safety" appears in the title of the major standard IEC 61508, and it is therefore used to refer to the part of total system safety that is dependent on the proper operation of active control and safety systems [37, 40]. IEC 61508 standards aim to guarantee that safety–critical systems are specified, designed, produced, installed, and operated in such a way that they fulfill their intended safety duties reliably. The purpose of these standards is to provide broad criteria and to act as a foundation for the creation of specific standards. The IEC 61508 standard is divided into five major stages [37, 39]:

1. ***Risk assessment***: The result is the formulation of the needed safety functions as well as the related reliability objectives.
2. ***Design and construction***: The end result is a safety–critical system made up of hardware and software components.
3. ***Planning for integration***: The main tasks include validation, operation, and maintenance.
4. ***Operation and maintenance***: When a modification is proposed, any change to the safety–critical systems should prompt a return to the most suitable life cycle phase.
5. ***Disposal***: It represents the end-of-life status of safety–critical systems.

### 7.3.2.2 Specific Standards

Specific standards have been established and tested using the IEC 61508 standard in a variety of industries, including process industry, mechanical systems, nuclear power plants, railway applications, and the automobile industry, among others [37, 39, 41]:

- ***Process industry***: Safety–critical systems in the process industry, including the oil and gas industry, are covered by IEC 61511, which is based on IEC 61508. When a safety–critical system is based on proven technology or technology whose design has been confirmed against the standards of IEC 61508, IEC 61511 is used. IEC 61511 does not cover the development of new technologies. As a result, IEC 61511 is often referred to as the end-user and system integrator standard, whereas IEC 61508 is referred to as the manufacturer's standard. To make the implementation of IEC 61508 and IEC 61511 easier, guidelines have been prepared. The following are two important guidelines [37]:

  - Guidelines for Safe and Reliable Instrumented Protective Systems published by the Center for Chemical Process Safety
  - Application of IEC 61508 and IEC 61511 in the Norwegian Petroleum Industry published by the Norwegian Oil and Gas Association.

- ***Machinery systems***: The EU Machinery Directive (EU-2006/42/EC, 2006) concerns machinery safety in Europe, with the first version being passed in 1989. The EU Machinery Directive specifies the basic health and safety criteria for the design and operation of machinery, allowing the particular aspects to be determined by harmonized standards. More information on various standards for Machinery systems could be found in Rausand's book [37].
- ***Nuclear industry***: Based on IEC 61508, the standard [42] was developed as a sector-specific standard for the nuclear power industry. An instrumentation and control (I&C) system is described in IEC 61513 as a "system, based on electrical and/or electronic and/or programmable electronic technology, performing I&C functions as well as servicing and monitoring activities connected to the system's operation."
- ***Automotive industry***: Under IEC 61508, ISO 26262 [43] was designed for the safety of road vehicle applications. It was also developed for electric and/or electronic systems in vehicles with a gross vehicle mass of up to 3500 kg. The standard consists of nine normative elements and a use guideline for ISO 26262.
- ***Railway transport***: Three European standards, EN 50126, EN 50128, and EN 50129, have been produced for railway transport with a scope equivalent to IEC 61508. Later, the three EN-norms were included in IEC-standards [37]:

  - IEC 62278 (EN 50126): Railway applications—The specification and demonstration of Reliability, Availability, Maintainability, and Safety (RAMS).
  - IEC 62279 (EN 50128): Railway applications—Communications, signaling, and processing systems—Software for railway control and protection systems.
  - IEC 62425 (EN 50129): Railway applications-Communication, signaling, and processing systems—safety-related electronic systems for signaling.

### 7.3.3 Safety Barrier and Life Cycle

Most risk studies use the phrase "safety barrier," which somewhat overlaps with our description of a safety–critical system. A safety barrier system can be either a technological technology or a concerted human and organizational effort. As a result, a safety barrier is not the same as a safety–critical system. A safety barrier, such as an emergency procedure, is not a safety–critical system. Safety barriers are frequently referred to as layers of protection or protective layers in the process industry, as shown in Fig. 7.4 [37]:

(a) ***Process design***: Applying design concepts that are fundamentally safe.
(b) ***Control***: Keeping the system in a normal (stable) condition by employing fundamental control functions, alerts, and operator reactions.

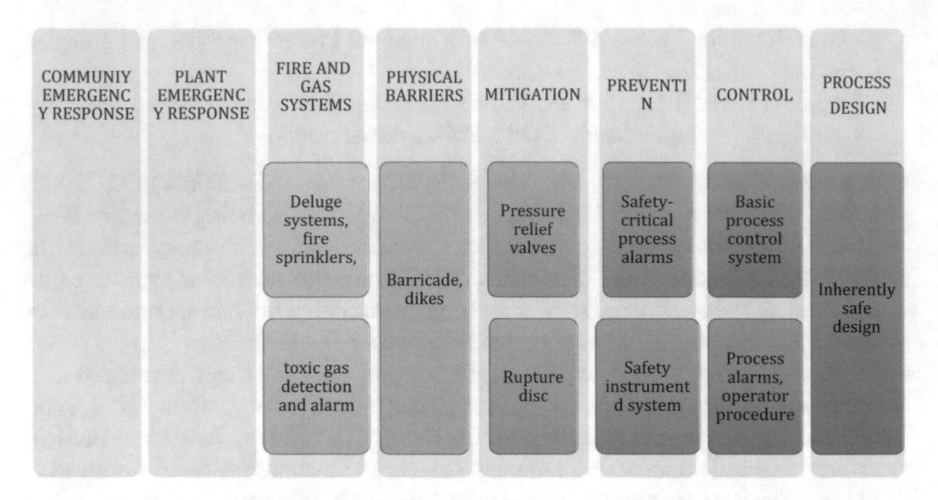


**Fig. 7.4** Protection layers (safety barrier) for process plants (Adapted from [37])

(c) ***Prevention***: Using safety-instrumented systems and safety–critical alarms to react to departures from the usual condition and thereby avoid an unwanted incident.
(d) ***Mitigation***: The use of safety-instrumented systems or functions provided by other technologies to lessen the effects of the undesirable event.
(e) ***Physical protection***: To improve mitigation, use permanent safety obstacles. Examples include the protection provided by dikes and barriers.
(f) ***Fire and gas detection and distinguishing***: As a third technique for mitigating the consequences of explosive gases and mixtures by preventing ignition and hence an accident.
(g) ***Emergency response***: Using a variety of methods to lessen the impact of the disaster, both locally and throughout the community.

In IEC 61508, a safety life cycle model was created, and by the time, various specialized standards had modified versions of this safety life cycle. There are six major phases in the safety life cycle model: (a) Preparation, (b) Analysis, (c) Planning and development, (d) Installation, (e) Operation and maintenance, and (f) Decommissioning [37]. This chapter focuses on the analysis phase of the safety life cycle model. IEC 61508 and IEC 61511 include information on further phases. The IEC 61508 describes a risk-based strategy to meeting the following objectives during the analysis phase [40]:

1. To recognize the undesirable occurrences that may affect the control systems
2. To identify the reasons and event sequences that can result in each undesirable occurrence
3. To determine the chain of events and the risk associated with each undesirable outcome
4. To define the requirements for risk reduction

5. To identify the safety functions required to accomplish the requisite risk reduction
6. To decide which of the safety functions should be used as a safety-instrumented function.

In addition to the previously described safety standards for risk analysis, numerous approaches for identifying and controlling hazards and undesirable occurrences can be applied, such as [40, 44, 45] hazard identification (HAZID), preliminary hazard analysis (PHA), hazard and operability analysis (HAZOP), structured what-if technique (SWIFT), failure modes, effects, analysis (FMEA), fault hazard assessment (FHA), fault tree analysis (FTA), and process hazard analysis (PHA). This chapter has focused on the FMEA approach for safety–critical systems, with the theory addressed in Sect. 7.3 and its application for safety–critical systems depicted in the following section for Automotive safety–critical systems (Sect. 7.3.4).

### *7.3.4 FMEA Implementation: Automotive Safety–Critical Systems*

FMEA procedures are used by many products and industries for their safety–critical systems. The systems discussed in this chapter are technical systems that may or may not require human operator intervention. The concepts and methods in this chapter can be used to examine the following safety–critical systems:

- ***Automotive industry***: Airbag systems, brakes, steering, electronic stability program systems.
- ***Process industry***: Emergency systems, fire and gas systems, gas burner management systems.
- ***Machinery systems***: Guard interlocking systems, emergency stop systems.
- ***Railway transport***: Signaling systems, automatic train stop systems.
- ***Nuclear power industry***: Turbine control systems, fire prevention systems.
- ***Medical devices***: Heart pacemakers, insulin pumps, electronic equipment used in surgery.

In order to implement the FMEA methodology, this chapter focuses on Automotive safety–critical systems, specifically Brake Oil Filling Machines (BOFMs) within assembly lines. The FMEA model's basic information for BOFMs was acquired from Soltanali et al. [15]. In fact, BOFMs are one of the safety–critical systems with semi-automatic capabilities. BOFMs ought to be reliable and safe from both operational and non-operational perspectives. First, because of the importance of speed rates in various operations, low reliability leads to an increase in operational costs, equipment breakdown, and, ultimately, assembly line downtime. According to the records, these systems are responsible for more than 43 percent of assembly line failures, which have been trending upward in recent years. Second, effective inspection and maintenance programs can improve the safety of operators and vehicle drivers by reducing the risk of unexpected events [12, 25]. A BOFM performs leakage tests by producing pressure and vacuum, as well as filling/charging and leveling various fluids in vehicle paths and pipes.

The process description (a) and outer and inner views (b) of a BOFM are depicted in Fig. 7.5. As shown in Fig. 7.5a, the system is comprised of six critical blocks: initialization, ready, pressure and vacuum, filling, process end, and lubrication [46]. The pressure supplement is handled by the initialization block; if the filling system tank is under pressure, the process will equalize/release the pressure. After that, the system is ready to begin the filling process (Ready block). The pressure block is used to inject air into the system and then check the pressure to make sure there are no leaks in the filing system. The vacuum block then performs the system's evacuation and checks for any vacuum leaks in order to maintain a proper vacuum level in the filling system. The filling block performs the fillings with various liquids and their leveling after setting the vacuum and pressure. Lubrication is performed during the filling process, which is provided through a lubrication tank, for continued operation of the rotary equipment, particularly pumps. Finally, the operator can unclamp the filling head and remove it from the vehicle (Process end block).

The results of FMEA for BOFM in automobile assembly line are displayed in Table 7.4 based on the worksheet in Fig. 7.2 and Tables 7.1, 7.2, and 7.3 and formula (7.1). According to the Geometric mean of four experts' judgments (one mechanical engineer, one electrical engineer, one process engineer, and one safety engineer), the total values of risk parameters (S, O, D) and RPN for the entire BOFM are 8.51, 6.27, 4.89, and 256.84, respectively, confirming that the Severity (S) parameter has the most effects on safety analysis in BOFM. As seen, the failure mode ($Fm_1$) of "Bearing failure affected by corrosive cause" related to filling pump, the failure

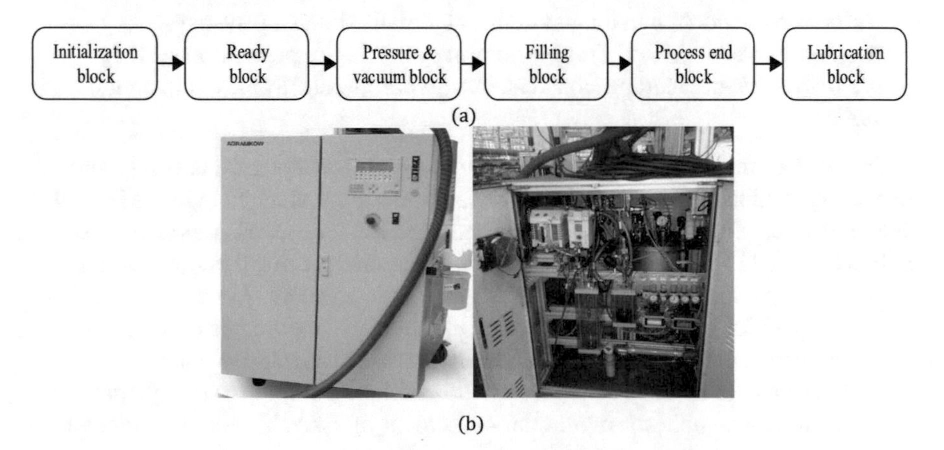


**Fig. 7.5** Process description (**a**) and outer and inner views (**b**) of BOFM in an Iranian automotive production line [15]

mode ($Fm_{12}$) of "Spring fails of pressure control valve" and "Failure and abrasion of activator" ($Fm_{13}$) related to breaker pressure set in Hydraulic-pneumatic, the failure mode ($Fm_{20}$) of "Leakage of head pipes" ($Fm_{17}$) related to head pipes and the failure mode of "Failure and leakage of Couplings" related to couplings in Filling headset, and the failure mode ($Fm_{21}$) of "Sensor's failure affected by more function and circuit confusion" related to sensors in Electronic circuit with more than (9 > ) rate have the highest Severity (S) values. It indicates that the most severe operational and non-operational consequences are associated with that of other components in BOFM. According to the RPN column in Table 7.4, the failure mode ($Fm_{19}$) "Failure of O-rings & seals affected by more function" associated to seals in filling head set with RPN = 632.49 has the highest risk potential of all BOFM breakdowns. Following that, the failure mode ($Fm_6$) of "Rotor fail impacted by more function" and the failure mode ($Fm_9$) of "Blade fail affected by more function" relate to vacuum pump circuit in the Hydraulic-pneumatic circuit with RPNs of 400.89 and 379.47 were assigned the highest RPN in BOFM. Furthermore, the failure modes ($Fm_{15}$) of "Valves failure effected by more function" and ($Fm_{16}$) of "Failure and abrasion of activator" related to valves in Hydraulic-pneumatic circuit, as well as the failure mode ($Fm_{17}$) of "Failure and leakage of Couplings" related to couplings in filling headset, had the highest RPN values. Based on the technical findings, it is possible to deduce that the majority of failures with the greatest FRPN value are associated with the filling head set and the hydraulic-pneumatic circuit. According to the filling headset, the operator's error might be attributed to maintenance staff's deficiencies in servicing and daily checks, as well as a failure to provide enough operator training. As a result, various training courses for maintenance personnel and fluid filling system operators should be considered in order to improve their performance and expand their experiences and skills. Furthermore, enhancing the technical components of the filing headset, such as employing a lighter head, may reduce personal mistakes and the ergonomic aspect would be barred from muscle and joint stresses. To decrease personal flaws, the $G_3$ Blue filling headset has been built with ergonomic advancement and weight reduction of up to 20% in mind. The key activities from the hydraulic-pneumatic circuit, notably for filling and vacuum pumps owing to high operations, are well-timed inspection and service.

## 7.4 Smart-FMEA Applied for Asset Digital Transformation

In the digital transformation era, smart-FMEA concept refers to a platform that is supported by advanced algorithms and technologies such as cloud computing, intelligent techniques (e.g., artificial intelligence, machine learning, neural networks, deep learning, reinforcement learning, etc.), Big data, or Internet of Things (IoT) platforms to support risk and reliability analysis as well as safety and maintenance management decisions. Furthermore, most previous studies have focused more on the current FMEAs' shortcomings and how to overcome them using uncertainty qualification methods (refer to Sect. 7.2.2), with less attention paid to the capability

**Table 7.4** FMEA worksheet for BOFM in an automotive assembly line

| Sub-system | Component | Functional failure | Failure modes (FM) | | Failure effects | S | O | D | RPN |
|---|---|---|---|---|---|---|---|---|---|
| Hydraulic-pneumatic circuit | Filling pump | Fluid filling failed | $Fm_1$ | Bearing failure affected by corrosive cause | Breakdown of filling pump and equipment | **9.74** | 2.45 | **7.20** | 171.79 |
| | | | $Fm_2$ | Electromotor failure affected by circuit faults | | 8.49 | 2.45 | **6.70** | 139.27 |
| | | | $Fm_3$ | Goring the wears | | 7.48 | 2.71 | 6.16 | 124.96 |
| | | | $Fm_4$ | Seals fail affected by more function | | 6.48 | 4.95 | 5.73 | 183.87 |
| | Vacuum pump | Vacuum supply failed | $Fm_5$ | Filter fail affected by more function | Breakdown of vacuum pump and equipment | 8.49 | 5.73 | 5.48 | 266.43 |
| | | | $Fm_6$ | Rotor fail affected by more function | | 9.00 | 9.02 | 4.95 | 400.89 |
| | | | $Fm_7$ | Fatigue and strain of spring affected by more pressure | | 8.21 | 6.70 | 5.96 | 327.63 |
| | | | $Fm_8$ | Electromotor failure affected by circuit faults | | 8.21 | 6.65 | 6.19 | 338.11 |
| | | | $Fm_9$ | Blade fail affected by more function | | 9.21 | 4.47 | 9.21 | 379.47 |
| | Fluid Pipes | Failure in air and fluid transfer | $Fm_{10}$ | Leakage and corrosion of pipes | Lead to leakage increase and fault in filling process | 5.18 | 2.71 | 3.13 | 43.95 |

(continued)

**Table 7.4** (continued)

| Sub-system | Component | Functional failure | Failure modes (FM) | | Failure effects | S | O | D | RPN |
|---|---|---|---|---|---|---|---|---|---|
| | Breaker pressure set | The actual pressure is not shown | $Fm_{11}$ | Excessive system pressure | Do not display the exact pressure. This issue leads to damage the pipes and valves | 8.74 | 6.96 | 4.95 | 301.20 |
| | | Pressure supply failed | $Fm_{12}$ | Spring fails of pressure control valve | Incorrect adjustment of circuit leads to pressure instability | **9.76** | 8.43 | 2.06 | 169.04 |
| | | | $Fm_{13}$ | Failure and abrasion of activator | Incorrect adjustment of circuit pressure that leads to pressure instability | **10.00** | 10.00 | 2.00 | 200.00 |
| | Valves | Improper close and open | $Fm_{14}$ | Failure and abrasion of spool valve | In addition to displaying the values, it can disrupt the process | 7.48 | 9.49 | 2.91 | 206.80 |
| | | | $Fm_{15}$ | Valve failure effected by more function | | 7.97 | 9.74 | 4.47 | 347.10 |
| | | Improper adjustment | $Fm_{16}$ | Failure and abrasion of activator | In addition to displaying the values, it can disrupt the process | 7.48 | 9.77 | 4.48 | 325.96 |
| Filling head set | Couplings | Fluid filling failure | $Fm_{17}$ | Failure and leakage of Couplings | Leaks in filling head interfere the process of filling and testing of fluid | 9.49 | 6.65 | 5.48 | 345.74 |
| | Mini-valves | | $Fm_{18}$ | Failure or leakage of mini-valves | | 8.49 | 7.71 | 4.23 | 276.59 |
| | Seals | | $Fm_{19}$ | Failure of O-rings and seals affected by more function | | 9.21 | 9.52 | 7.24 | 632.49 |

(continued)

**Table 7.4** (continued)

| Sub-system | Component | Functional failure | Failure modes (FM) | | Failure effects | S | O | D | RPN |
|---|---|---|---|---|---|---|---|---|---|
| | Head pipes | | Fm $_{20}$ | Leakage of head pipes | | 9.75 | 6.48 | 3.94 | 248.45 |
| Electronic circuit | Sensors | Detection of fluid, pressure failed | Fm $_{21}$ | Sensor's failure affected by more function and circuit confusion | Resulting in equipment fault and ultimately leading to disruption of production operations | 10.00 | 7.48 | 4.23 | 316.51 |
| | ABS | Failure in test brake paths | Fm $_{22}$ | Failure of conductor, cables, and main units such as bobbin and cores | There is no electronic connection to open the electric valves and hydraulic valves | 8.49 | 7.75 | 4.47 | 293.61 |
| | Starter | Fluid filling failed | Fm $_{23}$ | Starter failure affected by circuit confusion | There is no possibility of filling through the headset | 8.97 | 6.40 | 5.18 | 297.55 |

The bold values represent the maximum value of each risk parameter

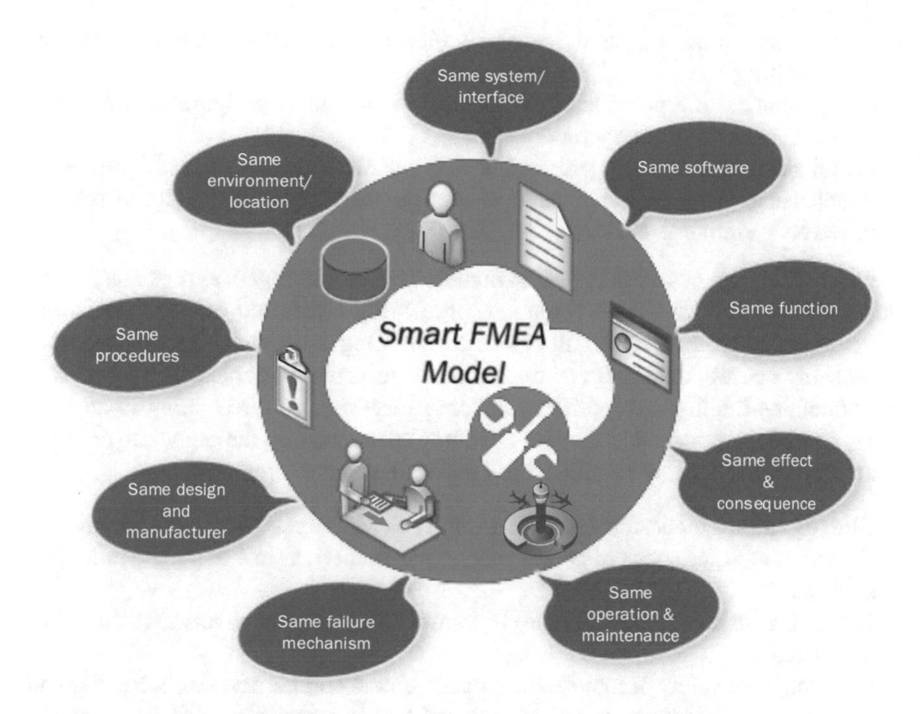


**Fig. 7.6** The concept of smart FMEA model

of FMEA models and how to improve them to meet the needs of modern industries like automation and digitalization. Figure 7.6 depicts the proposed smart-FMEA platform, which includes all potential technical, organizational, environmental, and operational factors to make maintenance and safety decisions more reliable. Some of these factors are covered in NASA's risk assessment manual [47]:

- Component type: e.g., motor-operated valve including any special design or construction
- Characteristics: component size and material, normally running, standby, etc.
- Component use: system isolation, parameter sensing, motive force, etc.
- Component manufacturer
- Component internal conditions: temperature range, normal flow rate, power requirements

Component boundaries and system interfaces: connections with other components, interlocks, etc.

- Component location name and/or location code
- Component external environmental conditions: e.g., temperature, radiation, vibration

- Component initial conditions: normally closed, normally open, energized, etc., and operating
- Component testing procedures and characteristics: test configuration or lineup, effect of test on system operation, etc.
- Component maintenance procedures and characteristics: planned, preventive maintenance frequency, maintenance configuration or lineup, effect of maintenance on system operation, etc.

Figure 7.7a displays a smart-FMEA model based on Intelligent approaches, which consists of three layers: input variables, processing layer, and output (prediction) layer. Input variables address all relevant technological, organizational, environmental, and operational issues, as well as other uncertain variables. The processing layer contains intelligent algorithms for pre-processing and evaluating input variables, as well as transferring them to the output part. Finally, the output layer of the smart-FMEA model may be utilized to achieve the following goals:

- Classify failure modes/components using critical analysis of risk-based models
- Classify the failure rates of comments/parts from high to low using the reliability analysis
- Divide the spare components into high, medium, and low levels that should be scheduled
- Determine the safety or maintenance management techniques, i.e., corrective (re-design, replacement, or repair) or preventative (time based or condition based).

Furthermore, this structure may be updated/upgraded with an IoT platform (Fig. 7.7b), which contains a cloudy layer (smart-FMEA platform, smart application, and smart database), a connectivity layer (platform/mode connectivity), and a physical layer (interconnection of software and hardware items).

## 7.5 Conclusion

Fault diagnosis and prognosis methodologies are critical in assessing system safety and reliability in digitalized environments. This chapter focuses on FMEA approach as a proactive diagnosis tool, as well as its advancements in identifying and mitigating adverse occurrences in high-risk businesses. It discusses several forms of FMEAs, including design-FMEA, process-FMEA, and system-FMEA. Furthermore, the notion of safety–critical systems and the use of FMEAs for risk and hazard analysis within such systems are presented. The existing disadvantages and limits of classical-FMEA theories are also surveyed in this chapter, as well as how they might be overcome by hybrid-FMEA models. Finally, the feasibility of developing smart-FMEA platforms in modern sectors, as well as their enrichment through advanced algorithms and technologies, is discussed in the context of Industry 4.0. It is worth noting that the smart-FMEA platform proposed in this study can be useful for automatically monitoring major risks and mitigating adverse consequences in high-risk

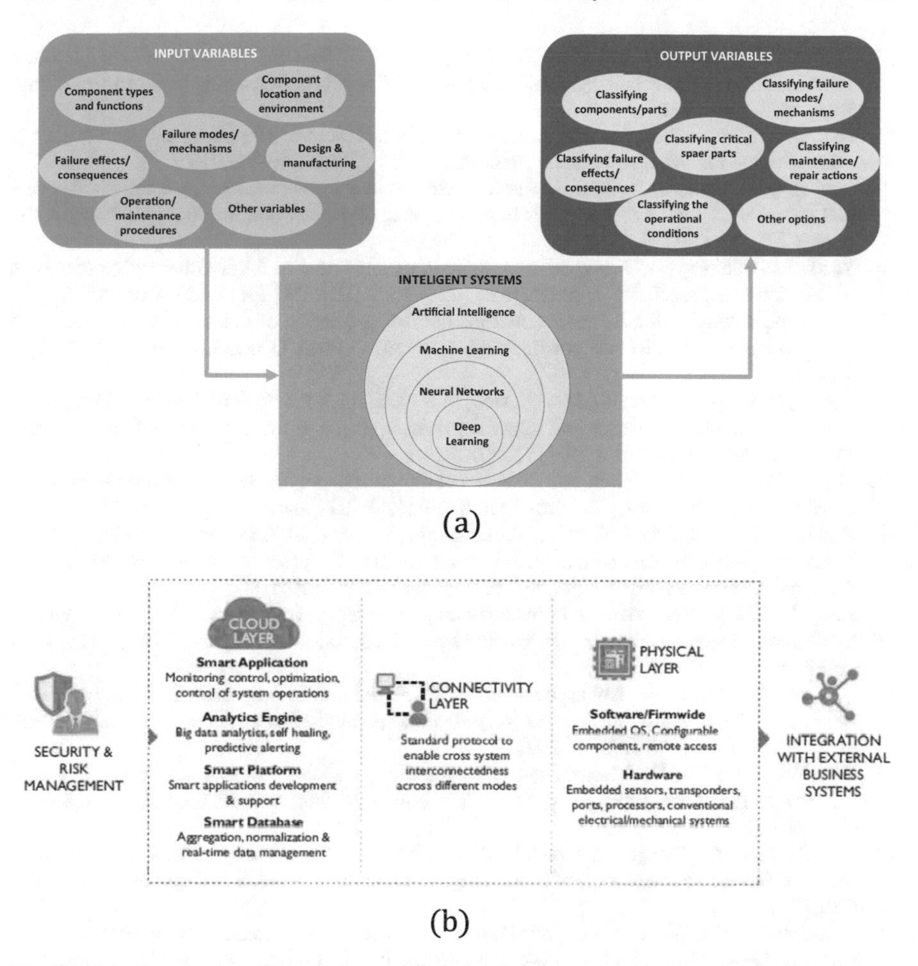


**Fig. 7.7** The proposed smart-FMEA based on **a** intelligent systems and **b** IoT platforms [48]

industries. As the current FMEA models struggle to meet the demands of the growing current digital transformation, more research on our proposed smart-FMEA platform with the capability of covering all potential operational and environmental issues in safety–critical systems is recommended.

## References

1. Farsi, M.A., Zio, E.: Industry 4.0: some challenges and opportunities for reliability engineering. Int. J. Reliab. Risk Safety: Theor. Appl. **2**(1), 23–34 (2019)
2. Lazarova-Molnar, S., Mohamed, N.: Reliability assessment in the context of industry 4.0: data as a game changer. Proc. Comput. Sci. **151**, 691–698 (2019)

3. Tseng, M.L., Tran, T.P.T., Ha, H.M., Bui, T.D., Lim, M.K.: Sustainable industrial and operation engineering trends and challenges toward Industry 4.0: a data driven analysis. J. Ind. Prod. Eng. **38**(8), 581–598 (2021)
4. Di Bona, G., Silvestri, A., Forcina, A., Petrillo, A.: Total efficient risk priority number (TERPN): a new method for risk assessment. J. Risk Res. **21**(11), 1384–1408 (2018)
5. Ilbahar, E., Kahraman, C., Cebi, S.: Risk assessment of renewable energy investments: a modified failure mode and effect analysis based on prospect theory and intuitionistic fuzzy AHP. Energy **239**, 121907 (2022)
6. Yazdi, M., Soltanali, H.: Knowledge acquisition development in failure diagnosis analysis as an interactive approach. Int. J. Interactive Des. Manuf. (IJIDeM) **13**(1), 193–210 (2019)
7. Das, A.K., Leung, C.K.: A fundamental method for prediction of failure of strain hardening cementitious composites without prior information. Cement Concr. Compos. **114**, 103745 (2020)
8. Djeziri, M.A., Benmoussa, S., Mouchaweh, M.S., Lughofer, E.: Fault diagnosis and prognosis based on physical knowledge and reliability data: application to MOS field-effect transistor. Microelectron. Reliab. **110**, 113682 (2020)
9. Vogl, G.W., Weiss, B.A., Helu, M.: A review of diagnostic and prognostic capabilities and best practices for manufacturing. J. Intell. Manuf. **30**(1), 79–95 (2019)
10. Alzghoul, A., Backe, B., Löfstrand, M., Byström, A., Liljedahl, B.: Comparing a knowledge-based and a data-driven method in querying data streams for system fault detection: a hydraulic drive system application. Comput. Ind. **65**(8), 1126–1135 (2014)
11. Jiang, Y., Yin, S.: Recursive total principle component regression-based fault detection and its application to vehicular cyber-physical systems. IEEE Trans. Industr. Inf. **14**(4), 1415–1423 (2017)
12. Soltanali, H., Rohani, A., Abbaspour-Fard, M.H., Farinha, J.T.: A comparative study of statistical and soft computing techniques for reliability prediction of automotive manufacturing. Appl. Soft Comput. **98**, 106738 (2021)
13. Cho, W.I., Lee, S.J.: Fault tree analysis as a quantitative hazard analysis with a novel method for estimating the fault probability of microbial contamination: a model food case study. Food Control **110**, 107019 (2020)
14. Jin, C., Ran, Y., Zhang, G.: Interval-valued q-rung orthopair fuzzy FMEA application to improve risk evaluation process of tool changing manipulator. Appl. Soft Comput. **104**, 107192 (2021)
15. Soltanali, H., Rohani, A., Abbaspour-Fard, M.H., Parida, A., Farinha, J.T.: Development of a risk-based maintenance decision making approach for automotive production line. Int. J. Syst. Assurance Eng. Manage. **11**(1), 236–251 (2020)
16. Zhang, G., Thai, V.V., Yuen, K.F., Loh, H.S., Zhou, Q.: Addressing the epistemic uncertainty in maritime accidents modelling using Bayesian network with interval probabilities. Saf. Sci. **102**, 211–225 (2018)
17. Filz, M.A., Langner, J.E.B., Herrmann, C., Thiede, S.: Data-driven failure mode and effect analysis (FMEA) to enhance maintenance planning. Comput. Ind. **129**, 103451 (2021)
18. Soltanali, H., Khojastehpour, M., Torres Farinha, J.: An improved risk and reliability framework-based maintenance planning for food processing systems. In: Quality Technology & Quantitative Management, pp. 1–23 (2022)
19. Yazdi, M., Daneshvar, S., Setareh, H.: An extension to fuzzy developed failure mode and effects analysis (FDFMEA) application for aircraft landing system. Saf. Sci. **98**, 113–123 (2017)
20. Cabanes, B., Hubac, S., Le Masson, P., Weil, B.: Improving reliability engineering in product development based on design theory: the case of FMEA in the semiconductor industry. Res. Eng. Design **32**(3), 309–329 (2021)
21. Huang, J., Xu, D.H., Liu, H.C., Song, M.S.: A new model for failure mode and effect analysis integrating linguistic Z-numbers and projection method. IEEE Trans. Fuzzy Syst. **29**(3), 530–538 (2019)
22. Dağsuyu, C., Göçmen, E., Narlı, M., Kokangül, A.: Classical and fuzzy FMEA risk analysis in a sterilization unit. Comput. Ind. Eng. **101**, 286–294 (2016)

23. Bartolomé, E., Benítez, P.: Failure mode and effect analysis (FMEA) to improve collaborative project-based learning: case study of a study and research path in mechanical engineering. Int. J. Mech. Eng. Educ. **50**(2), 291–325 (2022)
24. Silva, M.M., de Gusmão, A.P.H., Poleto, T., e Silva, L.C., Costa, A.P.C.S.: A multidimensional approach to information security risk management using FMEA and fuzzy theory. Int. J. Inf. Manage. **34**(6), 733–740 (2014)
25. Soltanali, H., Rohani, A., Tabasizadeh, M., Abbaspour-Fard, M.H., Parida, A.: An improved fuzzy inference system-based risk analysis approach with application to automotive production line. Neural Comput. Appl. **32**(14), 10573–10591 (2020)
26. AIAG: Potential Failure Mode and Effects Analysis (FMEA): Reference Manual, 4th edn. AIAG, Southfield, MI (2008)
27. AIAG and VDA.: Failure Mode and Effects Analysis—FMEA Handbook: Design FMEA, Process FMEA, Supplement FMEA for Monitoring and System Response. AIAG and VDA, Southfield, MI (2019)
28. Ford Motor Company: Failure Mode and Effects Analysis—FMEA Handbook (with Robustness Linkages). Version **4**, 2 (2011)
29. Stamatis, D.H.: Failure Mode and Effect Analysis: FMEA from Theory to Execution. ASQ Quality Press, Milwaukee (2003)
30. Bharathi, S.K., Vinodh, S., Gopi, N.: Development of software support for process FMEA: a case study. Int. J. Services Oper. Manage. **31**(4), 415–432 (2018)
31. Feng, X., Qian, Y., Li, Z., Wang, L., Wu, M.: Functional model-driven FMEA method and its system implementation. In: 2018 12th International Conference on Reliability, Maintainability, and Safety (ICRMS), pp. 345–350, IEEE (2018)
32. Haughey, B.: Product and process risk analysis and the impact on product safety, quality, and reliability. In: 2019 Annual Reliability and Maintainability Symposium (RAMS), pp. 1–5, IEEE (2019)
33. Catelani, M., Ciani, L., Galar, D., Guidi, G., Matucci, S., Patrizi, G.: FMECA assessment for railway safety-critical systems investigating a new risk threshold method. IEEE Access **9**, 86243–86253 (2021)
34. Liu, H.C.: FMEA using uncertainty theories and MCDM methods. In: FMEA using Uncertainty Theories and MCDM Methods, pp. 13–27. Springer, Singapore (2016)
35. Hassan, S., Wang, J., Kontovas, C., Bashir, M.: Modified FMEA hazard identification for cross-country petroleum pipeline using fuzzy rule base and approximate reasoning. J. Loss Prev. Process Ind. **74**, 104616 (2022)
36. Wu, X., Wu, J.: The risk priority number evaluation of FMEA analysis based on random uncertainty and fuzzy uncertainty. In: Complexity (2021)
37. Rausand, M.: Reliability of Safety-Critical Systems: Theory and Applications. Wiley, Hoboken, NJ (2014)
38. Knight, J.C.: Safety critical systems: challenges and directions. In: Proceedings of the 24th International Conference on Software Engineering, pp. 547–550 (2002)
39. IEC 61508: Functional Safety of Electrical/Electronic/Programmable Electronic Safety-Related Systems. Part 1-7. Geneva: International Electrotechnical Commission (2010)
40. Rausand, M.: Risk Assessment; Theory, Methods, and Applications. Wiley, Hoboken, NJ (2011)
41. IEC 61511: Functional Safety—Safety Instrumented Systems for the Process Industry. International Electrotechnical Commission, Geneva (2003)
42. IEC 61513: Nuclear Power Plants—Instrumentation and Control for Systems Important to Safety—General Requirements for Systems. International Electrotechnical Commission, Geneva (2004)
43. ISO 26262: Road Vehicles—Functional Safety. International Standardization Organization, Geneva (2011)
44. Dabous, S.A., Zadeh, T., Ibrahim, F.: A failure mode, effects and criticality analysis-based method for formwork assessment and selection in building construction. Int. J. Build. Pathol. Adaptation, (ahead-of-print) (2022)

45. Zhang, D., Li, Y., Li, Y., Shen, Z.: Service failure risk assessment and service improvement of self-service electric vehicle. Sustainability **14**(7), 3723 (2022)
46. AGRAMKOW Co.: Manual Instructions of Line-Side Brake Fluid Filling Equipment. Augustenborg Landevej 19DK-6400 Sønderborg, Denmark (2014). https://www.agramkow.com
47. Stamatelatos, M., Dezfuli, H., Apostolakis, G., Everline, C., Guarro, S., Mathias, D., Youngblood, R.: Probabilistic Risk Assessment Procedures Guide for NASA Managers and Practitioners (No. HQ-STI-11-213) (2011)
48. Anandavel, S.V.: Analysis of Manufacturing Processes According to FMEA Techniques and Implementation of IoT Systems to Prevent Process Failures (Doctoral dissertation, Politecnico di Torino) (2021)

**Hamzeh Soltanali** is a faculty member at Department of Industrial Engineering, Imam Hossein University (IHU), Iran. He was postdoctoral researcher at Ferdowsi University of Mashhad (FUM), Iran, and received his Ph.D. degree in reliability and maintenance engineering from FUM in early 2020. He was a Ph.D. visiting researcher at Luleå Tekniska Universitet (LTU), Division of Operation and Maintenance, Sweden, during 2017–2018. His research focuses on maintenance engineering, asset management, smart technologies, risk and failure analysis, and reliability (RAMS) engineering. He is passionate about utilizing uncertainty qualification and AI/machine learning techniques to overcome uncertainty and variability issues in such areas.

**Saeed Ramezani** is a faculty member at Department of Industrial Engineering, Imam Hossein University (IHU), Iran. He received his Ph.D. degree in Logistics and Maintenance Engineering from Iran University of Science and Technology (IUST), Iran. He was a Ph.D. visiting researcher at University of Seville, Spain. His research and works focus on asset and maintenance management, fault diagnosis and prognostics, predictive analytics, failure analysis, and reliability and safety assessment.

# Chapter 8
# Optimization of Redundancy Allocation Problem Using Quantum Particle Swarm Optimization Algorithm Under Uncertain Environment

**Rajesh Paramanik, Sanat Kumar Mahato, and Nabaranjan Bhattacharyee**

**Abstract** Reliability optimization of a redundancy allocation problem is an important area of research in the literature. The main purpose of this type of problem is to enhance the reliability of the system. In this paper, our target is to maximize the reliability of a redundancy allocation problem with time-dependent component failure rates considering the control parameters as trapezoidal fuzzy numbers. To compute the maximum system reliability of the series–parallel system softly, a novel quantum particle swarm optimization (QPSO) is used. In this QPSO algorithm, the decision variables are assigned by the position of the particles and the fitness value of each particle is evaluated using the reliability function related to this work. All the coding of QPSO algorithm is done in C++. Finally, a numerical example is solved to clarify the sensitivity of the proposed algorithm with respect to the crisp as well as the fuzzy atmospheres.

**Keywords** Redundancy allocation problem (RAP) · QPSO algorithm · Trapezoidal fuzzy number · Beta and uniform distribution method of crispification · Series–parallel system

R. Paramanik · S. K. Mahato (✉) · N. Bhattacharyee
Department of Mathematics, Sidho-Kanho-Birsha University, Purulia, West Bengal 723104, India
e-mail: sanatkmahato@gmail.com; sanatkr_mahato.math@skbu.ac.in

 
H. Garg (ed.), *Advances in Reliability, Failure and Risk Analysis*, Industrial and Applied Mathematics, https://doi.org/10.1007/978-981-19-9909-3_8

## 8.1 Introduction

The term reliability is very often used in our daily life. From the family of human beings to the industrial sectors or managements are deeply influenced by reliability. Reliability is a chance or probability of a system that happens continuously as the operation time of the system moves. After the Second World War, it becomes a challenge for the society that how to improve the reliability of a system in addition to different restrictions. It is seen that the researchers consider the components of a reliable system as constant in their corresponding researches. But it is not always good to take the reliability components as constants because reliability is time dependent in the existing systems.

Few works in connection with this topic have been done after considering time-varying components reliability of a system in the literature. Despite, so far, our knowledge goes, failures per interval of time of the associated components were not assumed in major part of the related topic in the existing literature. For smooth conduction of a reliable system, failure rates play an important role. It may be taken into account that major works related to this topic have been done by the researchers with good efforts after choosing fixed valued failure rate. Such types of reliability optimization problems are categorized as redundancy allocation problem (RAP). In case of time-varying failure rates along with redundant components, the optimization problem is termed as time-dependent reliability redundancy allocation problem (TDRRAP).

The main objective of redundancy allocation problem is to find the optimal or near-optimal number of redundant components in each subsystem subject to maximize the objective function under some constraints. Several types of reliability optimization problems with differently structured systems are solved to gain the number of redundant units in each subsystem by the renowned researchers [8, 14, 25, 26, 29, 47, 48]. These NP-hard problems are solved by differently designed evolutionary algorithms which are heuristic or meta-heuristic in nature [23, 24, 26, 29, 41, 42]. Heuristic technique by Nakagawa and Nakashima [35], reduced gradient method by Hwang et al. [18], and surrogate constraints algorithm by Nakagawa and Miyazaki [34] were implemented to find optimum value of reliability designed problems subject to different resource constraints. Tillman et al. [48], Sasaki and Shingai [46], Chern and Jan [9], Misra [31], and Park [38] reported differently coded soft computing algorithms to find the finer output of the objective function in their respective studies.

Misra and Sharma [30] and Kim and Yum [21] applied heuristic/meta-heuristic algorithms to obtain the outputs of the reliability allocation problems as well as the redundancy allocation problems. A multi-objective reliability optimization problem in which the objective function is defined based on the structure of the system in series form is solved by Huang [17]. The branch and bound technique worked well for redundancy allocation problems, according to Sun and Li [47]. To solve reliability optimization problems, Mahapatra and Roy (2006) adopted the fuzzy multi-objective mathematical programming technique. Mahapatra and Roy (2009) investigated triangular intuitionistic fuzzy numbers in the context of reliability optimization. Gupta et al. [14] employed the penalty technique in an interval setting to solve the redundancy allocation problem. Bhunia et al. [6], Mahato et al. [27], Ruigang et al. [40] described stochastic reliability optimization in an interval context. Sahoo et al. [43] offered a GA-based solution to reliability optimization problems. A series–parallel system's optimal redundancy is determined using the generalized fuzzy number by Mahapatra and Roy in 2011. For attaining an optimal solution in their proposed reliable systems, Sahoo et al. [28], Mahato et al. [41], Garg et al. 2014, Garg [12], Mahato et al. [26] and Sahoo et al. [45] have examined interval environments. Mahato et al. [29], Sahoo et al. [44] and Tillman et al. [49] used fuzzy atmospheres to solve reliability optimization problems utilizing GA and PSO as soft computing techniques.

By employing particle swarm optimization, Khalili-Damghani et al. [20] solved a multi-objective redundancy allocation problem. Garg and Rani [13] used PSO and intuitionistic fuzzy environments to find industrial system reliability. To solve a reliability problem with respect to series system in a fuzzy atmosphere, Sahoo et al. [42] used a GA. To handle nonlinear optimization problems, Jia et al. [19] created a unique attractor QPSO technique.

Various scholars as well as researchers [3–5, 10, 11, 36, 37] have recently investigated the uses of fuzzy and intuitionistic fuzzy environments to solve various types of reliability systems, such as series, series–parallel, and bridge systems.

In the preceding paras, it was observed that the researchers did not consider the time-dependent reliable components of the reliability optimization problems. Besides, they have taken the failure rate of the reliable components as constant values. Very few works are done on time-dependent reliability optimization problems. Mori and Ellingwood [32] explained the time-dependent system reliability analysis by adaptive importance sampling. An analysis on the optimization system reliability with respect to time value of money was drawn by Hamadani and Khorshidi [15]. Mori et al. [32] and Hamadani et al. [15] considered time-dependent reliability components for analyzing their proposed reliability optimization problems. For analyzing the time-dependent system reliability, Hu et al. [16] and Mourelatos et al. [33] utilized random field discretization technique and total probability theorem, respectively. Some reliability optimization problems are solved by Ardakan et al. [2] and Ahmadivala et al. [1] on considering time-dependent reliable components. Zafar and Wang [50] have implemented transferring learning for optimizing time-dependent reliability problem.

In this paper, we want to find the system reliability considering time-varying failure rates of the reliable components. The reliability of each component has to be diminished exponentially with respect to time. Trapezoidal fuzzy number is brought for making the environment more representative and handling the diverse situation. Also, as the components assume time-varying failure rates, the problem becomes more realistic. In this study, a series–parallel system with m stages is taken subject to some restrictions.

The rest of this study is framed as follows: In Sect. 8.2, the basic assumptions and symbols that are used throughout this work are kept. The definitions of fuzzy and trapezoidal fuzzy numbers with graphical representation are kept in Sect. 8.3. Also, the crispification technique for the considered fuzzy number is included in the same section. The formulation of the problem in crisp and imprecise environments is shown in Sect. 8.4. The solution methodology and the integration handling technique are included in Sects. 8.5 and 8.6, respectively. Section 8.7 describes the proposed soft computing techniques. For the clarification of our proposed environment and algorithm, numerical example is taken and sensitivities are drawn graphically in Sect. 8.8. Outcomes associated with this study are analyzed with the inclusion of Sect. 8.9. Section 8.10 concludes the whole work done in this paper with some future scopes.

## 8.2 Assumptions and Notation

### 8.2.1 Assumptions

In the entire work, we have considered the following assumptions:

- Series–parallel reliability system is taken.
- Redundant components are active and non-repairable.
- Each subsystem is comprised of identical components.
- All the reliability components are time varying.
- The failure rate of each component is linear function of time.
- System reliability is not dependent on the failure of components of the subsystems.
- Reliability components follow exponential distribution which diminishes as time goes.

### 8.2.2 Notation

| Symbols | Meanings |
|---|---|
| $r_i(t)$ | *Reliability of the* $i$-th *component* |

(continued)

(continued)

| Symbols | Meanings |
|---|---|
| $\tilde{P}$ | *Trapezoidal fuzzy number* |
| $\mu_{\tilde{P}}(v)$ | *Membership function w.r.t.* $\tilde{P}$ |
| $z_i$ | *Number of active redundant components in i*-th *subsystem* $(i = 1, 2, \ldots, m)$ |
| $z = (z_1, z_2, \ldots, z_m)$ | *Redundancy vector* |
| $T_s(z, \delta, t)$ | *System reliability function in crisp environment* |
| $\tilde{T}_s(z, \delta, t)$ | *System reliability function in fuzzy environment* |
| $a_j(z_1, z_2, \ldots, z_m)$ | *Crisp constraint's usability function* |
| $\tilde{a}_j(z_1, z_2, \ldots, z_m)$ | *Trapezoidal fuzzy constraint's usability function* |
| $q_j, \tilde{q}_j$ | *Crisp, Trapezoidal fuzzy requirement vectors* |
| $l_i, u_i$ | *Lower and upper bounds of* $z_i$ |
| $M$ | *Particles' size in QPSO* |
| $pbest_i$ | *i*-th *best particle* |
| $m_{best}$ | *Mean best position of j*-th *component at* $\tau$-th *iteration* |

# 8.3 Some Definitions

## *8.3.1 Fuzzy Number*

A fuzzy number $\left(\hat{A}\right)$ is a fuzzy set [51] which is both convex and normal. That is, a fuzzy number is a special case of a fuzzy set. The pictorial representation of a fuzzy number is given in Fig. 8.1.

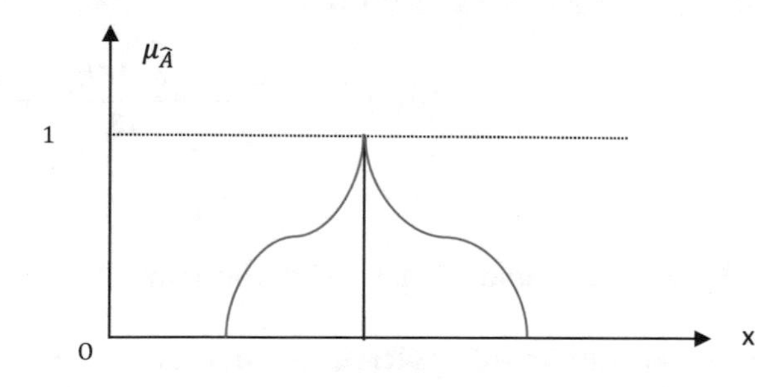


**Fig. 8.1** Membership function of $\hat{A}$

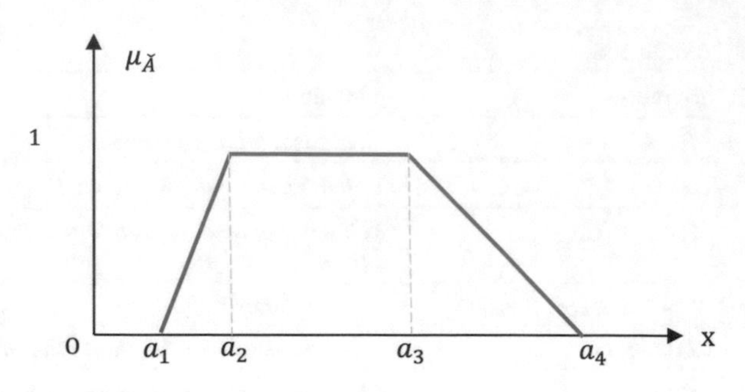


**Fig. 8.2** Trapezoidal fuzzy number

## 8.3.2 Trapezoidal Fuzzy Number (TrFN)

A trapezoidal fuzzy number [26] $\tilde{A}$ is a fuzzy set in R with the membership function $\left(\mu_{\tilde{A}}(x)\right)$ as given by Eq. (8.1).

$$\mu_{\tilde{A}}(x) = \begin{cases} \frac{x-a_1}{a_2-a_1}, \; for a_1 \le x \le a_2 \\ 1, \; for a_2 \le x \le a_3 \\ \frac{a_4-x}{a_4-a_3}, \; for a_3 \le x \le a_4 \\ 0, \; otherwise \end{cases} \tag{8.1}$$

where $a_1 \le a_2 \le a_3 \le a_4 \quad \forall x \in R$

This TrFN is denoted by $\tilde{A} = (a_1, a_2, a_3, a_4)$.

A trapezoidal fuzzy number is shown pictorially in Fig. 8.2.

## 8.3.3 Beta and Uniform Distribution Method of Crispification

According to Rahmani et al. [7, 39], the defuzzification of the fuzzy number $\tilde{A} = (a_1, a_2, a_3, a_4)$ is given by Eq. (8.2).

$$(Bet)_{\tilde{A}} = \frac{(2a_1 + 7a_2 + 7a_3 + 2a_4)}{18} \tag{8.2}$$

### 8.3.3.1 Derivation of Defuzzification Formula

First, we partition the interval $(a_1, a_4)$ into three subintervals $M_1 = (a_1, a_2)$, $M_2 = [a_2, a_3]$, and $M_3 = (a_3, a_4)$ so that $(a_1, a_4) = M_1 \cup M_2 \cup M_3$. After that, we consider two triangular fuzzy numbers $(a_1, a_2, a_2)$ and $(a_2, a_4, a_4)$ denoted by $\hat{B}_1$ and $\hat{B}_2$, respectively, with respect to the intervals $M_1$ and $M_2$. Now, applying beta distribution for the fuzzy numbers $\hat{B}_1$ and $\hat{B}_3$, the formulae of defuzzification become as $(Bet)_{\hat{B}_1} = \frac{(a_1+a_2+a_2)}{3}$ and $(Bet)_{\hat{B}_2} = \frac{(a_3+a_4+a_4)}{3}$. The crisp value of the interval $M_2$ is obtained by using the uniform distribution as $(Uni)_{\hat{B}_3} = \frac{(a_2+a_3)}{2}$. Thus, the combination of beta and uniform distribution yields the crispified value of the trapezoidal fuzzy number $\tilde{A}$ as the mean of $(Bet)_{\hat{B}_1}$, $(Bet)_{\hat{B}_2}$, and $(Uni)_{\hat{B}_3}$.

## 8.4 Formulation of the Problem

A series–parallel system (Fig. 8.3) with m number of subsystems is considered. The position of each subsystem is parallel in the series–parallel system. All the components connected to the system as well as subsystems are time dependent. Besides, the failure rates of each component are considered as a linear function of time. The problem thus formulated represents a reliability redundancy allocation problem. Our main purpose is to find the maximum value of the system reliability under some restrictions. The redundant components of the respective subsystems construct the required solution vector to the reliability redundancy allocation problem.

In crisp environment, the time-dependent RAP for series–parallel system is given as follows:

$$\begin{aligned}
&\text{Maximize } T_S(t) = \prod_{i=1}^{m}\left[1-\left(1-e^{-\int\limits_0^t \delta_i(k)\cdot dk}\right)^{z_i}\right]\\
&\text{subject to}\\
&a_j(z_1, z_2, \ldots, z_m) - q_j \leq 0,\ j = 1, 2, \ldots, n\\
&l_i \leq z_i \leq u_i,\ i = 1, 2, \ldots, m.
\end{aligned} \tag{8.3}$$

where $z = (z_1, z_2, \ldots, z_m)$ represents the vector of redundancy and $z_i (\geq 0)$ is an integer and $t \in (0, 100]$. Here, $r_i(t) = e^{-\int_0^t \delta_i(k).dk}$ is the reliability of the $i$-th

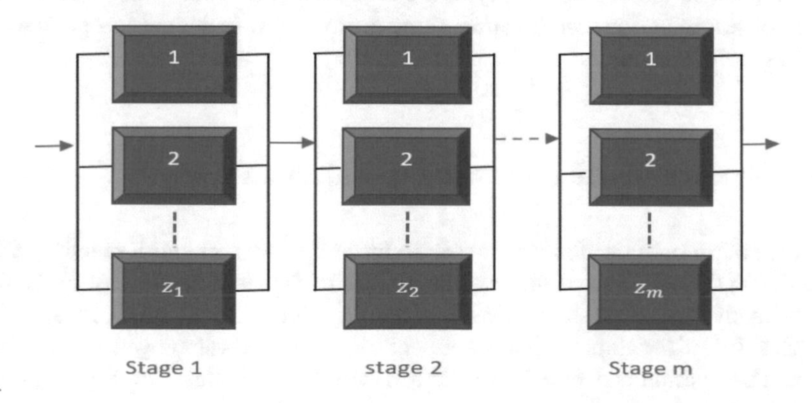


**Fig. 8.3** Series–parallel system with m stages

component of the reliable system and $\delta_i(t)$ is the time-dependent failure rate of the same component.

The corresponding fuzzified model using trapezoidal fuzzy number is given by Eq. (8.4)

$$\begin{aligned}
&\text{Maximize } T_S(t) = \prod_{i=1}^{m}\left[1 - \left(1 - e^{-\int_0^t \delta_i(k).dk}\right)^{z_i}\right] \\
&\text{subject to} \\
&\tilde{a}_j(z_1, z_2, \ldots, z_m) - \tilde{q}_j \le 0, j = 1, 2, \ldots, n \\
&l_i \le z_i \le u_i, i = 1, 2, \ldots, m.
\end{aligned} \tag{8.4}$$

where $z = (z_1, z_2, \ldots, z_m)$ represents the vector of redundancy and $z_i (\ge 0)$ is an integer and $t \in (0, 100]$. Here, $r_i(t) = e^{-\int_0^t \delta_i(k).dk}$ is the reliability of the $i$-th component of the reliable system and $\delta_i(t)$ is the time-dependent failure rate of the same component.

## 8.5 Solution Methodology

The considered reliability optimization problem is an integer nonlinear programing problem with some restrictions. Our main goal is to find the maximum of the system reliability $T_s(z, \delta, t)$ of the series–parallel system in the time interval (0, 100]. Simpson's 3/8-th method is employed to find the value of integral $\int_0^{100} T_S(t)dt$. Clearly, the maximum of $T_S(t)$ will agree with the maximum of the integral. Besides, Big-M penalty technique is implemented as tool for handling the constraints involved in the optimization problem. This method is very useful in obtaining unconstrained optimization problem from a given constrained optimization problem. On the other hand, in eradication of infeasibility of the solutions in addition to reduction of the region of searching for each iteration, there are very few techniques as good as the Big-M penalty technique.

## 8.6 Quantum Particle Swarm Optimization (QPSO)

Particle swarm optimization was first introduced by Eberhart and Kennedy (1995). Visualizing the activities of the swarms, school of fish, and flocking of birds, they got the motivation of the study of PSO. In this algorithm, the swarms are assigned as particles. To find the global optimal, a collection of swarms is employed in a searching space. The position and velocity of the $i$-th particle are updated by Eqs. (8.5) and (8.6), respectively.

$$X_{iD}^{\tau} = X_{iD}^{\tau-1} + V_{iD}^{\tau} \tag{8.5}$$

$$V_{iD}^{\tau} = V_{iD}^{\tau-1} + c_1 r_1 \left(\text{pbest}_{iD} - X_{iD}^{\tau-1}\right) + c_2 r_2 \left(\text{gbest}_{iD} - X_{iD}^{\tau-1}\right) \tag{8.6}$$

where $i = 1, 2, \ldots, M$; $c_1$ and $c_2$ are the learning factors and generally, $c_1 = c_2 = 1$; and $r_1$ and $r_2$ are the random numbers uniformly distributed between 0 and 1.

Later on, an inertial weight ($\omega$) is assigned with velocity in Eq. (8.6) to obtain better optimal value and it is given by Eq. (8.7).

$$V_{iD}^{\tau} = \omega V_{iD}^{\tau-1} + c_1 r_1 \left(\text{pbest}_{iD} - X_{iD}^{\tau-1}\right) + c_2 r_2 \left(\text{gbest}_{gD} - X_{iD}^{\tau-1}\right) \tag{8.7}$$

But this PSO algorithm is not always suitable in finding the global optimal solution of differently structured optimization problems in the existing literature. As a result, a developed version [4, 22] of this algorithm is launched which is called quantum particle swarm optimization (QPSO). The area of motivation of QPSO is quantum mechanics. In this algorithm, wave function $\psi(X, \tau)$ is used to specify the state of the particles. QPSO is able to search the global optimum in an infinite region with high convergency. On applying Monte Carlo stochastic simulation method, the position of the particle is defined by Eq. (8.8).

$$X_{iD} = p_{iD} \pm \frac{L}{2} \log\left({}^{1}/_{u}\right), u \sim U(0, 1) \tag{8.8}$$

where U (0,1) means a random number uniformly distributed between 0 and 1. $p_{iD}$ is the local attractor, and it can be defined as given by Eq. (8.9)

$$p_{iD} = \beta P_{iD} + (1 - \beta) P_{gD}, \beta \sim U(0, 1) \tag{8.9}$$

where $P_i = \left(P_{i1}, P_{i2}, \ldots, P_{iD}\right)$ is the best location of the $i$-th particle, $P_g = \left(P_{g1}, P_{g2}, \ldots, P_{gD}\right)$ is the best location of all the particles, and the parameter L can be evaluated from Eq. (8.10).

$$L = 2\alpha.|\text{mbest}_D - X_{iD}| \tag{8.10}$$

where mbest is the average optimal position of all the particles, and it can be computed by Eq. (8.11).

$$\text{mbest} = \frac{1}{M}\sum_{i}^{M} \text{pbest}_i = \left(\frac{1}{M}\sum_{i}^{M} \text{pbest}_{i,1}, \frac{1}{M}\sum_{i}^{M} \text{pbest}_{i,2}, \ldots, \frac{1}{M}\sum_{i}^{M} \text{pbest}_{i,d}\right) \tag{8.11}$$

Therefore,

$$X_{iD} = p_{iD} + \alpha|\text{mbest}_D - X_{iD}|\log\left(1/u\right), u \sim U(0, 1) \text{ if } \alpha > 0.5 \tag{8.12}$$

and

$$X_{iD} = p_{iD} - \alpha|\text{mbest}_D - X_{iD}|\log\left(1/u\right), u \sim U(0, 1) \text{ if } \alpha < 0.5 \tag{8.13}$$

where $\alpha$ is a parameter of the QPSO algorithm called the contraction–expansion coefficient.

The flowchart of the proposed QPSO algorithm is shown in Fig. 8.4.

## 8.7 Integration Handling Technique

Due to the complexity of integrands, implementation of traditional analytical method for finding the integrations associated to the proposed problems becomes a tedious work. To overcome this situation, we are interested to employ numerical technique. Simpson's 3/8-th rule is being used to find the integrations related to our considered problems. This integration formula is given by Eq. (8.14).

$$\int_a^b F(x)\mathrm{d}x = \frac{3l}{8}((v_0 + v_m) + 2 \times (v_3 + v_6 + \ldots) + 3 \times (v_1 + v_4 + \ldots) + 3 \times (v_2 + v_5 + \ldots)) \tag{8.14}$$

Here, $x_0, x_1, x_2, \ldots, x_m$ are the points at which the interval $[a, b]$ is divided into $m$ equal parts where $m$ is a multiple of three and $v_0, v_1, v_2, \ldots, v_m$ represents the corresponding ordinates, and $l$ represents the subintervals' length. There are 24 subintervals of equal length in the present exertion.

## 8.8 Numerical Example

To clarify our proposed QPSO algorithm, the following numerical example is considered with respect to crisp environment (Tables 8.1 and 8.2).

$$\text{Maximize } T_S(z, \delta, t) = \prod_{i=1}^{m}\left[1 - \left(1 - e^{-\int_0^t \delta_i(k).dk}\right)^{z_i}\right]$$

subject to

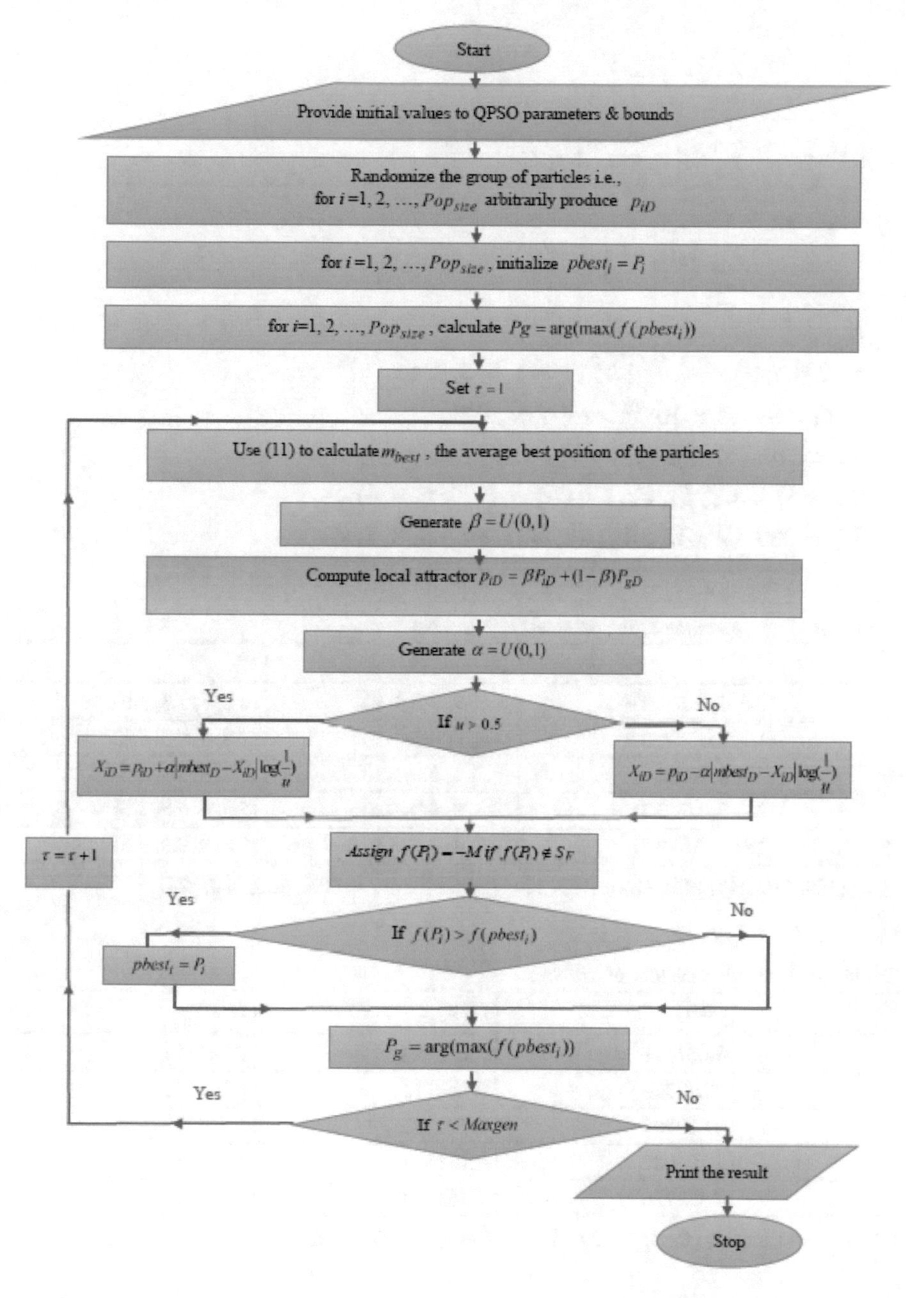


**Fig. 8.4** Flowchart of QPSO algorithm

$$\sum_{i=1}^{5} V_i z_i^2 \leq V_s$$

$$\sum_{i=1}^{5} C_i\left[z_i + e^{\frac{z_i}{4}}\right] \leq C_s$$

$$\sum_{i=1}^{5} W_i\left[z_i e^{\frac{z_i}{4}}\right] \leq W_s$$

where

$z_i \in \mathbb{N}, r_i(t) = e^{-\int_0^t \delta_i(k).dk}, i = 1, 2, 3, 4, 5;$

$\delta_i(t) = 99999 \times 10^{-10} \times t + 2 \times 10^{-7};$

$V_s = 110, (V_1, V_2, V_3, V_4, V_5) = (1, 2, 3, 4, 2);$

$C_s = 175, (C_1, C_2, C_3, C_4, C_5) = (7, 7, 5, 9, 4);$

$W_s = 200, (W_1, W_2, W_3, W_4, W_5) = (7, 8, 8, 6, 9).$

**Table 8.1** Representation of the parameters as TrFNs

| $i$ | $\tilde{V}_i$ | $\tilde{C}_i$ | $\tilde{W}_i$ |
|---|---|---|---|
| 1 | (0.8, 1, 1.1, 1.4) | (6.8, 7, 7.3, 7.6) | (6.8, 7, 7.3, 7.6) |
| 2 | (1.7, 2.0, 2.2, 2.3) | (7.8, 8, 8.3, 8.6) | (7.8, 8, 8.3, 8.6) |
| 3 | (2.6, 3, 3.2, 3.5) | (5.8, 5, 5.3, 5.6) | (7.8, 8, 8.3, 8.6) |
| 4 | (2.7, 3, 3.2, 3.7) | (8.8, 9, 9.3, 9.6) | (5.8, 6, 6.3, 6.6) |
| 5 | (1.4, 2, 2.3, 2.7) | (3.8, 4, 4.3, 4.6) | (8.8, 9, 9.3, 9.6) |

$\widetilde{V_s} = (104, 110, 115, 123)$; $\widetilde{C_s} = (169, 174, 180, 186)$; $\widetilde{W_s} = (194, 200, 210, 221)$

**Table 8.2** Defuzzified values of TrFNs

| $i$ | $(Bet)_{\tilde{V}_i}$ | $(Bet)_{\tilde{C}_i}$ | $(Bet)_{\tilde{W}_i}$ |
|---|---|---|---|
| 1 | 1.061111 | 7.161111 | 7.161111 |
| 2 | 2.077778 | 8.161111 | 8.161111 |
| 3 | 3.088889 | 5.272222 | 8.161111 |
| 4 | 3.100000 | 9.161111 | 6.161111 |
| 5 | 2.127778 | 4.161111 | 9.161111 |

$(Bet)_{\tilde{V}_s} = 112.7222$; $(Bet)_{\tilde{C}_s} = 177.1111$; $(Bet)_{\tilde{W}_s} = 205.5556$

## 8.9 Result Analysis

We offered a numerical example in the preceding section to demonstrate our strategy for optimizing the system reliability with appropriate redundancy component allocations. We ran 30 independent runs of the QPSO algorithms in C++ on a notebook with an Intel CORE i3 10$^{th}$ generation CPU and 4 GB RAM running in Linux and gathered the results to find the best and worst found system reliability, as well as their average value, standard deviation, and running time. For the considered example, the population size and maximum number of generations taken in the QPSO algorithm are 80 and 150, respectively. The other parameters of the QPSO algorithm are $\alpha_1, \alpha_2, \beta_1$ and $\beta_2$. For the proposed QPSO $\alpha_1 = 1.5, \alpha_2 = 0.5, \beta_1 = 1.2$ and $\beta_2 = 2$ The comparative results have been presented in Tables 8.3 and 8.4. The most important thing is that the problem obtains maximum system reliability in case of trapezoidal environment. Also, from the above statistical data table for system reliability, we see that our proposed QPSO algorithm gives better outcomes for the considered example in trapezoidal fuzzy atmosphere in comparison with crisp environment. Figures 8.5, 8.6, 8.7, 8.8, 8.9, 8.10, 8.11 and 8.12 were drawn to visualize the behavior of the system reliability with respect to different swarm parameters in different environments. From these figures, it is clear that the system reliability increases significantly with the increment of the maximum number of generations as well as the population size of the swarm.

## 8.10 Conclusions and Future Scopes

In this paper, we look at a more realistic and practical RAP that takes into account time-dependent reliabilities for components as an exponential function. The soft computing technique of QPSO was used to solve the numerical problem in diverse techniques and forms, including crisp and fuzzy. Our main goal was to obtain the maximum system reliability subject to different resource limitations in the entire work. We were able to find significant results on maximizing the system reliability in comparison with the works done previously. Throughout this work, we have seen that the fuzzy environment produces a remarkable result for the considered example.

For further study, one may consider other imprecise environments like neutrosophic, type-2 fuzzy, and nonlinear intuitionistic fuzzy for finding maximum of the mission design life in association with differently designed reliability systems.

**Table 8.3** Comparison of the best-obtained results with existing method

| Algorithms | Environments | Redundancy vector ($z_i$) | Reliability components ($r_i$) | System reliability ($T_S$) |
|---|---|---|---|---|
| Proposed QPSO | **Crisp** | **(3, 3, 3, 3, 3)** | 0.999138,0.999138,0.999138,0.999138,0.999138 | **0.9956961704** |
| | **Trapezoidal fuzzy** | **(4, 3, 3, 3, 3)** | 0.999918,0.999138,0.999138,0.999138,0.999138 | **0.9964736657** |
| [4] | Triangular fuzzy | (2, 2, 2, 2, 2) | 0.991901,0.997501,0.993602,0.977499,0.995103 | 0.9562648909 |
| | Crisp | (2, 2, 2, 2, 2) | 0.991900,0.997500,0.993600,0.977500,0.995100 | 0.9562601906 |

**Table 8.4** Statistical data of system reliability

| Algorithms | Environments | Best | Worst | Average | SD | CPU (time) |
|---|---|---|---|---|---|---|
| Proposed QPSO | Crisp | **0.9956961704** | 0.988371979 | 0.994882371 | 0.002301771 | 0.113 |
| | Trapezoidal fuzzy | **0.9964736657** | 0.996473657 | 0.996473657 | 0 | 0.130 |
| [4] | Triangular fuzzy | 0.9562648909 | Not found | Not found | Not found | 0.071150 |
| | Crisp | 0.9562601906 | Not found | Not found | Not found | 0.068281 |

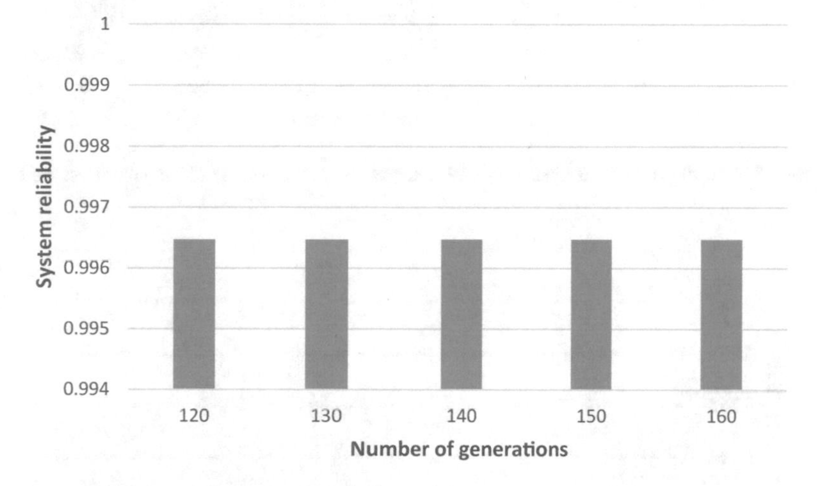


**Fig. 8.5** Sensitivity of system reliability with respect to maxgen (fuzzy environment)

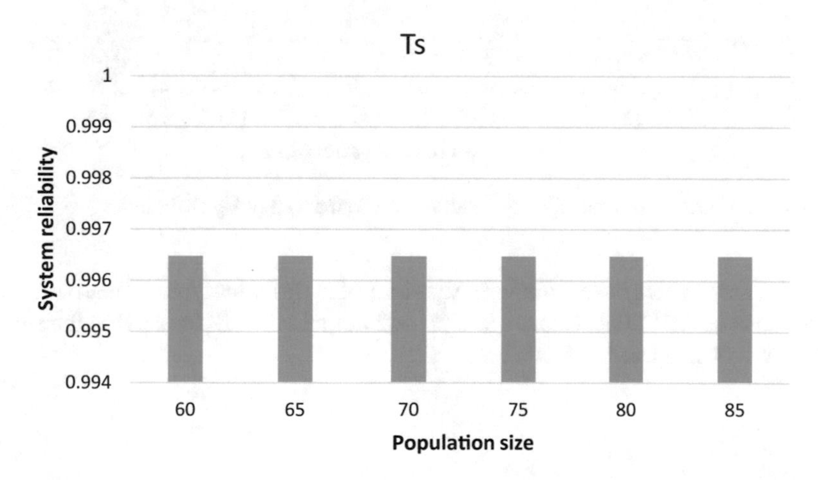


**Fig. 8.6** Sensitivity of system reliability with respect to population size (fuzzy environment)

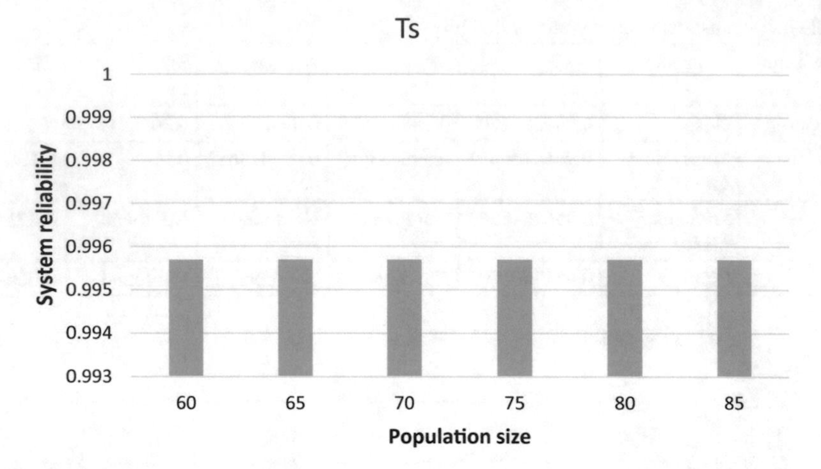


**Fig. 8.7** Sensitivity of system reliability with respect to population size (crisp environment)

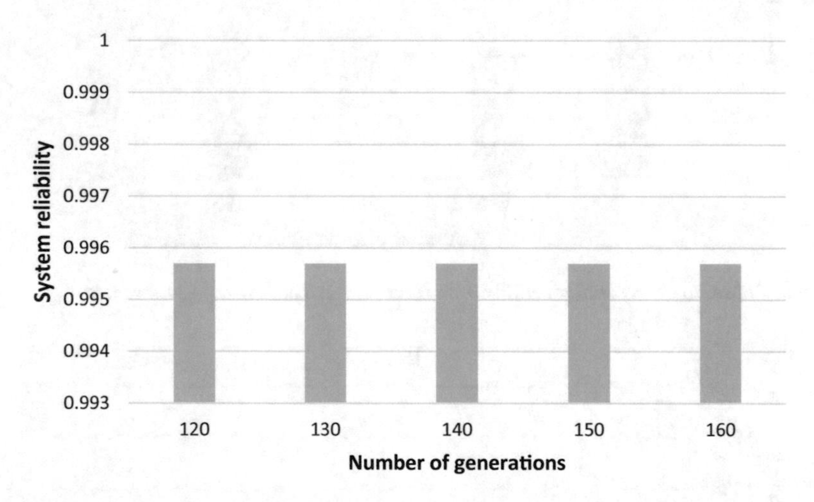


**Fig. 8.8** Sensitivity of system reliability with respect to maxgen (crisp environment)

Also, the researchers have a platform to implement different types of heuristic techniques such as ABC, GA, Cuckoo search, and neural network for solving the similar problems as considered in this work.

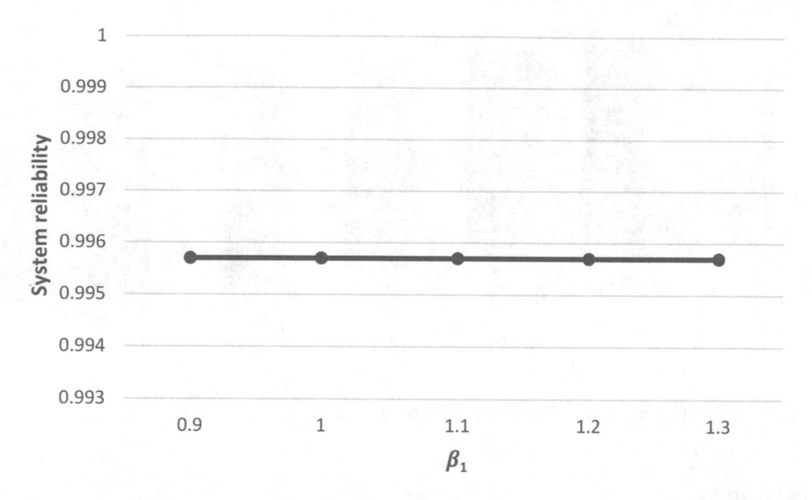


**Fig. 8.9** Sensitivity of system reliability with respect to $\beta_1$

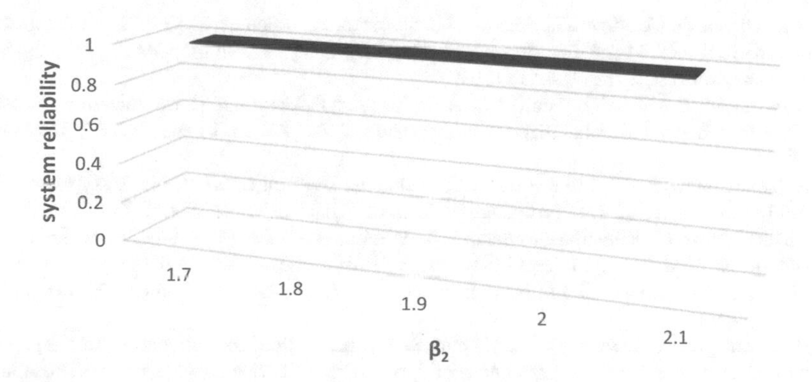


**Fig. 8.10** Sensitivity of system reliability with respect to $\beta_2$

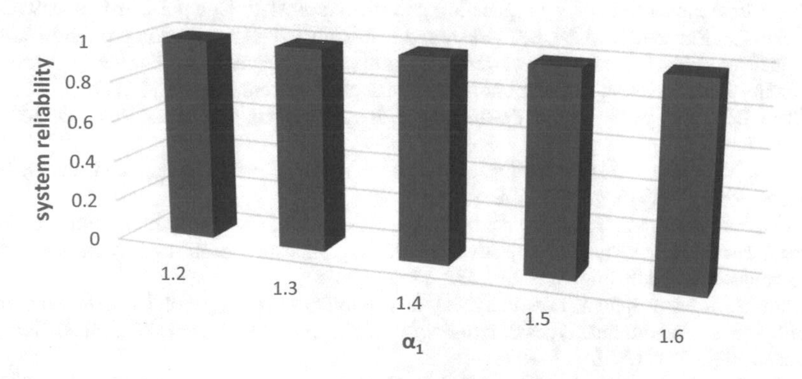


**Fig. 8.11** Sensitivity of system reliability with respect to $\alpha_1$

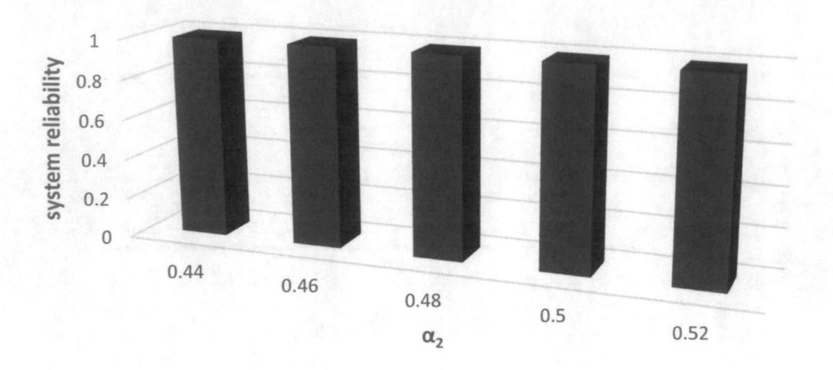


**Fig. 8.12** Sensitivity of system reliability with respect to $\alpha_2$

# References


1. Ahmadivala, M., Mattrand, C., Gayton, N., Dumasa, A., Yalamas, T., Orcesi, A.: Application of AK-SYS method for time-dependent reliability analysis. 24e'me Congre's Franc¸ais de Me´canique Brest, 26 au 30 Aou^t (2019)
2. Ardakan, M.A., Mirzaei, Z., Hamadani, A.Z., Elsayed, E.A.: Reliability optimization by considering time dependent reliability for components. Qual. Reliab. Engng. Int. **33**, 1641–1654 (2017)
3. Baloui Jamkhaneh, E.: System reliability using generalized intuitionistic fuzzy exponential lifetime distribution. Int. J. Soft. Comput. Eng. (IJSCE) **7**, 2231–2307 (2017)
4. Bhattacharyee, N., Kumar, N., Mahato, S.K., Bhunia, A.K.: Development of a blended particle swarm optimization to optimize mission design life of a series–parallel reliable system with time dependent component reliabilities in imprecise environments. Soft. Comput. **25**(17), 11745–11761 (2021)
5. Bhattacharyee, N., Paramanik, R., Mahato, S.: Optimal redundancy allocation for the problem with chance constraints in fuzzy and intuitionistic fuzzy environments using soft computing technique. Ann. Opt Theor. Pract. **3**(2), 25–47 (2020)
6. Bhunia, A.K., Sahoo, L., Roy, D.: Reliability stochastic optimization for a series system with interval component reliability via genetic algorithm. Appl. Math. Comput. **216**, 929–939 (2010)
7. Biswas, P., Pramanik, S., Giri, B.C.: Value and ambiguity index based ranking method of single-valued trapezoidal neutrosophic numbers and its application to multi-attribute decision making. Neutrosophic Sets Syst. **12** (2016). https://digitalrepository.unm.edu/nss_journal/vol12/iss1/17
8. Chen, D.X.: Fuzzy reliability of the bridge circuit system. Syst. Eng-Theor. Pract. **11**, 109–112 (1977)
9. Chern, M.S., Jan, R.H.: Reliability optimization problems with multiple constraints. IEEE Trans. Reliab. **35**(4), 431–436 (1986)
10. Dolatshahi-Zand, A., Khalili-Damghani, K.: Design of SCADA water resource management control center by a bi-objective redundancy allocation problem and particle swarm optimization. Reliab. Eng. Syst. Saf. **133**, 11–21 (2015)
11. Garg, H.: A novel approach for analyzing the reliability of series-parallel system using credibility theory and different types of intuitionistic fuzzy numbers. J. Braz. Soc. Mech. Sci. Eng. **38**(3), 1021–1035 (2016)
12. Garg, H.: Performance analysis of an industrial system using soft computing based hybridized technique. J. Braz. Soc. Mech. Sci. Eng. **39**(4), 1441–1451 (2017)
13. Garg, H., Rani, M.: An approach for reliability analysis of industrial systems using PSO and IFS technique. ISA Trans. Elsevier **52**(6), 701–710 (2013)

14. Gupta, R., Bhunia, A.K., Roy, D.: A GA based penalty function technique for solving constrained redundancy allocation problem of series system with interval valued reliabilities of components. J. Comput. Appl. Math. **232**(2), 275–284 (2009)
15. Hamadani, A.Z., Khorshidi, H.A.: System reliability optimization using time value of money. Int. J. Adv. Manuf. Technol. **66**(14), 97–106 (2013)
16. Hu, Z., Mahadevan, S.: Time-dependent system reliability analysis using random field discretization. ASME J. Mech. Des. **137**(10), 101404 (2015). https://doi.org/10.1115/1.403133
17. Huang, H.: Fuzzy multi-objective optimization decision-making of reliability of series system. Microelectron. Reliab. **37**(3), 447–449 (1996)
18. Hwang, C.L., Tillman, F.A., Kuo, W.: Reliability optimization by generalized Lagrangian-function based and reduced-gradient methods. IEEE Trans. Reliab. **28**, 316–319 (1979)
19. Jia, P., Duan, S., Yan, J.: An enhanced quantum-behaved particle swarm optimization based on a novel computing way of local attractor. Information **6**, 633–649 (2015). https://doi.org/10.3390/info6040633
20. Khalili-Damghani, K., Abtahi, A.R., Tavana, M.: A new multi objective particle swarm optimization method for solving reliability redundancy allocation problems. Reliab. Eng. Syst. Saf. **111**, 58–75 (2013)
21. Kim, J.H., Yum, B.J.: A heuristic method for solving redundancy optimization problems in complex systems. IEEE Trans. Reliab. **42**(4), 572–578 (1993)
22. Kumar, N., Mahato, S.K., Bhunia, A.K.: A new QPSO based hybrid algorithm for constrained optimization problems via tournamenting process. Soft. Comput. **24**, 11365–11379 (2020). https://doi.org/10.1007/s00500-019-04601-3
23. Kumar, N., Shaikh, A.A., Mahato, S.K., Bhunia, A.K.: Applications of new hybrid algorithm based on advanced cuckoo search and adaptive Gaussian quantum behaved particle swarm optimization in solving ordinary differential equations. Expert Syst. Appl. **172**, 114646 (2021)
24. Kumar, N., Shaikh, A.A., Mahato, S.K., Bhunia, A.K.: Development of some techniques for solving system of linear and nonlinear equations via hybrid algorithm. Expert Syst. e12669 (2021b)
25. Mahapatra, G.S., Roy, T.K.: Optimal redundancy allocation in series-parallel system using generalized fuzzy number. Tamsui Oxford J. Inf. Math Sci. **27**(1), 1–20 (2011)
26. Mahato, S.K., Bhattacharyee, N., Paramanik, R.: Fuzzy reliability redundancy optimization with signed distance method for defuzzification using genetic algorithm. Int. J. Oper. Res. (IJOR) **37**(3), 307–323 (2020)
27. Mahato, S.K., Bhunia, A.K.: Reliability Optimization in Fuzzy and Interval Environments: Applications of Genetic Algorithm in Reliability Optimization in Crisp, Stochastic, Fuzzy & Interval Environments. LAP LAMBERT Academic Publishing (2016)
28. Mahato, S.K., Sahoo, L., Bhunia, A.K.: Reliability-redundancy optimization problem with interval valued reliabilities of components via genetic algorithm. J. Inf. Comput. Sci. (UK) **7**(4), 284–295 (2012)
29. Mahato, S.K., Sahoo, L., Bhunia, A.K.: Effects of defuzzification methods in redundancy allocation problem with fuzzy valued reliabilities via genetic algorithm. Int. J. Inf. Comput. Sci. **2**(6), 106–115 (2013)
30. Misra, B., Sharma, U.: An efficient algorithm to solve integer programming problems arising in system-reliability design. IEEE Trans. Reliab. **40**(1), 81–91 (1991)
31. Misra, K.: On optimal reliability design: a review. Syst. Sci. **12**(4), 5–30 (1986)
32. Mori, Y., Ellingwood, B.R.: Time-dependent system reliability analysis by adaptive importance sampling. Struct. Safety **12**(1), 59–73 (1993)
33. Mourelatos, Z.P., Majcher, M., Pandey, V., Baseski, I.: Time dependent reliability analysis using the total probability theorem. ASME J. Mech. Des. **137**(3), 031405 (2015). https://doi.org/10.1115/1.4029326
34. Nakagawa, Y., Miyazaki, S.: Surrogate constraints algorithm for reliability optimization problems with two constraints. IEEE Trans. Reliab. **30**(2), 181–184 (1981)
35. Nakagawa, Y., Nakashima, K.: A heuristic method for determining optimal reliability allocation. IEEE Trans. Reliab. **26**(3), 156–161 (1977)

36. Paramanik, R., Kumar, N., Mahato, S.K.: Solution for the optimality of an intuitionistic fuzzy redundancy allocation problem for complex system using Yager's ranking method of defuzzification with soft computation. Int. J. Syst. Assurance Eng. Manage. 1–10 (2021)
37. Paramanik, R., Mahato, S.K., Bhattacharyee, N., Supakar, P., Sarkar, B.: Multiple constrained reliability-redundancy optimization under triangular intuitionistic fuzziness using a genetic algorithm. In: Reliability Management and Engineering, pp. 205–232. CRC Press (2020)
38. Park, K.: Fuzzy apportionment of system reliability. IEEE Trans. Reliab. **36**(1), 129–132 (1987)
39. Rahmani, A., Hosseinzadeh Lotfi, F., Rostamy-Malkhalifeh, M., Allahviranloo, T.: New method for defuzzification and ranking of fuzzy numbers based on the statistical beta distribution. Hindawi Publishing Corporation Adv. Fuzzy Syst. (2016). https://doi.org/10.1155/2016/6945184
40. Ruigang, Y., Wenzhao, L., Guangli, Z., Yuzhen, L., Weichen, J.: Interval non-probabilistic time-dependent reliability analysis of boom crane structures. J. Mech. Sci. Tech. **35**(2), 535–544 (2021)
41. Sahoo, L., Bhunia, A.K., Kapur, P.K.: Genetic algorithm based multi-objective reliability optimization in interval environment. Comput. Ind. Eng. **62**(1), 152–160 (2012)
42. Sahoo, L., Bhunia, A.K., Mahato, S.K.: Optimization of system reliability for series system with fuzzy component reliabilities by genetic algorithm. J. Uncertain Syst. **8**(2), 136–148 (2014)
43. Sahoo, L., Bhunia, A.K., Roy, D.: A genetic algorithm-based reliability redundancy optimization for interval valued reliabilities of components. J. Appl. Quant. Methods **5**(2), 270–287 (2010)
44. Sahoo, L., Bhunia, A.K., Roy, D.: Reliability optimization with high and low-level redundancies in interval environment via genetic algorithm. Int. J. Syst. Assur. Eng. Manage. (2013). https://doi.org/10.1007/s13198-013-0199-9
45. Sahoo, L., Mahato, S.K., Bhunia, A.K.: Optimization of system reliability in the imprecise environment via genetic algorithm. Int. J. Swarm Intell. Res. **13**(1), 1–21 (2022)
46. Sasaki, R.O.T., Shingai, S.: A new technique to optimize system reliability. IEEE Trans. Reliab. **32**, 175–182 (1983)
47. Sun, X.L., Li, D.: Optimization condition and branch and bound algorithm for constrained redundancy optimization in series system. Optim. Eng. **3**(1), 53–65 (2002)
48. Tillman, F.A., Hwang, C.L., Kuo, W.: Optimization of System Reliability. Marcel Dekker, New York (1980)
49. Xi, M., Sun, J., Xu, W.: An improved quantum-behaved particle swarm optimization algorithm with weighted mean best position. Appl. Math. Comput. **205**, 751–759 (2008)
50. Zafar, T., Wang, Z.: Time-dependent reliability prediction using transfer learning. Struct. Multidiscip. Optim. **62**, 147–158 (2020). https://doi.org/10.1007/s00158-019-02475-5
51. Zadeh, L.A.: Fuzzy sets. Info Ctrl. **8**(3), 338–352 (1965)

**Rajesh Paramanik** is Research Scholar in the Department of Mathematics, Sidho-Kanho-Birsha University, Purulia. He did his M.Sc. in Mathematics from the same university in 2015. His research area of interest is reliability optimization.

**Dr. Sanat Kumar Mahato** is Professor and Former Head of the Department of Mathematics, Sidho-Kanho-Birsha University, PuruliaWest Bengal, India. Prof. Mahato has completed his master's in Applied Mathematics in 2004 from the University of Burdwan, West Bengal. He has been awarded the Ph.D. degree in Mathematics by The University of Burdwan in August 2014. Prof. Mahato has published more than 50 research papers in different peer-reviewed international journals, two books, several book chapters, and also a monograph in LAMBERT Academic Publication. His area of research work includes Application of Genetic Algorithm in inventory, reliability optimization, interval analysis and its application, particle swarm optimization, Fluid Mechanics, Biomathematics, etc. Presently, eight research scholars are working under his guidance and four have been awarded.

**Dr. Nabaranjan Bhattacharyee** has been Research Scholar in the Department of Mathematics, Sidho-Kanho-Birsha University, Purulia. He is serving as Teacher of Mathematics in a school since 2004. He did his M.Sc. in Applied Mathematics from Burdwan University in 2004. He has been awarded Ph.D. degree in Mathematics in 2021 from Sidho-Kanho-Birsha University. His research area is reliability optimization, soft computing, and optimization under uncertainty.

# Chapter 9
# Resilience: Business Sustainability Based on Risk Management

**Mohsen Imeni and Seyyed Ahmad Edalatpanah**

**Abstract** Businesses must be resilient to withstand many occurrences. Companies of all sizes can deal with adversity and possible danger in this manner. When things are unclear, it is critical to have a strong risk management strategy in place. In times of crisis, risk management enables businesses to remain adaptable and robust while avoiding any hurried or erroneous action. In the wake of recent crises like COVID-19, it is clear that risk management must be taken seriously. In the current epidemic, businesses worldwide have become more vulnerable due to the absence of appropriate risk management implementation (not the risk management method itself). Despite this, there is little knowledge about how resilience is related to risk management. Theoretically, it is also crucial to understand how these can impact corporate performance. Therefore, businesses with sufficient knowledge of resilience and risk management can be expected to protect their shareholders and customers against an unplanned disruption. Firms can deal with sustainable development and risk management by using the concepts of resilience, robustness, and antifragility. Therefore, in this chapter, to help businesses, resilience and risk management concepts were introduced, and the relationship between these two variables was explained. The findings can help enterprises to adopt good practices for proper planning and risk management, given the degree of resilience of that business. Its implications can help enterprises to adopt appropriate policies and provide valuable insights to help them develop risk management and resilience capacities to prevent and respond to related disasters.

**Keywords** Risk management · Risk · Resilience · Robustness · Antifragility

M. Imeni
Department of Accounting, Ayandegan Institute of Higher Education, Tonekabon, Iran

S. A. Edalatpanah (✉)
Department of Applied Mathematics, Ayandegan Institute of Higher Education, Tonekabon, Iran
e-mail: saedalatpanah@gmail.com; s.a.edalatpanah@aihe.ac.ir



H. Garg (ed.), *Advances in Reliability, Failure and Risk Analysis*, Industrial and Applied Mathematics, https://doi.org/10.1007/978-981-19-9909-3_9

## 9.1 Introduction

Small- and medium-sized enterprises (SMEs) have faced several challenges and risks over the last decade. One of the most challenging business conditions in recent decades has been created by intense global competition, economic uncertainty, rapid technological change, and growing customer demand [18]. In today's world, risk management is no longer focused on reducing vulnerability but rather on strengthening resilience. An individual, system, or community's resilience to risk and uncertainty is fundamentally determined by its ability to respond to disturbances, surprises, and changes [45]. As a result, the application of the concept of resilient societies and the ways to create and strengthen them have become more widely used [15]. Economic resilience, given its dynamic and forward-looking nature, can be more effective in increasing the economy's ability to adapt to risks. Economic resilience means identifying ways and behaviors that increase the capacity to resilience external shocks or adverse effects. Alternatively, resilience seeks to reduce the probability of failure or losses of economic risks, and it is before and after the occurrence of shocks [77]. Paying attention to ambiguity and uncertainty is essential [33].

One of the critical features mentioned for a resistance economy is the resilience of the economy. Accordingly, a definition of economic resilience is "the capacity or ability of the economy to maintain performance and optimal allocation of resources in the face of economic uncertainty." Resilience consists of two tangible sources (such as internal resources) and an intangible source (for instance, strong leadership and fast decision-making) [11]. Liu et al. [39] believe that internal resources help companies adapt to external crises and opportunities to improve corporate business performance potentially. Several recent natural and economic events have demonstrated the vulnerability of countries of all levels of development to disasters. Like terrorism and natural disasters, COVID-19 threatened managers' lives, emotions, and rationality [25, 28, 72]. From about 350 in 1980 to almost 1000 yearly in 2014, these high-risk events have steadily increased around the world. Economic losses have risen from about $50 billion in the 1980s to about $250 billion in the past decade [72].

Radović-Marković et al. [54] argue that the concept of resilience shows how a country's economy can return to its previous level, based on this concept. However, they say a country's economic resilience is impossible unless small- and medium-sized businesses resist the adverse effects. Therefore, many consider it two sides of the same coin. On the other hand, the risks of "new" forms such as terrorism, the COVID-19 pandemic, the financial sector's recent economic collapse, and the ensuing global economic crisis can all be borderless in nature [65]. A negative economic cycle characterizes SMEs during COVID-19, which is strongly impacted by the pandemic [2, 5]. Barbosa [4] believes that due to the need for creating resilient ecosystems, resilience is a new research opportunity. Building resilience through risk-informed sustainable development is essential to generating sustainable and resilient communities [2]. Resilience can correct the structure and tools to function in the face of stress, change, and uncertainty. Understanding the risk perspective correctly and identifying the best place to own and manage those risks are essential. Understanding how interrelationships between system components affect the performance of the system and strengthening the components that address those risks are essential [72].

There are many similarities and points of convergence between risk and resilience, according to Mitchell and Harris [45]. In their view, risk assessment and management are key to business resilience, and monitoring risks is a necessary condition for business resilience. Risk management approaches and resilience, however, are often viewed as independent variables in the literature because "Resilience thinking challenges the widely held notions about stability and resistance to change in risk management around the world" [6, 43].

Hence, this study provides a definition and explanation of the concept of resilience at different levels of the economy and to express risk and risk management. In this way, the knowledge gap between risk management and resilience will be reduced. Therefore, businesses with sufficient knowledge of resilience and risk management can be expected to protect their shareholders and customers against an unplanned disruption. Meanwhile, understanding concepts such as resilience, robustness, and antifragility can help firms remain stable in crises and risks.

There are several ways in which this study contributes to the literature on resilience and risk management. First, concepts such as resilience and risk management have not yet been poorly addressed empirically. By identifying different types of resilience and levels of risk, this study increases theoretical and managerial knowledge. Second, in the present study, other concepts such as antifragility and robustness were introduced in future research, these three concepts can be used more accurately in management research, but now the dominant concept is resilience. Third, the findings can help businesses adopt good practices for proper planning and risk management, given the degree of resilience of that business. Businesses can benefit from its implications by adopting appropriate policies and constructing risk management and resilience capacities in order to prevent and respond to disasters.

## 9.2 Definition of Resilience, Robustness, and Antifragility

As we examine the definitions of resilience in the planning, environmental, psychology, engineering, organizational behavior, sociology, and economic fields for the past forty years, we are able to draw a comparison between them [59]. Holling first used the term in 1973. The term "resilience" is used to mean "going back in time," which is derived from the root "resilio" [16], and the equivalent of the word resilience means the ability to recover, rapid recovery, change, elasticity, and buffer and elasticity [47].

The root of resilience (resiliency) is derived from the science of physics and means jumping backward. Holling [30] defined resilience as the system's stability to sudden changes and ability to absorb shocks while maintaining past relationships between parameters and variables in the same state. Finally, [30] defines "the ability of systems to absorb change … as well as to survive" as a definition of resilience. In addition, he sometimes refers to resilience as "buffer capacity." Pimm [53] describes resilience as the speed at which a system returns to equilibrium. In addition, [51] defined resilience as relative: "the gap between current and critical loads." In [23], resilience is defined as the ability of a system to adapt, grow, and survive in changing conditions.

On the other hand, organizations can face three consequences after disasters: (1) declining performance and subsequent improvement (i.e., resilience), (2) insensitivity to uncertainty (i.e., robustness), and (3) upside gain (i.e., antifragility). In exploring these distinct outcomes, [49] transfer knowledge from the uncertainty, risk, and system theory literature into organizational resilience. Distinguishing the differences between these different outcomes can clarify our insight of enterprise answers at all times and in any situation with adversity.

The most crucial area of this concept is robustness. These include "abilities that aim to stabilize enterprise in the time of the disorders" and "abilities that aim to reduce the effect of the disorders on performance" [68]. Because if enterprises cannot maintain their strength in adverse times, they cannot recover from the disruption. In words, any enterprise that lacks the ability to have some robustness is likely to suffer from subpar performance and consequent failure [49]. Thus, for example, during the COVID-19 epidemic, online companies could withstand the crisis (i.e., were robust), while audit firms were initially damaged and then recovered (i.e., were resilient).

Robustness is the ability to maintain and resist adverse effects [20, 67]. Systems with the necessary robustness can resist or absorb pressure [20] or withstand and absorb strain and maintain their performance [19]. Robust systems often change their states to keep up performance [38]. While uncommon or unforeseen affairs can reveal vulnerabilities, and as a result, they create situations that they are unable to handle, and eventually firms fail [9].

The notions of fragility and antifragility were introduced by Taleb [70]. It means "things that gain from disorder." The definition of an antifragile system is as follows:

Some systems profit from shock. That is, when volatility and disorder are present, they grow. In other words, they are interested in adventure, risk, and uncertainty. It is, however, impossible to find a word to describe the careful inverse of fragile. So we call it antifragile. There is no such thing as resilience or robustness when it comes to antifragility. Taleb [70] believes that in times of disorder and shock, strength and resilience cannot help organizations and individuals, while antifragility benefits them. Based on the above, it can be said that any disorder or shock is not necessarily harmful [7]. It should be noted, however, that necessarily is not gainful for all volatility and disorder [7].

Of course, robustness has a limited capacity to absorb disruptions without subsequently disrupting performance [30]. In contrast, resilient systems can return to their previous acceptable levels after the reduced performance [75], enabling business resumption in prior performance levels [63]. It is complicated to differentiate between robustness and resilience. Hillmann and Guenther [29] argue that there is an invisible boundary between the two concepts. Because resilient organizations, in a way, need robustness to be able to withstand pressure [36]. Resilience is about performance recovery after it has been reduced. As antifragility is defined as an increase in performance in the face of adversity [70], as a result, antifragility differs from resilience [55].

## *9.2.1 Different Levels of Resilience*

Rose and Krausmann [60] expressed the operational criteria of resilience in two categories: Direct Static Economic Resilience (DSER) and Total Static Economic Resilience (TSER). Economic resilience on a direct static basis refers to the level of a business or industry (micro and meso), which is based on the evaluation of "partial equilibrium" or the performance of an enterprise or household. Basically, total static economic resilience (TSER) refers to the macro-level of economics. Ideally, it would encompass all price-quantity interactions in an economy, referred to as "general equilibrium." In addition, resilience can be assessed behaviorally. Researchers face three problems in the area of resilience. Measures of resilience, including those that violate rational behavior, need to be identified at the conceptual level. This may cause challenges to the model of individual, group, and community behavior on an operational level. At the practical level, data collection on resilience, particularly to determine models, is difficult [58].

Finally, [59] expresses resilience on the following levels: microeconomic (households or businesses), meso-economic (markets or industries), and macroeconomic (an economic entity that includes all economic entities and their interactions) (Fig. 9.1).

### 9.2.1.1 Resilience at the Microeconomic Level

The purpose of this section is to show how economic production theory can be used to analyze economic resilience at the micro-level. A business's ability to produce profits from different inputs is represented by this abstract model. A framework called Computable General Equilibrium (CGE) is viewed in economics as a set of integrated (macro-level) supply chains and deals with how businesses interact in supply chains (meso-level). Businesses' performance remains the focus of this approach.

Resilience in business has two aspects. Customer-side resilience creates through disruption (quantity and timing) in the delivery of inputs and to utilize the resources available in businesses and households. Also, it has a relationship with static resilience. For example, resilience is primarily a demand-side issue in a particular period, meaning the existence of a specific fixed capital or any fundamental disruption in the supply of input. In contrast, supply-side resilience is related to providing output to customers. In addition to system redundancy, dynamic resilience usually involves repairing or constructing critical inputs. Supplier efforts or capital repairs are entirely separate from customer resilience, which is the responsibility of input providers [59].

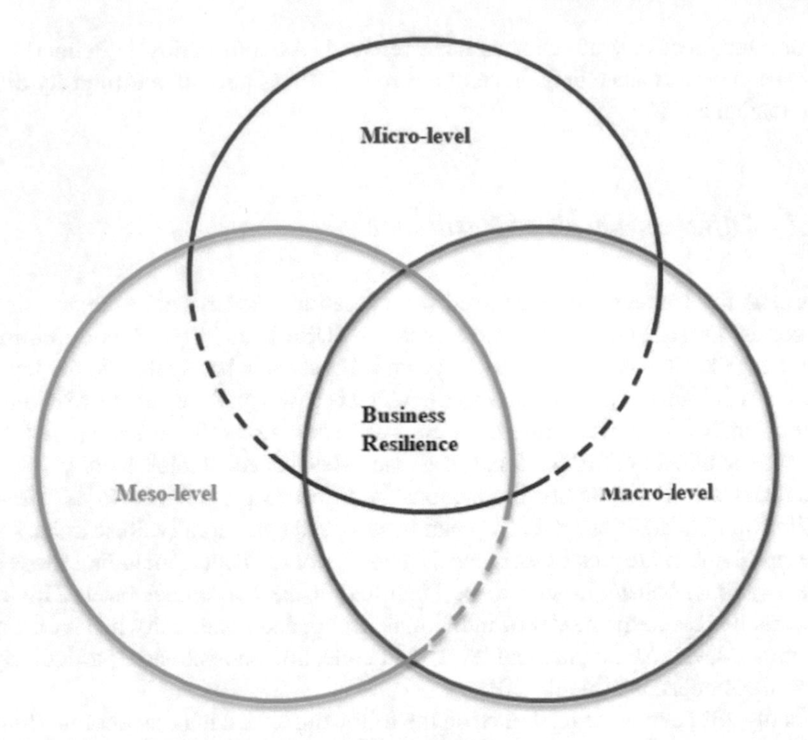


**Fig. 9.1** Business resilience based on different levels [64]

From an economic point of view, economic resilience relates to the ability of the economy to withstand shocks and return to its original path; however, this ability also depends on the level of vulnerability of the economy [8]. Economic resilience means identifying ways and behaviors that increase the capacity to deal with external shocks or adverse effects. On the other hand, resilience seeks to reduce the likelihood of failure or loss of economic risks before and after shocks [77]. Due to its dynamic and forward-looking nature, the economic resilience approach can have been more effective in adapting to risks.

#### 9.2.1.2 Resilience at the Meso- and Macroeconomic Level

From a meso-economic perspective, resilience is an option to strengthen the market or parts of it. Examples are pricing mechanisms, aggregation of resources and industry information, and various types of infrastructure. Most researchers in the field of economic believe that the intrinsic resilience of market prices, which acts as an "invisible hand" to direct resources to better allocation, should be considered as a disaster. Rose [59] believes that the market is likely similar to its buildings, and human disasters will be damaged. Under normal market failure conditions, two alternatives to some (or all) economies are presented similarly, public goods, and market power:

**Fig. 9.2** Meso-level resilience (the business level) [13]

(1) It is replacing judgments or scheduling with significant costs and higher than its implementation.
(2) Both approaches, such as improving information, are considered resilient to strengthen the market.

Connelly [13] explained resilience at the meso-level (the level of businesses) shown in Figure 9.2. Connelly [13] believes that a business's risk and governance management are the most critical elements.

From a macroeconomic perspective, resilience is affected by sector interactions. Martin and Sunley [42] reason that the resilience of macroeconomics is not just a function of resilience measures at the microeconomic level (individual business or household). All companies and markets have an impact on it (at the meso-economic level).

After understanding resilience, robustness, and antifragility concepts and getting familiar with different levels of resilience, the concepts of risk and its management will be discussed. In addition, the relationship between resilience and risk management will be explained.

## 9.3 Risk, Risk Management, and Business Resilience

Around 1200 in Venice, "risk" was the first word to appear and be used in European languages to resolve uncertainty. Other terms, such as "uncertainty," appeared much later [37]. In general, risk refers to the possibility of loss or unfavorable outcomes associated with an action. From [31] point of view, risk is a situation in which the result of activities is accompanied by uncertainty. The Institute of Internal Auditors

(IIA) describes risk as uncertainty about the occurrence of an event that could influence the goal achievement. Having uncertainty means not knowing what will happen in the future—risk increases with uncertainty [14]. The goal of risk management for a manager is to maximize expected returns subject to the risks and tolerances of the organization. Crane et al. [14] argue that taking risks allows you to make money. In other words, there would be no return if there were no risks. Positive or negative consequences may result from the risk, or simply uncertainty may result. For an organization, thus, the risks associated with opportunity and loss or the existence of uncertainty may be taken into account. Each risk has characteristics requiring exceptional management or analysis [32]. Therefore, attention to risk and its management has become more important in businesses.

Pike et al. [52] have divided the risks into two categories with internal and external drivers. There are two kinds of risks associated with external drivers: financial risks (risks associated with accounting standards, foreign exchange, interest rates, and customer credit) and market environment risks (risks associated with competition, customer demand, economic conditions, technology developments, and legal requirements). As well as risks associated with internal incentives, there are risks associated with controls and control environments, liquidity, investment, fraud, accounting information systems, and human resources. Moeller [46] also divides the risks that must be managed into strategic, operational, financial, and information risks. Therefore, a business must address the various risks in achieving its targets.

The enterprise is a dynamic combination of organized resources to achieve a set of purposes and missions. Therefore, defining these goals is vital to any enterprise's management. In these circumstances, it is clear that the unavailability of all or part of a particular resource can prevent the organization from achieving its goals. Reasons for this "lack of access" to resources include the occurrence of risks or unidentified "accidental" events. In this context, the purpose of the risk management process can be defined as the availability, under any set of conditions, of resources at a level consistent with the enterprise's core objectives. Then, to achieve an operational definition of the risk management goal, a closer look at the organization's goals is necessary [12]. Risk management is the process used by the board of directors and senior executives to identify events that affect the business [57]. Risk management has become an essential issue for all companies today, so it is used in decision-making in the categories of CEO and middle managers [10]. Risk management theory shows the reduction of various accounting costs that help improve the company's performance [76].

The relationship between risk management and performance has also been debated by researchers for a long time, mainly because the relationship between risk management and value in imperfect and inefficient markets has not yet been established [24]. Past studies of corporate risk management have concluded that organizations may be able to improve their performance by adopting a dependent view of risk management [22]. Iwedi et al. [35] studies have shown that risk management aims to maximize shareholder value. Studies by Ashari and Krismiaji Krismiaji [3] and Shad et al. [61] also show the effect of risk management on performance and return on equity.

One of the dimensions of risk management is industry competition. An essential feature of the competition is that there is more than one firm in the market, and this feature makes firms comparable in terms of performance. Comparability of firms allows investors to choose firms with optimal performance for investment. In particular, Shatnawi's [62] studies have shown that corporate risk management involves a combination of managing threats and strategic risks based on corporate policies. In this sense, the risk management process not only becomes a tool for preventing and managing the impact of destructive events on the business but also becomes a force for seeing an opportunity [69]. The three stages of risk management are as follows: (1) identification of risks, (2) risk assessment, and (3) risk reduction [21], and higher-ability managers can improve performance [34]. Enterprise risk management is made up of eight parts: control environment, targeting, event identification, risk assessment, risk response, control activities, communication, and monitoring [57]. Merna and Al-Thani [44] believe the risk management process should be dynamic and regularly reviewed.

Risk management plays a central role in resilience [41]. Risk management purpose is to manage all the uncertainties that may interfere with the goals and missions of the organization and to ensure the survival of the organization in any situation (environmental and economic) that it may face. Risk analysis is fundamental to preventing business failure, including risk assessment and management [50]. Park et al. [50] believe risk analysis is impossible where risks are unknown.

Conversely, the goal of resilience is to build the capacity to overcome disorders or stress while maintaining the functions needed for survival and possibly progress [40]. Somers [66] argues that there is more to resilience than just surviving. In the event of a disaster, this means identifying potential risks and taking precautions to ensure the organization's progress.

The main areas of resilience can be categorized as follows: organizational, operational, financial, technological, and business resilience. Since the objective of businesses is sustainability, they must understand the risks they will face in the future and be prepared for those risks. They can achieve this goal through adaptive and mitigation measures. Resilience plays a significant role in this case because resilience demonstrates the ability of businesses to cope with expected and unpredictable events. To this end, it is imperative for companies to identify, evaluate, and plan for future risks that they may face.

Torabi et al. [71] argument measures such as risk reduction and process reengineering can help in proper risk control in traditional risk management. The resilience concept, along with the risk concept, has been considered by researchers in recent years. Hence, different perspectives have been expressed in this regard [56]: As a complementary approach to risk management, resilience is viewed from the first perspective. In this perspective, it is believed that traditional risk management systems in times of crisis have not been able to respond appropriately, so resilience can be considered a new and appropriate way to manage crises. As seen from the second perspective, resilience and risk are completely separate concepts. In this perspective, it is believed that businesses will move toward risk management or increased resilience in times of crisis. Lastly, both concepts are regarded as unifying elements

**Table 9.1** Perspectives on resilience and risk management [50]

| | Risk management | Resilience |
|---|---|---|
| Design principles | Maintain the status quo by avoiding transformative change; minimize failure risks | Adaptability to change (e.g., changing paths, if not destinations) without permanent loss of function. Acknowledging unknown hazards. To reduce the possibility of a larger system experiencing permanent loss of function, intentional failure may be allowed at the subsystem level |
| Design objectives | Minimizing failure probability, albeit with rare catastrophic consequences | Reducing the consequences of failure, although they may occur more often and may require a more rapid recovery |
| Design strategies | Armoring, resistance, strengthening, redundancy, oversizing, isolation | Diversity, cohesion, adaptability, renewability, flexibility, innovation, regrowth, and transformation |
| Relation to sustainability | Security, longevity | Recovery, innovation, renewal |
| Mechanisms of coordinating response | Coordination of efforts is facilitated by centrally located, hierarchically organized decision structures | Local conditions are responded to by decentralized, autonomous agents |
| Modes of analysis | Analysis of identified hazards using quantitative (probability-based) and semiquantitative (scenario-based) techniques in the context of utility theory (i.e., costs and benefits) | Analyzing scenarios with unknown causes and possible consequences |

from the third perspective. In terms of logic, resilience and risk are quite different, but they both aim to increase access to resources for a longer period of time [56]. The resilience approach requires preparation for the unexpected. Risk analysis, on the other hand, assumes that risks are identifiable. Table 9.1 compares the perspectives of risk management and resilience.

## 9.4 Conclusion and Discussion

The business environment is constantly changing and full of risk [1]. By ignoring the invisible nexus between business and their environment, businesses miss out on many new sustainable development opportunities that may prohibit businesses from collapsing [48]. Moore and Manring [48] argue that business plans should articulate the opportunities and constraints of changing social and environmental conditions. If businesses, tiny and medium-sized businesses, cannot adequately cope with possible crises and problems, they cannot survive and are doomed [26]. Therefore, companies must take steps to make themselves more competitive to survive and increase their competitiveness. Varmazyari and Imani [74] believe that the resilience of the country's economy (at the macro-level) and businesses (at the micro- and meso-level) can make the economy resistant to external and internal shocks.

This concept increases the ability of businesses to cope with abrupt and unpredictable changes and shocks (internal and external factors). Herbane [27] points out that businesses mainly suffer from a lack of resources such as liquidity to meet market and customers' needs. Van Gils [73] also enumerates factors such as raw material supply and financial needs in this regard. This school of thought includes perspectives based on management risk-focused strategies.

Based on the above, if businesses at the micro- and meso-levels want to resilience possible crises such as sanctions and coronavirus which affect their economic conditions, they must understand their strengths and weaknesses. Analysis of strengths and weaknesses can affect the future and survival of such businesses because businesses can increase their competitiveness and resilience through resource and cost management (liquidity, financing, etc.). In addition, recognizing the risks and proper coverage of these risks can be very important in the continuity of businesses. Understanding threats and opportunities allow businesses to use various financial tools (using the concept of financial engineering and risk management) to counter these threats. Dahles and Susilowati [17] argue that companies have been able to survive (and even grow some) emerging crises by using flexible expertise (in any situation), diversifying, and combining different sources of revenue within and across sectors.

**Conflicts of Interest** The authors declare that they have no conflicts of interest regarding the publication of this paper.

**Funding** The authors funded the research.

## References

1. Aleksić, A., Stefanović, M., Arsovski, S., Tadić, D.: An assessment of organizational resilience potential in SMEs of the process industry, a fuzzy approach. J. Loss Prev. Process Ind. **26**(6), 1238–1245 (2013)
2. Arcese, G., Traverso, M.: Sustainability and resilience assessment in the pandemic emergency. Symphonya. Emerging Issues Manage. (symphonya.unicusano.it) (2), 99–117 (2021).https://doi.org/10.4468/2021.2.09arcese.traverso
3. Ashari, S., Krismiaji, K.: Audit committee characteristics and financial performance: Indonesian evidence. Equity **22**(2), 139 (2020)
4. Barbosa, M.W.: Uncovering research streams on agri-food supply chain management: a bibliometric study: Global Food Security **28**, 100517 (2021). https://doi.org/10.1016/j.gfs.2021.100517
5. Bellandi, M.: Some notes on the impacts of Covid-19 on Italian SME productive systems. Symphonya. Emerging Issues Manage. (symphonya.unicusano.it) (2), 63–72 (2020). https://doi.org/10.4468/2020.2.07bellandi

6. Berkes, F.: Understanding uncertainty and reducing vulnerability: lessons from resilience thinking. Nat. Hazards **41**(2), 283–295 (2007)
7. Blečić, I., Cecchini, A.: Antifragile planning. Plan. Theor. **19**(2), 172–192 (2020)
8. Briguglio, L., Cordina, G., Farrugia, N., Vella, S.: Economic vulnerability and resilience: concepts and measurements. Oxf. Dev. Stud. **37**(3), 229–247 (2009)
9. Carlson, J.M., Doyle, J.: Complexity and robustness. Proc. Natl. Acad. Sci. **99**, 2538–2545 (2002)
10. Castro, L.M., Gulías, V.M., Abalde, C., Santiago Jorge, J.: Managing the risks of risk management. J. Decis. Syst. **17**(4), 501–521 (2008)
11. Chowdhury, M., Prayag, G., Orchiston, C., Spector, S.: Postdisaster social capital, adaptive resilience and business performance of tourism organizations in Christchurch, New Zealand. J. Travel Res. **58**(7), 1209–1226 (2019)
12. Condamin, L., Louisot, J.P., Naım, P.: Risk Quantification-Management, Diagnosis and Hedging. Wiley, New York (2006)
13. Connelly, C.: From disaster recovery to business resilience. In: Enterprise Risk Management Seminar Institute of Actuaries of Australia (2011)
14. Crane, L., Gantz, G., Isaacs, S., Jose, D., Sharp, R.: Introduction to risk management: understanding agricultural risks: production, marketing, financial, legal, human Resources, 2nd edn. Extension Risk Management Education and Risk Management Agency (2013)
15. Cutter, S.L., Barnes, L., Berry, M., Burton, C., Evans, E., Tate, E., Webb, J.: A place-based model for understanding community resilience to natural disasters. Glob. Environ. Chang. **18**(4), 598–606 (2008)
16. Dadashpoor, H., Adeli, Z.: Measuring the amount of regional resilience in Qazvin urban region. J. Emergency Manage. **4**(2), 73–84 (2016). (In Persian)
17. Dahles, H., Susilowati, T.P.: Business resilience in times of growth and crisis. Ann. Tour. Res. **51**, 34–50 (2015)
18. Demmer, W.A., Vickery, S.K., Calantone, R.: Engendering resilience in small-and medium-sized enterprises (SMEs): a case study of Demmer Corporation. Int. J. Prod. Res. **49**(18), 5395–5413 (2011)
19. Dubey, R., Gunasekaran, A., Childe, S.J., Papadopoulos, T., Blome, C., Luo, Z.: Antecedents of resilient supply chains: an empirical study. IEEE Trans. Eng. Manage. **66**(1), 8–19 (2017)
20. Durach, C.F., Wieland, A., Machuca, J.A.: Antecedents and dimensions of supply chain robustness: a systematic literature review. Int. J. Phys. Distrib. Logist. Manag. **45**, 118–137 (2015)
21. Dwyer, A., Zoppou, C., Nielsen, O., Day, S., Roberts, S.: Quantifying Social Vulnerability: A Methodology for Identifying those at Risk to Natural Hazards (2004)
22. Ellul, A., Yerramilli, V.: Stronger risk controls, lower risk: evidence from US bank holding companies. J. Financ. **68**(5), 1757–1803 (2013)
23. Fiksel, J.: Sustainability and resilience: toward a systems approach. Sustain.: Sci. Pract. Policy **2**(2), 14–21 (2006)
24. Florio, C., Leoni, G.: Enterprise risk management and firm performance: the Italian case. Br. Account. Rev. **49**(1), 56–74 (2017)
25. George, G., Howard-Grenville, J., Joshi, A., Tihanyi, L.: Understanding and tackling societal grand challenges through management research. Acad. Manag. J. **59**(6), 1880–1895 (2016)
26. Hamel, G., Välikangas, L.: The quest for resilience. Harv. Bus. Rev. **81**, 52–63 (2003)
27. Herbane, B.: Small business research: time for a crisis-based view. Int. Small Bus. J. **28**(1), 43–64 (2010)
28. Heredia, J., Rubiños, C., Vega, W., Heredia, W., Flores, A.: New strategies to explain organizational resilience on the firms: a cross-countries configurations approach. Sustainability **14**(3), 1612 (2022). https://doi.org/10.3390/su14031612
29. Hillmann, J., Guenther, E.: Organizational resilience: a valuable construct for management research? Int. J. Manag. Rev. **23**, 7–44 (2021)
30. Holling, C.S.: Resilience and stability of ecological systems. Ann. Rev. Ecol. Syst. **4**(1), 1–23 (1973)

31. Holton, G.A.: Defining risk. Financial Anal. J. **60**(6), 19–25 (2004)
32. Hopkin, P.: Fundamentals of Risk Management: Understanding, Evaluating and Implementing Effective Risk Management. Kogan Page Publishers (2018)
33. Imeni, M.: Fuzzy logic in accounting and auditing. J. Fuzzy Extension Appl. **1**(1), 69–75 (2020)
34. Imeni, M., Fallah, M., Edalatpanah, S.A.: The effect of managerial ability on earnings classification shifting and agency cost of Iranian listed companies. In: Discrete Dynamics in Nature and Society (2021)
35. Iwedi, M., Anderson, O.E., Barisua, P.S., Zaagha, S.A.: Enterprise risk management practice and shareholders value: evidence from selected quoted firms in Nigeria. Green Finance **2**(2), 197–211 (2020)
36. Kantur, D., İşeri-Say, A.: Organizational resilience: a conceptual integrative framework. J. Manag. Organ. **18**, 762–773 (2012)
37. Kast, R., Lapied, A.: Economics and Finance of Risk and of the Future. Wiley, New York (2006)
38. Kitano, H.: Biological robustness. Nat. Rev. Genet. **5**, 826–837 (2004)
39. Liu, C.L., Shang, K.C., Lirn, T.C., Lai, K.H., Lun, Y.V.: Supply chain resilience, firm performance, and management policies in the liner shipping industry. Transp. Res. Part A: Policy Pract. **110**, 202–219 (2018)
40. Louisot, J.P.: Risk and/or resilience management. Risk Governance Control: Financial Markets Inst. **5**(2), 84–91 (2015)
41. Mamaghani, E.J., Medini, K.: Resilience, agility and risk management in production ramp-up. Proc. CIRP **103**, 37–41 (2021)
42. Martin, R., Sunley, P.: On the notion of regional economic resilience: conceptualization and explanation. J. Econ. Geogr. **15**(1), 1–42 (2015)
43. Martinelli, E., Dellanoce, F., Carozza, G.: Business resilience and risk management during the Covid-19 pandemic: the Amadori case-study. Sinergie Italian J. Manage. **39**(3), 123–139 (2021)
44. Merna, T., Al-Thani, F.F.: Corporate Risk Management. Wiley, New York (2008)
45. Mitchell, T., Harris, K.: Resilience: a risk management approach. In: ODI Background Note, pp. 1–7 (2012)
46. Moeller, R.R.: Brink's Modern Internal Auditing: A Common Body of Knowledge. Wiley, New York (2009)
47. Mohammadi, T., Shakeri, A., Taghavi, M., Ahmadi, M.: Explaining the concepts, dimensions and components of economic resilience. Basij Strategic Stud. **20**(75), 89–120 (2017). (In Persian)
48. Moore, S.B., Manring, S.L.: Strategy development in small and medium sized enterprises for sustainability and increased value creation. J. Clean. Prod. **17**(2), 276–282 (2009)
49. Munoz, A., Billsberry, J., Ambrosini, V.: Resilience, robustness, and antifragility: towards an appreciation of distinct organizational responses to adversity. Int. J. Manag. Rev. **24**(2), 181–187 (2022)
50. Park, J., Seager, T.P., Rao, P.S.C., Convertino, M., Linkov, I.: Integrating risk and resilience approaches to catastrophe management in engineering systems. Risk Anal. **33**(3), 356–367 (2013)
51. Perrings, C.: Resilience and sustainable development. Environ. Dev. Econ. **11**(4), 417–427 (2006)
52. Pike, R., Neale, B., Linsley, P.: Corporate Finance and Investment: Decisions and Strategies, 7th edn. Pearson Education Ltd (2012)
53. Pimm, S.L.: The Balance of Nature?: Ecological Issues in the Conservation of Species and Communities. University of Chicago Press (1991)
54. Radović-Marković, M., Shoaib Farooq, M., Marković, D.: Strengthening the resilience of small and medium-sized enterprises. In: Review of Applied Socio-economic Research, pp. 345–356 (2017)
55. Ramezani, J., Camarinha-Matos, L.M.: Approaches for resilience and antifragility in collaborative business ecosystems. Technol. Forecast. Soc. Chang. **151**, 1–26 (2020)

56. Rezaei Soufi, H., Esfahanipour, A., Akbarpour Shirazi, M.: Risk reduction through enhancing risk management by resilience. Int. J. Disaster Risk Reduction **64**, 102497 (2021). https://doi.org/10.1016/j.ijdrr.2021.102497
57. Romney, M.B., Steinbart, P.J.: Accounting Information Systems, 12th edn. Pearson Education Ltd (2012)
58. Rose, A.: Defining and measuring economic resilience to disasters. Disaster Prevention Manage.: Int. J. **13**, 307–314 (2004)
59. Rose, A.: Defining and Measuring Economic Resilience from a Societal, Environmental and Security Perspective. Springer, Heidelberg (2017)
60. Rose, A., Krausmann, E.: An economic framework for the development of a resilience index for business recovery. Int. J. Disaster Risk Reduction **5**, 73–83 (2013)
61. Shad, M.K., Lai, F.W., Fatt, C.L., Klemeš, J.J., Bokhari, A.: Integrating sustainability reporting into enterprise risk management and its relationship with business performance: a conceptual framework. J. Clean. Prod. **208**, 415–425 (2019)
62. Shatnawi, S., Hanefah, M., Adaa, A., Eldaia, M.: The moderating effect of enterprise risk management on the relationship between audit committee characteristics and corporate performance: a conceptual case of Jordan. Int. J. Acad. Res. Bus. Soc. Sci. **9**(5), 177–194 (2019)
63. Sheffi, Y.: Building a resilient supply chain. Harv. Bus. Rev. **1**(8), 1–4 (2005)
64. Skouloudis, A., Tsalis, T., Nikolaou, I., Evangelinos, K., Leal Filho, W.: Small & medium-sized enterprises, organizational resilience capacity and flash floods: insights from a literature review. Sustainability **12**(18), 7437 (2020)
65. Smith, D., Fischbacher, M.: The changing nature of risk and risk management: the challenge of borders, uncertainty and resilience. Risk Manage. **11**(1), 1–12 (2009)
66. Somers, S.: Measuring resilience potential: an adaptive strategy for organizational crisis planning. J. Contingencies Crisis Manage. **17**(1), 12–23 (2009)
67. Sorourkhah. A., Edalatpanah. S.A.: Using a combination of Matrix Approach to Robustness Analysis (MARA) and Fuzzy DEMATEL-Based ANP (FDANP) to choose the best decision. Int. J. Math. Eng. Manage. Sci. **7**(1), 68–80 (2022). https://doi.org/10.33889/IJMEMS.2022.7.1.005
68. Sorourkhah. A., Babaie-Kafaki. S., Azar. A., Shafiei Nikabadi. M.: A fuzzy-weighted approach to the problem of selecting the right strategy using the robustness analysis (case study: Iran automotive industry). Fuzzy Inf. Eng. **11**(1), 39–53 (2019). https://doi.org/10.1080/16168658.2021.1886811
69. Spikin, I.C.: Risk management theory: the integrated perspective and its application in the public sector. Estado, Gobierno y Gestión Pública **21**, 89–126 (2013)
70. Taleb, N.N.: Antifragile: How to Live in a World We Don't Understand, vol. 3. Allen Lane, London (2012)
71. Torabi, S.A., Giahi, R., Sahebjamnia, N.: An enhanced risk assessment framework for business continuity management systems. Saf. Sci. **89**, 201–218 (2016)
72. Van Der Vegt, G.S., Essens, P., Wahlström, M., George, G.: Managing risk and resilience. Acad. Manag. J. **58**(4), 971–980 (2015)
73. Van Gils, A.: Management and governance in Dutch SMEs. Eur. Manage. J. **23**(5), 583–589 (2005)
74. Varmazyari, H., Imani, B.: Analyzing resilience of rural businesses in Malekan County. J. Entrepreneurship Dev. **10**(1), 181–200 (2017). (In Persian)
75. Walker, B., Holling, C.S., Carpenter, S., Kinzig, A.: Resilience, adaptability and transformability in social–ecological systems. Ecol. Soc. **9**, 1–10 (2004)
76. Yang, S., Ishtiaq, M., Anwar, M.: Enterprise risk management practices and firm performance, the mediating role of competitive advantage and the moderating role of financial literacy. J. Risk Financial Manage. **11**(3), 35 (2018). https://doi.org/10.3390/jrfm11030035
77. Zaman, G., Vasile, V.: Conceptual framework of economic resilience and vulnerability at national and regional levels. Romanian J. Econ. **39**(2), 48 (2014)

**Mohsen Imeni** received a B.Sc. (2005) and M.Sc. (2011) in Accounting from Iran. Then, he received a Ph.D. degree in accounting from Islamic Azad University of Rasht Branch, Iran, in 2018. He is an assistant professor in department of accounting at the Ayandegan Institute of Higher Education. He is the author of 3 textbooks and more than 30 publications, and advisory board of 2 international journals, such as the Management Decision (Emerald) and Discrete Dynamics In Nature and Society (Hindawi). He also is a reviewer of journals of the Emerald and the Hindawi. Dr. Imeni's research interests are earnings management, political economy, and auditing.

**Seyyed Ahmad Edalatpanah** received the Ph.D. degree in applied mathematics from the University of Guilan, Rasht, Iran. He is currently working as the chief of R&D at the Ayandegan Institute of Higher Education, Iran. He is also an academic member of Guilan University and the Islamic Azad University of Iran. His fields of interest are uncertainty, fuzzy mathematics, numerical linear algebra, soft computing, and optimization. He has published over 150 journal and conference proceedings papers in the above research areas. He serves on the editorial boards of several international journals. He is also the editor in chief of the International Journal of Research in Industrial Engineering.

# Chapter 10
# Reliability Analysis of Process Systems Using Intuitionistic Fuzzy Set Theory

**Mohammad Yazdi, Sohag Kabir, Mohit Kumar, Ibrahim Ghafir, and Farhana Islam**

**Abstract** In different engineering processes, the reliability of systems is increasingly evaluated to ensure that the safety–critical process systems will operate within their expected operational boundary for a certain mission time without failure. Different methodologies used for reliability analysis of process systems include Failure Mode and Effect Analysis (FMEA), Fault Tree Analysis (FTA), and Bayesian Networks (BN). Although these approaches have their own procedures for evaluating system reliability, they rely on exact failure data of systems' components for reliability evaluation. Nevertheless, obtaining exact failure data for complex systems can be difficult due to the complex behavior of their components, and the unavailability of precise and adequate information about such components. To tackle the data uncertainty issue, this chapter proposes a framework by combining intuitionistic fuzzy set theory and expert elicitation that enables the reliability assessment of process systems using FTA. Moreover, to model the statistical dependencies between events, we use

M. Yazdi (✉)
Faculty of Engineering and Applied Science, Memorial University of Newfoundland, St. John's, Newfoundland A1B 3X5, Canada
e-mail: myazdi@mun.ca; mohammad.yazdi@mq.edu.au

School of Engineering, Macquarie University, Sydney, Australia

S. Kabir · I. Ghafir
Department of Computer Science, University of Bradford, Bradford BD7 1DP, UK
e-mail: s.kabir2@bradford.ac.uk

I. Ghafir
e-mail: i.ghafir@bradford.ac.uk

M. Kumar
Department of Mathematics, Institute of Infrastructure Technology Research and Management (IITRAM), Ahmedabad, Gujarat 380026, India
e-mail: mohitkumar@iitram.ac.in

F. Islam
Department of Education, Bangabandhu Sheikh Mujibur Rahman Digital University, Kaliakair, Gazipur 1750, Bangladesh
e-mail: farhana@edu.bdu.ac.bd



H. Garg (ed.), *Advances in Reliability, Failure and Risk Analysis*, Industrial and Applied Mathematics, https://doi.org/10.1007/978-981-19-9909-3_10

the BN for robust probabilistic inference about system reliability under different uncertainties. The efficiency of the framework is demonstrated through application to a real-world system and comparison of the results of analysis produced by the existing approaches.

**Keywords** Process safety · Intuitionistic fuzzy set · Reliability · Bayesian networks · Expert elicitation · Decision-making

## 10.1 Introduction

Chemical process industries are one of the most hazardous sectors where the potential of occurrence of serious undesirable events, rare accidents, mishaps, or near misses is significant. Such unexpected events can directly or indirectly cause serious injuries like loss of life, serious and immutable environmental damage, loss of material and equipment assets, and decrease the forgotten factor as the reputation of the company. Fire and explosion, the release of toxic, and hazardous materials are common examples of the abovementioned events [1]. Catastrophic accidents such as the Piper Alpha fire and explosion in 1988, BP explosion in 2005, and Deepwater Horizon tragedy in 2010 reveal the tragic effects of major accidents in the chemical process industry [2]. Thus, the prediction of the occurrence of unexpected events and subsequent consequences has a high necessity to assure the safe operation of the system and to prevent the upcoming occurrence of similar events. In this regard, safety and risk analysis can help to prevent the occurrence of unwanted events and develop operational mitigation actions [3]. Several qualitative and quantitative methods, including fault tree analysis (FTA), event tree analysis (ETA), failure mode and effect analysis (FMEA), hazard and operability study (HAZOP), and risk matrix, have been widely used in the risk analysis of chemical process industries. Among the available techniques, FTA is a well-established technique, which can graphically describe the relationships between the cause and effects of different events in the form of Basic Events (BEs), Intermediates Events (IEs), and Top Event (TE). FTA can provide both qualitative and quantitative analysis by presenting undesired events and giving probabilistic analysis from root causes to the consequence [4].

FTA uses the probabilities of BEs (located at bottom of the tree) as quantitative input to calculate the probability of the undesired event as TE (located at the top of the tree). Therefore, the probability of all BEs as crisp values or probability density functions (PDF) is required for quantitative analysis [5]. However, in the real-world industry, because of the lack of knowledge and missing data or systematic bias, the availability of all necessary data cannot be guaranteed. Thus, collecting data from varieties of sources having different features such as dissimilar operating environments, industrial sectors, and experts from diverse backgrounds is an important solution, which has been widely used to obtain the known probability. In addition, even with consideration of exact probabilities or PDFs, intrinsic uncertainties may remain because of different failure modes, lack of knowledge of the mechanism of the failure process, and ambiguity of system experiences. Therefore, a robust method is required for calculating the probability of BEs and addressing the uncertainty among the data collection and analysis procedure [6, 7].

Experts' knowledge has been used to obtain the BEs probability when objective data are limited, incomplete, imprecise, or unknown [8]. The fuzzy set theory (FST) introduced by Zadeh [9] has been demonstrated to be effective and efficient in data uncertainty handling and computing the probability of BEs utilizing multi-expert opinions. The previous studies generally used FST to acquire the probability of BEs from impression and subjectivity in expert judgment. For example, Yazdi and Kabir [10] proposed a framework to obtain the known failure rates from the reliability data handbook and the unknown failure rate according to the experts' opinions. Due to the elicitation procedure considering the unavailability of sufficient data, fuzzy set theory is used to transform linguistic expressions provided by experts into fuzzy numbers. Subsequently, fuzzy possibility, crisp possibility, and failure probability of each BEs are calculated. The risk matrix analysis framework proposed by Yan et al. [11] considered potential risk influences such as controllability, manageability, criticality, and uncertainty. The likelihood in the risk matrix has been calculated by obtaining the probability of the TE of a fault tree. In the TE probability computation process, the probabilities of the BEs of the fault tree have been obtained through expert elicitation. The analytical hierarchy process (AHP) is utilized to improve the accuracy of the failure probability data by minimizing the subjective biases of the experts by quantifying their weightings. Yazdi and Kabir [12] revised Yan et al.'s methodology as a new framework using fuzzy AHP and similarity aggregation procedure (SAM) in the fuzzy environment to cope with available ambiguities of identified BEs. All mentioned papers used a combination of FST and multi-expert knowledge to approximate the BEs' probabilities. However, the FST suffers from several shortages. The one worth mentioning is related to the uncertainty or hesitation about the degree of membership. The FST cannot include the hesitation in the membership functions. In this regard, Atanassov [13] extended conventional fuzzy set to propose the intuitionistic fuzzy set (IFS), in which non-membership degrees and hesitation margin groups have been included with the membership degrees. The IFS data are more complete than the conventional fuzzy data that considers membership function only [14]. In another example, it is demonstrated the use of IFSs to handle uncertainties in FMEA [15]. Yazdi [16] utilized IFS and specifically intuitionistic fuzzy numbers (IFNs) to develop a conventional risk matrix.

To the best of the authors' knowledge, limited research has been conducted to combine IFNs and multi-expert knowledge to address the issues of data uncertainty in FTA. For instance, Shu et al. [17] utilized IFNs to analyze the failure behavior of the printed circuit board assembly. A vague FTA approach has been proposed [18] by integrating experts' judgment into the analysis to calculate the fault interval of system components. Afterward, for fuzzy reliability evaluation of a "*liquefied natural gas terminal emergency shutdown system*", Cheng et al. [19] used IFS with FTA.

The weakest t-norm-based IFS has been used with FTA [20] to evaluate system reliability. Recently, Kabir et al. [21] have utilized IFS for dynamic reliability analysis.

On the other hand, traditional FTA as well as fuzzy FTA are well known to have a static structure and cannot consider the variation of risk due to the dynamic behavior of the system. In addition, BEs are assumed to be independent in both methods and they are considered to have binary states—failed and non-failed, whereas, in practice, events can be in more than two states. Moreover, the effects of common cause failure (CCF) in the reliability of systems are usually not considered in traditional FTA. Such mentioned issues are commonly named as model uncertainty in risk analysis [22]. Thus, model uncertainty is recognized as a considerable limitation of risk analysis methods. In this regard, a dependency coefficient method is introduced by Ferdous et al. [23] to evaluate the interdependencies of BEs in static FT. The joint likelihood function in the hierarchical Bayesian network is developed [24] to consider the interdependencies among BEs in conventional FT. Besides, Hashemi et al. [25] used the copula function technique to evaluate and model the interdependencies of BEs to improve uncertainty analysis.

Bayesian networks (BN) have become a popular method, which has been widely used to incorporate a variety of information types such as extrapolated data, experts' judgment, or partially related data in risk analysis of process industries [26, 27]. Kabir and Papadopoulos [22] provided a review of the applications of BNs in reliability and risk assessment areas. Examples of such applications include risk analysis of fire and explosion [28, 29], leakage [30, 31], human error [32–34], maintenance activity [35, 36], and offshore and drilling operations [37–39] utilized BN as a probabilistic interface tool for reasoning under uncertainty. BN used a chain rule or *d-separation* to represent the causal relationships between a set of variables (in case of FTA is BEs) considering the dependencies [40]. BN is also able to cope with the limitations of conventional FTA as well as having a flexible structure. Several scholars have used BN in parallel with FTA and addressed the shortages of the conventional FTA by mapping FT into the corresponding BN [41–44]. Because of the modeling flexibility provided by BN, the interdependencies of BEs can be effortlessly modeled by using BN. BN can also model multiple states for BEs and common cause failure (CCF) scenarios. Furthermore, to deal with the model uncertainty, BN can perform the probability updating mechanism using Bayes' theorem by adding new information about the system over time.

The novelty and contribution of this work are utilizing the advantages of IFNs over traditional FST to evaluate the TE probability of an FT. Besides, this chapter adopts BN to allow dynamic risk assessment under uncertainty, where the BEs' probabilities are calculated based on the combination of subjective opinions and IFNs, and BN is used to take into account the interdependencies of BEs as well as CCF. The rest of the chapter is organized as follows. In Sect. 10.2, the uncertainty sources in chemical process industries are reviewed. A short overview of the IFS theory is presented in Sect. 10.2. In Sect. 10.3, the proposed methodology is described. Section 10.4 demonstrates the feasibility and efficiency of the proposed approach via a numerical example with sensitivity analysis. Lastly, the concluding statements are presented in Sect. 10.5.

## 10.2 Background

### *10.2.1 Uncertainty Sources in Chemical Process Industries*

The term uncertainty is widely used with a different meaning in the literature on risk analysis. Several scholars claimed that uncertainty is equal to risk about the future and accordingly risk is equal to uncertainty. Others stated that uncertainty and risk are from two different schools and it has not been complicated to each other [40]. In this chapter, it is assumed the terms risk and uncertainty are two different concepts. There exist two distinct concepts of uncertainty in chemical process industries including (i) uncertainty caused by physical unpredictability (aleatory uncertainty) and (ii) uncertainty caused by insufficient knowledge (epistemic uncertainty) [45, 46].

The existence of aleatory and epistemic uncertainties in risk analysis of chemical process industries implies that the probability of numerous risk factors cannot be measured in an appropriate way when they are ambiguous and unknown. Referring to aleatory uncertainty, the random behavior of some parameters in a system or its environment should be stated such as inconsistency in weather conditions and experimental data variability for BEs in FT. In contrast, epistemic uncertainty is related to fuzziness, vagueness, or imprecision regarding the quality of chemical process safety, particularly in the accident scenario identification and consequence modeling. In reality, it is difficult to reduce aleatory uncertainty because of the intrinsic nature of a system, whereas it is possible to reduce epistemic uncertainty when more knowledge about the system is available over time. More information about the characteristics of aleatory and epistemic uncertainties can be found in [47]. This study concentrates on epistemic uncertainty.

During analysis, a certain explanation or assumption about the models leads to model uncertainty. Moreover, mathematical and other analytical tools are utilized to reduce properties of interest, ranging from structural, stochastic, human behavior, accident, evacuation, dispersion model, etc. This study concentrates on the model uncertainty caused by the independence assumptions among BEs in FTA. Thus, the modeling capability of BN is used to assess the dependency among events to address the abovementioned issue.

Parameter uncertainties are caused by the imprecisions and inaccuracies in the input data used in the process safety analysis. These uncertainties are intrinsic due to the imperfect nature of the available data, and the analysis process requires to be based on partial knowledge. Nonetheless, it is believed that parameter uncertainty is the easiest one to be quantified [48]. In the literature, to cope with parameter uncertainty, it is commonly expressed by PDFs and Monte Carlo simulation-based probability theory [49–51]. However, as mentioned earlier, PDFs are rarely easy to obtain. In this chapter, IFNs are utilized to deal with parameter uncertainty, where the probabilities of BEs are treated as IFNs that are derived from multi-experts' knowledge.

### 10.2.2 IFS Theory

The concept of the classical fuzzy sets has been generalized by Atanassov [13] into IFS through the introduction of a non-membership function $\upsilon_{\tilde{A}}(x)$ indicating the evidence against $x \in X$ along with the membership value $\mu_{\tilde{A}}(x)$ indicating evidence for $x \in X$ and this admits an aspect of indeterminacy.

An IFS $\tilde{A}$ in the universe of discourse $X$ is given by

$$\tilde{A} = \{\langle x, \mu_{\tilde{A}}(x), \upsilon_{\tilde{A}}(x)\rangle : x \in X\} \tag{10.1}$$

where $\mu_{\tilde{A}} : X \to [0, 1]$ and $\upsilon_{\tilde{A}} : X \to [0, 1]$ are membership and non-membership functions, respectively, where

$$0 \le \mu_{\tilde{A}^i}(x) + \upsilon_{\tilde{A}^i}(x) \le 1, \forall x \in X \tag{10.2}$$

For every value $x \in X$, the values $\mu_{\tilde{A}}(x)$ and $\upsilon_{\tilde{A}}(x)$ represent, respectively, the degree of membership and degree of non-membership to $\tilde{A} \subseteq X$ Moreover, the uncertainty level or hesitation degree of the membership of $x$ in $\tilde{A}$ is denoted as:

$$\pi_{\tilde{A}}(x) = 1 - \mu_{\tilde{A}}(x) - \upsilon_{\tilde{A}}(x) \tag{10.3}$$

If $\pi_{\tilde{A}}(x) = 0, \forall x \in X$, then the IFS becomes a classical fuzzy set.

If the membership and non-membership functions of an IFS $\tilde{A}$ (i.e., $\mu_{\tilde{A}}(x)$ and $\upsilon(_{\tilde{A}}(x))$ satisfy the following conditions given by Eqs. (10.4) and (10.5), then $\tilde{A}$ in $X$ is considered as IF-convex

$$\mu_{\tilde{A}}(\lambda x_1 + (1-\lambda)x_2) \ge \min(\mu_{\tilde{A}}(x_1), \mu_{\tilde{A}}(x_2)) \forall x_1, x_2 \in X, 0 \le \lambda \le 1. \tag{10.4}$$

$$\upsilon_{\tilde{A}}(\lambda x_1 + (1-\lambda)x_2) \le \max(\upsilon_{\tilde{A}}(x_1), \upsilon_{\tilde{A}}(x_2)) \forall x_1, x_2 \in X, 0 \le \lambda \le 1. \tag{10.5}$$

If there exist at least two points $x_1, x_2 \in X$ such that $\mu_{\tilde{A}}(x_1) = 1$ and $\upsilon_{\tilde{A}}(x_2) = 1$, then the IFS $\tilde{A}$ in X is considered as IF-normal [52]:

An IFS $\tilde{A} = \{\langle x, \mu_{\tilde{A}}(x), \upsilon_{\tilde{A}}(x)\rangle : x \in R\}$ is called an IFN if

(i) $\tilde{A}$ is IF-normal and IF-convex.
(ii) $\mu_{\tilde{A}}(x)$ is an upper and $\upsilon_{\tilde{A}}(x)$ is a lower semi-continuous.
(iii) $Supp\tilde{A} = \{x \in X : \upsilon_{\tilde{A}}(x) < 1\}$ is bounded (see Fig. 10.1).

A Triangular-IFN is an IFN given by

$$\mu_{\tilde{A}}(x) = \begin{cases} \frac{x-a_1}{a_2-a_1}, & a_1 \le x \le a_2 \\ \frac{a_3-x}{a_3-a_2}, & a_2 \le x \le a_3 \\ 0, & \text{otherwise} \end{cases} \tag{10.6}$$

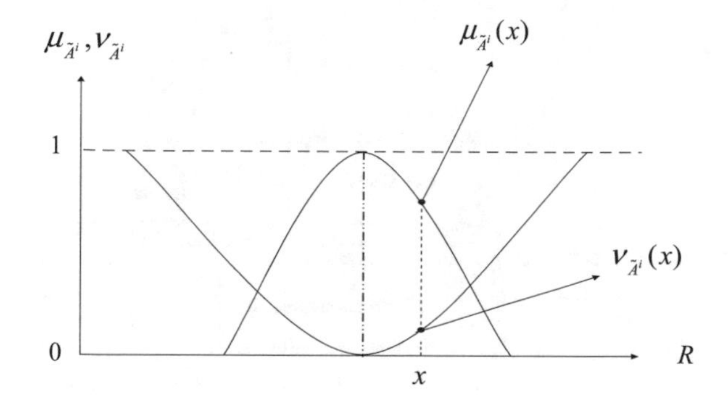


**Fig. 10.1** Graphical representation of IFNs

and

$$\upsilon_{\tilde{A}}(x) = \begin{cases} \frac{a_2 - x}{a_2 - a_1'}, & a_1' \le x \le a_2 \\ \frac{x - a_2}{a_3' - a_2}, & a_2 \le x \le a_3' \\ 1, & \text{otherwise} \end{cases} \tag{10.7}$$

where $a_1' \le a_1 \le a_2 \le a_3 \le a_3'$. This TIFN is denoted by $\tilde{A} = \left(a_1, a_2, a_3; a_1', a_2, a_3'\right)$.

## 10.3 Material and Method

To introduce the methodology developed in this chapter, this section briefly describes the framework as can be seen in Fig. 10.2.

### *10.3.1 Hazard Analysis*

There are numerous methods available for hazard analysis in different types of industrial sectors. The initial step of all hazard analysis methods is identifying all possible hazards. Therefore, well understanding of process function has a high necessity for this purpose. All information about a process system should be collected to understand its functionality appropriately. Then, any hazards which have enough potential to destroy the industrial equipment, surrounding environment, or harm to the public should be considered [40]. HAZOP technique is based on the brainstorming method that has enough capability to recognize hazardous systems and sub-systems by employing a group of specialists, commonly a third-party company. Thus, this study considers the outcome of HAZOP as a highly probable and severe event. In fact, the HAZOP study is commonly in conducted process-based industries to identify the deviation as a pre-step fault tree analysis. However, considering the inherent

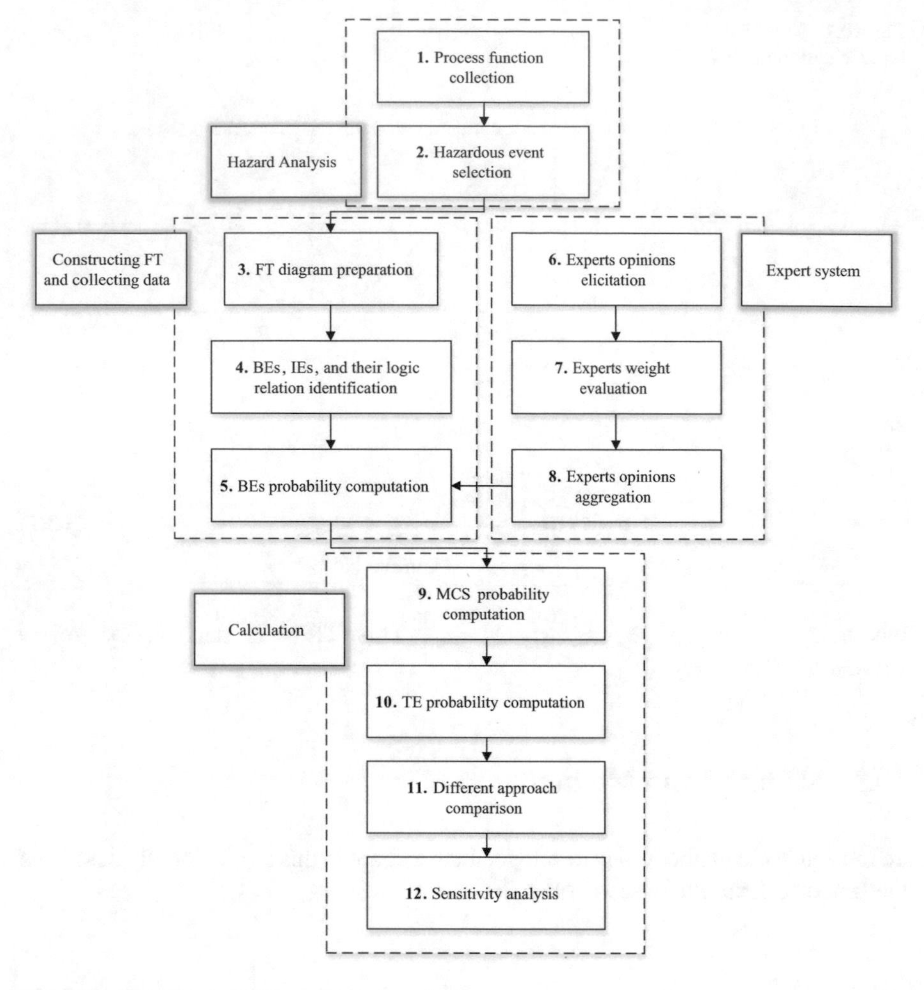


**Fig. 10.2** The structure of the proposed method

features process, FMEA or other types of risk assessment method can also be carried out.

## *10.3.2 Developing a Fault Tree and Collecting Data*

After identifying an event as the TE of a fault tree, the rest of the tree is developed from top to bottom in a downward direction. It should be noted that further analysis of the FT is performed based on the TE. Therefore, the TE of the FT must be chosen appropriately for further analysis. The TE is commonly specified as an accident or hazardous event which can potentially be a cause of asset loss or harm to the public. After finalizing the development of an FT, the BEs that are put at the bottom level of the tree (leaves) should be identified to facilitate further analysis. The logic relationship between BEs, IEs, and TE is defined using Boolean OR and AND gates.

The reliability data such as the ones from OREDA [53] can be used to obtain the failure rate of known BEs. Nevertheless, when there is a difficulty in using a reliability handbook to obtain failure rates of rare events with unknown or limited failure data, three popular methods, including expert judgment, extrapolation, and statistical methods, can be utilized to estimate the failure rates [54]. The statistical method estimates the failure rates by estimating the failure probabilities by performing a short test on the practical data. In addition, statistical methods can be distinguished with deterministic methods, which are suitable where observations are precisely reproducible or are expected to be in this manner. The extrapolation method denotes the utilization of a predicting model, equal condition, or the available reliability data sources. The expert judgment method directly calculates probabilities based on experts' opinions on the occurrence of BEs. This study employs the expert judgment method to estimate BEs' occurrence probability. In this regard, a combination of subjective opinions expressed by experts and IFNs can help assessors to deal with the uncertainty that may arise during the analysis. In the following subsection, the procedure of using an expert system is presented.

### *10.3.3 Use of the Expert System*

Expert systems are convenient to use in quantitative analysis models in circumstances where the available situations make it difficult or even more impossible to make enough observation to quantify the models using real data. Thus, expert systems are commonly used to approximate the model parameter under ambiguous conditions. Expert systems can also be used to improve the estimation, which is gained from real data.

An expert provides his/her judgment about a subject based on knowledge and experience according to his/her background. Thus, an employed expert will require to respond to a predefined set of questions related to a subject, which can include personal information, probabilities, rating, weighting factor, uncertainty estimation, and so on. The experts' opinions can be collected during an eliciting. An important issue related to the elicitation process is that experts' opinions should not be used instead of rigorous reliability and risk analysis approaches, whereas it can be used to supplement them where reliability and risk analytical approaches are inconsistent or inappropriate.

#### 10.3.3.1 Experts' Opinion Elicitation

Due to the increased complexity of systems and the subjective nature of expert judgment, no officially renowned approach has been developed for treating expert opinion. Once the elicitation process is finished, opinions are analyzed by combining them to obtain an aggregated result to be used in the reliability analysis. Clemen and Winkler [55] divided the elicitation and aggregation processes into two categories—behavioral and mathematical methods. Behavioral methods aim to create some sort of group agreement between the employed experts. While, in mathematical methods, the experts expressed their opinion about an uncertain quantity in the form of subjective probabilities. Afterward, suitable mathematical methods are used to combine these opinions. The rationale behind using mathematical approaches for the processing of experts' opinions was provided in [56, 57]. Hence, in this study, one of the mathematical methods is used to analyze experts' opinions.

According to [58], probability can be considered as a numerical representation of uncertainty because it offers a way to quantify the likelihood of occurrence of an event. Therefore, it is much easier for the employed experts to use linguistic expressions like high probable, low probable, and so on to express their opinions. Three elicitation methods that have been widely used for subjective analysis are *Indirect*, *Direct*, and *Delphi*. The basis of the *Indirect* method is to utilize the betting rates of experts to reach a point of indifference between obtainable choices according to an issue. The *Direct* method is the direct estimation of the degree of confidence of an expert on some subject. The *Delphi* technique is the first organized tool for methodologically collecting opinions on a specific subject using a cautiously defined ordered set of questionnaires mixed with summarized information and feedback resulting from previously received responses [59, 60]. The selection of each method for a particular purpose should fulfill the rational consensus principles such as accountability and fairness. In this study, among the abovementioned methods, *Delphi*, because of having enough capacity for expert opinion elicitation, is selected for eliciting process.

#### 10.3.3.2 Experts Weighting Evaluation

Once the experts' opinion elicitation process is completed, the expert weighting calculation is started. This step is necessary because, in real life, each employed expert has a different weight according to his/her experience and background. Thus, to obtain realistic results for the probability of each BE, the weight (importance of the judgment outcome) of the employed experts should be identified. There are many methods such as simple averaging besides many unmethodical techniques that may be used for giving specific weighting to the experts. However, they cannot diminish subjective bias and help domain experts to carry out the eliciting procedure in an effective way.

AHP (analytical hierarchy process) introduced by Saaty [61] is a widely used process in multi-criteria decision-making. This process breaks large decision problems into smaller ones and then uses a hierarchy of decision layers to handle the complexity of the problems. This allows focusing on a smaller set of the decision at a time. There exist criticism regarding AHP's use of lopsided judgmental scales and its inability to appropriately reflect the characteristic uncertainty and imprecision of pair comparisons [62]. The verbal statements provided by the decision-makers in AHP could be unclear. Moreover, they regularly would choose to provide their preferences as oral expressions instead of numerical quantities and the type of pair comparisons used cannot properly reflect their decisions about priorities [63–66]. The abovementioned shortages represent that in most cases, the nature of decision-making is full of ambiguities and complexities, and accordingly it is denoted that most decisions are made in a fuzzy environment.

Let $O = \{o_1, o_2, \ldots, o_n\}$ is a set of objects and $W = \{w_1, w_2, \ldots, w_m\}$ is a set of goals. Therefore, the extent analysis values for $m$ goals for each object can be denoted as:

$$M_{gi}^{1}, M_{gi}^{2}, \ldots, M_{gi}^{m} \quad i = 1, 2, \ldots n \tag{10.8}$$

where each of $M_{gi}^{m}$ is a triangular fuzzy set.

*Step 1.* The fuzzy synthetic extent concerning the $i$-th object is denoted as:

$$\sum_{j=1}^{m} \mathrm{M}_{gi}^{j} \otimes \left[ \sum_{i=1}^{n} \sum_{j=1}^{m} \mathrm{M}_{gi}^{j} \right]^{-1} \tag{10.9}$$

To get $\sum_{j=1}^{m} \mathrm{M}_{gi}^{j}$ the fuzzy addition operation of $m$ extent analysis values for a particular matrix is achieved as:

$$\sum_{j=1}^{m} \mathrm{M}_{gi}^{j} = \left( \sum_{j=1}^{m} l_j, \sum_{j=1}^{m} m_j, \sum_{j=1}^{m} u_j \right) \tag{10.10}$$

and afterward, the inverse of the vector is calculated as follows:

$$\left[ \sum_{i=1}^{n} \sum_{j=1}^{m} \mathrm{M}_{gi}^{j} \right]^{-1} = \left( \frac{1}{\sum_{j=1}^{m} l_j}, \frac{1}{\sum_{j=1}^{m} m_j}, \frac{1}{\sum_{j=1}^{m} u_j} \right) \tag{10.11}$$

Step 2. The degree of likelihood of $\mathrm{M}_2 = (l_2, m_2, u_2) \geq \mathrm{M}_1 = (l_1, m_1, u_1)$ is calculated as:

$$V(M_2 \geq M_1) = \sup_{y \geq x} \left[ \min\left( \mu_{M_1}(x), \mu_{M_2}(y) \right) \right] \tag{10.12}$$

It can be represented by Eq. (10.13).

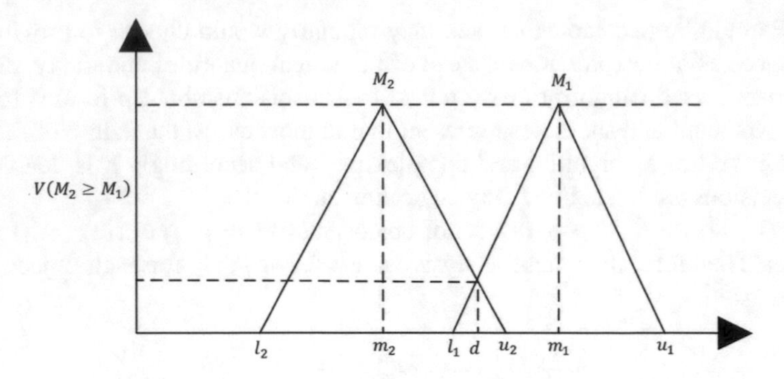


**Fig. 10.3** The intersection between $M_1$ and $M_2$

$$V(M_2 \geq M_1) = hgt(M_1 \cap M_2) = \mu_{M_2}(d) = \begin{cases} 1, & \text{if } m_2 \geq m_1 \\ 0, & \text{if } l_2 \geq u_1 \\ \frac{l_1 \geq u_1}{(m_2 - u_2) - (m_1 - u_1)}, & \text{otherwise} \end{cases} \tag{10.13}$$

As seen in Fig. 10.3, $d$ is the highest intersection point between $\mu_{M_1}$ and $\mu_{M_2}$.

*Step 3*. The degree of likelihood that a convex fuzzy number is greater than $k$ convex fuzzy $M_i (i = 1, 2, \ldots, k)$ numbers can be obtained by:

$$\begin{aligned} V(M \geq M_1, M_2, \ldots, M_k) &= V[(M \geq M_2) \text{ and } (M \geq M_1) \text{and } \ldots \text{ and } (M \geq M_k)] \\ &= \min V(M \geq M_i), i = 1, 2, 3, \ldots k \end{aligned} \tag{10.14}$$

Suppose that $d'(A_i) = \min V(S_i \geq S_k)$ for $k = 1, 2, \ldots, n; k \neq i$. Now, the given weight vector is denoted by:

$$W' = (d'(A_1), d'(A_2), \ldots, d'(A_n))^{\mathrm{T}} \tag{10.15}$$

where $A_i (i = 1, 2, \ldots, n)$ are $n$ elements.

*Step 4*. Using normalization, the normalized weight vectors are:

$$W_{\text{FAHP}} = d(A_1), d(A_2), \ldots, d(A_n))^T \tag{10.16}$$

where $W$ is a non-fuzzy number.

The fuzzy linguistic variables are used to allow experts to provide their subjective opinions reflecting nine-point essential scale. In this chapter, the linguistic variables and their equivalent fuzzy numbers are used.

#### 10.3.3.3 Experts' Opinion Aggregation

The experts' opinion aggregation process can be completed in three phases including (i) obtaining linguistic terms from experts describing the likelihood of occurrence of BEs, (ii) mapping linguistic variables into the corresponding fuzzy numbers, and (iii) applying an aggregation process under fuzzy environment.

Firstly, the engaged experts provided their judgements about the likelihood of occurrence of each BE in the fault tree. Their opinions can be obtained in the form of linguistic variables represented as IFNs.

As experts may have dissimilar opinions about a subject due to having a different level of experience, background, and expertise, it is essential to aggregate multi-expert opinions to reach an agreement. Different kinds of aggregation methods like the arithmetic averaging method and similarity aggregation method (SAM) can be utilized for this purpose. However, Yazdi and Zarei [56] pointed out the benefits of such methods in the context of fuzzy FTA. It is concluded that SAM has enough capability for this purpose. Therefore, an extension of SAM as described in [67] is used in this chapter for the aggregation of IFNs. The SAM method contains the following steps.

*Step A. Mapping of linguistic variables into equivalent IFNs*:

After each expert, $E_k (k = 1, 2, \ldots, n)$ provides his/her judgment about the occurrence possibility of each BE in the form of linguistic variables; accordingly, it is transformed into the equivalent IFNs.

*Step B. Degree of similarity computation*:

The similarity $S_{uv}\left(\tilde{A}_u, \tilde{A}_v\right)$ between the opinions $\tilde{A}_u$ and $\tilde{A}_v$ of experts $E_u$ and $E_v$ is evaluated as:

$$S_{uv}\left(\tilde{A}_u, \tilde{A}_v\right) = \begin{cases} \frac{EV_u}{EV_v} if EV_u \leq EV_v \\ \frac{EV_v}{EV_u} if EV_v \leq EV_u \end{cases} \tag{10.17}$$

where $S_{uv}\left(\tilde{A}_u, \tilde{A}_v\right) \in [0, 1]$ is the function to measure similarity, where $\tilde{A}_u$ and $\tilde{A}_v$ are two regular intuitionistic fuzzy numbers, $EV_u$ and $EV_v$ are the expectancy evaluation for $\tilde{A}_u$ and $\tilde{A}_v$. The $EV$ of a triangular IFN $\tilde{A} = \left(a, b, c; a', b', c'\right)$ is calculated as:

$$EV\left(\tilde{A}\right) = \frac{\left(a + a'\right) + 4b + (c + c')}{8} \tag{10.18}$$

A *similarity matrix* $(SM)$ for $m$ experts is defined as:

$$SM = \begin{bmatrix} 1 & s_{12} & s_{13} \cdots & s_{1m} \\ s_{21} & 1 & s_{23} \cdots & s_{2m} \\ \vdots & \vdots & \vdots \ddots & \vdots \\ s_{m1} & s_{m2} & s_{m3} \ldots & 1 \end{bmatrix} \tag{10.19}$$

where $S_{uv} = s\left(\tilde{A}_u, \tilde{A}_v\right)$, if $u = v$ then $S_{uv} = 1$.

*Step C. Degree of agreement computation:*

The average agreement degree $AA(E_m)$ for each expert is calculated as

$$\mathrm{AA}(E_m) = \frac{1}{m-1} \sum_{\substack{v=1 \\ v \neq 1}}^{m} S_{uv} \tag{10.20}$$

where $m = 1, 2, \ldots, n$.

*Step D. The relative agreement computation*:

The $\mathrm{RAD}(E_m)$ is the relative agreement degree, which can be calculated as:

$$\mathrm{RAD}(E_m) = \frac{\mathrm{AA}(E_m)}{\sum_{u=1}^{n} \mathrm{AA}(E_n)} \tag{10.21}$$

where $m = 1, 2, \ldots, n$.

*Step E. Consensus degree computation*:

The aggregation weight ($w_m$) of an expert $E_m$ is computed using $\mathrm{RAD}(E_m)$, and the weight of each expert ($W_{\mathrm{FAHP}}$) is obtained by FAHP as follows.

$$w_m = \alpha \odot W_{\mathrm{FAHP}}(E_m) + (1 - \alpha) \odot \mathrm{RAD}\ (E_m) \tag{10.22}$$

where $\alpha (0 \leq \alpha \leq 1)$ is the weighting factor also known as a relaxation factor that can be assigned to $W_{\mathrm{FAHP}}(E_m)$ $\mathrm{RAD}(E_m)$ to define their relative importance.

*Step F. Aggregated result computation*:

The aggregated result for each basic event can be computed as:

$$\tilde{P}_j = \sum_{i=1}^{n} w_m \otimes \tilde{P}_{ij} \tag{10.23}$$

where $\tilde{P}_j$ is the aggregated possibility of basic event $j$ in the form of IFNs.

So far, the aggregation possibility of each BE based on IFNs is computed. In the next section, the procedure of TE computation is explained.

### 10.3.4 Calculation of Probability of TE

Once the occurrence possibilities of all BEs are obtained, these values are translated into the equivalent probabilities using the following equation introduced by [68]:

$$\text{FP} = \begin{cases} 1/10^k & \text{FPS} \neq 0 \\ 0 & \text{FPS} = 0 \end{cases} \tag{10.24}$$

where FP and FPS represent failure probability and failure possibility, respectively, and

$$k = 2.301 \times [(1 - \text{FPS})/\text{FPS}]^{1/3} \tag{10.25}$$

Once the intuitionistic fuzzy failure probabilities of the BEs are obtained, they are used to calculate the IF probability of the TE. Intuitionistic fuzzy arithmetic operations are adopted to evaluate the probabilities of the minimal cut sets of the FT and the same for the TE probability.

A set of minimal cut sets of a fault tree can be denoted as:

$$S = C_i : i = 1, 2, \ldots, m \tag{10.26}$$

where $C_i$ is the $i$-th minimal cut set of order $k$ and is denoted as $C_i = e_1.e_2 \ldots e_k$.

Let the probability $\tilde{P}_j$ of event $e_j : i = 1, 2, \ldots, n$ be characterized by triangular IFNs $\left(a_j, b_j, c_j; a'_j, b_j, c'_j\right)$, then the failure probability of $\tilde{P}_{C_i}$ of the minimal cut set $C_i$ is estimated using the following expressions.

$$\begin{aligned} \tilde{P}_{C_i} &= \text{AND}\left(\tilde{P}_1, \tilde{P}_2, \ldots, \tilde{P}_k\right) = \tilde{P}_1 \otimes \tilde{P}_2 \otimes \ldots \otimes \tilde{P}_k \\ &= \left(\prod_{j=1}^{n} a_j, \prod_{j=1}^{n} b_j, \prod_{j=1}^{n} c_j; \prod_{j=1}^{n} a'_j \prod_{j=1}^{n} b_j, \prod_{j=1}^{n} c'_j\right) \end{aligned} \tag{10.27}$$

As the TE of an FT is represented by an OR gate, the failure probability of the TE can be calculated using the following equation:

$$\begin{aligned} \tilde{P}_{C_i} =& \text{OR}\left(\tilde{P}_{c1}, \tilde{P}_{c2}, \ldots, \tilde{P}_{cm}\right) = 1\ominus\left(1\ominus \tilde{P}_{c1}\right) \otimes \left(1\ominus \tilde{P}_{c2}\right) \otimes \ldots \otimes \left(1\ominus \tilde{P}_{cm}\right) \\ =& \left(1 - \prod_{j=1}^{n}(1 - a_j), 1 - \prod_{j=1}^{n}(1 - b_j), 1 - \prod_{j=1}^{n}(1 - c_j);\right. \\ & \left. 1 - \prod_{j=1}^{n}(1 - a'_j), 1 - \prod_{j=1}^{n}(1 - b_j), 1 - \prod_{j=1}^{n}(1 - c'_j)\right) \end{aligned} \tag{10.28}$$

where $\tilde{P}_{C_1}, \tilde{P}_{C_2}, \ldots, \tilde{P}_{C_m}$ denoted the failure probabilities of all MCSs $C_i : i = 1, 2, \ldots, m$.

Through IF-defuzzification process an IFN can be converted to a single scalar quantity. The failure probability of the TE obtained as triangular IFN $\tilde{A} = \left(a, b, c; \acute{a}, b, \acute{c}\right)$ can be defuzzified as follows.

$$X = \frac{1}{3}\left[\frac{(c' - a')(b - 2c' - 2a') + (c - a)(a + b + c) + 3(c'^2 - a'^2)}{c' - a' + c - a}\right] \tag{10.29}$$

### 10.3.5 Different Approach Comparison

To understand the efficiency of the proposed model, the results are compared with the common approaches. Firstly, conventional FFTA based on the FST which is widely used in different engineering applications is applied. Then, an approach based on the integration of the BN and FST which was introduced in [12] is utilized.

As mentioned in the literature, the procedure of conventional FFTA is utilizing triangular or trapezoidal fuzzy numbers for the probability expression of all BEs in FT. Then, fuzzy arithmetic operations are utilized to compute the TE probability in terms of a fuzzy number.

In the second approach, after [69] that compared conventional FTA and BN, many studies have been performed by mapping FT into the corresponding BN for different applications. A list of such works can be found in literature [70], which makes use of the advantages of multi-expert opinions and FST for uncertainty handling in the data and BN for modeling dependency between events. According to their approach, the probability of each BE is computed in five key steps as collecting experts' opinions in qualitative terms, fuzzification, aggregation, defuzzification, and probability computation. Once the probability of each BEs is obtained, then FT is mapped into the corresponding BN. According to the Bayes theorem, the TE probability can be calculated as follows.

In a BN, the joint probability distribution of a set of variables can be denoted using the conditional dependency of variables and chain rules as follows:

$$P(U) = \prod_{i=1}^{n} P(X_i | X_{i+1}, \ldots X_n) \tag{10.30}$$

where $U = \{X_1, X_2, \ldots, X_n\}$ and $X_{i+1}$ is the parent of $X_i$. Consequently, the probability of $X_i$ can be calculated as:

$$P(X_i) = \sum_{U \setminus X_i} P(U) \tag{10.31}$$

Using Bayes theorem as seen in Eq. (10.32), the prior probability of an event ($E$) can be updated.

$$P(U|E) = \frac{P(U \cap E)}{P(E)} = \frac{P(U \cap E)}{\sum_U P(U \cap E)} \tag{10.32}$$

To get further details, readers can refer to [71].

### 10.3.6 Sensitivity Analysis

Once the relative competency of each expert's opinion is predicted, it is better to determine the consensus coefficient. Thus, the decision-maker needs to allocate a proper value for the relaxation factor $\alpha$ in Eq. (10.22); otherwise, sensitivity analysis (SA) should be performed to evaluate the reliability of the system when $\alpha$ has been given different values ranging from 0 to 1. In this study, the relaxation factor is considered as 0.5 to give equal weights to both factors on the right side of the Eq. (10.22). However, to identify the sensitivity of the BEs, we have performed the sensitivity analysis by varying the values of $\alpha$. This helped to understand which of the BEs are more sensitive to uncertainty.

Using BIM, the criticality of an event is identified as follows:

$$\text{BIM}(\text{BE}_i) = P(\text{Top Event}|P(\text{BE}_i) = 1) - P(\text{Top Event}|P(\text{BE}_i) = 0) \tag{10.33}$$

As seen in the above equation, the criticality of the basic event $\text{BE}_i$ is computed by taking the difference between the top event probabilities when the $\text{BE}_i$ is assumed to have occurred and non-occurred, respectively.

## 10.4 Application to the Case Study

The developed methodology is applied to the risk analysis of an ethylene oxide (EO) production plant that is a component of an ethylene transportation line to demonstrate its effectiveness. The detail of the system is shown in Fig. 10.4. A prior study performed on the abovementioned system by [72] identified the most hazardous components of the system, including the ethylene oxide storage and reaction unit, ethylene oxide distillation column, transportation line, and ethylene re-boiler. It was recommended that further risk assessment is essential for the declared units. Therefore, Khan and Haddara [73] found optimal maintenance in the above case study using a risk-based maintenance method. Additionally, the ethylene transportation line component was recognized as the third key hazard in the available units. In this regard, [12] applied their proposed approach to EO Transportation line as a case study.

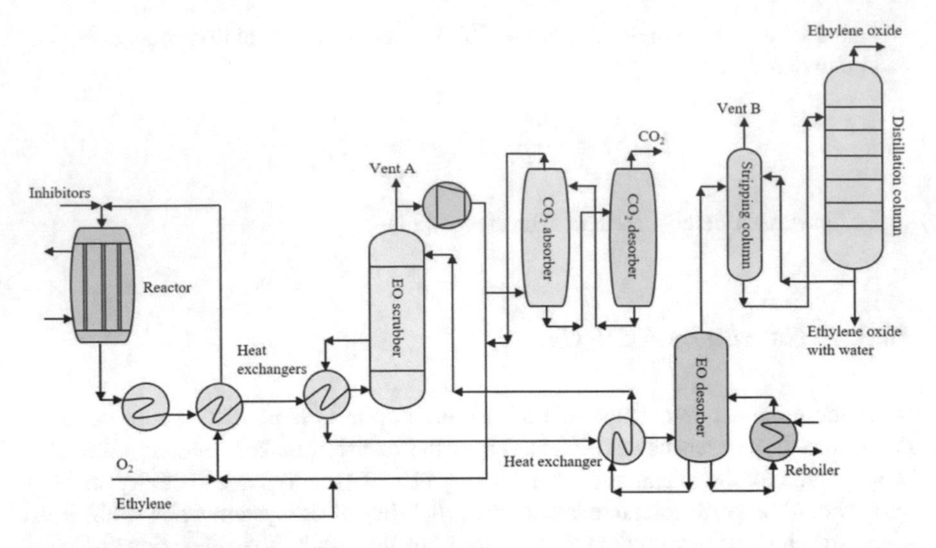


**Fig. 10.4** Schematic diagram of the EO plant [72]

## *10.4.1 Probabilistic Risk Assessment*

An ignition of vapor cloud that may lead to a fireball is selected as the TE of the FT. The developed FT is shown in Fig. 10.5. As seen in the fault tree, there are 25 BEs (represented as circles) and details of these BEs are presented in Table 10.1. To compute the occurrence probability of each BE, the heterogeneous group of experts used in [12] has been used in this chapter. Using the Delphi method, employed experts were asked to provide their judgements in relevant linguistic terms. The weights of experts have been computed using the FAHP method, and the calculated weights of experts 1, 2, 3, and 4 are 0.249, 0.126, 0.495, and 0.128, respectively [12].

To show the aggregation procedure of expert's judgment; consider the case of BE24 (Corrosion) as an example. Concerning the characterization of IFNs, the linguistic variables, obtained from four experts, are categorized as "L", "M", "FH", and "M". The detailed computation of aggregation for BE24 is shown in Table 10.2. The aggregated results for all BEs are presented in Table 10.3.

To calculate the TE of the FT of Fig. 10.5, it was qualitatively analyzed to obtain 102 MCSs. Each of the MCSs is a combination of a number of BEs that can cause the TE. Using the Eqs. (10.22), (10.23) and the IF-probabilities of the BEs from Table 10.3, the TE probability as IFN is calculated as: {3.296E-11, 8.270E-10, 1.132E-08, 1.804E-11, 8.270E-10, 1.922E-08}. After defuzzification, the crisp probability of the TE obtained is 5.715E-09. We have also used the crisp probabilities of the BEs (see the last column of Table 10.3) to evaluate the TE probability and the value obtained was 1.620E-09. As can be seen, this value is close to the value obtained through the defuzzification of the IF-probabilities.

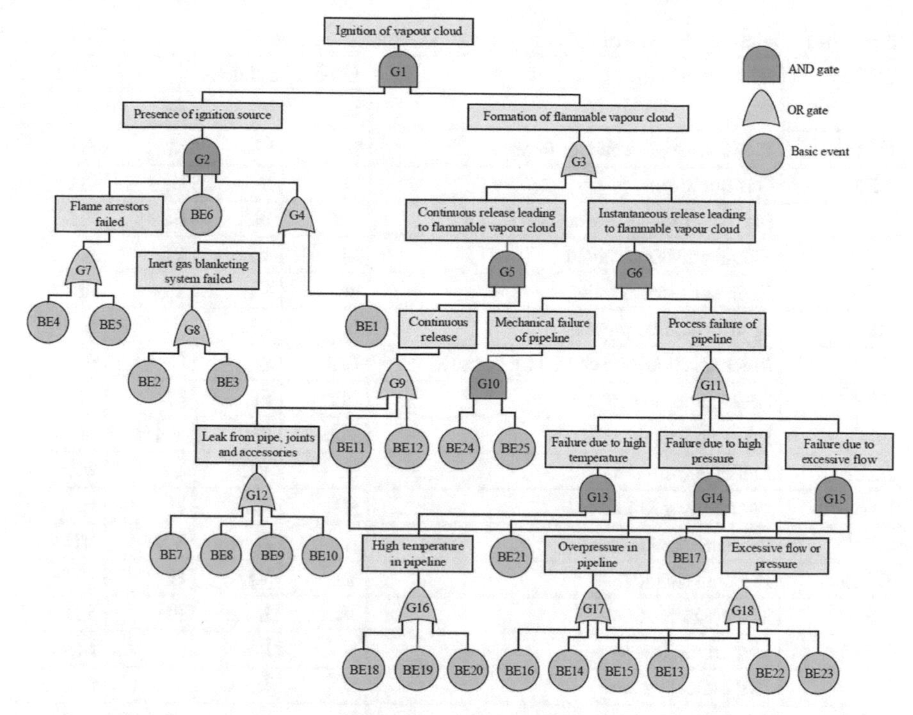


**Fig. 10.5** FT for the ethylene transportation line (reworked and modified from [12])

According to step 11 of the framework shown in Fig. 10.2, the TE probability has been evaluated using the BN-based approach for comparison of the result. Figure 10.6 shows the BN model of the FT illustrated in Fig. 10.5. In this BN, the prior probabilities of the root nodes are specified based on the crisp probabilities of the BEs as shown in Table 10.3. Conversely, the conditional probabilities of nodes representing logic gates are characterized according to the specification of the gates. After running a query on this BN model, the probability of TE obtained was 1.576E-9, which is quite close to the value of TE probability calculated by the algebraic formulation.

## *10.4.2 Sensitivity Analysis*

As discussed in Sect. 3.6, a SA can be applied to show the validity of the proposed method, as well as highlight some features of the method. By varying the value of $\alpha$ from 0 to 1, the probability of each BE is computed. Accordingly, the TE probability is estimated using BN. The probabilities of all BEs based on the corresponding value of $\alpha$ are provided in Table 10.4.

**Table 10.1** Details of the BEs of FT of Fig. 10.5 and experts' opinions

| FT Tag | BEs Description | Experts' opinions | | | |
|---|---|---|---|---|---|
| | | E1 | E2 | E3 | E4 |
| BE1 | Flammable gas detector failed | M | FL | M | FH |
| BE2 | Gas out of run | H | L | L | FH |
| BE3 | Inert gas release mechanism failed | FH | FL | FL | FH |
| BE4 | Flame arrestor A failed | M | VL | VL | FH |
| BE5 | Flame arrestor B failed | M | VL | VL | FH |
| BE6 | Ignition source present | H | VL | VL | H |
| BE7 | Mechanical failure caused by corrosion | FL | L | L | FL |
| BE8 | Leakage in two valves | FL | FL | FL | FL |
| BE9 | Leakage from four bends | FH | FH | VH | FL |
| BE10 | Leakage from ten joints | M | VL | VL | M |
| BE11 | Flow sensor failed | H | FL | H | H |
| BE12 | Pressure sensor failed | L | L | VH | VH |
| BE13 | Pipeline chocked | VL | VL | H | VL |
| BE14 | Valve chocked | VL | L | FH | VL |
| BE15 | High inlet flow | M | L | L | M |
| BE16 | High inlet pressure | M | L | L | M |
| BE17 | Pressure controller/trip failed | FL | FL | FL | L |
| BE18 | High inlet temperature | FL | VL | VL | FL |
| BE19 | External heat source present | M | L | L | H |
| BE20 | Side reaction | FL | FL | FL | L |
| BE21 | Temperature controller/trip failed | FL | L | L | L |
| BE22 | Phase change | VL | L | L | L |
| BE23 | Valves fail open | VL | FL | FL | FH |
| BE24 | Corrosion | L | M | FH | M |
| BE25 | Mechanical damage | L | M | M | L |

VL: Very Low
L: Low
FL: Fairly Low
M: Medium
FH: Fairly High
H: High
VH: Very High

It should be added that the sensitivity analysis assists experts to allocate priorities and make it flexible to perform the risk assessment. Figure 10.7 shows the results of the sensitivity analysis.

The SA specifies that the estimated probability for all the basic events is not pretty sensitive to the variations in the value of $\alpha$. Using different values of $\alpha$ ranging from

**Table 10.2** Aggregation calculations for the BE24

| | | |
|---|---|---|
| Expert 1 (L) | (0.07,0.13,0.19;0.06,0.13,0.20) | |
| Expert 2 (M) | (0.35,0.50,0.65;0.32,0.50,0.68) | |
| Expert 3 (FH) | (0.62,0.73,0.82;0.61,0.73,0.85) | |
| Expert 4 (M) | (0.35,0.50,0.65;0.32,0.50,0.68) | |
| S(E1&E2) | 0.260 | $EV_1 = \frac{0.07+0.19+0.06+0.20+4\times0.13}{8} = 0.13 \geq EV_2 = \frac{0.35+0.68+0.65+0.32+4\times0.50}{8} = 0.50 \rightarrow \frac{EV_1}{EV_2} = 0.260$ |
| S(E1&E3) | 0.179 | |
| S(E1&E4) | 0.260 | |
| S(E2&E3) | 0.687 | |
| S(E2&E4) | 1.000 | |
| S(E3&E4) | 0.687 | |
| AA(E1) | 0.233 | $AA(E_m) = \frac{1}{m-1} \sum_{v=1, v \neq 1}^{m} S_{uv}$ $1/(4-1)(0.260 + 0.179 + 0.260) = 0.233$ |
| AA(E2) | 0.649 | |
| AA(E3) | 0.518 | |
| AA(E4) | 0.649 | |
| RA(E1) | 0.114 | $RAD(E_m) = \frac{AA(E_m)}{\sum_{u=1}^{m} AA(E_m)}$ $0.233/(0.649 + 0.233 + 0.518 + 0.649) = 0.114$ |
| RA(E2) | 0.317 | |
| RA(E3) | 0.253 | |
| RA(E4) | 0.317 | |
| CC(E1) | 0.181 | $\alpha \cdot W_{FAHP}(E_m) + (1-\alpha) \cdot RAD(E_m) = 0.5 \times 0.114 + 0.5 \times 0.249 = 0.181$ |
| CC(E2) | 0.221 | |
| CC(E3) | 0.374 | |
| CC(E4) | 0.222 | |
| Aggregation for BE24 $\tilde{P}_j = \sum_{i=1}^{n} w_m \otimes \tilde{P}_{ij}$ | | |

$$
\begin{aligned}
&= 0.181 \otimes (0.07, 0.13, 0.19; 0.06, 0.13, 0.20) \oplus 0.221 \\
&\quad \otimes (0.35, 0.50, 0.65; 0.32, 0.50, 0.68) \oplus 0.374 \\
&\quad \otimes (0.62, 0.73, 0.82; 0.61, 0.73, 0.85) \oplus 0.222 \\
&\quad \otimes (0.35, 0.50, 0.65; 0.32, 0.50, 0.68) \\
&= (0.400, 0.518, 0.629; 0.381, 0.518, 0.656)
\end{aligned}
$$

0 to 1, we can see that the risk probability of only 4 of the 25 basic events (16%) is quite different and these BEs are BE4, BE9, BE11, and BE20. Therefore, in this study, the differences between the rankings concerning different $\alpha$ values are low.

In addition, choosing an adequate value of $\alpha$ illustrates an important role in the top event probability computation. The value of $\alpha$ can have an effect on the probability of each BE and accordingly top event. Thus, the value of $\alpha$ should be allocated taking into account the following issues. As an initial subject, decision-makers can consult any existing historical data from similar operation conditions and risk assessment, which have received feedback from them earlier. Next, using a questionnaire or

**Table 10.3** The fuzzy and crisp failure data of all BEs

| FT tag | Intuitionistic fuzzy failure possibilities | Intuitionistic fuzzy failure probabilities (FPs) | Defuzzified FPs |
|---|---|---|---|
| BE1 | (0.369,0.504,0.635;0.344,0.504,0.664) | (1.777E-3, 5.149E-3, 1.225E-2, 1.408E-3, 5.149E-3, 1.465E-2) | 6.770E-3 |
| BE2 | (0.347,0.416,0.482;0.335,0.416,0.498) | (1.448E-3, 2.662E-3, 4.394E-3, 1.283E-3, 2.662E-3, 4.929E-3) | 2.903E-3 |
| BE3 | (0.367,0.471,0.567;0.352,0.471,0.591) | (1.742E-3, 4.070E-3, 7.882E-3, 1.507E-3, 4.070E-3, 9.238E-3) | 4.773E-3 |
| BE4 | (0.184,0.261,0.335;0.176,0.261,0.347) | (1.667E-4, 5.587E-4, 1.286E-3, 1.404E-4, 5.587E-4, 1.446E-3) | 6.946E-4 |
| BE5 | (0.184,0.261,0.335;0.176,0.261,0.347) | (1.667E-4, 5.587E-4, 1.286E-3, 1.404E-4, 5.587E-4, 1.446E-3) | 6.946E-4 |
| BE6 | (0.355,0.404,0.453;0.346,0.404,0.461) | (1.560E-3, 2.400E-3, 3.536E-3, 1.435E-3, 2.400E-3, 3.778E-3) | 2.520E-3 |
| BE7 | (0.114,0.191,0.269;0.099,0.191,0.283) | (2.747E-5, 1.902E-4, 6.132E-4, 1.596E-5, 1.902E-4, 7.307E-4) | 2.964E-4 |

(continued)

**Table 10.3** (continued)

| FT tag | Intuitionistic fuzzy failure possibilities | Intuitionistic fuzzy failure probabilities (FPs) | Defuzzified FPs |
|---|---|---|---|
| BE8 | (0.170,0.270,0.370;0.150,0.270,0.390) | (1.244E-4, 6.208E-4, 1.782E-3, 7.872E-5, 6.208E-4, 2.126E-3) | 8.973E-4 |
| BE9 | (0.560,0.669,0.760; 0.549,0.669,0.789) | (7.535E-3, 1.511E-2, 2.709E-2, 6.990E-3, 1.511E-2, 3.283E-2) | 1.756E-2 |
| BE10 | (0.249,0.371,0.492;0.224,0.371,0.517) | (4.721E-4, 1.797E-3, 4.738E-3, 3.319E-4, 1.797E-3, 5.622E-3) | 2.473E-3 |
| BE11 | (0.731,0.796,0.861;0.711,0.796,0.881) | (2.244E-2, 3.450E-2, 5.569E-2, 1.975E-2, 3.450E-2, 6.576E-2) | 3.897E-2 |
| BE12 | (0.499,0.555,0.611;0.487,0.555,0.623) | (4.949E-3, 7.275E-3, 1.049E-2, 4.548E-3, 7.275E-3, 1.131E-2) | 7.649E-3 |
| BE13 | (0.209,0.255,0.300;0.204,0.255,0.305) | (2.611E-4, 5.102E-4, 8.837E-4, 2.393E-4, 5.102E-4, 9.357E-4) | 5.570E-4 |
| BE14 | (0.189,0.252,0.310;0.184,0.252,0.320) | (1.812E-4, 4.929E-4, 9.849E-4, 1.662E-4, 4.929E-4, 1.098E-3) | 5.706E-4 |

(continued)

**Table 10.3** (continued)

| FT tag | Intuitionistic fuzzy failure possibilities | Intuitionistic fuzzy failure probabilities (FPs) | Defuzzified FPs |
|---|---|---|---|
| BE15 | (0.193,0.292,0.392;0.174,0.292,0.410) | (1.953E-4, 8.116E-4, 2.161E-3, 1.356E-4, 8.116E-4, 2.530E-3) | 1.113E-3 |
| BE16 | (0.193,0.292,0.392;0.174,0.292,0.410) | (1.953E-4, 8.116E-4, 2.161E-3, 1.356E-4, 8.116E-4, 2.530E-3) | 1.113E-3 |
| BE17 | (0.155,0.249,0.343;0.137,0.249,0.362) | (8.983E-5, 4.762E-4, 1.394E-3, 5.594E-5, 4.762E-4, 1.662E-3) | 6.965E-4 |
| BE18 | (0.075,0.141,0.207;0.066,0.141,0.216) | (4.699E-3, 6.242E-5, 2.513E-4, 2.676E-6, 6.242E-5, 2.903E-4) | 1.128E-4 |
| BE19 | (0.252,0.333,0.415;0.235,0.333,0.431) | (4.912E-4, 1.259E-3, 2.621E-3, 3.909E-3, 1.259E-3, 2.988E-3) | 1.506E-3 |
| BE20 | (0.155,0.249,0.343;0.137,0.249,0.362) | (8.983E-5, 4.762E-4, 1.394E-3, 5.594E-5, 4.762E-4, 1.662E-3) | 6.965E-4 |
| BE21 | (0.091,0.159,0.227; 0.078,0.159,0.239) | (1.083E-5, 9.719E-5, 3.445E-4, 5.882E-6, 9.719E-5, 4.113E-4) | 1.622E-4 |

(continued)

**Table 10.3** (continued)

| FT tag | Intuitionistic fuzzy failure possibilities | Intuitionistic fuzzy failure probabilities (FPs) | Defuzzified FPs |
|---|---|---|---|
| BE22 | (0.057,0.113,0.170;0.049,0.113,0.178) | (1.382E-6, 2.708E-5, 1.239E-4, 6.512E-7, 2.708E-5, 1.467E-4) | 5.480E-05 |
| BE23 | (0.213,0.305,0.393;0.198,0.305,0.411) | (2.780E-4, 9.344E-4, 2.189E-3, 2.153E-4, 9.344E-4, 2.551E-3) | 1.189E-3 |
| BE24 | (0.400,0.518,0.629;0.381,0.518,0.656) | (2.319E-3, 5.685E-3, 1.179E-2, 1.971E-3, 5.685E-3, 1.393E-2) | 6.932E-3 |
| BE25 | (0.227,0.337,0.448;0.206,0.337,0.469) | (3.446E-4, 1.312E-3, 3.404E-3, 2.454E-4, 1.312E-3, 3.993E-3) | 1.777E-3 |

other available methods, the value of $\alpha$ can be obtained based on the decision-makers' opinions. If a decision-maker has a high confidence regarding his/her judgment about the probability of basic events, the value of $\alpha$ can be set to a higher value, on the contrary, a smaller value can be assigned to $\alpha$. Finally, the value of $\alpha$ can be assigned according to a realistic circumstance, meaning that the value of $\alpha$ should be allocated a higher value when it is easy to get the consensus of decision-makers' judgements on the probability of basic events or when the appropriately selected decision-makers are present.

The above SA illustrates that the presented model can offer vital data to analysts and other involved parties in the risk assessment process. Accordingly, the probability of the top event is computed by varying the value of $\alpha$.

According to the new estimated probability of BEs, the probability of the TE is also updated and provided in Table 10.5.

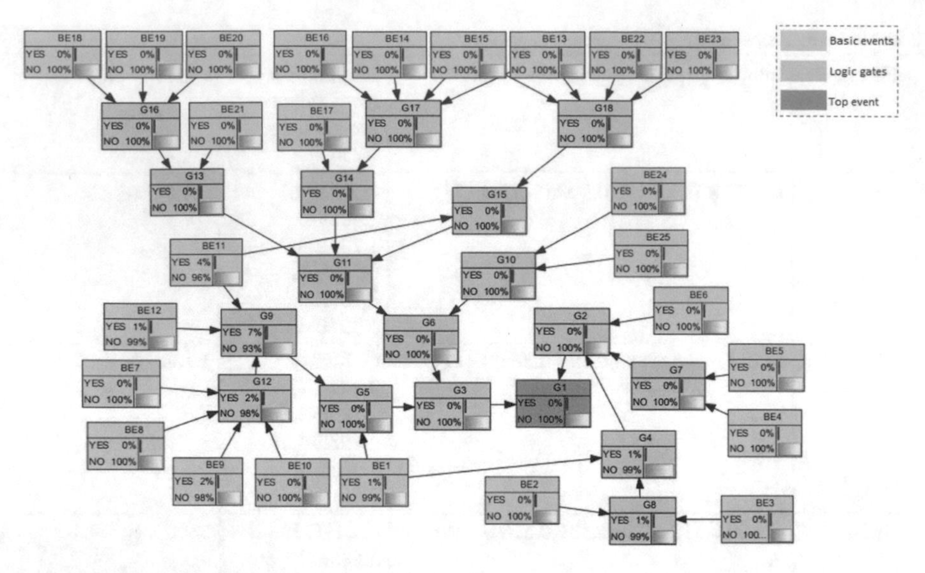


**Fig. 10.6** BN model of the FT of Fig. 10.5

## *10.4.3 Identification of Critical BEs and Corrective Actions for the Most Critical BEs*

As we all know, one of the important outputs of FTA and correspondingly BN is recognizing the critical basic events. Based on this recognition, decision-makers can provide corrective and/or preventive actions to reduce the probability of critical basic events. As a result, the TE probability will be reduced; subsequently, the probability reduction will lead to improved performance of the system.

By following the criticality calculation approach shown in Sect. 3.6, the criticality of the BEs is estimated and conveyed in Table 10.6. As seen in the table, Flame arrestor A failed (BE4), Flame arrestor B failed (BE5), Ignition source present (BE6), Flammable gas detector fail (BE1), Flow sensor failed (BE11), and Leak from bends (four bends) (BE9) are recognized to be the most critical events (in the descending order of criticality), which are also recognized as top six critical events in [12]. This chapter provides corrective actions for the first five critical basic events because in the realistic case, the system cannot apply any interpretative actions to all BEs. The existing control measures for the aforementioned BEs can fall into the process safety management system since the construction of the complex plant. However, the performance of the control measures needs to be upgraded based on all requirements and changed after a couple of years.

Several control measures as corrective actions are recommended for the critical basic events. It is believed that any corrective actions need to satisfy the three main criteria as (i) it should have acceptable efficiency, (ii) it should fall into the acceptable economic perspective, and (iii) the recommended corrective actions should be

**Table 10.4** The probability of BEs based on different relaxation factor

| FT tag | $\alpha = 0$ | $\alpha = 0.1$ | $\alpha = 0.2$ | $\alpha = 0.3$ | $\alpha = 0.4$ | $\alpha = 0.6$ | $\alpha = 0.7$ | $\alpha = 0.8$ | $\alpha = 0.9$ | $\alpha = 1$ |
|---|---|---|---|---|---|---|---|---|---|---|
| BE1 | 0.006874 | 0.006853 | 0.006832 | 0.006811 | 0.006791 | 0.006750 | 0.006729 | 0.006709 | 0.006689 | 0.006670 |
| BE2 | 0.003553 | 0.003414 | 0.003280 | 0.003151 | 0.003025 | 0.002785 | 0.002670 | 0.002560 | 0.002452 | 0.002349 |
| BE3 | 0.005818 | 0.005594 | 0.005379 | 0.005170 | 0.004968 | 0.004584 | 0.004402 | 0.004226 | 0.004055 | 0.003891 |
| BE4 | 0.000855 | 0.000821 | 0.004156 | 0.000756 | 0.000725 | 0.000665 | 0.000637 | 0.000609 | 0.000583 | 0.026624 |
| BE5 | 0.000855 | 0.000821 | 0.003314 | 0.000756 | 0.000725 | 0.000665 | 0.000637 | 0.000609 | 0.000583 | 0.000557 |
| BE6 | 0.003766 | 0.003486 | 0.003221 | 0.002973 | 0.002739 | 0.002314 | 0.002121 | 0.001940 | 0.001770 | 0.001612 |
| BE7 | 0.000342 | 0.000333 | 0.000323 | 0.000314 | 0.000305 | 0.000288 | 0.000279 | 0.000271 | 0.000263 | 0.000085 |
| BE8 | 0.000900 | 0.000900 | 0.000900 | 0.000900 | 0.000900 | 0.000897 | 0.000897 | 0.000897 | 0.000897 | 0.000897 |
| BE9 | 0.023861 | 0.024767 | 0.025714 | 0.026705 | 0.027740 | 0.017586 | 0.031150 | 0.008067 | 0.033712 | 0.035091 |
| BE10 | 0.000861 | 0.000804 | 0.000750 | 0.000699 | 0.000650 | 0.002410 | 0.000518 | 0.000478 | 0.000441 | 0.000405 |
| BE11 | 0.039908 | 0.039718 | 0.039530 | 0.039342 | 0.039156 | 0.038787 | 0.038604 | 0.038423 | 0.038242 | 0.073468 |
| BE12 | 0.007058 | 0.007549 | 0.008070 | 0.008623 | 0.009210 | 0.010493 | 0.011196 | 0.011943 | 0.012738 | 0.013584 |
| BE13 | 0.000006 | 0.004124 | 0.000074 | 0.000168 | 0.000323 | 0.000887 | 0.001822 | 0.001923 | 0.002682 | 0.003648 |
| BE14 | 0.000043 | 0.000286 | 0.000156 | 0.000255 | 0.000391 | 0.000801 | 0.001089 | 0.001445 | 0.001878 | 0.002399 |
| BE15 | 0.001421 | 0.001355 | 0.001291 | 0.001229 | 0.001170 | 0.001057 | 0.001004 | 0.000953 | 0.000903 | 0.000856 |
| BE16 | 0.001421 | 0.001355 | 0.001291 | 0.001229 | 0.001170 | 0.001057 | 0.001004 | 0.000953 | 0.000903 | 0.000856 |
| BE17 | 0.000677 | 0.000681 | 0.000685 | 0.000689 | 0.000693 | 0.000700 | 0.000704 | 0.000708 | 0.000712 | 0.000716 |

(continued)

**Table 10.4** (continued)

| FT tag | $\alpha = 0$ | $\alpha = 0.1$ | $\alpha = 0.2$ | $\alpha = 0.3$ | $\alpha = 0.4$ | $\alpha = 0.6$ | $\alpha = 0.7$ | $\alpha = 0.8$ | $\alpha = 0.9$ | $\alpha = 1$ |
|---|---|---|---|---|---|---|---|---|---|---|
| BE18 | 0.000153 | 0.000144 | 0.000136 | 0.000128 | 0.000120 | 0.000106 | 0.000099 | 0.000092 | 0.000263 | 0.000080 |
| BE19 | 0.001742 | 0.001693 | 0.001644 | 0.001459 | 0.001551 | 0.001462 | 0.001418 | 0.001376 | 0.000903 | 0.001295 |
| BE20 | 0.000677 | 0.000681 | 0.000685 | 0.000689 | 0.000693 | 0.000700 | 0.000704 | 0.000708 | 0.015788 | 0.000716 |
| BE21 | 0.000143 | 0.000147 | 0.000151 | 0.000154 | 0.000158 | 0.001463 | 0.000170 | 0.000174 | 0.000178 | 0.000183 |
| BE22 | 0.000065 | 0.000063 | 0.000061 | 0.000059 | 0.000057 | 0.000934 | 0.000051 | 0.000049 | 0.000048 | 0.000046 |
| BE23 | 0.001660 | 0.001557 | 0.001458 | 0.001364 | 0.001274 | 0.001107 | 0.001030 | 0.000957 | 0.000888 | 0.000822 |
| BE24 | 0.007154 | 0.007107 | 0.007061 | 0.007017 | 0.006974 | 0.006892 | 0.006853 | 0.006816 | 0.006780 | 0.006745 |
| BE25 | 0.001421 | 0.001487 | 0.001556 | 0.001627 | 0.001701 | 0.003900 | 0.001936 | 0.002020 | 0.002106 | 0.002195 |

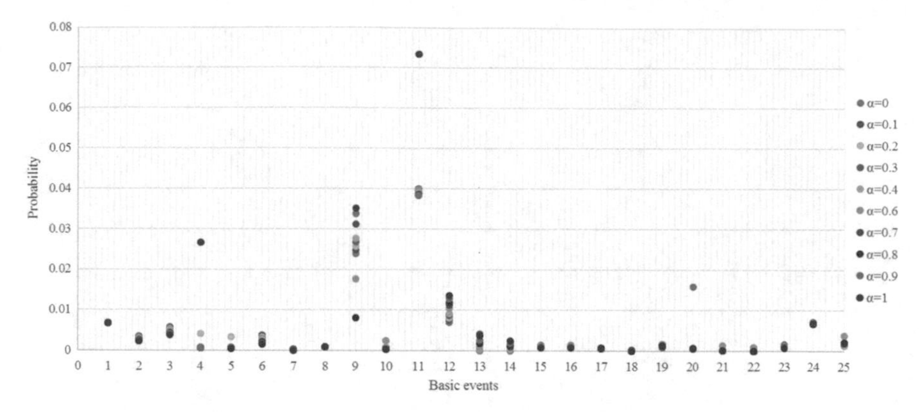


**Fig. 10.7** The probability of basic events based on the variation of $\alpha$

environmentally friendly. Keeping these criteria in our mind, to control the Flame arrestors A and B, increasing the number of inspection can effectively reduce the probability of failure. In addition, cleaning, as an important part of the Flame arrestor maintenance procedure, is required to be continuously considered. The Ignition source present event can properly be eliminated by providing natural or in some specific cases of a fireproof ventilation system. The ventilation system has been widely used and accepted method in the oil and gas industries. It can prevent smoke and fire propagation through the air ducts even in case of fire. To reduce the failure probability of a flammable gas detector, one possible and applicable way is using an updated version of a gas detector. The flammable gas detector may fail due to some identical causes. These causes also need to be identified. Thus, the failure can be eliminated only and only by some simple modifications. According to this, continual maintenance to preserve the detector in operational conditions is recommended. To deal with another critical basic event as "Flow sensor failed", a potential acceptable solution is by introducing redundancy, i.e., changing the current system into the parallel one by adding one more sensor. In this case, one sensor is operating and the second sensor is in a standby mode. In case of the failure of the operating sensor, the standby sensor can take over the operational responsibility of the failed sensor, thus preventing the failure. Finally, the "Leak from bends" is controlled by bare-eye inspection. To cope with this failure, electrical testing such as voltage and resistance measurement, physical testing like drop test, bending test, and pull test can be applied. Also, such visual inspection including optical microscope and X-ray microscope is also possible to be used.

Adding to this, the risk assessment is a continuous procedure to improve the safety performance of the studied system. Therefore, continuous review and revision must be taken into account.

**Table 10.5** The probability of the top event based on different values of $\alpha$

| Value of $\alpha$ | $\alpha = 0$ | $\alpha = 0.1$ | $\alpha = 0.2$ | $\alpha = 0.3$ | $\alpha = 0.4$ | $\alpha = 0.6$ | $\alpha = 0.7$ | $\alpha = 0.8$ | $\alpha = 0.9$ | $\alpha = 1$ |
|---|---|---|---|---|---|---|---|---|---|---|
| TE probability | 3.15E-09 | 2.84E-09 | 1.20E-08 | 2.30E-09 | 2.05E-09 | 1.44E-09 | 1.46E-09 | 9.40E-10 | 1.15E-09 | 3.33E-08 |

**Table 10.6** Criticality ranking of the BEs of the FT of Fig. 10.5

| Basic events | BIM | Rank | Basic events | BIM | Rank |
|---|---|---|---|---|---|
| BE4 | 1.134E-06 | 1 | BE23 | 1.280E-14 | 13 |
| BE5 | 1.134E-06 | 1 | BE22 | 1.270E-14 | 14 |
| BE6 | 6.252E-07 | 2 | BE24 | 5.600E-15 | 15 |
| BE1 | 2.327E-07 | 3 | BE2 | 5.000E-15 | 16 |
| BE11 | 2.301E-08 | 4 | BE3 | 5.000E-15 | 16 |
| BE9 | 2.251E-08 | 5 | BE17 | 1.900E-15 | 17 |
| BE12 | 2.229E-08 | 6 | BE21 | 1.300E-15 | 18 |
| BE10 | 2.217E-08 | 7 | BE14 | 4.000E-16 | 19 |
| BE8 | 2.214E-08 | 8 | BE16 | 4.000E-16 | 19 |
| BE7 | 2.212E-08 | 9 | BE18 | 1.000E-16 | 20 |
| BE25 | 2.190E-14 | 10 | BE19 | 1.000E-16 | 20 |
| BE15 | 1.320E-14 | 11 | BE20 | 1.000E-16 | 20 |
| BE13 | 1.310E-14 | 12 | | | |

## 10.5 Conclusion

This chapter presents a framework for FTA and BN-based reliability analysis of process systems using IFS theory where there exists uncertainty with the availability of precise failure data. The proposed approach enables the gathering of uncertain data by combining IFS theory with expert elicitation. The IFS theory differs from the traditional fuzzy set theory in the sense that it considers both the membership and non-membership of an element in the set. Therefore, the utilization of the IFS theory would allow us to model situations where a varying level of confidence is associated with the fuzziness of numerical data. Therefore, by using IFS theory together with expert judgment as presented in this chapter, the analysts would get increased flexibility while expressing failure data in the form of fuzzy numbers.

The sensitivity analysis performed within the proposed framework would help the analysts to determine the events that are more sensitive to uncertainty, thus allowing to make informed decision to improve the data quality of the associated events. Furthermore, the criticality analysis of the events followed by the recommendation of corrective actions would greatly help to increase the reliability of the studied system. The efficiency of the proposed framework has been verified by applying it to a practical system. The experimentations illustrate that the IFS-based method offers a valuable way of reliability assessment of process systems when the fuzzy failure data of system components cannot be defined with high confidence. It should be added that, as a direction for future works, the same approach can be integrated using much more advanced fuzzy set theory such as but not limited PFS.

## References


1. Mannan, S., Lees, F.P., Lees' Loss Prevention in the Process Industries: Hazard Identification, Assessment, and Control. Elsevier Butterworth-Heinemann (2005)
2. Amyotte, P.R., Berger, S., Edwards, D.W., Gupta, J.P., Hendershot, D.C., Khan, F.I., Mannan, M.S., Willey, R.J.: Why are major accidents still occurring? Process Saf. Prog. **35**, 253–257 (2016). https://doi.org/10.1002/prs.11795
3. Khan, F.I., Abbasi, S.A.: Risk analysis of a typical chemical industry using ORA procedure. J. Loss Prev. Process Ind. **14**, 43–59 (2000). https://doi.org/10.1016/S0950-4230(00)00006-1
4. Kabir, S.: An overview of fault tree analysis and its application in model based dependability analysis. Expert Syst. Appl. **77**, 114–135 (2017). https://doi.org/10.1016/j.eswa.2017.01.058
5. Markowski, A.S., Mannan, M.S., Bigoszewska, A.: Fuzzy logic for process safety analysis. J. Loss Prev. Process Ind. **22**, 695–702 (2009). https://doi.org/10.1016/j.jlp.2008.11.011
6. Omidvar, M., Zarei, E., Ramavandi, B., Yazdi, M.: Fuzzy Bow-Tie Analysis: Concepts, Review, and Application BT—Linguistic Methods Under Fuzzy Information in System Safety and Reliability Analysis. In: Yazdi, M. (ed.) Springer International Publishing, Cham, pp. 13–51 (2022). https://doi.org/10.1007/978-3-030-93352-4_3
7. Yazdi, M., Adumene, S., Zarei, E.: Introducing a Probabilistic-Based Hybrid Model (Fuzzy-BWM-Bayesian Network) to Assess the Quality Index of a Medical Service BT—Linguistic Methods Under Fuzzy Information in System Safety and Reliability Analysis. In: Yazdi, M. (ed.) Springer International Publishing, Cham, pp. 171–183 (2022). https://doi.org/10.1007/978-3-030-93352-4_8
8. Cooke, R.: Experts in Uncertainty: Opinion and Subjective Probability in Science. Oxford University Press (1991)
9. Zadeh, L.: Fuzzy sets. Inf. Control. **8**, 338–353 (1965)
10. Yazdi, M., Kabir, S.: Fuzzy evidence theory and Bayesian networks for process systems risk analysis. Hum. Ecol. Risk Assess. **26**, 57–86 (2020). https://doi.org/10.1080/10807039.2018.1493679
11. Wang, D., Zhang, Y., Jia, X., Jiang, P., Guo, B.: Handling uncertainties in fault tree analysis by a hybrid probabilistic-possibilistic framework. Qual. Reliab. Eng. Int. **32**, 1137–1148 (2016). https://doi.org/10.1002/qre.1821
12. Yazdi, M., Kabir, S.: A fuzzy Bayesian network approach for risk analysis in process industries. Process Saf. Environ. Prot. **111** (2017). https://doi.org/10.1016/j.psep.2017.08.015
13. Atanassov, K.T.: Intuitionistic fuzzy sets. Fuzzy Sets Syst. **20**, 87–96 (1986). https://doi.org/10.1016/S0165-0114(86)80034-3
14. Govindan, K., Khodaverdi, R., Vafadarnikjoo, A.: Intuitionistic fuzzy based DEMATEL method for developing green practices and performances in a green supply chain. Expert Syst. Appl. **42**, 7207–7220 (2015). https://doi.org/10.1016/j.eswa.2015.04.030
15. Sayyadi Tooranloo, H., Sadat Ayatollah, A.: A model for failure mode and effects analysis based on intuitionistic fuzzy approach. Appl. Soft Comput. J. **49**, 238–247 (2016). https://doi.org/10.1016/j.asoc.2016.07.047
16. Yazdi, M.: Risk assessment based on novel intuitionistic fuzzy-hybrid-modified TOPSIS approach. Saf. Sci. **110**, 438–448 (2018). https://doi.org/10.1016/j.ssci.2018.03.005
17. Ming-Hung, S., Ching-Hsue, C., Chang, J.-R.: Using intuitionistic fuzzy sets for fault-tree analysis on printed circuit board. Assembly **46**, 2139–2148 (2006). https://doi.org/10.1016/j.microrel.2006.01.007
18. Chang, J.R., Chang, K.H., Liao, S.H., Cheng, C.H.: The reliability of general vague fault-tree analysis on weapon systems fault diagnosis. Soft Comput. **10**, 531–542 (2006). https://doi.org/10.1007/s00500-005-0483-y
19. Cheng, S.R., Lin, B., Hsu, B.M., Shu, M.H.: Fault-tree analysis for liquefied natural gas terminal emergency shutdown system. Expert Syst. Appl. **36**, 11918–11924 (2009). https://doi.org/10.1016/j.eswa.2009.04.011

20. Kumar, M., Yadav, S.P.: The weakest t-norm based intuitionistic fuzzy fault-tree analysis to evaluate system reliability. ISA Trans. **51**, 531–538 (2012). https://doi.org/10.1016/j.isatra.2012.01.004
21. Kabir, S., Walker, M., Papadopoulos, Y.: Dynamic system safety analysis in HiP-HOPS with Petri nets and Bayesian networks. Saf. Sci. **105**, 55–70 (2018). https://doi.org/10.1016/j.ssci.2018.02.001
22. Kabir, S., Papadopoulos, Y.: Applications of Bayesian networks and Petri nets in safety, reliability, and risk assessments: a review. Saf. Sci. **115**, 154–175 (2019). https://doi.org/10.1016/j.ssci.2019.02.009
23. Ferdous, R., Khan, F., Veitch, B., Amyotte, P.R.: Methodology for computer aided fuzzy fault tree analysis. Process Saf. Environ. Prot. **87**, 217–226 (2009). https://doi.org/10.1016/j.psep.2009.04.004
24. Yu, H., Khan, F., Veitch, B.: A flexible hierarchical Bayesian modeling technique for risk analysis of major accidents. Risk Anal. **37**, 1668–1682 (2017). https://doi.org/10.1111/risa.12736
25. Hashemi, S.J., Khan, F., Ahmed, S.; Multivariate probabilistic safety analysis of process facilities using the Copula Bayesian network model. Comput. Chem. Eng. **93**, 128–142 (2016). https://doi.org/10.1016/j.compchemeng.2016.06.011
26. Yazdi, M., Nikfar, F., Nasrabadi, M.: Failure probability analysis by employing fuzzy fault tree analysis. Int. J. Syst. Assur. Eng. Manag. **8**, 1177–1193 (2017). https://doi.org/10.1007/s13198-017-0583-y
27. Yazdi, M.: Footprint of knowledge acquisition improvement in failure diagnosis analysis. Qual. Reliab. Eng. Int. 405–422 (2018). https://doi.org/10.1002/qre.2408
28. Wang, Y.F., Qin, T., Li, B., Sun, X.F., Li, Y.L.: Fire probability prediction of offshore platform based on dynamic Bayesian network. Ocean Eng. **145**, 112–123 (2017). https://doi.org/10.1016/j.oceaneng.2017.08.035
29. Xin, P., Khan, F., Ahmed, S.: Dynamic hazard identification and scenario mapping using Bayesian network. Process Saf. Environ. Prot. **105**, 143–155 (2017). https://doi.org/10.1016/j.psep.2016.11.003
30. Liu, X., Zheng, J., Fu, J., Nie, Z., Chen, G.: Optimal inspection planning of corroded pipelines using BN and GA. J. Pet. Sci. Eng. **163**, 546–555 (2018). https://doi.org/10.1016/j.petrol.2018.01.030
31. Yan, F., Xu, K., Yao, X., Li, Y.: Fuzzy Bayesian network-bow-tie analysis of gas leakage during biomass gasification. PLoS ONE **11**, e0160045 (2016). https://doi.org/10.1371/journal.pone.0160045
32. Pasman, H., Rogers, W.: The bumpy road to better risk control: a Tour d'Horizon of new concepts and ideas. J. Loss Prev. Process Ind. **35**, 366–376 (2015). https://doi.org/10.1016/j.jlp.2014.12.003
33. Naderpour, M., Lu, J., Zhang, G.: An abnormal situation modeling method to assist operators in safety-critical systems. Reliab. Eng. Syst. Saf. **133**, 33–47 (2015). https://doi.org/10.1016/j.ress.2014.08.003
34. Ren, J., Jenkinson, I., Wang, J., Xu, D.L., Yang, J.B.: A methodology to model causal relationships on offshore safety assessment focusing on human and organizational factors. J. Safety Res. **39**, 87–100 (2008). https://doi.org/10.1016/j.jsr.2007.09.009
35. Cai, B., Zhang, Y., Wang, H., Liu, Y., Ji, R., Gao, C., Kong, X., Liu, J.: Resilience evaluation methodology of engineering systems with dynamic-Bayesian-network-based degradation and maintenance. Reliab. Eng. Syst. Saf. **209**, 107464 (2021). https://doi.org/10.1016/j.ress.2021.107464
36. Li, X., Zhu, H., Chen, G., Zhang, R.: Optimal maintenance strategy for corroded subsea pipelines. J. Loss Prev. Process Ind. **49**, 145–154 (2017). https://doi.org/10.1016/j.jlp.2017.06.019
37. Abimbola, M., Khan, F., Khakzad, N., Butt, S.: Safety and risk analysis of managed pressure drilling operation using Bayesian network. Saf. Sci. **76**, 133–144 (2015). https://doi.org/10.1016/j.ssci.2015.01.010

38. Khakzad, N., Khan, F., Amyotte, P.: Quantitative risk analysis of offshore drilling operations: a Bayesian approach. Saf. Sci. **57**, 108–117 (2013). https://doi.org/10.1016/j.ssci.2013.01.022
39. Khakzad, N., Khan, F., Amyotte, P.: Quantitative risk analysis of offshore drilling operations: a Bayesian approach. Saf. Sci. **57**, 108–117 (2013). https://doi.org/10.1016/j.ssci.2013.01.022
40. Rausand, M., Haugen, S.: Risk Assessment: Theory, Methods, and Applications. Wiley, Hoboken (2020)
41. Mohammadfam, I., Zarei, E., Yazdi, M., Gholamizadeh, K.: Quantitative risk analysis on rail transportation of hazardous materials. Math. Probl. Eng. **2022**, 6162829 (2022). https://doi.org/10.1155/2022/6162829
42. Adumene, S., Adedigba, S., Khan, F., Zendehboudi, S.: An integrated dynamic failure assessment model for offshore components under microbiologically influenced corrosion. Ocean Eng. **218**, 108082 (2020). https://doi.org/10.1016/j.oceaneng.2020.108082
43. Adumene, S., Okwu, M., Yazdi, M., Afenyo, M., Islam, R., Orji, C.U., Obeng, F., Goerlandt, F.: Dynamic logistics disruption risk model for offshore supply vessel operations in Arctic waters. Marit. Transp. Res. **2**, 100039 (2021). https://doi.org/10.1016/j.martra.2021.100039
44. Li, F., Wang, W., Dubljevic, S., Khan, F., Xu, J., Yi, J.: Analysis on accident-causing factors of urban buried gas pipeline network by combining DEMATEL, ISM and BN methods. J. Loss Prev. Process Ind. **61**, 49–57 (2019). https://doi.org/10.1016/j.jlp.2019.06.001
45. Yazdi, M., Kabir, S., Walker, M.: Uncertainty handling in fault tree based risk assessment: state of the art and future perspectives. Process. Saf. Environ. Prot. **131**, 89–104 (2019). https://doi.org/10.1016/j.psep.2019.09.003
46. Yazdi, M., Golilarz, N.A., Adesina, K.A., Nedjati, A.: Probabilistic risk analysis of process systems considering epistemic and aleatory uncertainties: a comparison study. Int. J. Uncertainty Fuzziness Knowl.-Based Syst. **29**, 181–207 (2021). https://doi.org/10.1142/S0218488521500098
47. Verma, A.K., Srividya, A., Karanki, D.R.: Reliability and Safety Engineering. Springer London (2010). https://doi.org/10.1007/978-1-84996-232-2
48. Markowski, A.S., Mannan, M.S.: Fuzzy risk matrix. J. Hazard. Mater. **159**, 152–157 (2008). https://doi.org/10.1016/j.jhazmat.2008.03.055
49. Nadjafi, M., Farsi, M.A., Jabbari, H.: Reliability analysis of multi-state emergency detection system using simulation approach based on fuzzy failure rate. Int. J. Syst. Assur. Eng. Manag. **8**, 532–541 (2016). https://doi.org/10.1007/s13198-016-0563-7
50. Abdo, H., Flaus, J.-M.: Monte Carlo simulation to solve fuzzy dynamic fault tree*. IFAC-PapersOnLine **49**, 1886–1891 (2016). https://doi.org/10.1016/j.ifacol.2016.07.905
51. Abdo, H., Flaus, J.M., Masse, F.: Uncertainty quantification in risk assessment—representation, propagation and treatment approaches: application to atmospheric dispersion modeling. J. Loss Prev. Process Ind. **49**, 551–571 (2017). https://doi.org/10.1016/j.jlp.2017.05.015
52. Garg, H.: A novel approach for analyzing the behavior of industrial systems using weakest t-norm and intuitionistic fuzzy set theory. ISA Trans. **53**, 1199–1208 (2014). https://doi.org/10.1016/j.isatra.2014.03.014
53. OREDA: Offshore Reliability Data Handbook, 4th edn. Trondheim (2015)
54. Preyssl, C.: Safety risk assessment and management-the ESA approach. Reliab. Eng. Syst. Saf. **49**, 303–309 (1995). https://doi.org/10.1016/0951-8320(95)00047-6
55. Clemen, R.T., Winkler, R.L.: Combining probability distributions from experts in risk analysis. Risk Anal. **19**, 155–156 (1999). https://doi.org/10.1023/A:1006917509560
56. Yazdi, M., Zarei, E.: Uncertainty handling in the safety risk analysis: an integrated approach based on fuzzy fault tree analysis. J. Fail. Anal. Prev. **18**, 392–404 (2018). https://doi.org/10.1007/s11668-018-0421-9
57. Yazdi, M., Korhan, O., Daneshvar, S.: Application of fuzzy fault tree analysis based on modified fuzzy AHP and fuzzy TOPSIS for fire and explosion in the process industry. Int. J. Occup. Saf. Ergon. **26**, 319–335 (2020)

58. Berni, R.: Quality and reliability in top-event estimation: quantitative fault tree analysis in case of dependent events. Commun. Stat. Theor. Methods **41**, 3138–3149 (2012). https://doi.org/10.1080/03610926.2011.621574
59. Yazdi, M., Khan, F., Abbassi, R., Rusli, R.; Improved DEMATEL methodology for effective safety management decision-making. Saf. Sci. **127**, 104705 (2020). https://doi.org/10.1016/j.ssci.2020.104705
60. Jiang, G.-J., Chen, H.-X., Sun, H.-H., Yazdi, M., Nedjati, A., Adesina, K.A.: An improved multi-criteria emergency decision-making method in environmental disasters. Soft Comput. (2021). https://doi.org/10.1007/s00500-021-05826-x
61. Saaty, T.L.: Decision Making with Dependence and Feedback: The Analytic Network Process: The Organization and Prioritization of Complexity. RWS Publications (1996)
62. Deng, H.: Multicriteria analysis with fuzzy pairwise comparison. Int. J. Approx. Reason. **21**, 215–231 (1999). https://doi.org/10.1016/S0888-613X(99)00025-0
63. Rezaei, J.: Best-worst multi-criteria decision-making method. Omega (United Kingdom) **53**, 49–57 (2015). https://doi.org/10.1016/j.omega.2014.11.009
64. Rezaei, J.: Best-worst multi-criteria decision-making method: Some properties and a linear model. Omega (United Kingdom) **64**, 126–130 (2016). https://doi.org/10.1016/j.omega.2015.12.001
65. Li, H., Guo, J.-Y., Yazdi, M., Nedjati, A., Adesina, K.A.: Supportive emergency decision-making model towards sustainable development with fuzzy expert system. Neural Comput. Appl. **33**, 15619–15637 (2021). https://doi.org/10.1007/s00521-021-06183-4
66. Yazdi, M.: Acquiring and sharing tacit knowledge in failure diagnosis analysis using intuitionistic and pythagorean assessments. J. Fail. Anal. Prev. **19**, 369–386 (2019). https://doi.org/10.1007/s11668-019-00599-w
67. Kabir, S., Geok, T.K., Kumar, M., Yazdi, M., Hossain, F.: A method for temporal fault tree analysis using intuitionistic fuzzy set and expert elicitation. IEEE Access **8** (2020). https://doi.org/10.1109/ACCESS.2019.2961953
68. Onisawa, T.: An application of fuzzy concepts to modelling of reliability analysis. Fuzzy Sets Syst. **37**, 267–286 (1990). https://doi.org/10.1016/0165-0114(90)90026-3
69. Khakzad, N., Khan, F., Amyotte, P.: Safety analysis in process facilities: comparison of fault tree and Bayesian network approaches. Reliab. Eng. Syst. Saf. **96**, 925–932 (2011). https://doi.org/10.1016/j.ress.2011.03.012
70. Yazdi, M.: A review paper to examine the validity of Bayesian network to build rational consensus in subjective probabilistic failure analysis. Int. J. Syst. Assur. Eng. Manag. **10**, 1–18 (2019). https://doi.org/10.1007/s13198-018-00757-7
71. Jensen, F.V., Nielsen, T.D.: Bayesian Networks and Decision Graphs (2007). https://doi.org/10.1007/978-0-387-68282-2
72. Khan, F.I., Husain, T., Abbasi, S.A.: Design and evaluation of safety measures using a newly proposed methodology "SCAP." J. Loss Prev. Process Ind. **15**, 129–146 (2002). https://doi.org/10.1016/S0950-4230(01)00026-2
73. Khan, F.I., Haddara, M.: Risk-based maintenance (RBM): a new approach for process plant inspection and maintenance. Process Saf. Prog. **23**, 252–265 (2004). https://doi.org/10.1002/prs.10010

**Mohammad Yazdi** Ph.D., is a senior researcher at the Macquarie University, Australia. He received the B.Sc. degree in process safety engineering from the Petroleum University of Technology, Abadan, Iran, in 2012, and the M.Sc. degree in industrial engineering from Eastern Mediterranean University, Famagusta, Cyprus, in 2017, and Dual Ph.D. from Memorial University of Newfoundland, Canada and Macquarie University, Australia. Before undertaking the aforementioned career, he served as a Safety Expert and an Auditor in the oil and gas industry, from 2012 to 2016. He has been collaborating as a project, operation, and asset integrity management

Researcher and a Consultant in both academia and industrial sectors. His research mainly focuses on risk assessment.

**Sohag Kabir** received the Ph.D. degree in computer science and the M.Sc. degree in embedded systems from the University of Hull, UK, in 2016 and 2012, respectively. He is currently working as Assistant Professor in the Department of Computer Science at the University of Bradford, UK, where he is the program leader for the M.Sc. Internet of Things and M.Sc. Big Data Science and Technology programs. He is Fellow of the Higher Education Academy (FHEA). From 2017 to 2019, he was Research Associate in the Dependable Intelligent Systems (DEIS) Research Group at the University of Hull. His research interests include safety and reliability analysis, autonomous systems, Internet of Things, artificial intelligence, and fault-tolerant computing.

**Mohit Kumar** received the M.Sc. degree in Mathematics from Chaudhary Charan Singh University, Meerut, India, in 2006. He received the Ph.D. degree in Applied Mathematics with specialization "System Reliability Studies in Intuitionistic Fuzzy Environment" from Indian Institute of Technology, Roorkee, India in 2013. From August 2012 to June 2014, he was Assistant Professor in Department of Mathematics, Amity University, Noida, India. He is currently Assistant Professor of Mathematics with Institute of Infrastructure Technology Research, and Management, Ahmedabad, India. He has authored or co-authored more than 25 research papers in journals, book chapters, and conference proceedings. His research interests include fuzzy system reliability analysis, fuzzy multi-criteria decision-making, fuzzy statistics, fuzzy optimization, and intuitionistic fuzzy set theory. He is IEEE Member and Reviewer of many journals.

**Ibrahim Ghafir** is Assistant Professor in Computer Science at Faculty of Engineering and Informatics, University of Bradford. He received his Ph.D. in Computer Science (Cyber Security) from Manchester Metropolitan University, UK. Ibrahim also has another Ph.D. in Informatics (Computer Systems and Technologies) from Durham University, UK, and Masaryk University, Czech Republic (European Doctorate). He is Fellow of the Higher Education Academy (FHEA). Prior to joining Bradford, Ibrahim worked as Research Associate in Cyber Security at Loughborough University, UK, from 2017 to 2019. Before Loughborough, he worked with Masaryk University as a Research Assistant from 2013 to 2015. Dr. Ghafir's research interests include network security, intrusion detection systems, wireless communication, Internet of Things, artificial intelligence, and machine learning.

**Farhana Islam** received her B.Sc. and M.Sc. degrees in Information Technology from Jahangirnagar University, Bangladesh, in 2015 and 2016, respectively. Currently, she is working as Lecturer in the Department of Education at Bangabandhu Sheikh Mujibur Rahman Digital University, Bangladesh (BDU). Before joining BDU, she worked as Assistant Lecturer at Gono Bishwabidyalay. Her research interests include IoT, machine learning, big data, image processing, and wireless communication.

# Chapter 11
# Smart Systems Risk Management in IoT-Based Supply Chain

**Hamed Nozari and Seyyed Ahmad Edalatpanah**

**Abstract** The Internet of Things (IoT) offers a modern solution in which the boundaries between real and digital realms are gradually blurred by the continuous transformation of each physical device into a smart object. Each of these intelligent objects plays a role in different realms of life, but at the same time leads to new challenges. In order to develop industries and make them smarter using transformational technologies, there are risks that slow down or prevent progress. As the risk increases in the processes, the task of managing and controlling the project becomes more difficult. Many of the failures that occur in business processes are due to risk and instability in the environment and within the supply chain structure. Therefore, a comprehensive quantitative relationship that can measure supply chain risk and take into account all dimensions of risk has not yet been proposed. IoT-based intelligent supply chains have always been studied as one of the high-risk sectors due to the presence of the Internet and network and huge data flow. This study examines and prioritizes the risk of implementing smart systems in IoT-based supply chains that have been prioritized using a nonlinear fuzzy approach. The results show that lack of knowledge and lack of maintenance of technical infrastructure is one of the most important risk factors in smart food chains, and for sustainable and efficient development, special attention should be paid to the risks resulting from these deficiencies.

**Keywords** Risk management · Smart system risk · Cybersecurity · IoT-based supply chain

H. Nozari
Department of Industrial Engineering, Iran University of Science and Technology, Tehran, Iran
e-mail: ham.nozari.eng@iauctb.ac.ir

S. A. Edalatpanah (✉)
Department of Applied Mathematics, Ayandegan Institute of Higher Education, Tonekabon, Iran
e-mail: saedalatpanah@gmail.com

 
H. Garg (ed.), *Advances in Reliability, Failure and Risk Analysis*, Industrial and Applied Mathematics, https://doi.org/10.1007/978-981-19-9909-3_11

## 11.1 Introduction

Due to the dramatic global changes in today's world and due to the advances in information technology and its applications in various industries and services, companies have a variety of capabilities such as establishing diverse communications, combining capabilities and competencies, fast and diverse transfers, and activities. The supply chain industry is one of the most complex, and at the same time, the most influential industries in each country in terms of providing facilities and facilities related to the production, warehousing, supply, and distribution of goods and services [1]. A successful organization can be effective in maintaining its financial and human capital by having a targeted risk management program and minimize potential losses. With risk management, in addition to preventing an accident and incurring damages, you can also reduce its negative effects in the event of an accident. In fact, to create a relationship between producers and consumers, there are several processes that, with the expansion of the scope and volume of activities in this field, the realization of productivity-related indicators in creating this relationship becomes very complex [2].

In addition to being a transformative technology in various industries and businesses, the Internet of Things has shown its ability in the key processes of the supply chain. Professional tools based on the Internet of Things as well as forecasting and monitoring tools help managers to improve the operational capability of the organization's professional distribution processes and increase transparency in organizational decisions. Therefore, in this era, more than ever, the benefits of using the Internet of Things in the supply chain are evident. The Internet of Things adds many features and exclusive features to traditional supply chains, which increases the reliability and activity of the supply chain [3]. Smart objects will have the ability to manage the supply chain with the help of analyzing the market situation, the situation of competitors, the interests of customers, etc. However, security and privacy concerns need to be kept in mind as well. As the number of tools and devices connected to the Internet increases, the possibility of attacking them will increase, and in such circumstances, creating a relative balance between providing security and using the tools and facilities of the Internet of Things will become a necessity [4].

The interface that connects IoT systems and their devices creates a mechanism for spreading risk and creating danger and damage on physical, social, and economic scales. Risk assessment methods are a complex process that requires the consideration of various factors. Also, the interpretation and assessment of risk may change depending on the scope of the work, which should be considered in the new risk assessment solution as a periodic assessment according to the significant changes in IoT environments.

Therefore, risk management is about evaluating and selecting strategies for evaluating network-based risks in intelligent systems, such as intelligent supply chains, for the employer and its shareholders to maximize the return on investment. Applying risk management in different stages of supply chain activities from purchasing raw materials to distributing sales can be useful and effective in reducing risks and increasing technical knowledge. Increasing the use of risk management means that employers can add and manage risk management and supply chain uncertainties to gain technical knowledge about all IoT-based intelligent processes to save time and money. Risk management can be considered as a process for maintaining assets, power, and security, and value engineering is value management that is an integral part of risk management. Integrating value engineering and risk management creates synergies. This synergy, when implemented in cybersecurity, can affect cybersecurity. IoT-based supply chain cyber-risk management includes identifying security risks and vulnerabilities, identifying actions, and implementing comprehensive security solutions to ensure the security of an organization's assets [5].

Accordingly, some organizations that use IoT data in their supply chain, in addition to business concerns, should also consider legal concerns in their cybersecurity risk management system. For this reason, cybersecurity, in order to increase its effectiveness, must have a layered approach to further protect the important assets of organizations such as data about the organization, business partners, and customers. The reason for this is that the consequences of an intrusion can do more harm to the organization than the intrusion itself [6].

In this regard, over the years, studies have been conducted regarding risk management based on the Internet of Things. Lee [7] investigated cyber-risks based on the Internet of Things and categorized previous studies in this field. Zakaria et al. [8] presented a model for analyzing risks based on the Internet of Things in the medical industry. Rabelo et al. [9] investigated and analyzed the effects of risk management based on the Internet of Things in order to increase business resilience. Zhou et al. [10] investigated the effects of reset management based on big data from the Internet of Things and presented a framework for this management.

This chapter examines risk management in IoT-based intelligent supply chain systems. Therefore, after identifying the strengths and weaknesses of the two management techniques, the necessity of combining them has been investigated, and nonlinear decision-making methods have been performed to analyze risk management in the use of intelligent systems in supply chains. This study provides a realistic insight for supply chain practitioners and activists about the structure of Internet of Things systems, the use cases of these digital systems and the changes related to the processes resulting from these technologies. In addition, concepts, benefits, and limitations related to the factors involved are identified that can be used to redesign sustainable and secure supply chains. Therefore, a correct understanding of the risks based on digital technologies can help the organization to act correctly in the face of them and reduce the number of risks, and this is very necessary for the survival of the organization.

## 11.2 IoT-Based Supply Chain

The Internet of Things network is very wide. In various sources, the number of active devices in the Internet of Things network is estimated up to several billions.

These objects connected to the network can collect a large amount of information. If this massive data is properly processed, reliable knowledge can be obtained. This knowledge can help a lot in supply chain management. The smart objects of the future have the ability to analyze human behavior using a variety of digital developments such as artificial intelligence systems, deep learning and with the help of data from the Internet of Things technology, and with the help of designing a variety of prediction models, analysis to do detailed processes for complex supply chain processes [7]. By linking things to information technology through smart devices, the entire supply chain process from supply to production and distribution can be optimized, and the entire product life cycle can be controlled. By tagging items and contents, more information can be obtained about the status of the workshop, the location of the production machinery. Useful tag information can be used as input data to generate refined applications and improve logistics. Self-organization and intelligent production solutions can be identified alongside design items [8].

Applications of the Internet of Things in the supply chain include the following:

- ***Reduce logistics costs and improve distribution efficiency***

Using the Internet of Things, you can monitor all processes from the supply chain to warehousing, handling, and delivery. This can identify which parts need improvement. On the other hand, unnecessary exchange costs are eliminated due to the effectiveness of information exchange in digital supply chain processes. By improving the condition of different departments and creating more coordination among them, you can increase the productivity and profit in addition to reducing the collection costs [9].

- ***Improve supply chain information and supply and demand balance***

Since the Internet of Things has been added to logistics, the ability to track, review, monitor, and analyze statistics has greatly increased. The Internet of Things has improved the work of transporting goods and increased the health of cargoes by providing proper information to all stakeholders. In the Internet of Things, sensors can be used for specific products to check different points such as altitude, temperature, and humidity. The information sharing mechanism across the supply chain can enable manufacturers, distributors, and retailers to access market supply and demand information in time. Information flow eventually balances supply and demand and prevents product price fluctuations [10].

- ***Ensure product and product security by controlling each part of the supply chain***

Professional and intelligent monitoring and control of the process in the supply chain and logistics by examining various stages in the supply chain from production to warehousing and packaging increases the security and speed of processing in the process and reduces costs. The positive point of using the Internet of Things in the chain processes is increasing the amount of supervision and control over the processes and most importantly the supervision over the fleet and drivers. Building an IoT infrastructure helps supply chain managers manage and optimize costs intelligently by providing them with real-time information [11].

- ***Promote and improve the production of quality products on a large scale and industrial development***

Traditional products are often small in scale; products and goods often go through a long cycle from production to consumption, and information is exchanged slowly during this cycle; the supply chain is inefficient, and products often come to market with many problems. The Internet of Things connects every part of the supply chain, and information is transmitted in a timely, accurate, and efficient manner. Manufacturers, shippers, wholesalers, and retailers of the supply chain can receive and analyze market information correctly and ultimately make the right decisions. When supply and demand information is shared in the market, everyone from manufacturers to retailers will benefit, and logistics companies can consolidate resources, ultimately reducing operating costs and increasing the scale of the business. In the long run, this will help build reputable brands and industrial development [12].

- ***Improve the business of retailers***

IoT can offer several benefits in retail and supply chain management. For example, by equipping shelves with equipment such as RFIDs, a retailer can manage the needs of their goods properly. Also, if major manufacturers are aware of the needs of retailers, they can better manage their products and control the market situation. So, by collecting submitted information from retailers' needs, they can optimize their products. IoT can create huge potential for product warehousing in retail. According to annual statistics, about 0.4% of retailers' sales are lost due to shortages. Therefore, this technology is able to play a major role in the production and supply chain of goods by displaying the moment of inventory and sales of each product [13].

Figure 11.1 schematically shows an IoT-based supply chain.

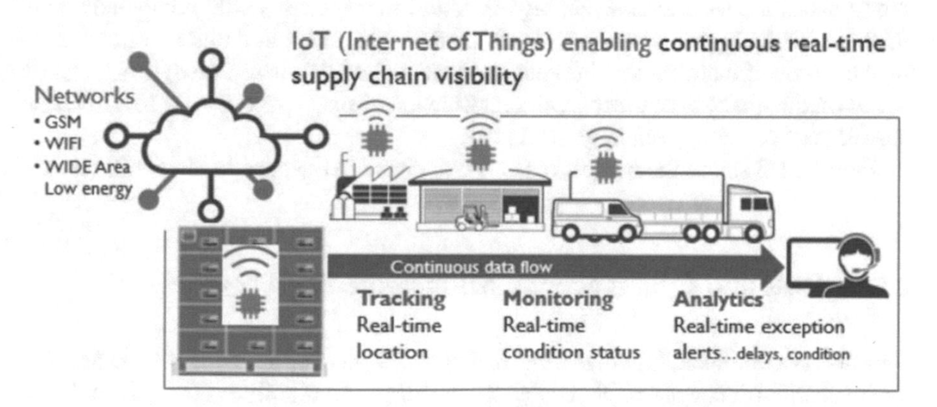


**Fig. 11.1** IoT-based supply chain

## 11.3 Internet of Things and Risks

The Internet of Things (IoT) in organizations is growing rapidly, paving the way for new and innovative approaches and services in all industries, including manufacturing. As a result, new cybersecurity risks are emerging. IoT devices are vulnerable to moderate or high-intensity attacks. This provides a good environment for hacking, and companies will be at risk of stealing personal and confidential information on the network.

The Internet of Things relies heavily on networks and wireless communications to provide connectivity for smart devices, yet their openness makes wireless communications vulnerable to security threats and risks. The risks in the IoT system are critical, and any interruptions or deviations can lead to costly changes or life-threatening challenges. Therefore, security and trust in IoT systems are important [14].

The common way to analyze such challenges is to conduct cyber-risk assessments that seek to identify critical assets and threats that organizations face. In this method, after prioritizing the risks identified in the supply chain processes, appropriate solutions are selected to effectively face these risks. Only, the use of pre-existing risk assessment methods for the Internet of Things may prevent the detection of new risks in this ecosystem [15].

IoT risk assessment and management mechanisms are needed to identify, evaluate, and prioritize risks in organizational assets and operations. There are several basic concepts in risk assessment that include assets, vulnerabilities, threats, the likelihood of an attack, and cyber-damage. IoT risk assessment and management mechanisms are needed to identify, evaluate, and prioritize risks in organizational assets and operations. There are several basic concepts in risk assessment that include assets, vulnerabilities, threats, the likelihood of an attack, and cyber-damage.

Framework-based mechanisms are based on a set of assumptions, concepts, values, and actions that make up the method of observing reality. Framework-based mechanisms are used to analyze, organize, and manage risks with completely new methods. At its highest level, risk-based modeling is a criterion that requires statistical analysis of many factors in order to assess risks. Risk-based models, which are based on the analysis of assets, vulnerabilities, and risks, have a view to assessing their impact on the system in general [16].

Figure 11.2 shows the biggest reasons that cause IoT security vulnerabilities.

## 11.4 Risk and Cybersecurity Management Process

The first step in deciding what form of risk to expect for a business is to use the cybersecurity framework implemented for the business area. Once the organization's data is provided to the organization in a mapped form, managers can make better decisions about how to control and manage the data and create strategies to reduce the risk effects associated with the data. It can be concluded that even with a strong

**Fig. 11.2** Five biggest reasons that cause IoT security vulnerabilities [17]

security training and culture, sensitive organizational information such as data stored in spreadsheet rows, or notes in employee presentation files, or as topics in a long email, can easily and randomly from an organization [18].

In most cases, monitoring the organization to find sensitive information in which no information is exchanged in the organization, as well as deleting any data that does not belong to its storage location, greatly reduces the risk of accidental loss of sensitive information. Once the expected risk pattern has been identified, all of the organization's technology infrastructure will be tested to determine the baseline status for the current risk and what the organization needs to do to move from the current state to the expected risk state. As long as preventive measures are taken in the organization to find potential dangers and threats, the probability of danger and victimization at the time of the attack decreases [6].

Small security vulnerabilities can cause major damage to network systems. The intrusion into a worthless area of the organization's network can lead to unauthorized access to important systems and more sensitive information [19].

The only way to create a 100% secure system is to make sure the system is not accessible. This is impractical at best. Excessive restrictions on network systems and access to data required by users may make it more difficult for authorized personnel to do business. On the other hand, if authorized users find that they cannot access the systems or data they need to do their job, they may seek solutions that could compromise the organization's information. The following has been suggested to enhance the quality characteristics of risk management:

- *Advanced encryption*:

Data encryption is not a new feature for databases. But today, encryption must be implemented according to a specific strategy and systematically to protect the organization's data against external and internal threats. This includes access control based on the organizational role of individuals, standard encryption, advanced key management systems, segregation based on individual tasks, and advanced encryption algorithms that dramatically reduce information disclosure.

Although data encryption is useful to prevent it from being stolen against external threats, it is a little weak against internal data theft. Because people who have access to sensitive information do not necessarily have enough information to decrypt it. Therefore, organizations must also protect data that is transferred from the organization's sensitive systems using portable media such as external drives and other portable devices [20].

- *Correction and review*:

Organizations need to balance data protection with the ability to share. Correction and review enables the organization to share information with minimal hassle and by hiding sensitive information such as letters and numbers when searching and updating [21].

- *Element-level security*

Because correction and review play a very important role in organizations, organizations need to be able to do this based on the role of employees at the level of specific characteristics or based on the roles of employees in the organization. Organizations must be able to implement and enforce custom rules as well as out-of-the-box rules [22].

- *Cybersecurity solutions and risk management services*

Ideally, organizations implement a comprehensive security structure including a combination of different technologies such as firewalls, endpoint security, intrusion prevention systems, threat intelligence systems, and access control systems in their network. To reach this point, organizations may want to focus on risk assessment services to provide a comprehensive assessment and solution to ensure that their security budgets are optimally spent [23].

The most important criteria were then examined by the Minister of Security Standards for risk management in the IoT-based supply chain.

## 11.5 Smart Systems Risk Management in IoT-Based Supply Chain

In this section, using literature review and published studies, the most important risk criteria in the smart supply chain were identified and summarized. For this purpose, in the first step, the research question was formulated. In this study, the research questions are as follows: What are the risks affecting the implementation of an IoT-based digital intelligent supply chain? And what classifications do they have? In the same step, the statistical population and the search period were determined. To determine the statistical population, articles indexed in Scopus, Google Scholar, and ISI Web of Science indexing databases were considered. Also, due to the new concept of smart supply chain, indexed articles between 2016 and 2022 were considered as time periods.

Among all the risks that an intelligent supply chain can face, according to experts, two criteria and seven risks were considered as the most important risks of the intelligent supply chain, which can be seen in Fig. 11.3, the risk-related decision tree.

As shown in Fig. 11.3, the risks of an intelligent supply chain are shown in two sections, design risks, and implementation risks. The following is a description of these risks:

- ***Technical knowledge risks***

Technical knowledge risks are one of the most important and effective risks in the development process. Technological risk refers to a company's inability to fully understand or accurately predict certain technical aspects of the environment that are relevant to new product development projects. Therefore, identifying, reducing, and

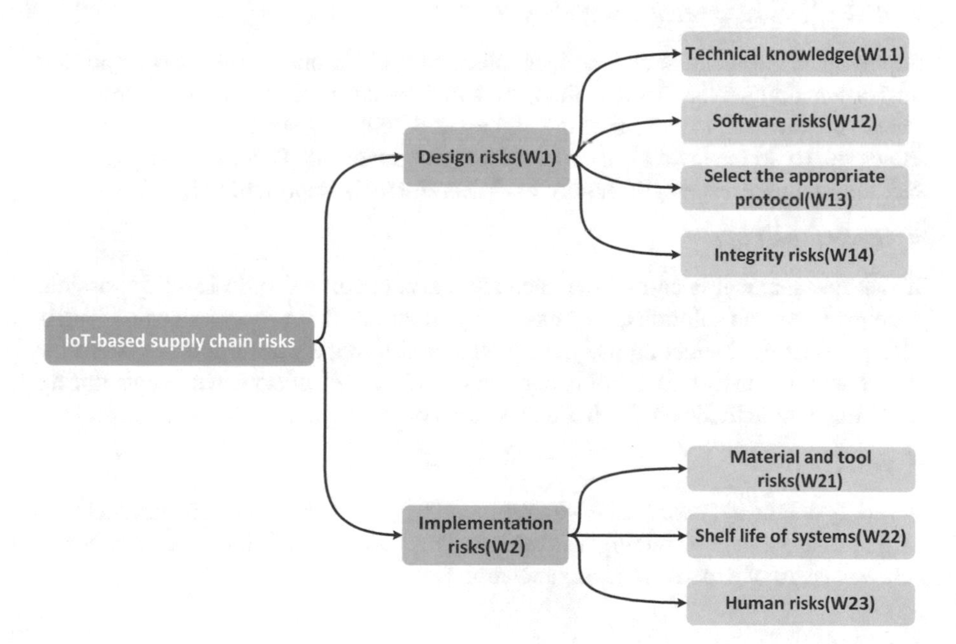


**Fig. 11.3** IoT-based supply chain risks

managing these risks will help to increase the success rate of the product development process in the production sector in the supply chain [24–26].

- ***Software risks***

Software risks refer to errors that occur in software systems related to supply chain intelligence [25, 26].

- ***Select the appropriate protocol***

Risk management in an organization is a process that is implemented continuously. To implement this process, a risk management system must be established in the organization. This system includes risk strategy, risk architecture, and risk management protocols. Risk management protocols are the guidelines, methods, standards, and tools by which the risk management process is implemented in the organization. Risk management protocols define the step-by-step steps of the risk management process and provide specific tools for each role for each step [27].

- ***Integrity risk***

The information must be complete, comprehensive, and accurate, and not tampered with at the time of entry, processing, and storage. The risk of information integration is associated with the risk of manipulating information, reducing its accuracy and precision, and removing or adding unrelated parts [28].

- ***Risk of choosing materials and tools***

Supplier selection and supply of technological tools is one of the most important decision-making issues in this field, in which many qualitative and quantitative factors are involved to determine the highest capability. As supply chain complexity increases, so do the level of uncertainty and risk involved. Therefore, supply chain risk management, especially supply risk assessment, is important [27].

- ***System shelf life***

If risk management is carried out accurately and continuously to identify possible problems and find solutions, they can easily monitor other processes such as organizing, planning, budgeting and cost control and find the most optimal system for the system. It can also largely prevent the occurrence of unexpected events during the lifetime of activities in the basic processes of the organization [24, 25].

- ***Human risks***

A variety of human factors can have a profound impact on the results and productivity of risk management of intelligent systems. This risk is due to the lack of effective role and share of experts in the organization [29].

## 11.6 Quantitative Assessment of IoT-Based Supply Chain Risks

This research seeks to manage the risks of an intelligent digital supply chain based on IoT technology with a quantitative approach. In this study, in order to investigate the risks, first, using the literature on the subject and the opinions of active experts, 7 criteria were extracted as the most important risks. In this study, Cronbach's alpha method was used to assess the reliability of the relevant questionnaires using SPSS software. A nonlinear hierarchical analysis method has been used to analyze and evaluate the risks of the intelligent supply chain. This method is known as the Mikhailov method [30]. The quantitative analysis method used in this research is described below. The research method is shown in Fig. 11.4.

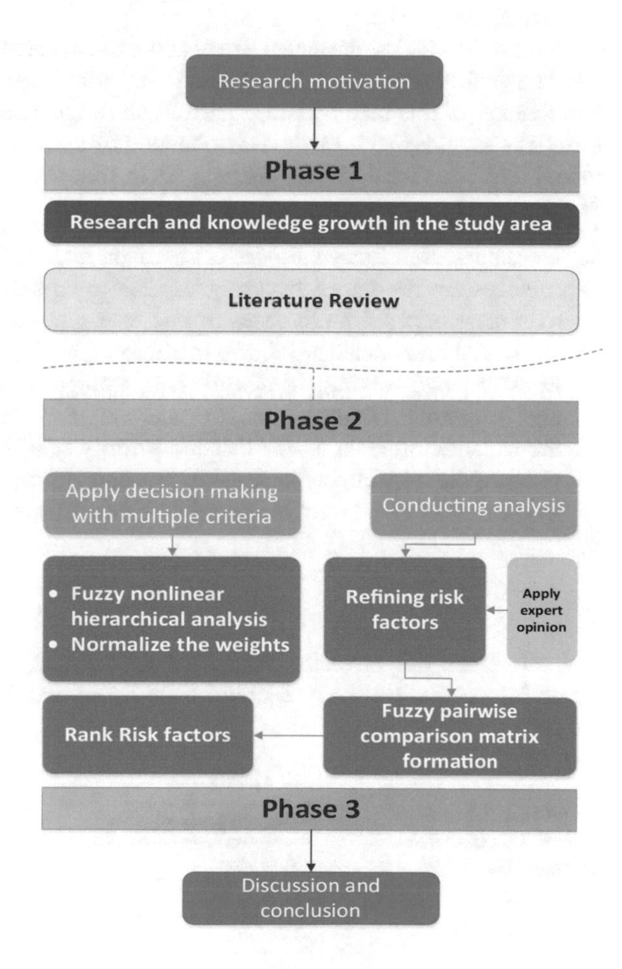


**Fig. 11.4** Research method

### *11.6.1 Mikhailov Ranking Method*

Due to the vague and imprecise nature of data and information in real life and making decisions based on this inaccurate and uncertain information, the modeling process of many phenomena may not be done properly and efficiently. In order to solve the ambiguity and inaccuracy of individual judgments, the theory of fuzzy sets was proposed to introduce linguistic conditions (phrases) in the decision-making process. The traditional AHP (early version of AHP), which has been used so far, requires accurate judgments. But, due to the complexity and uncertainty of real decision-making issues, it is often unrealistic or even impossible to provide accurate judgments. Therefore, it is much more realistic and practical if this possibility can be provided to the decision maker to use inaccurate judgments using fuzzy logic instead of accurate judgments.

In 2004, Mikhailov presented a new approach to calculating weights in the fuzzy AHP method, and he called this method fuzzy prioritization. One of the most important features of this method is the calculation of the compatibility rate in the fuzzy state. The weights in this method are obtained from solving a nonlinear optimization model. The steps for using this method are as follows:

1. The development of the hierarchical structure is as shown in Fig. 11.1.
2. Development of fuzzy pairwise comparison matrix: Fuzzy judgment agreement matrices are developed by using experts' opinions. Fuzzy numbers are used to express experts' preferences in this research. Linguistic variables and their associated fuzzy scale are shown in Table 11.1.
3. Modeling and solving: In this method, triangular fuzzy numbers are used to analyze pairwise comparisons. Certain weight vectors $w = (w_1, w_2, \ldots, w_n)$ are extracted in such a way that the priority rate is approximately within the range of the initial fuzzy judgments. In other words, the weights are determined in such a way that the following relationship is established.

$$l_{ij} \leq \frac{w_i}{w_j} \leq u_{ij} \tag{11.1}$$

Each certain weight vector ($w$) has a degree in fuzzy inequalities as given above. These inequalities can be measured by the linear membership function of relation (11.2):

**Table 11.1** Linguistic variables for pairwise comparisons

| Linguistic variable | Triangular fuzzy scale |
|---|---|
| Very low | (1,2,3) |
| Low | (2,3,4) |
| Medium | (3,4,5) |
| High | (4,5,6) |
| Very high | (5,6,7) |

$$\mu_{ij}\left(\frac{w_i}{w_j}\right) = \begin{cases} \frac{(w_i/w_j)-l_{ij}}{m_{ij}-l_{ij}} & \frac{w_i}{w_j} \le m_{ij} \\ \frac{u_{ij}-(w_i/w_j)}{u_{ij}-m_{ij}} & \frac{w_i}{w_j} \le m_{ij} \end{cases} \tag{11.2}$$

According to the specific form of the membership functions, the fuzzy ranking problem is represented as a nonlinear optimization problem as shown in Eq. (11.3).

$$\begin{aligned} & \max \ \lambda \\ & \text{Subject to :} \\ & (m_{ij} - l_{ij})\lambda w_j - w_i + l_{ij} w_j \le 0 \\ & (u_{ij} - m_{ij})\lambda w_j + w_i - u_{ij} w_j \le 0 \\ & i = 1, 2, \ldots, n-1, \ j = 2, 3, \ldots, n, \ j > i, \\ & \sum_{k=1}^{n} w_k = 1 \quad w_k > 0, \quad k = 1, 2, \ldots, n \end{aligned} \tag{11.3}$$

Given the nonlinearity of Eq. 11.3, it is obvious that it cannot be solved without the use of software. Therefore, LINGO software has been used to solve the models created in this research. Positive optimal values for the index (objective function) indicate that all weight ratios are completely true in the initial judgment.

### 11.6.2 Research Findings

The main stages in analyzing and prioritizing and managing supply chain risks based on the Internet of Things and other transformative technologies in this research have two parts; these two parts are as follows:

1. Compilation of the matrix of pairwise comparisons based on the integration of experts' opinions.
2. Applying the nonlinear model to prioritize the risks of the digital supply chain based on the Internet of Things and obtain the weight of each of the risks

The pairwise comparison tables obtained from the opinions of the experts are shown in Tables 11.2, 11.3 and 11.4. These tables have been used for calculations by the Mikhailov method.

By placing the data from Tables 11.2, 11.3 and 11.4 in the nonlinear model (11.3) as a model providing weights and rankings based on hierarchical analysis and model

**Table 11.2** Matrix of pairwise comparisons for technological challenges of two general categories

| | W1 | | | W2 | | |
|---|---|---|---|---|---|---|
| W1 | – | – | – | – | – | – |
| W2 | 1.5 | 2.47 | 3.25 | – | – | – |

**Table 11.3** Matrix of pairwise comparisons for design risks

| | W11 | | | W12 | | | W13 | | | W14 | | |
|---|---|---|---|---|---|---|---|---|---|---|---|---|
| W11 | – | – | – | – | – | – | – | – | – | – | – | – |
| W12 | 3.1 | 4.2 | 5.1 | – | – | – | – | – | – | – | – | – |
| W13 | 2.1 | 2.8 | 4.7 | 2.3 | 3.1 | 4.2 | – | – | – | – | – | – |
| W14 | 3.1 | 3.5 | 5.4 | 3.1 | 3.5 | 4.5 | 2.1 | 2.45 | 3.21 | – | – | – |

**Table 11.4** Paired comparison matrix for implementation risks

| | W21 | | | W22 | | | W23 | | |
|---|---|---|---|---|---|---|---|---|---|
| W21 | – | – | – | – | – | – | – | – | – |
| W22 | 2.1 | 2.7 | 3.8 | – | – | – | – | – | – |
| W23 | 1.5 | 1.75 | 2.5 | 3.1 | 3.95 | 5.12 | – | – | – |

equation using LINGO software, the weight and rank of each of the risks can be acquired general dimensions as well as in exclusive categories. The computational results related to the solution of the nonlinear model for general and individual batches of chrysanthemums are shown in Tables 11.5, 11.6 and 11.7.

A positive value for the compatibility index indicates the acceptable compatibility of the matrices. After obtaining the weights of the general categories and the weights

**Table 11.5** Weight and ranking of the main categories

| Category | Code | Weight | Rank | Objective function ($\lambda$ ) |
|---|---|---|---|---|
| Design risks | W1 | 0.678225 | 1 | 0.5251 |
| Implementation risks | W2 | 0.323775 | 2 | |

**Table 11.6** Weight and ranking of design risks

| Risk | Code | Weight | Rank | Objective function ($\lambda$ ) |
|---|---|---|---|---|
| Technical knowledge | W11 | 0.397849 | 1 | 0.4214 |
| Software risks | W12 | 0.241141 | 2 | |
| Select the appropriate protocol | W13 | 0.181998 | 3 | |
| Integrity risks | W14 | 0.180306 | 4 | |

**Table 11.7** Weight and ranking and implementation risks

| Risk | Code | Weight | Rank | Objective function ($\lambda$ ) |
|---|---|---|---|---|
| Material and tool risks | W21 | 0.373887 | 1 | 0.2374 |
| Shelf life of systems | W22 | 0.281210 | 3 | |
| Human risks | W23 | 0.347111 | 2 | |

**Table 11.8** Normal weight and IoT-based digital supply chain risk rating

| Category | Weight | Risk | Weight | Normalized weight | Rank |
|---|---|---|---|---|---|
| Design risks | 0.678225 | Technical knowledge | 0.397849 | 0.269831 | 1 |
| | | Software risks | 0.241141 | 0.163548 | 2 |
| | | Select the appropriate protocol | 0.181998 | 0.123436 | 3 |
| | | Integrity risks | 0.180306 | 0.122288 | 4 |
| Implementation risks | 0.323775 | Material and tool risks | 0.373887 | 0.121055 | 5 |
| | | Shelf life of systems | 0.281210 | 0.091049 | 7 |
| | | Human risks | 0.347111 | 0.112386 | 6 |

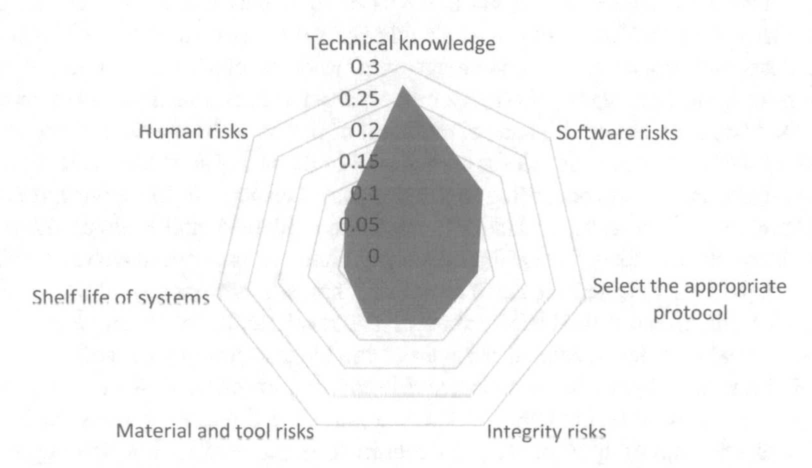


**Fig. 11.5** IoT-based supply chain risk assessment

in the specific categories, we can normalize the weights to obtain the total weight regardless of the category as well as their overall rank. The normalized computational results are shown in Table 11.8.

As shown in Table 11.8, technological risks are one of the most important risks of implementing an object-based supply chain in the age of evolving technologies. These results are shown in Fig. 11.5.

## 11.7 Conclusion

Today, due to the growth of the competitive environment in the supply chain management sector, we are witnessing the increasing efforts of organizations to increase efficiency and increase the performance of the supply chain. The intelligence of the supply chain under the influence of disruptive technologies such as the Internet of

Things and artificial intelligence at different levels can have a high impact on the overall performance of the supply chain and the optimal response of the organization to continuous changes in the business environment in the wide network of local and global supply chains. But, the presence of Internet and network technology undoubtedly offers many challenges for these intelligent processes. In order to manage these risks, identifying and analyzing these risks can be very important. Before launching a risk management system in cybersecurity, organizations should identify their assets that need protection and prioritize them based on value. The Internet of Things means the possibility of communicating all objects with each other, along with their identification and discovery under a unified network. Its main purpose is to share the information in each object among other objects related to it. The Internet of Things environment consists of heterogeneous devices that continuously exchange information and are present everywhere, but at the same time, they lead to new challenges. The interface that connects IoT systems and their devices creates a mechanism to spread risk and create danger and damage on physical, social, and economic scales. Risk assessment methods are a complex process that requires consideration of various factors. Also, the interpretation and assessment of risk may change depending on the scope of the work, which should be considered in the new risk assessment solution as a periodic assessment according to the significant changes in IoT environments.

Therefore, in this study, an attempt was made to identify and evaluate the most important risks affecting the IoT-based supply chain. In order to evaluate the risks of the supply chain based on the Internet of Things, a nonlinear ranking method called the Mikhailov method has been used. The results show that technological and technological knowledge-related risks are of the highest importance, and therefore, creating a technological environment and increasing knowledge of evolving technologies and raising technology levels can reduce the threatening risks. Software risks and selection of appropriate protocols in the next degree, they are important and should be given special attention. The calculation results are in line with previous researches regarding risk factors in smart supply chains and confirm them.

## References


1. Ogbuke, N.J., Yusuf, Y.Y., Dharma, K., Mercangoz, B.A.: Big data supply chain analytics: ethical, privacy and security challenges posed to business, industries and society. Prod. Plann. Control **33**(2–3), 123–137 (2022)
2. Nozari, H., Fallah, M., Szmelter-Jarosz, A., Krzemiński, M.: Analysis of security criteria for IoT-based supply chain: a case study of FMCG industries. Central Eur. Manage. J. **29**(4) (2021)
3. Lee, K., Romzi, P., Hanaysha, J., Alzoubi, H., Alshurideh, M.: Investigating the impact of benefits and challenges of IOT adoption on supply chain performance and organizational performance: an empirical study in Malaysia. Uncertain Supply Chain Manage. **10**(2), 537–550 (2022)
4. Nozari, H., Fallah, M., Kazemipoor, H., Najafi, S.E.: Big data analysis of IoT-based supply chain management considering FMCG industries. Бизнес-информатика **15**(1 eng) (2021)
5. Hasan, T.A.S.M., Sabah, S., Haque, R.U., Daria, A., Rasool, A., Jiang, Q.: Towards convergence of IoT and blockchain for secure supply chain transaction. Symmetry **2022**(14), 64 (2022)

6. Wang, L., Wang, Y.: Supply chain financial service management system based on block chain IoT data sharing and edge computing. Alex. Eng. J. **61**(1), 147–158 (2022)
7. Fallah, M., Sadeghi, M.E., Nozari, H.: Quantitative analysis of the applied parts of Internet of Things technology in Iran: an opportunity for economic leapfrogging through technological development. Sci. Technol. Policy Lett. **11**(4) (2021)
8. Nozari, H., Fallah, M., Szmelter-Jarosz, A.: A conceptual framework of green smart IoT-based supply chain management. Int. J. Res. Ind. Eng. **10**(1), 22–34 (2021)
9. Ju, C.: Research on the challenges and strategies of enterprises in reverse logistics cost control under B2C mode. In: Advances in Smart Vehicular Technology, Transportation, Communication and Applications, pp. 401–410. Springer, Singapore (2022)
10. Rajaguru, R., Matanda, M.J., Zhang, W.: Supply chain finance in enhancing supply-oriented and demand-oriented performance capabilities–moderating role of perceived partner opportunism. J. Bus. Ind. Marketing (2022)
11. Bhushan, B., Kumar, A., Katiyar, L.: Security magnification in supply chain management using blockchain technology. In: Blockchain Technologies for Sustainability, pp. 47–70. Springer, Singapore (2022)
12. Zheng, J., Hu, H.: Study on food quality and safety management model based on industrial agglomeration theory. Acta Agriculturae Scandinavica, Section B—Soil Plant Sci. 1–11 (2022)
13. Purohit, S.: Review of Adoption of Technology Towards Intelligence Operation and Customer Services in Indian Petro Retailing (2022)
14. Nozari, H., Ghahremani-Nahr, J., Fallah, M., Szmelter-Jarosz, A.: Assessment of cyber risks in an IoT-based supply chain using a fuzzy decision-making method. Int. J. Innov. Manage. Econ. Soc. Sci. **2**(1) (2022)
15. Nozari, H., Szmelter-Jarosz, A., Ghahremani-Nahr, J.: Analysis of the challenges of Artificial Intelligence of Things (AIoT) for the smart supply chain (case study: FMCG industries). Sensors **22**(8), 2931 (2022)
16. Selvan, S., Mahinderjit Singh, M.: Adaptive contextual risk-based model to tackle confidentiality-based attacks in fog-IoT paradigm. Computers **11**(2), 16 (2022)
17. Joshi, N.: Allerin (2016) [Online]. Available: https://www.allerin.com/blog/security-risks-and-challenges-to-iot-devices. Accessed 04 2022
18. Broday, E.E., da Silva, M.C.G.: The role of internet of things (IoT) in the assessment and communication of indoor environmental quality (IEQ) in buildings: a review. In: Smart and Sustainable Built Environment (2022)
19. Ge, Y., Zhu, Q.: Accountability and Insurance in IoT Supply Chain. arXiv preprint arXiv:2201.11855 (2022)
20. Chandrashekhar, R.V., Visumathi, J., Anandaraj, A.P.: Advanced lightweight encryption algorithm for android (IoT) devices. In: 2022 International Conference on Advances in Computing, Communication and Applied Informatics (ACCAI), pp. 1–5, IEEE (2022)
21. Georgiou, G.: Do correctional authorities treat all offenders equally? Evaluating the use of a risk assessment instrument. Int. Rev. Law Econ. **69**, 106033 (2022)
22. Sobana, S., Prabha, S.K., Seeranguravar, T., Sudha, S.: Securing future autonomous applications using cyber-physical systems and the Internet of Things. In: Handbook of Research of Internet of Things and Cyber-Physical Systems, pp. 81–148. Apple Academic Press (2022)
23. Rahmani, K.R., Rana, M.S., Hossan, M.A., Wadeed, W.M.: Lightweight cyber security for decision support in information security risk assessment. Eur. J. Electrical Eng. Comput. Sci. **6**(1), 24–31 (2022)
24. Hafiani, M., Maslouhi, M., El Abbadi, L.: Supply Chain Risks: A Review Study
25. Mouloudi, L., Evrard Samuel, K.: Critical materials assessment: a key factor for supply chain risk management. Supply Chain Forum: Int. J., 1–15 (Taylor & Francis) (2022)
26. Rajaeifar, M.A., Ghadimi, P., Raugei, M., Wu, Y., Heidrich, O.: Challenges and recent developments in supply and value chains of electric vehicle batteries: A sustainability perspective. Resour. Conserv. Recycl. **180**, 106144 (2022)
27. Mu, W.: Analysis and warning model of logistics risks of cross-border E-commerce. In: Discrete Dynamics in Nature and Society (2022)

28. Cahill, A.G., Samano, P.S.G.: Prioritizing stewardship of decommissioned onshore oil and gas wells in the United Kingdom based on risk factors associated with potential long-term integrity. Int. J. Greenhouse Gas Control **114**, 103560 (2022)
29. Powell, W., Foth, M., Cao, S., Natanelov, V.: Garbage in garbage out: the precarious link between IoT and blockchain in food supply chains. J. Ind. Inf. Integr. **25**, 100261 (2022)
30. Mikhailov, L.: A fuzzy programming method for deriving priorities in the analytic hierarchy process. J. Oper. Res. Soc. **51**(3), 341–349 (2000)

**Hamed Nozari** is a research assistant in Industrial engineering at the Iran University of Science and Technology. He holds a Ph.D. in Industrial Engineering with a focus on Production Management and Planning and PostDoc in Industrial Engineering from the Iran University of Science and Technology. He has taught various courses in the field of Industrial Engineering and has published many books and papers as well. Now he is a researcher in the field of digital developments and smart systems and optimization.

**Seyyed Ahmad Edalatpanah** is Associate Professor at the Ayandegan Institute of Higher Education, Tonekabon, Iran. S. A. Edalatpanah received his Ph.D. degree in Applied Mathematics from the University of Guilan, Rasht, Iran. He is currently working as the Chief of R&D at the Ayandegan Institute of Higher Education, Iran. He is also an academic member of Guilan University and the Islamic Azad University of Iran. Dr. Edalatpana's fields of interest include numerical computations, operational research, uncertainty, fuzzy set and its extensions, numerical linear algebra, soft computing, and optimization. He has published over 150 journal and conference proceedings papers in the above research areas. He serves on the editorial boards of several international journals. He is also the Director-in-Charge of the Journal of Fuzzy Extension & Applications.

# Chapter 12
# Risk and Reliability Analysis in the Era of Digital Transformation

**Fatemeh Afsharnia**

**Abstract** Evolution of Industry 4.0 and the integration of the digital, physical, and human worlds, reliability and safety engineering must evolve in order to address the challenges currently and in the future. This chapter aimed to describe the application of digital transformation in the reliability engineering and risk analysis. In this chapter, the principle of digital transformation is introduced as well as some of the opportunities and challenges in reliability engineering. New directions for research in system modeling, big data analysis, health management, cyber-physical system, human–machine interaction, uncertainty, jointly optimization, communication, and interfaces are proposed. Various topics may be investigated individually, however, we present here a perspective on safety and reliability analysis in the era of digital transformation that would be suitable for discussion and consideration by scientists interested in this topic. The digital transformation combines software and systems engineering to build and run large-scale, massively distributed, fault-tolerant systems.

**Keywords** Reliability · Risk · Cyber-physical system · IoT · Big data · Digital

## Abbreviations

| Notation | Main acronyms |
|---|---|
| CPS | Cyber-physical systems |
| IoT | Internet of Things |
| IT | Information technology |
| ML | Machine learning |
| ANN | Artificial neural networks |
| AI | Artificial intelligence |
| ITU | International Telecommunication Union |
| FTA | Fault tree analysis |

F. Afsharnia (✉)
Department of Agricultural Machinery and Mechanization Engineering, Agricultural Sciences and Natural Resources University of Khuzestan, Ahvaz, Iran
e-mail: phd.afsharnia@asnrukh.ac.ir; afsharniaf@yahoo.com



H. Garg (ed.), *Advances in Reliability, Failure and Risk Analysis*, Industrial and Applied Mathematics, https://doi.org/10.1007/978-981-19-9909-3_12

| | |
|---|---|
| FMEA | Failure modes and effects analysis |
| HAZOP | Hazard and operability methodology |
| MBE | Model-based engineering |
| GTST-MLD | Goal tree-success tree and master logic diagram |
| STPA | System theoretic process analysis |
| CIA | Confidentiality, integrity, and availability |

## 12.1 Introduction

Global digitization can improve reliability and reduce costs, so maintenance managers need to be more ambitious in their move toward digital maintenance [1]. With the digitization of maintenance operations and reliability in heavy industries, it is expected that the availability of company assets will increase by 5–15% and their repair and maintenance costs will also decrease by 18–25%.

The decision-making processes that support maintenance and reliability operations may be sped up and standardized with the aid of new digital technologies. For instance, reliability teams may plan and manage repair or replacement decisions throughout the lifecycles of individual assets or whole fleets with the use of digital asset management systems. On the other hand, new digital technologies can assist teams in selecting the best maintenance strategy (e.g., run-to-fail, scheduled preventative maintenance, or condition-based maintenance) for each equipment, as well as they can promote reliability-centered maintenance [2].

With the advent of the Internet and the widespread use of information technology, manufacturing industry has been impacted by digital information technology. As the digital, physical, and human worlds increasingly integrate, the industry undergoes deep transformation, and emergence of the Fourth Industrial Revolution called Industry 4.0. This technology offers opportunities for factories to be used as open platforms and distributed systems, where they can operate faster, more efficiently, and with a more flexible and resilient supply chain [3].

Based on the change in the manufacturing environments and the increasing competition among companies, we need a new concept to define and build manufacturing factories. This is because the future industrial factory must work as a flexible, resilient, and affordable system. To illustrate this new concept more clearly, the past Industrial Revolutions are examined in this section [4].

First Industrial Revolution used steam power to cause major changes in industries in the eighteenth century. Second Industrial Revolution was made possible by electric power and assembly lines. During the Third Industrial Revolution, computers and information technology became integral parts of manufacturing as well as computer-aided systems. A major feature of the Fourth Industrial Revolution is the strong use of automation and data exchange in manufacturing. Cyber-physical systems, the Internet of Things (IoT), 3D printing, digital twinge, advanced analytics, and cloud

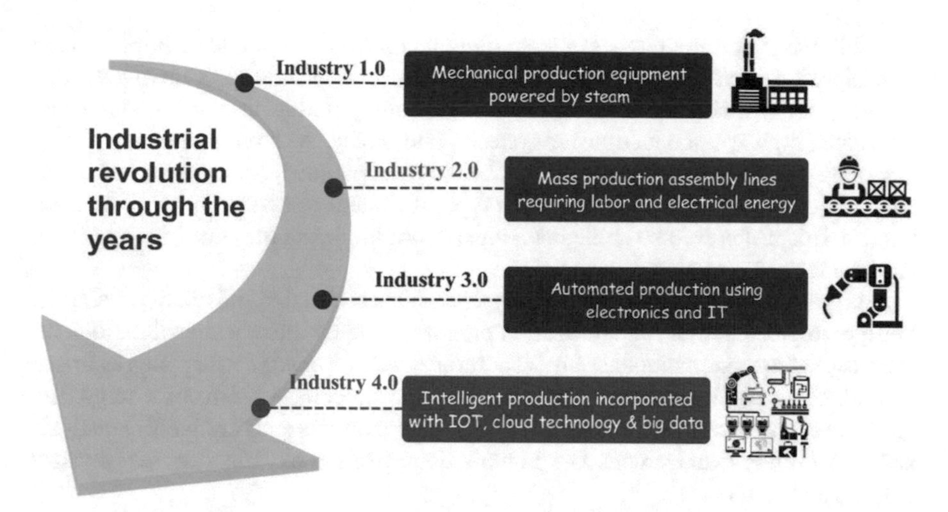


**Fig. 12.1** Overview of industrial revolutions over time [7]

computing, among others are used in the new systems [5]. An illustration of the transformation takes place is shown in Fig. 12.1.

In order to fully utilize digitalization, cyber-physical system integration and intelligent control, industrial systems must process digitalization and implement cyber-physical system integration. In addition to efficiency increases, supporting systems need to be integrated into the main system, such as maintenance, logistics, and supply chain. We deal with a smart system that consists of many systems with dynamic structure. By changing the manufacturing environment, the system speed and flexibility increase. Therefore, smart manufacturing and Industry 4.0 investment have been increasing rapidly and several countries have focused on this subject. A variety of research methods have also been used to introduce and analyze Industry 4.0 and smart manufacturing systems as well [6].

In order to implement Industry 4.0, several fundamental requirements must be met [8]:

- Integrated enterprise systems and interoperability
- An organization that is distributed
- A model-based approach to monitoring and controlling
- Environments and systems that are heterogeneous
- A dynamic and open structure
- Teamwork and collaboration
- Human-to-machine integration and interoperability
- The ability to scale, be agile, and be fault-tolerant
- A network of interdependence
- Collaborative manufacturing platforms that are service-oriented
- Decision support systems based on data-driven analysis, modeling, control, and learning.

Additionally, different types of technological innovations should be implemented to establish a smart factory [9–11]. There are several technologies involved, such as software, advanced collaborative robotics, configurations that are modular and adaptable, high-speed data transfer systems, and others. As a prerequisite to a fully smart system, we need a smart supply chain, smart maintenance system, and smart labor. From a technical standpoint, this type of system presents a number of challenges. This has prevented some companies from implementing this idea and there is still a long way to go.

It is well known that manufacturing systems must be reliable and readily available. Design, implementation, and utilization processes should include considerations for security, safety, and maintainability. Therefore, when a smart factory idea is investigated, these challenges and opportunities must be considered from a reliability engineering point of view. The rest of the chapter explores smart reliability analysis and smart safety management based on big data, Internet of Things, cyber-physical system, and so on.

## 12.2 Reliability Analysis

### *12.2.1 Big Data and Data Processing*

Intelligent systems incorporate advanced instruments and facilities to collect and analyze data at different phases in the life cycle of a product, such as raw materials, machine operations, facility logistics, quality control, product use, and warranty duration. This data plays a crucial role in smart systems, and big data empowers companies to develop more flexible and effective strategies to compete on the market. It is imperative to store and analyze the data collected from manufacturing systems. As industrial development progressed and technology was integrated with manufacturing, as well as the use of computerized systems, data is collected and stored on a machine. In recent years, the capabilities of information technology have rapidly grown up and advanced technologies (e.g., big data analytics, Internet of Things, cloud computing, and artificial intelligence) are becoming more prevalent in industrial and business systems. By integrating IT with systems, a new paradigm is created called Industry 4.0. A similar pattern of data evolution can probably also be assumed for other systems; Fig. 12.2 illustrates how data evolved in manufacturing systems.

The big data collected must be processed and applied in order for system performance to improve. Different types of parameters with different quality and forms are contained in this data due to the use of different sensors and sources. Various types of data may be collected, including video, voice, electronic signal, image, and others, and these should be preprocessed, processed, and analyzed before they can be applied. Data from crude sources is not valuable and may also contain noisy data, therefore, the data should be converted into specific information content and

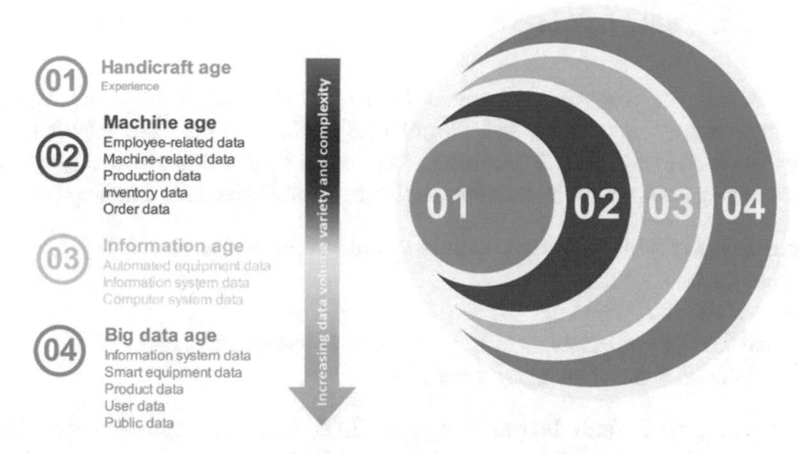


**Fig. 12.2** History of data volume variety and complexity in manufacturing systems [12]

context that users can directly understand. In order to achieve this, we need advanced methods such as cloud computing, neural networks, and deep learning.

In recent decades, cloud-based big data processing technology has been studied as an interesting topic, and various computer models are planned based on different platforms and focuses, such as stream-based, batch-based, directed acyclic graphs-based, graph-based, interactive and visual processing.

Neural networks are a powerful tool in reliability engineering, particularly for predicting how long equipment will be usable. Although artificial neural networks (ANN) are beneficial for data processing, deep learning is more effective. Reduced operating expenses, improved productivity and reduced downtime, keeping up with changing customer demand, improved visibility, and extracting more value from operations for worldwide competitiveness are all points of interest in deep learning.

As computing techniques and data processing have advanced, computer-aided engineering systems and design methods have improved, for instance, different kinds of failure in the system are now modeled and evaluated by simulations. In the utilization stage, this capability provides a greater understanding of failure mechanisms and how to avoid them. Using these capabilities, a reliability engineer can optimize the predictability of a new product in the design phase. Conversely, designers apply artificial intelligence (AI) and deep learning to their own design processes and to new products. It will be a challenge for engineers to use these tools in their own designs in order to optimize final designs more quickly [13]. The dynamic behavior of the system is another challenge in big data processing. System modeling requires the use of a model that adapts to the age, degradation behavior, and condition of the system. Because the system can be influenced by the data collected in real time, the pre-defined model could be changed. The topic of model updating is therefore an interesting one in this field [3].

### 12.2.2 Internet of Things

IoT in maintenance program can help increase safety, reliability, efficiency, connectivity, and communication [14, 15]. Figure 12.3 depicts the increase growth of IoT devices from 2015 to 2025. The production capacity of a manufacturing plant is reduced during equipment breakdowns. IoT-based predictive maintenance could:

- Increase the reliability and availability of equipment and machines;
- Reduce costs;
- Improve uptime;
- Reduce the risks of safety, health, environment, and quality; and
- Extend the lifetime of an aging asset [16].

By identifying a fault before it occurs, IoT predictive maintenance allows machines to be maintained in advance. A machine's condition can be monitored in real time by Internet of Things maintenance systems. The data is analyzed by software to create performance reports. The architecture of IoT-based predictive maintenance is illustrated in Fig. 12.4.

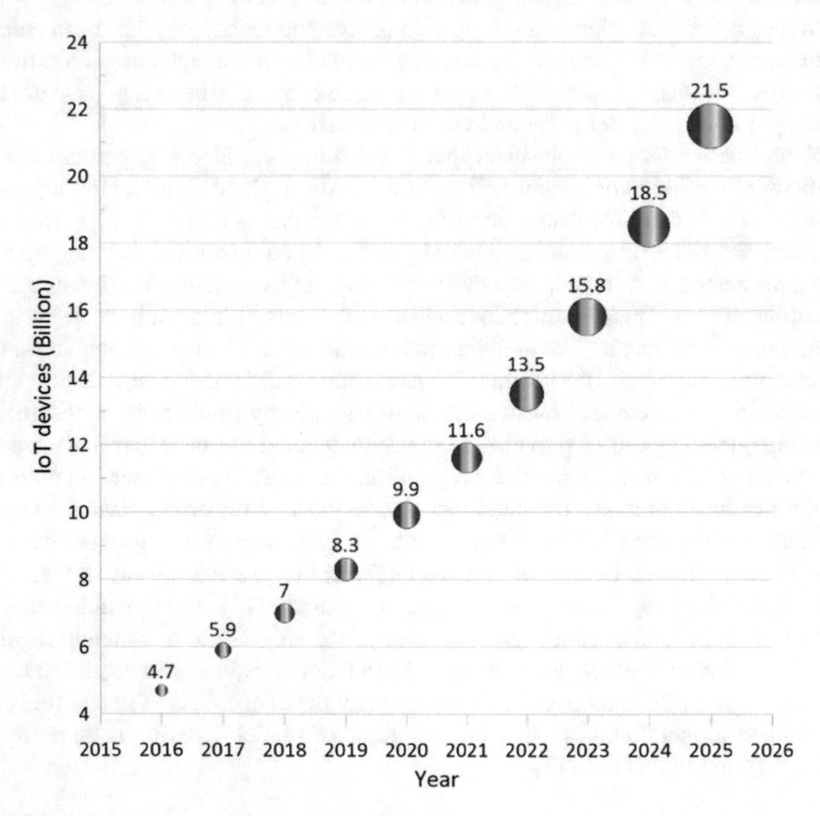


**Fig. 12.3** IoT devices growth during 2015–2025 [17]

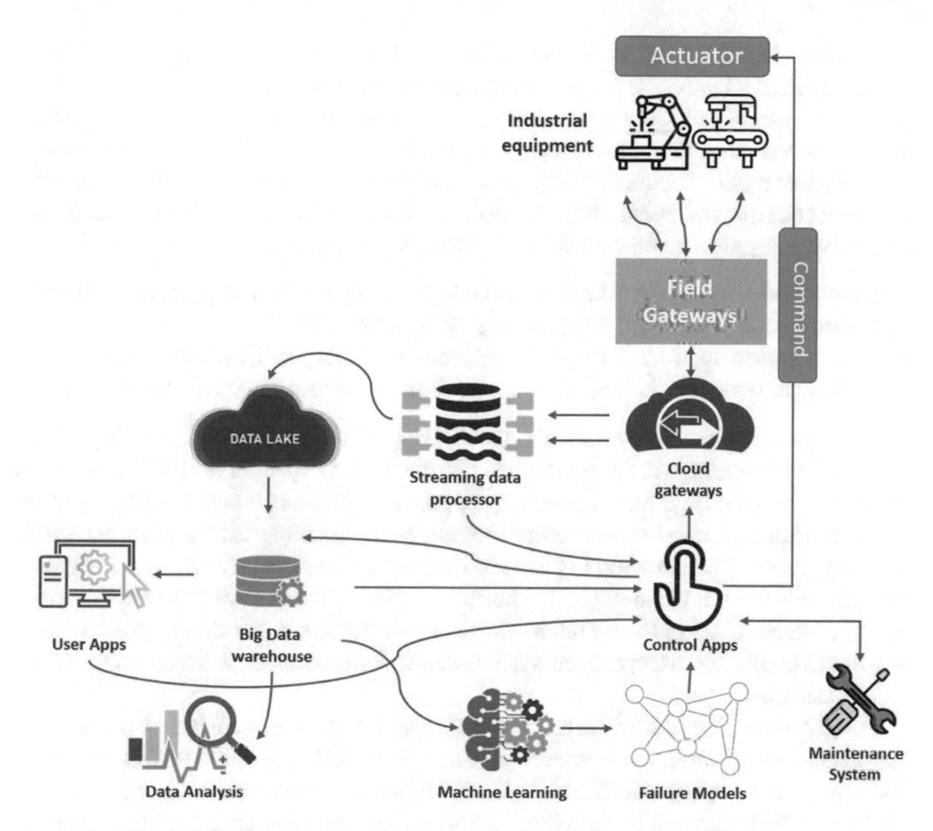


**Fig. 12.4** IoT-based predictive maintenance architecture

Identifying key factors that determine equipment's health is necessary before proceeding to technical details. As soon as these variables are determined, equipment is outfitted with sensors to collect information about them and send it to the cloud for processing. Gateways are required to transfer sensor data to the cloud—it cannot pass directly. In field gateways, the data is filtered and preprocessed. Connecting various gateways via various protocols is possible with a cloud gateway, which enables data transmission and ensures secure data transmission. Streaming data processors then receive the sensor data that was entered into the cloud part. Data lakes are used to store data streams and to transmit them quickly and efficiently to data storage, enabling continuous flow of data. The data collected by sensors is stored in a data lake. Currently, the data is raw, so it may contain inaccurate or erroneous information. It is displayed as a collection of measurements taken at the corresponding time by a number of sets of sensor. In order to gain insight into the health of the equipment, the data is loaded into a big data warehouse. It contains vibration, temperature, and other parameters measured at a corresponding time and contextual information about equipment' locations, types, dates, etc.

Machine learning (ML) algorithms are used to analyze the data after it has been prepared. Machine learning algorithms are used to detect abnormal patterns in datasets and reveal hidden correlations. Predictive models take into account the patterns in the data. Predictive models are built, trained, and then applied to diagnose whether a fault occurs in an equipment, identify the weak points of equipment, or predict equipment' remaining useful life. Predictive models which are used for predictive equipment maintenance may follow two approaches:

- **Regression approach**: These models indicate how many days/cycles remain before an equipment will reach the end of its useful life.
- **Classification approach**: Using this approach, we can predict whether equipment is likely to fault and determines whether their properties are lower than usual.

The update of predictive models usually occurs once a month, and then they are tested for accuracy. If the result does not match the expected one, it is changed, retrained, and tested again until it works properly. A significant amount of exploratory analytics should be performed before moving on to machine learning. In machine learning datasets, data analysis is used to detect relationships, trends, and insights. Furthermore, several technological assumptions are evaluated during the exploratory analytics stage to aid in the selection of the best-fit machine learning algorithm. An IoT-based predictive maintenance system can inform users of a likely equipment failure using user apps.

For instance, Fig. 12.5 illustrates the implementation of IoT-based predictive maintenance in a production line. Sometimes, physical inspections of production line equipment require personnel to enter dangerous environments to inspect the facilities, which may not be possible. Factories may use IoT-based predictive maintenance to anticipate possible breakdowns and boost the productivity of highly essential equipment. The solution measures temperature, vibration levels, and the other equipment's properties, with sensors deployed throughout the equipment. The system collects real-time sensor data and sends it to the cloud for analysis, prediction, and assessment [18].

### *12.2.3 Cyber-Physical System*

Cyber-physical systems (CPS) are intelligent systems that include engineered networks with the ability to interact with physical and computational components (based on algorithms). These systems are highly interconnected and integrated, providing new functions to improve and enhance the quality of life and leading to technological advances in critical areas such as personal health care, emergency response, traffic flow management, smart manufacturing, national security and defense, and produce and consume energy. Currently, in addition to CPS, there are many other words and phrases that describe similar or related systems and concepts, such as Industrial Internet, Internet of Things (IoT), Machine-to-Machine (M2M),

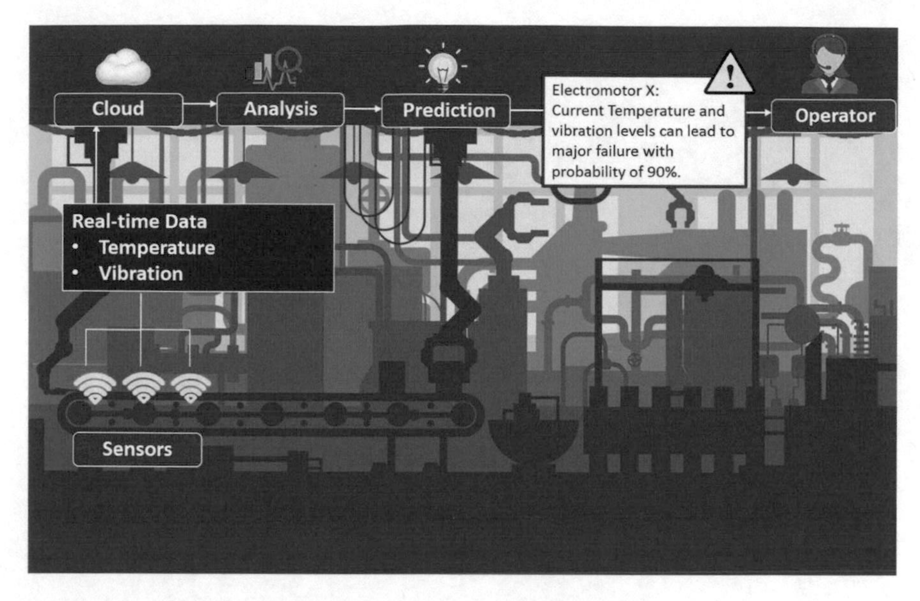


**Fig. 12.5** IoT-based predictive maintenance in a production line

smart cities, and so on. There is a lot of overlap between these concepts, especially between CPS and IoT, as they are sometimes used interchangeably (Fig. 12.6). In 2013, the International Telecommunication Union (ITU) defined the Internet of Things in a recommendation as follows:

> A global social information infrastructure created by the interconnection (physical and virtual) of objects, based on existing and evolving information and communication technologies, with the ability to work with each other and enable the provision of advanced services.

The true value of the Internet of Things is determined when the data generated by sensors, devices, machines, and terminals of the Internet of Things can be received, interpreted, and processed through predicted systems, and finally, the necessary commands given to the appropriate operators. In other words, the true value of the Internet of Things for manufacturers lies in the analysis that results from the cyber-physical models of machines and systems. In the fourth generation industry, the systems that can add value to the Internet of Things are cyber-physical systems (CPS). Objects in the IoT include physical world objects (physical assets) and virtual world objects, i.e., information. When the IoT is integrated with sensors and actuators, the resulting technology becomes an example of more general systems, such as cyber-physical systems, which include technologies such as smart grids, smart homes, smart transportation, and smart cities. The cyber-physical system is an interface between the human world and the cyber sphere, enabling the data collected by the system to be transformed into operational information and, ultimately, to optimize processes by interacting with physical assets.

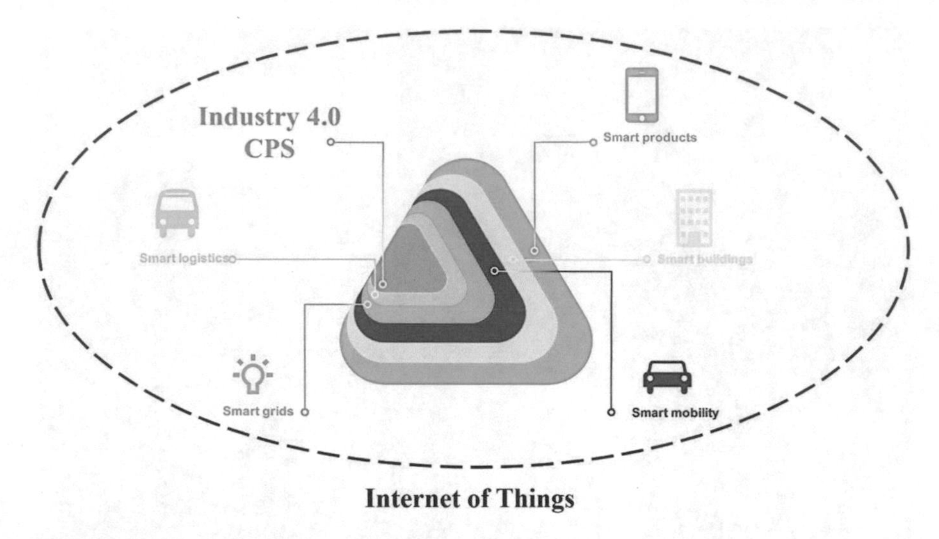


**Fig. 12.6** Internet of Things and cyber-physical systems

When data is collected from physical assets using sensors embedded in them using IoT technology, large volumes of data are generated and made available. Unfortunately, existing technologies are not enough to categorize and manage this huge amount of data that is generated daily. In addition, analytical methods and algorithms are not mature enough to use this large amount of data and have not grown enough to be able to intelligently and efficiently process and analyze all generated data. This is considered as a big data challenge.

From the similarity of IoT and CPS in their use of network, Internet, and sensors, it can be concluded that they are different definitions of a common concept. Despite this similarity, IoT and CPS are not the same thing. The conversion of data into information or tasks has placed a special emphasis on fault detection and prediction. For instance, the use of nonlinear data analysis methods in robotic applications and the application of multiple baselines to achieve a health machine model that analyzes data related to vibration, temperature, and torque and diagnoses the faults of their axis.

To meet the needs of the cyber surface, it is necessary to use historical records and algorithms that are learned over time to obtain reliable information of the health and estimated life of machines. As the machine has several decreases in performance, the development of health monitoring algorithms based on historical data is important. Although analytical methods for practical applications in industry are complex, life prediction methods need to respond to changes in operating conditions and the impact of maintenance operations on life estimation. The cyber provides more reliable information about the health status of the machines compared to the information obtained from the traditional method of condition monitoring. In the traditional condition monitoring, the condition of the machine is compared to the condition at start-up or the ideal condition, which is called the "baseline", and the health status of the machine is determined by their differences and the trend of changes.

As a perception-level cascading system, it must include decision-making algorithms and support systems that are able to suggest appropriate maintenance and production measures through the use of condition-based maintenance and predictive maintenance in the form of CPS based on the health of monitored machines and their reliability value. Currently, there is no mature and fully integrated system that combines machine health with decision-making processes in a way that reflects the true values of machine health. Therefore, for many industries, achieving the level of perception is a major challenge. For example, according to studies based on "alternatives theory [19]" and by estimating the remaining life of a physical asset (which is the output of the health monitoring system), the appropriate time for the maintenance and repair operations can be decided. Alternatives theory is an idea that has been used for many years to buy and sell a fixed asset item at the end of its useful life or before. At this time, the amount of information that needs to be processed is so large and beyond the capacity of human decision makers that it is necessary to first provide decision-making systems with various options to operational staff, engineers, or maintenance staff in order for them to make the final decision. The studies showed that current technologies in practice cannot adequately give machines the ability to self-adjust or self-configure, and there are many research opportunities for the development of this aspect of CPS. For instance, although much work has been done to control vibrations and unbalance of machines, to neutralize the effect of chatter on rolling racks, or to control machine tools, there is a long way to go before automatic rotating machines can be configured. Nevertheless, knowing the capabilities of cyber-physical systems allows for the development of a promising design approach for CPS-based maintenance applications. Interconnectivity, which was covered in the previous section, gives access to a wealth of data. However, having access to data alone does not offer a major benefit. Therefore, managing, classifying, and processing data so that PHM algorithms may further analyze it requires a robust and flexible technique. This approach has to be comprehensive enough to fully take use of cyber-physical systems' benefits.

Lee and Bagheri [20] suggested the "Time Machine Methodology for Cyber-Physical Systems" that is a methodical approach and being used to deploy CPS in maintenance applications. This strategy is in charge of correctly arranging the data that is already accessible in a big data environment so that it is ready for use in PHM algorithms and that every single asset in the fleet has a time machine record, which represents a type of digital. This cyber twin's approach is to gather and clean up data in preparation for future use. Other information that is taken from the cyber side includes sensory data as well as installation history, operating parameters, system configuration, maintenance events, and others. The stability of the cyber model over time is its most significant benefit. The actual asset will eventually collapse, but its digital duplicate will continue to maintain its data indefinitely. The schematic representation of CPS-based maintenance strategy is depicted in Fig. 12.7.

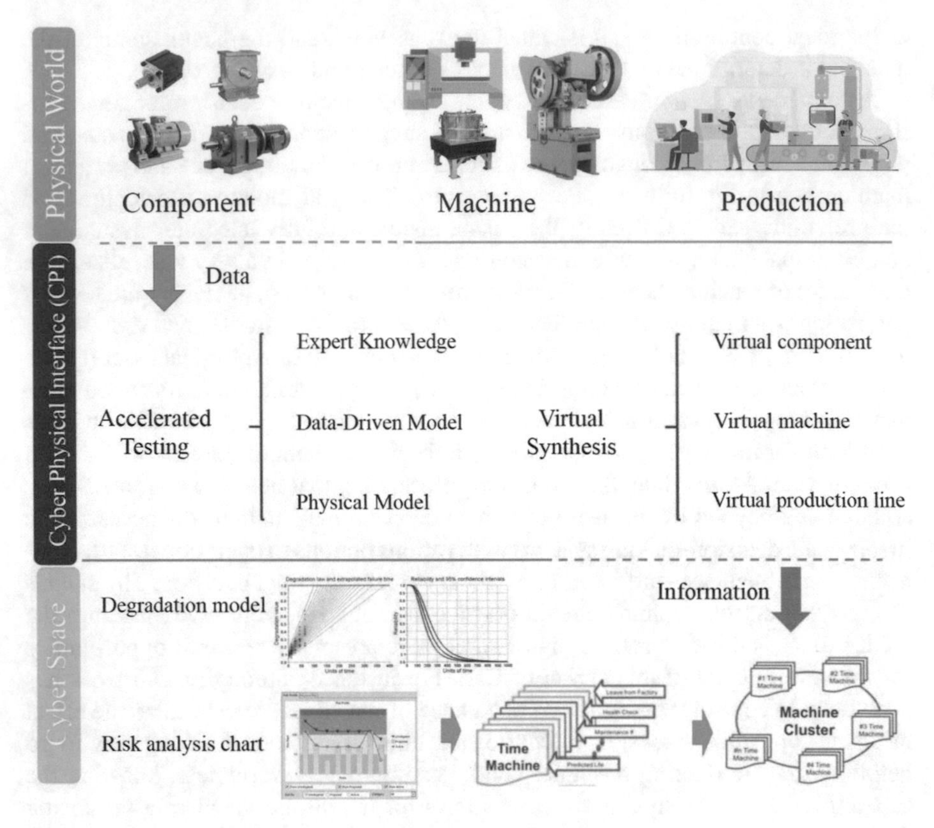


**Fig. 12.7** CPS-based maintenance strategy [20]

## 12.3 Assessment of Safety Risks

An assessment and management of risk is focused on identifying assets, analyzing vulnerabilities, and evaluating and estimating damages that could occur. Generally, risk assessment can be roughly divided into qualitative and quantitative aspects. Quantitative assessment is based heavily on expert experience, while qualitative entails calculating the exact risk value of the system. There have been many methods of assessing safety risk to date; below are some typical technologies for safety.

### *12.3.1 Big Data*

The big data mainly contains the five aspects in detail which include basic theories of safety big data, big data-driven safety management, big data-driven risk assessment and forecasting, big data application platform and design scheme in safety management, and big data-related technology developments in safety management [21]. The

**Table 12.1** Security objectives in CPS versus IT systems in order of priority

| Priority | CPS | IT systems |
|---|---|---|
| Low | Confidentiality | Availability |
| Medium | Integrity | Integrity |
| High | Availability | Confidentiality |

application of big data in the field of safety science precedes its theoretical studies without a doubt [22].

### 12.3.2 *Cyber-Physical System*

Cyber-physical system places great importance on risk assessment and management. When CPS was first developed, system designers gave more consideration to safety [23]. As a result of interactions between the environment and the control system, the control system itself, and the control system and authorized users, safety risks may occur. The CIA triad, which is commonly known as the three basic security objectives (confidentiality, integrity, and availability) in CPS and IT systems, represents the fundamental security objectives [23–25]. In contrast with traditional IT systems, CPS places the highest priority on availability. According to Table 12.1, these fundamental objectives are important for both CPS and IT systems, but their priorities are different [26, 27]. The goal of availability and safety is to keep the system under a pre-defined and acceptable threshold [28].

CPS safety risk assessment methods have been developed in many ways, some examples are Fault tree analysis (FTA), Failure modes and effects analysis (FMEA), Hazard and operability methodology (HAZOP), Model-based engineering (MBE), Goal tree-success tree and master logic diagram (GTST-MLD), system theoretic process analysis (STPA), and Temporary Structures Monitoring [29].

From the foregoing, it appears that the application of digital tools will help to:

- Accurately calculate reliability due to online condition monitoring;
- Improved productivity of staff and reduced human labor;
- Efficient maintenance management;
- Better use of equipment and assets;
- Cost-effective operation;
- Improved work safety and reduce risk;
- Reduce the machine stoppage; and
- Reduce the costs related to major repairs.

## 12.4 Conclusion

In the fourth generation of industry, big data, the Internet of Things, cyber-physical system, and quick response to change provide an opportunity for reliability engineering to improve system reliability. Additionally, complexity increases, interconnections and dependencies between components, dynamic behavior, and advanced components, such as CPSs and sensors, make reliability engineering challenges for designers. It is necessary to update traditional methods and to develop new frameworks for reliability, risk, safety, and security.

Besides that, by using IoT-based predictive maintenance, equipment life can be extended by 30%, time-based maintenance can be eliminated, and equipment downtime decreased by 50%. However, a well-thought-out architecture with an emphasis on machine learning is required for a mature and dependable predictive maintenance system.

In this chapter, the application of new methods and tools such as big data and data processing, IoT, and cyber-physical system was described to analyze the reliability and risk of equipment. For future research, it seems necessary that the advantages and benefit–cost analysis of digital tools are compared to traditional tools and methods.

Our suggestion is that managers don't limit themselves to using a specific mode of digital tools, but think about how advanced digital analytics techniques can transform their maintenance and reliability system. This means constantly looking for opportunities to improve the use of data and user-centered design principles, in order to digitize processes. Sustained efficiency requires a combination of new digital tools, changes in asset strategy, and improves reliability performance.

## References


1. Sergi, D., Ucal Sari, I.: Prioritization of public services for digitalization using fuzzy Z-AHP and fuzzy Z-WASPAS. Complex Intelligent Syst. **7**(2), 841–856 (2021)
2. Bradbury, S., Carpizo, B., Gentzel, M., Horah, D., Thibert, J.: Digitally Enabled Reliability: Beyond Predictive Maintenance. McKinsey and Company (2018)
3. Zio, E.: Some challenges and opportunities in reliability engineering. IEEE Trans. Reliab. Inst. Electrical Electronics Eng. **65**(4), 1769–1782 (2016)
4. Farsi, M.A., Zio, E.: Industry 4.0: some challenges and opportunities for reliability engineering. Int. J. Reliab. Risk Safety: Theor. Appl. **2**(1), 23–34 (2019)
5. Muhuri, P.K., Shukla, A.K., Abraham, A.: Industry 4.0: a bibliometric analysis and detailed overview. Eng. Appl. Artif. Intell. **78**, 218–235 (2019)
6. Shahbakhsh, M., Emad, G.R., Cahoon, S.: Industrial revolutions and transition of the maritime industry: the case of Seafarer's role in autonomous shipping. Asian J. Shipping Logistics **38**(1), 10–18 (2022)
7. Torres, M.B., Gallego-García, D., Gallego-García, S., García-García, M.: Development of a business assessment and diagnosis tool that considers the impact of the human factor during industrial revolutions. Sustainability **14**(2), 940 (2022)
8. Panetto, H., Iung, B., Ivanov, D., Weichhart, G., Wang, X.: Challenges for the cyber-physical manufacturing enterprises of the future. In: Annual Reviews in Control (2019)

9. Alcácer, V., Cruz-Machado, V.: Scanning the Industry 4.0: A Literature Review on Technologies for Manufacturing Systems, Engineering Science and Technology, An International Journal, In Press (2019)
10. Vaidya, S., Ambad, P., Bhosle, S.: Industry 4.0—a glimpse. Proc. Manuf. **20**, 233–238 (2018)
11. Rauch, E., Linder, C., Dallasega, P.: Anthropocentric perspective of production before and within Industry 4.0. In: Computers & Industrial Engineering, Published Online (2019). https://doi.org/10.1016/j.cie.2019.01.018
12. Tao, F., Qi, Q., Liu, A., Kusiak, A.: Data-driven smart manufacturing. J. Manuf. Syst. **48**, 157–169 (2018)
13. Albright, B.: Deep Learning and Design Engineering, addressed by (2019) https://www.digitalengineering247.com/article/deep-learning-and-designengineering
14. Kolar, D., Lisjak, D., Curman, M., Pająk, M.: Condition monitoring of rotary machinery using industrial IOT framework: step to smart maintenance. Tehnički glasnik **16**(3), 343–352 (2022)
15. Fuenmayor, E., Parra, C., González-Prida, V., Crespo, A., Kristjanpoller, F., Viveros, P.: Calculating the optimal frequency of maintenance for the improvement of risk management: plausible models for the integration of cloud and IoT. In: IoT and Cloud Computing for Societal Good, pp. 209–219. Springer, Cham (2022)
16. Dehbashi, N., SeyyedHosseini, M., Yazdian-Varjani, A.: IoT based condition monitoring and control of induction motor using raspberry pi. In: 2022 13th Power Electronics, Drive Systems, and Technologies Conference (PEDSTC), pp. 134–138 (2022)
17. Singh, R., Sharma, R., Akram, S.V., Gehlot, A., Buddhi, D., Malik, P.K., Arya, R.: Highway 4.0: Digitalization of highways for vulnerable road safety development with intelligent IoT sensors and machine learning. Safety Sci. **143**, 105407 (2021)
18. Killeen, P., Ding, B., Kiringa, I., Yeap, T.: IoT-based predictive maintenance for fleet management. Proc. Comput. Sci. **151**, 607–613 (2019)
19. Vogel, J.: The new relevant alternatives theory. Philos. Perspect. **13**, 155–180 (1999)
20. Lee, J., Bagheri, B.: Cyber-physical systems in future maintenance. In: 9th WCEAM Research Papers, pp. 299–305. Springer, Cham (2015)
21. Wang, B., Wang, Y.: Big data in safety management: an overview. Saf. Sci. **143**, 105414 (2021)
22. Wang, B., Wu, C.: Study on the innovation research of safety science based on the safety big data. Sci. Technol. Manag. Res. 37–43 (2017)
23. Stouffer, K., Falco, J., Scarfone, K.: Guide to industrial control systems (ICS) security. NIST Spec. Publ. **800**(82), 29–32 (2011)
24. Dzung, D., Naedele, M., Von Hoff, T.P., et al.: Security for industrial communication systems. Proc. IEEE **93**(6), 1152–1177 (2005)
25. ISO/IEC 27001: Information Technology Security Techniques Information Security Management Systems—Requirements (2013)
26. Cheminod, M., Durante, L., Valenzano, A.: Review of security issues in industrial networks. IEEE Trans. Ind. Inf. **9**(1), 277–293 (2013)
27. Peng, Y., Lu, T., Liu, J., et al.: Cyber-physical system risk assessment. In: Proceedings of the 9th International Conference Intelligent Information Hiding and Multimedia Signal Processing, IIH-MSP 2013, Beijing, China, October 2013, 16–18 (2013)
28. Lyu, X., Ding, Y., Yang, S.H.: Safety and security risk assessment in cyber-physical systems. IET Cyber-Phys. Syst.: Theor. Appl. **4**(3), 221–232 (2019)
29. Yuan, X., Anumba, C.J.: Cyber-physical systems for temporary structures monitoring. In: Cyber-Physical Systems in the Built Environment, pp. 107–138 (2020)

**Fatemeh Afsharnia** received her BS, MS, and Ph.D. degree in agricultural mechanization engineering form the Agricultural Sciences and Natural Resources University of Khuzestan, Ahvaz, Iran. Dr. Fatemeh Afsharnia is a member of the Iranian Maintenance Association (IRMA) and has published over 30 publications in several journals. Moreover, she has been working as a reviewer in peer-review journals such as Measurement (Elsevier), International Journal of Quality and Reliability Management (Emerald), Journal of the Brazilian Society of Mechanical Sciences and Engineering (Springer), etc. Her main research interests are Reliability Engineering and Analysis, Maintenance Management, Maintenance Planning, Regression Modeling, Reliability Theory, Risk Assessment and Analysis, Safety Management and Engineering, Probabilistic Risk Analysis, Maintenance optimization, Condition-based Maintenance, etc.

# Chapter 13 Qualitative Analysis Method for Evaluation of Risk and Failures in Wind Power Plants: A Case Study of Turkey

**İbrahim Yilmaz and Emre Caliskan**

**Abstract** Due to rising worldwide energy demand and growing worries about environmental issues such as climate change and global warming, renewable energy resources have attracted a lot of attention in recent years. Renewable energy is defined as energy sources that provide energy via natural processes and can regenerate noticeably faster than the depletion rate of the resources consumed. Wind energy is being more widely preferred to satisfy the community's demand. The increasing growth of the wind power business necessitates improved equipment sustainability and reliability. As a result, the issue of reliability is critical for large-scale wind turbines which provide electricity to the national energy system. Wind farms have been built in numerous regions around Turkey in recent years and are currently being built. The objective of this study is to provide a decision-making model for ranking the risk and failures that wind power plants may during electricity production. In the proposed model, four risk and failure sources are examined under five evaluation criteria. In order to reflect evaluation criteria more comprehensively, fuzzy MCDM methodology is used to determine the most and least risk and failure sources that power plants may face during their operations. In this context, the VIKOR method is applied to determine the weights of the criteria and the ratings of the risk according to each criterion with qualitative data. VIKOR method is extended with the concept of an intuitionistic fuzzy set to define accurately the vague and imprecise situations which are defined qualitatively. This study could be considered as one of the first attempts to evaluate the integration of emerging risk and failure analysis in wind power plants with qualitative datasets. The results of this study imply that the most important risk and failure factors depend on the uncontrollable conditions or stochastic nature of the events such as unstable weather conditions.

İ. Yilmaz (✉)
Department of Industrial Engineering, School of Engineering and Natural Sciences, Ankara Yıldırım Beyazıt University, Ankara 06010, Turkey
e-mail: iyilmaz@ybu.edu.tr

E. Caliskan
Department of Industrial Engineering, School of Engineering, Gazi University, Ankara 06570, Turkey



H. Garg (ed.), *Advances in Reliability, Failure and Risk Analysis*, Industrial and Applied Mathematics, https://doi.org/10.1007/978-981-19-9909-3_13

**Keywords** Risk and failure evaluation · Renewable energy · Multi-criteria decision-making · Fuzzy logic

## 13.1 Introduction

The demand for energy in Turkey has been steadily increasing over the previous few decades, and this trend is expected to continue in the future. For example, it can be seen that consumption has increased by 5.1% per year in the ten years from 2006 to 2015. The percentage change in Turkey's annual electricity consumption since 1971 compared to previous years is given in Fig. 13.1. As shown in Fig. 13.1, while the biggest increase in consumption was in 1976 with 18.4%, the biggest decrease was experienced in 2009 with a decrease of 2%. Since 1971, electricity consumption has decreased only in 2001 and 2009 compared to the previous year. When the average is taken from 1971 to 2015, it is seen that consumption increased by 8.0% each year [1].

Renewable energy is defined as the type of energy that comes from the natural environment continuously or is obtained from sources that are accessed repeatedly. Renewable energy sources are accepted as sustainable energies as well as being obtained from natural sources such as solar, wind, hydraulic, biomass, geothermal, and wave energy [2].

As one of the renewable energy sources, wind power turbines are preferred by communities to satisfy the electricity demand of customers. The sun's heat is absorbed at varied rates by the earth's surface, which is made upon various types of

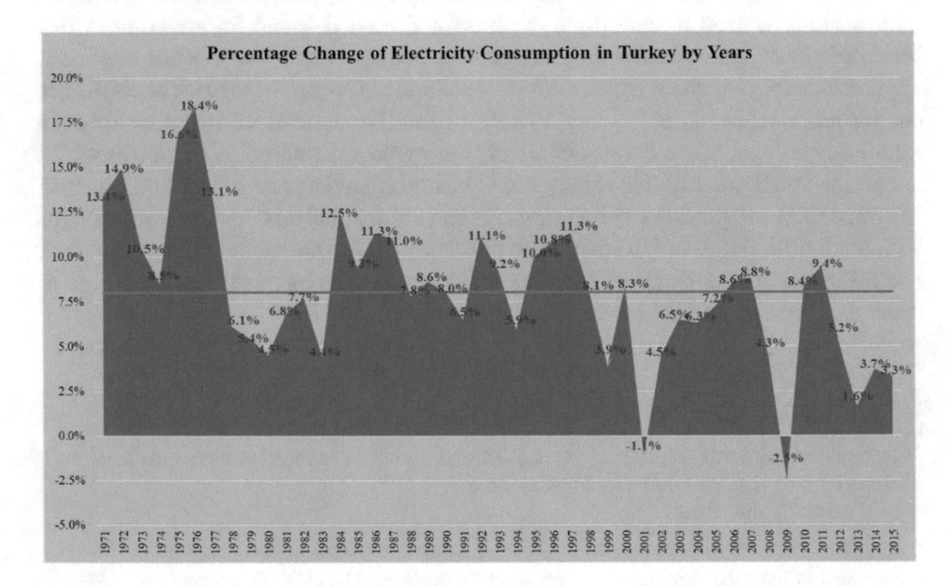


**Fig. 13.1** Percentage change in Turkey's annual electricity consumption from 1971 to 2015

land and ocean. The daily wind cycle is one example of this inconsistent warmth. Wind energy is mostly utilized to create power; therefore, all over the world with the increasing number of wind energy power plants being constructed. Wind energy accounts for 2% of the world's electricity production [3].

Wind power facilities are growing in Turkey similar to other countries. It is stated in the Turkish Ministry of Energy and Natural Resources' Wind Yearly Report that wind energy is the second most preferred source among the renewable energy sources. As of December 2020, Turkey's wind power capacity is calculated as 8832 MW. Wind energy has a share of 8.09% in the total electricity production, and the change in installed power over the years is shown in Fig. 13.2 [4].

It is both economically and ecologically unsustainable to meet expanding energy demand with typical carbon-fueled power plants. Traditional carbon-fueled power plant technologies are transforming into renewable energy plants when energy efficiency and climate change are taken into consideration. Due to this transformation, communities are becoming engaged in renewable energy sources. Also, renewable energy sources are becoming more affordable and plentiful every day.

According to the Turkish Wind Energy Association's annual report, there are 273 wind power plants in which 3983 wind turbines are already installed in Turkey. These wind power plants produce 9.84% of produced energy in Turkey. In addition, 20 projects are continuing around Turkey. However, some risk factors affect the electricity production efficiency in wind power plants. One of the most important risks that renewable energy risk management seeks to deal with is the randomness and uncertainty of the production of these resources. Wind energy is also a resource where this risk is high due to its variability and unpredictability. In the literature, six main risks are determined as technical, economic, environmental, social, political, and supply chain issues. These risks are extended to risk factors as shown in Table 13.2.

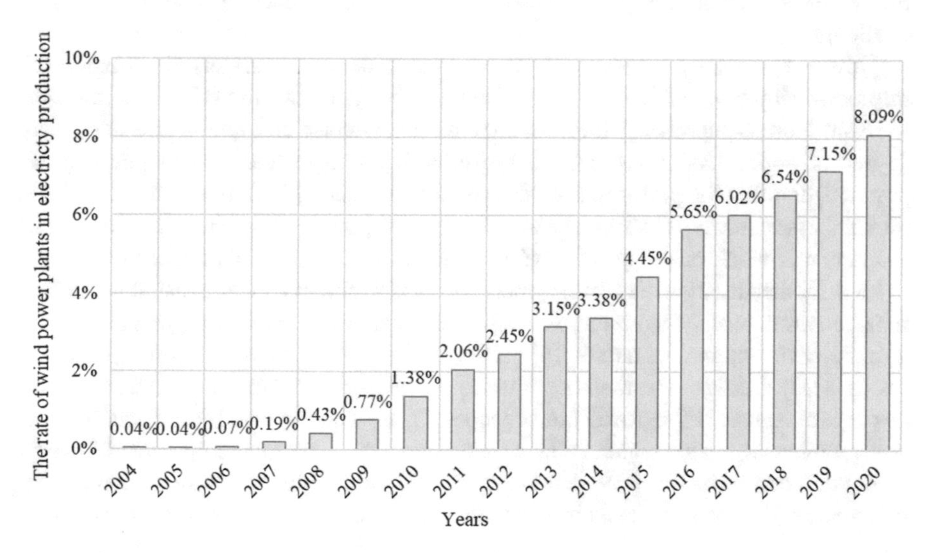


**Fig. 13.2** Rate of wind power electricity in electricity production

The risk factors could be collected in four groups (1) delay of production ($R_1$), (2) downtime ($R_2$), (3) unstable weather conditions ($R_3$), and (4) changes in government policy ($R_4$). If these risk factors cause the failure of production in wind power plants. The outcomes of the failure could be an effect (1) life quality of residents ($C_1$), (2) production sustainability of power plants ($C_2$), (3) operational and maintenance difficulties in power plants ($C_3$), (4) working conditions of power plant workers ($C_4$), and (5) environmental issues ($C_5$).

In this chapter, the four risk factors ($R_1$–$R_4$) regarding the efficiency of electricity production are evaluated under five criteria ($C_1$–$C_5$). Four experts evaluate the decisions to deal with verbal factors from literature and renewable energy experts under conflicting and unmeasurable 5 criteria. In this context, this research is one of the first research regarding risk and failure analysis in wind power plants with qualitative datasets. Due to the difficulty of precisely determining the weights of the criteria and the ratings of the risk factors according to each criterion with qualitative data, VIKOR method is applied in this chapter.

Due to the existence of conflicting and competing considerations such as quality and price, decision-makers commonly encounter complicated multiple criteria decision-making (MCDM) challenges. MCDM has the potential to enhance all engineering decision-making processes, from design to manufacturing in terms of reliability engineering, but it is also advantageous for applications in various sectors, where product and competition and competitive advantage are frequently attained by just very small improvements in performance. Opricovic (1998) proposes the VIKOR method to take care of such issues. VIKOR is a useful tool for dealing with discrete multi-criteria situations involving incompatible and incommensurable criteria. VIKOR could be applied to rank a set of alternatives to help decision-makers to reach a decision [5]. Thus, the VIKOR approach's distinctive features and capabilities have been successfully utilized in a variety of real-world decision-making challenges.

However, the crisp numbers do not enable to define accurately the vague and imprecise situations which are defined qualitatively. Fuzzy logic is used in order to make more comprehensive evaluations with imprecise and qualitative data. The degree of importance of the risks is ranked using fuzzy logic for this purpose. In practice, defining the performance of the alternatives is difficult due to the decision-makers' assessments, which can be inaccurate or imprecise under any circumstances. In many real-world challenges, it is difficult to gain sufficient and precise knowledge to identify the situation. Therefore, most MCDM problems are suggested under the fuzzy environment in the literature. To describe a situation more comprehensively in a fuzzy environment, Atanassov [6] proposes the concept of an intuitionistic fuzzy set (IFS). IFS covers the membership degree and non-membership degree with a hesitation degree. IFS extends the concept of type-1 fuzzy sets by introducing the non-membership value which is the complement of the membership value of every element. Therefore, IFS provides an additional possibility to express incomplete information,they may be used to better characterize many real-world issues under uncertain and limited information [7]. Therefore, this research aims to fill the gap in literature by applying IF-VIKOR to evaluate the risk and failures in wind power plants.

The remainder of this chapter is presented as follows: Section 13.2 provides a related literature review of risk and failure analysis applications related to renewable energy power plants and integrated with MCDM. Section 13.3 shows the proposed methodology with the intuitionistic fuzzy VIKOR (IF-VIKOR) method. Section 13.4 presents a case study to rank the risk factors in wind power plant. This chapter is completed with conclusions and future works which are given in Section 13.5.

## 13.2 Literature Review

This section presents general information on risk management studies related to renewable energy sources and provides an overview of risk factors and risk management studies that have already been conducted for wind energy in particular. The methodologies used in this area have been also discussed. The issue of risk management in the renewable energy sector has been investigated from different viewpoints in the literature.

### *13.2.1 Risk Management in the Renewable Energy*

Risk management is a process that involves assessing, developing a strategy, and mitigating the effects of uncertainties (*risks*) through the activities of a project. Related academic renewable energy studies are mostly risk analyses including risk identification and strategy development studies conducted at the initial investment. A comprehensive study on the handling of risk factors retrospectively evaluated 11,504 news reports about renewable energy-based incidents [8]. They used the ontology-based Bayesian network method to determine the accident risk factors in this news and the relationship between them. Indicated in that study, risks were identified from the literature review as renewable energy systemic risks, ecosystems risks, and social system risks according to falling, injury, or fatality incidents. Other methods such as FMEA have been also used as a method for determining risk factors [9]. In their studies, the researchers considered five alternative renewable resources, namely hydropower, solar energy, wind energy, geothermal energy, and biomass, and identified the risk factors associated with them, such as cost risks, political risks, technological risks, environmental risks, and construction-management risks. Hashemizadeh et al. [10] applied the TODIM method to determine the risks associated with investments in renewable resources and grouped these factors into economic, technical, environmental, social, and political. Somi et al. [11] proposed a new framework for identifying risks in renewable energy projects. In this structure, they combined case-based reasoning and fuzzy logic. They applied the proposed methodology in the case of the

**Table 13.1** Summary of types and definitions of risks

| Risk type | Definitions |
|---|---|
| Economic/financial risk | Risk relates to cost structures, currency fluctuations, access to credit sources, national economic conditions |
| Technical risk | Risks arise from design flaws, uncertainties in the use, implementation and efficiency of renewable energy technologies, the extent of progress in different countries/industries, replicability of technologies due to proprietary rights, patents |
| Social risk | Risk arises from difficulties in social/local communication, social acceptance, well-being of organizations' stakeholders (including occupational health and safety) |
| Environmental risk | Risk refers to the impact on environment such as damage to biodiversity (fatality of animal species), as well as the risk of being affected by nature (i.e., corrosion) or disasters |
| Political risk | Risk arises from changes in policies related to the environmental or financial investment, national stability, approach of governments to project implementation |
| Regulatory risk | Risk refers to the uncertainty that regulatory authorities might change existing laws and regulations such as incentive or tax laws |
| Supply chain risk | Risk relates to disruptions of supply of materials and information, transport issues, logistics complexity of components like blades, as well as delivery issues because of the variation and uncertainty in power load |

onshore wind investment project. While the literature mostly discusses all renewable energy sources, some studies focus on a specific energy source. For instance, Yatim et al. [12] conducted a study that identified key risks to the biomass industry in Malaysia. They summarized the associated risks as regulatory, financing, technology, supply chain and feedstock, business, and social and environmental risks. Types of risks and definitions in the literature are listed in Table 13.1.

Renewable energy investments depend on the geographical location of production and the market. Therefore, case studies on risk management for geographic regions can be found in the literature. Abba et al. [13] examined the risks and methodologies considered in renewable energy risk assessment and mitigation for developed and emerging economies, focusing on sub-Saharan African countries. The qualitative methods considered in the study were compared with other methods, and it was found that these methods can assess technical and economic risks in a similar way to social, political, and policy risks. In another study addressing the risk relationship between renewable energy sources and geography, Hansen et al. [14] evaluated the sustainability risks of forest-based bioenergy in Scandinavian Countries. Ranganath et al. [15] identified the risks associated with solar power projects in India and developed a methodology for risk analysis using fuzzy TOPSIS which is an MCDM technique. Aquila et al. [16] conducted a study to identify the uncertainties and risks

**Table 13.2** Main risk factors in wind energy facilities extended from [27]

| Risk type | Risk factors |
|---|---|
| Technical | Equipment/facility damage |
| | Transportation |
| | Inexperienced engineering teams |
| | Interfaces<br>Foundation failure<br>Maintenance issues<br>Corrosion issues |
| Economical | Finance |
| | Contract |
| | Market |
| | Business interruption<br>High O/M cost |
| Environmental | Land/sea conditions |
| | Weather (i.e., extreme waves, wind) |
| | Seabed/geological conditions<br>Biohazards (i.e., bird fatality) |
| Social | Personnel safety (i.e., falls, drowning)<br>Working at height<br>Workers' transportation (i.e., boat, road, helicopter) |
| Politics | Regulatory changes<br>Changes in state policy |
| Supply chain | Supply chain bottlenecks<br>Delay of power generation<br>Impact of the cost of raw material |

affecting wind energy investments in the Brazilian market. They used simulations to analyze the financial behavior of the investment project under the identified risks. Value-at-risk (VaR) method was also used for the analysis of risk management. To predict extreme situations that may occur in the pricing mechanism and manage risk, Hagfors et al. [17] studied the impact of key players. The impact of renewable energy sources, solar and wind, on extreme price developments was investigated taking Germany as an example and found that the probability models identified can be used in price-based risk management. Guerrero-Liquet et al. [18] consider the risk assessment and management of renewable energy facilities in the Dominican Republic as a decision-making problem. The conclusions were drawn using the cause-effect method and the SWOT method, by using the knowledge of the experts, and the risks were prioritized using the AHP method. Some studies consider Turkey in particular. Kul et al. [19] stated that the Turkish government encourages renewable energy investments, while there are inherent risks in investing in and developing such projects. Therefore, the authors provided an MCDM-based three-stage decision framework related to the risk factors of the aforementioned investments in Turkey.

They also suggested the evaluation of strategies to manage those factors. Apak et al. [20] examined financial risk management tools that meet the needs of the renewable energy sector. A comparative analysis was presented by looking at the European Union and Turkey together.

### *13.2.2 Risk Management in Wind Power Plants*

The risk factors that apply to renewable energy sources are also valid for wind power generation. There are risks such as costs, lack of capacity, and thus intermittent power generation at the operational stage in wind power plants. The technical risks of wind energy also include the cost of materials used in power plants. In their study, Kanamura et al. [21] performed an econometric analysis for the future markets of these materials, taking this risk into account. They showed the practical application of their proposed theoretical economic model in the case of illiquidity. Kumar et al. [22] stated that risks arising from the technical characteristics of the parts required for the wind turbine, as well as maintainability and servicing, are a natural risk in power generation and are related to supply chain risks. The researchers focused on operational risks and methods to improve wind farm reliability and availability. One of the most important risks that renewable energy risk management seeks to deal with is the randomness and uncertainty of the production of these resources. Wind energy is also a resource where this risk is high due to its variability and unpredictability. Soroudi et al. [23] proposed a mixed-integer nonlinear multi-objective programming model. In addition to minimizing risk levels, they considered the minimization of energy supply costs as a second objective. They used conditional VaR (CVaR) to estimate uncertainty as a risk management tool.

In the case of wind energy, the risks vary depending on the location of the power plants, i.e., onshore or offshore. Somi et al. [11] presented the risk breakdown matrix for the onshore wind farm. Considering each work package level of projects, this pioneering study identified 169 risk factors affecting the onshore wind farm. Onshore and offshore wind turbines also pose environmental risks. Considering these environmental risks, Macrander et al. [24] highlighted the impacts of offshore wind farms on marine creatures in the regions where they are located. Astiaso-Garcia and Bruschi [25] considered onshore wind farms for their proposed risk assessment tool to prioritize occupational health and safety standards. Gatzert and Kosub [26] also highlighted the risks and offered risk management solutions by separating investments in onshore and offshore wind farms in the European market. In contrast, studies that have taken the risk assessment of offshore wind energy projects into account have been few in number. Chou et al. [27] analyzed and evaluated the risks associated with the construction and use of these plants in their study. Accordingly, offshore wind projects continue to be at the preliminary stage, particularly in Asia. Researchers conducted a risk assessment study for Taiwan and identified risk factors from the literature. An extended list of risk factors related to onshore/offshore wind facilities has been demonstrated in Table 13.2 based on Chou et al. [27].

### 13.2.3 Risk Management Methods in Related Literature

Risk analysis, assessment, and mitigation—collectively referred to as risk management—use a variety of methods that may be qualitative, quantitative, semi-quantitative, or a combination of these methods. Figure 13.3 shows the methods preferred in the literature. Qualitative risk methods are subjective assessments used when sufficient data are not available. These types of methods include literature reviews, interviews with experts, etc. Quantitative risk management methods mainly use statistical data and probabilities to assess the level of risk [13]. Simulation methods such as Monte Carlo simulation, agent-based modeling, portfolio optimization methods, VaR methods, and mathematical programming—especially stochastic programming—are the most commonly encountered quantitative methods in the literature. On the other hand, semi-quantitative methods combine the advantages of both qualitative and quantitative methods. These methods use expert interviews and quantitative techniques to convert to numerical values. Therefore, such methods have the flexibility to consider statistical and non-statistical risks [28].

As shown in Fig. 13.3, there are many methods to evaluate or analyze the risk and failures related to renewable energy production. However, there is limited research focused on the qualitative evolution of risk factors of wind power plants under an imprecise and vague environment. The application of IF-VIKOR could be one of the first attempts to evaluate the risk and failures of wind power plants under imprecise and vague conditions.

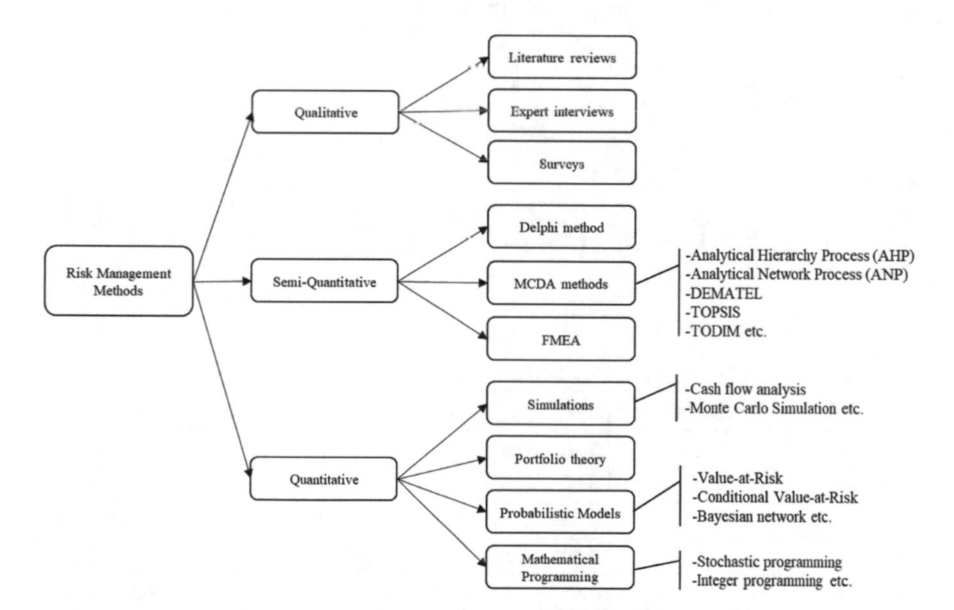


**Fig. 13.3** Overview of risk management methods in relevant studies

## 13.3 Methodology

This research combines IFS concepts and calculations which are shown in Atanassov [6] and VIKOR method steps which are shown in Opricovic and Tzeng [5]. The risk factors ($R_i$) are accepted as (1) delay of production ($R_1$), (2) downtime ($R_2$), (3) unstable weather conditions ($R_3$), and (4) changes in government policy ($R_4$). These risk factors are evaluated under possible results when these risk factors cause the failure of wind power plants. The evaluation criteria ($C_j$) are defined as (1) life quality of residents ($C_1$), (2) production sustainability of power plants ($C_2$), (3) operational and maintenance difficulties in power plants ($C_3$), (4) working conditions of power plant workers ($C_4$), and (5) environmental issues ($C_5$). The ratings of $R_i$ and the subjective weights of the criteria are given in Table 13.3.

It is assumed that there are 4 decision-makers ($\text{DM}_k, k = 1, \ldots, 4$) and 4 alternatives ($R_i, i = 1, \ldots, 4$) and 5 evaluation criteria $\left(C_j, j = 1, \ldots, 5\right)$. During criteria evaluation, specified a weight $\lambda_k \in (0, 1]$ and $\sum_{k=1}^{K} \upsilon_i = 1$ assigned to each DM show the relative importance of DMs while raking the alternatives. Therefore, the IF-VIKOR method steps are shown as follows:

**Step 1**. Collect the evaluation views of all DMs for each alternative under each criterion as IFN then aggregate all DMs' opinions.

Assume that $r_{ij}^k = (\mu_{ij}^k, \upsilon_{ij}^k)$ represents the evaluation of $R_i$ with respect to the $C_j$ as the IFN. Then aggregate IFN rating $r_{ij}$ of alternatives regarding each criterion can be calculated by SIFWA operator as follows:

$$r_{ij} = \text{SIFWA}\left(r_{ij}^1, r_{ij}^2, \ldots, r_{ij}^K\right) = \sum_{k=1}^{K} \lambda_k r_{ij}^k$$
$$= \left( \frac{\Pi_{k=1}^{K}\left(\mu_{ij}^k\right)^{\lambda_k}}{\Pi_{k=1}^{K}\left(\mu_{ij}^k\right)^{\lambda_k} + \Pi_{k=1}^{K}\left(1-\mu_{ij}^k\right)^{\lambda_k}}, \frac{\Pi_{k=1}^{K}\left(\upsilon_{ij}^k\right)^{\lambda_k}}{\Pi_{k=1}^{K}\left(\upsilon_{ij}^k\right)^{\lambda_k} + \Pi_{k=1}^{K}\left(1-\upsilon_{ij}^k\right)^{\lambda_k}} \right) \tag{13.1}$$

**Table 13.3** Linguistic terms for each alternatives and criteria

| Linguistic terms for risk factors | IFN | | Linguistic terms for criteria | IFN | |
|---|---|---|---|---|---|
| Very poor (VP) | 0.10 | 0.90 | Very low (VL) | 0.15 | 0.80 |
| Poor (P) | 0.20 | 0.65 | Low (L) | 0.25 | 0.65 |
| Moderately poor (MP) | 0.35 | 0.55 | Medium low (ML) | 0.40 | 0.50 |
| Fair (F) | 0.50 | 0.50 | Medium (M) | 0.50 | 0.50 |
| Moderately good (MG) | 0.65 | 0.25 | Medium high (MH) | 0.60 | 0.30 |
| Good (G) | 0.80 | 0.05 | High (H) | 0.75 | 0.15 |
| Very good (VG) | 0.90 | 0.10 | Very high (VH) | 0.90 | 0.05 |

After each $R_i$ evaluated to under $C_j$, a group decision matrix $R$ is created as follows:

$$R = \begin{bmatrix} r_{11} & \cdots & r_{1n} \\ \vdots & \ddots & \vdots \\ r_{m1} & \cdots & r_{mn} \end{bmatrix} \tag{13.2}$$

**Step 2**. Assign the subjective weights for each criterion.

The fuzzy weight of each $C_j$ is defines as $\omega_j^k = (\mu_j^k, \upsilon_j^k)$ by DM$_k$. Then, the importance weight of the $j$th criterion $\omega_j = (\mu_j, \upsilon_j)$ calculated as follows:

$$\omega_j = \text{SIFWA}\left(\omega_j^1, \omega_j^2, \ldots, \omega_j^K\right) = \sum_{k=1}^{K} \lambda_k \omega_j^k$$
$$= \left( \frac{\Pi_{k=1}^{K}\left(\mu_j^k\right)^{\lambda_k}}{\Pi_{k=1}^{K}\left(\mu_j^k\right)^{\lambda_k} + \Pi_{k=1}^{K}\left(1-\mu_{ij}^k\right)^{\lambda_k}}, \frac{\Pi_{k=1}^{K}\left(\upsilon_j^k\right)^{\lambda_k}}{\Pi_{k=1}^{K}\left(\upsilon_j^k\right)^{\lambda_k} + \Pi_{k=1}^{K}\left(1-\upsilon_j^k\right)^{\lambda_k}} \right) \tag{13.3}$$

After each $\omega_j$ is calculated, normalized subjective weight of each criterion, $\omega_j^s$, is calculated as follows:

$$\omega_j^s = \left( \frac{\mu_j + \pi_j \frac{\mu_j}{\mu_j + \upsilon_j}}{\sum_{j=1}^{J}\left(\mu_j + \pi_j \frac{\mu_j}{\mu_j + \upsilon_j}\right)} \right) \tag{13.4}$$

where $\pi_j$ denotes the hesitation degree, $\pi_j = 1 - \mu_j - \upsilon_j$.

**Step 3**. Determine the IF positive and negative ideal solutions, $f_j^* = \left(\mu_j^*, \upsilon_j^*\right)$ and $f_j^- = \left(\mu_j^-, \upsilon_j^-\right)$, respectively, according to the following rule

$$\begin{aligned} f_j^- &= \{\min r_{ij} \text{ for benefit criteria, } \max r_{ij} \text{ for cost criteria}\} \\ f_j^* &= \{\max r_{ij} \text{ for benefit criteria, } \min r_{ij} \text{ for cost criteria}\} \end{aligned} \tag{13.5}$$

**Step 4.** Calculate the normalized intuitionistic fuzzy distances, $\widehat{d_{ij}}$

$$\widehat{d_{ij}} = \frac{d\left(f_j^*, r_{ij}\right)}{d\left(f_j^-, r_{ij}\right)} \tag{13.6}$$

where

$$d\left(f_j^*, r_{ij}\right) = \frac{|\mu_j^* - \mu_{ij}| + |\upsilon_j^* - \upsilon_{ij}|}{4} + \frac{\max\left(|\mu_j^* - \mu_{ij}| + \left|\upsilon_j^* - \upsilon_{ij}\right|\right)}{2}$$

$$d\left(f_j^-, r_{ij}\right) = \frac{|\mu_j^- - \mu_j| + |\upsilon_j^- - \upsilon_j|}{4} + \frac{\max\left(|\mu_j^- - \mu_j| + \left|\upsilon_j^- - \upsilon_j\right|\right)}{2} \tag{13.7}$$

**Step 5**. Calculate $\tau_i$ and $\varphi_i$ values as follow

$$\tau_i = \sum_{j=1}^{J} \omega_s^j \widehat{d_{ij}} \tag{13.8}$$

$$\varphi_i = \max\left(\omega_j^s, \widehat{d_{ij}}\right) \tag{13.9}$$

**Step 6.** Calculate VIKOR index $Qi$

$$Qi = \upsilon \frac{\tau_i - \tau^*}{\tau_i - \tau^-} - (1 - \upsilon) \frac{\varphi_i - \varphi^*}{\varphi_i - \varphi^-} \tag{13.10}$$

where $\tau^* = \min \tau_i$, $\tau^- = \max \tau_i$, $\varphi^* = \min \varphi_i$, $\varphi^- = \max \varphi_i$ and $\upsilon$ is introduced as weight of strategy of the majority of criteria.

**Step 7**. Rank the alternatives with increasing order regarding $Qi$, $\tau i$, and, $\varphi i$.

**Step 8**. Recommend a compromise solution based on the Acceptable Advantage and Acceptable Stability rules which are defined as follow.

i. Acceptable Advantage: $\mathrm{Q}\left(A^{(2)}\right) - \mathrm{Q}\left(A^{(1)}\right) - \geq \frac{1}{m-1}$ where $A^{(2)}$ show the second location in the increasing order of $Qi$
ii. Acceptable Stability: $A^{(1)}$ must be the first location in the increasing order of $\varphi_i \mathrm{or} \tau_i$

If the first criterion is not fulfilled, we may argue that the $A^{(1)}$ and $A^{(2)}$ alternatives are similar solutions. This situation defines that $A^{(2)}$ does not have an advantage over $A^{(1)}$. When the second rule is not met, there is no stability in decision-making. Therefore, the positions of $A^{(1)}$ and $A^{(2)}$ are in the same location in the increasing order of alternatives.

## 13.4 Case Scenario: Evaluation of Wind Power Plants in Turkey

Renewable energy plants are getting the attention of communities to meet residents' energy demands. However, renewable energy sources are instable and unreliable. For example, wind or solar energy sources depend on the stochastic nature of weather condition. For this reason, renewable energy sources have risks in terms of production variability and unpredictability. Six main risks are determined in the literature, and these risks are extended to risk factors as shown in Table 13.2. The risk factors could be collected in four groups (1) delay of production ($R_1$), (2) downtime ($R_2$), (3) unstable weather conditions ($R_3$), and (4) changes in government policy ($R_4$). If these risk factors cause the failure of production in wind power plants, the outcomes of the failure could be an effect (1) life quality of residents ($C_1$), (2) production sustainability of power plants ($C_2$), (3) operational and maintenance difficulties in power plants ($C_3$), (4) working conditions of power plant workers ($C_4$), and (5) environmental issues ($C_5$).

The data used in the case scenario is collected from four DMs. $DM_1$ and $DM_2$ are reliability engineers; on the other hand, $DM_3$ and $DM_4$ are expert reliability engineers who are at least 10 years of experience in wind power plants. All DMs are working for the same company in Turkey, and their opinions are taken simultaneously and independently by applying a survey. The survey is prepared based on the literature review that is shown in Sect. 13.2. However, the data does not reflect a private institution's or organization's official opinion. The DMs' weights are assumed as $\lambda_1 = 0.20$, $\lambda_2 = 0.20$, $\lambda_3 = 0.30$, and $\lambda_4 = 0.30$. DMs have applied the linguistic terms shown in Table 13.3 to evaluate the risk factor under the defined criteria. The DMs' opinions on risk factors and the criteria weights are shown in Tables 13.4 and 13.5, respectively.

The proposed IF-VIKOR method is applied to evaluate risk factors in the wind power plants as follows:

**Step 1**. The decision matrix is created by using Eq. (13.1), and the criteria weights ($\omega_j$) are calculated by Eq. (13.3), which are shown in Table 13.6.

**Step2**: Normalized fuzzy decision matrix and weight of criteria are calculated by using Eq. (13.4), which are shown in Table 13.7.

**Step 3**. The positive and negative ideal solutions are calculated by using Eq. (13.5) regarding cost criteria ($C_1$–$C_5$) (Table 13.8).

**Step 4**. Calculate the normalized intuitionistic fuzzy differences $\widehat{d_{ij}}$ (Table 13.9).

**Step 5.** $\tau_i$ and $\psi_i$ are calculated by using Eqs. (13.8)–(13.9), respectively, as shown in Table 13.10.

**Step 6.** VIKOR index $Qi$ is calculated by using Eq. (13.10) as shown in Table 13.11.

**Step 7:** The rank of risk factors in wind power plants with increasing order based on $\tau_i$, $\psi_i$, and $Q_i$ (Table 13.12).

In this research, the failure factors of wind power plants are evaluated based on the defined criteria. The components of each failure components assumed as risk factors. The results IF-VIKOR method which is derived from $\tau_i$, $\psi_i$, and $Q_i$ are analyzed based on the acceptable advantage and acceptable stability rules; the most important barrier is defined as the unstable weather conditions ($R_3$). Also, the other risk factors are ordered based on the importance level as downtime ($R_2$), changes in government policy ($R_4$), and delay of production ($R_1$). Thus, the risk factors in wind power plants are ranked as $R_3 > R_2 > R_4 > R_1$ under the five criteria ($C_1$–$C_5$).

**Table 13.4** DMs' opinions on the four alternatives

| Criteria | Decision-makers | Risk factors | | | |
|---|---|---|---|---|---|
| | | $R_1$ | $R_2$ | $R_3$ | $R_4$ |
| $C_1$ | $DM_1$ | MG | VG | MP | MG |
| | $DM_2$ | VG | VG | MP | MP |
| | $DM_3$ | VG | MG | MP | MP |
| | $DM_4$ | MG | VG | MP | MG |
| $C_2$ | $DM_1$ | MG | MP | MP | MG |
| | $DM_2$ | MP | MG | MG | VG |
| | $DM_3$ | MG | MP | MF | MG |
| | $DM_4$ | MP | MG | MF | VG |
| $C_3$ | $DM_1$ | MG | MP | MP | MG |
| | $DM_2$ | MP | MG | MP | VG |
| | $DM_3$ | MG | MG | MP | VG |
| | $DM_4$ | MP | MP | MP | MG |
| $C_4$ | $DM_1$ | MP | VG | MG | MG |
| | $DM_2$ | MP | VG | MP | VG |
| | $DM_3$ | MG | MP | MP | MP |
| | $DM_4$ | MP | VG | MP | MG |
| $C_5$ | $DM_1$ | MG | MP | VG | VG |
| | $DM_2$ | MG | MP | VG | MG |
| | $DM_3$ | MP | VG | MG | VG |
| | $DM_4$ | MP | MG | MG | VG |

**Table 13.5** DMs opinion on the weights of criteria

| Criteria | Decision-makers | | | |
|---|---|---|---|---|
| | $DM_1$ | $DM_2$ | $DM_3$ | $DM_4$ |
| $C_1$ | M | MH | MH | M |
| $C_2$ | M | VH | M | M |
| $C_3$ | MH | H | MH | M |
| $C_4$ | M | VH | VH | VH |
| $C_5$ | M | MH | ML | ML |

As a managerial implication of the results, it is recommended to give more focus on the most important risk factor unstable weather conditions. However, it is known that the weather conditions are very stochastic and hard to predict and prevent from them. Therefore, managers should focus on both the specific risk factors they can control and the overall risk potential in wind power plants. [5, 8, 13]. On the other hand, from the practical implications, determining the risk factors under the fuzzy

**Table 13.6** Fuzzy decision matrix and assessment of criteria weight

| | $C_1$ | | $C_2$ | | $C_3$ | | $C_4$ | | $C_5$ | |
|---|---|---|---|---|---|---|---|---|---|---|
| $R_1$ | (0.3500 | 0.5500) | (0.4082 | 0.4852) | (0.3933 | 0.5351) | (0.4082 | 0.4852) | (0.7774 | 0.1768) |
| $R_2$ | (0.8486 | 0.1338) | (0.5000 | 0.3896) | (0.5000 | 0.3896) | (0.7945 | 0.1858) | (0.6450 | 0.2873) |
| $R_3$ | (0.5000 | 0.3896) | (0.8035 | 0.1614) | (0.8035 | 0.1614) | (0.6372 | 0.2832) | (0.8678 | 0.1216) |
| $R_4$ | (0.7622 | 0.1333) | (0.5000 | 0.3896) | (0.5000 | 0.3896) | (0.4384 | 0.4529) | (0.4691 | 0.4209) |
| $\omega_j$ | (0.5505 | 0.3956) | (0.6591 | 0.2925) | (0.8529 | 0.0866) | (0.5605 | 0.3756) | (0.4495 | 0.5000) |

**Table 13.7** Normalized fuzzy decision matrix and weight of criteria

| | $C_1$ | $C_2$ | $C_3$ | $C_4$ | $C_5$ |
|---|---|---|---|---|---|
| $R_1$ | 0.0000 | 0.0000 | 0.0000 | 0.0000 | 0.7816 |
| $R_2$ | 0.9997 | 0.2508 | 0.3385 | 1.0000 | 0.4423 |
| $R_3$ | 0.3300 | 1.0000 | 1.0000 | 0.6096 | 1.0000 |
| $R_4$ | 0.8692 | 0.2508 | 0.3385 | 0.0873 | 0.0000 |
| $\omega_j$ | 0.1759 | 0.2094 | 0.2745 | 0.1810 | 0.1592 |

**Table 13.8** Positive and negative ideal solutions

| | Min | | Min | | Min | | Min | | Min | |
|---|---|---|---|---|---|---|---|---|---|---|
| | $C_1$ | | $C_2$ | | $C_3$ | | $C_4$ | | $C_5$ | |
| $f_j^*$ | (0.3500 | 0.5500) | (0.4082 | 0.4852) | (0.3933 | 0.5351) | (0.4082 | 0.4852) | (0.4691 | 0.4209) |
| $f_j^-$ | (0.8486 | 0.1333) | (0.8035 | 0.1614) | (0.8035 | 0.1614) | (0.7945 | 0.1858) | (0.8678 | 0.1216) |

**Table 13.9** Normalized intuitionistic fuzzy differences

| | $C_1$ | $C_2$ | $C_3$ | $C_4$ | $C_5$ |
|---|---|---|---|---|---|
| $R_1$ | 0.0000 | 0.0000 | 0.0000 | 0.0000 | 0.1244 |
| $R_2$ | 0.1759 | 0.0525 | 0.0929 | 0.1810 | 0.0704 |
| $R_3$ | 0.0581 | 0.2094 | 0.2745 | 0.1104 | 0.1592 |
| $R_4$ | 0.1529 | 0.0525 | 0.0929 | 0.0158 | 0.0000 |

**Table 13.10** $\tau_i$ and $\psi_i$ values

| | $R_1$ | $R_2$ | $R_3$ | $R_4$ |
|---|---|---|---|---|
| $\tau_i$ | 0.1244 | 0.5727 | 0.8115 | 0.3141 |
| $\psi_i$ | 0.1244 | 0.1810 | 0.2745 | 0.1529 |

**Table 13.11** VIKOR index $Qi$ values

| | $R_1$ | $R_2$ | $R_3$ | $R_4$ |
|---|---|---|---|---|
| $Q_i$ | 0.0000 | 0.5699 | 1.0000 | 0.2502 |

**Table 13.12** Rank of risk factors in wind power plants based on $Qi$ values

| | $R_1$ | $R_2$ | $R_3$ | $R_4$ |
|---|---|---|---|---|
| $\tau_i$ | 4 | 2 | 1 | 3 |
| $\psi_i$ | 4 | 2 | 1 | 3 |
| $Q_i$ | 4 | 2 | 1 | 3 |

environment could the research one step ahead. Since the presence of failure could cause more cost, the IF-VIKOR captures a board frame to define vague decisions [16, 19]. Therefore, fuzzy logic and multi-criteria decision-making offer different research topics with various managerial and practical implications.

## 13.5 Conclusion and Future Research

This chapter addresses a qualitative IF-VIKOR method to evaluate the risk factors of wind power plants. The increasing demand for energy needs sustainable and reliable energy production. While satisfying the resident's energy demands, environmental issues are getting the attention of the communities. For these reasons, communities try to increase their renewable energy production rate in overall energy production. The communities are turning to renewable energy sources such as solar and wind power. However, the reliability and sustainability issues could not eliminate in renewable energy sources due to the stochastic nature of the weather conditions. Such a situation causes an increase in the possibility of risk and failure during energy production. Therefore, the objective of this chapter is to provide a decision-making model for ranking the risk and failures that wind power plants may during electricity production. Four risk and failure causes are investigated in the proposed model using numerical and verbal components and five assessment criteria. To reflect evaluation criteria more comprehensively, IF-VIKOR methodology is applied to define the highest and lowest risk and failure sources that power plants may face during their operations. IF-VIKOR approach is proposed to evaluate the most and least risk and failure sources that power plants may experience throughout their operations. The proposed approach is applied to a case scenario in which there are 4 risk factors and 5 criteria. In the case scenario, four decision-makers evaluated the risk factor under the determined criteria. The results imply that the most important risk factor is unstable weather conditions. The remaining risk factors are ranked in importance as follows downtime of the power plant, changes in government policy, and delay in production. According to the results of this chapter, it is clear that risk factors in wind power plants cannot be eliminated; therefore, policymakers should take into consideration the weather conditions as the first priority.

Future research could focus on implementing various weights of criteria to examine the suggested model's sensitivity. To decide the limitations for the weight of the criteria in the best way possible, an optimization model could be used. To determine the subjective weight of criterion, the proposed IF-VIKOR method could be combined with other MCDM methods. In addition, Intuitionistic, Pythagorean, Neutrosophic, or different fuzzy logic types can be used to increase the functionality of the suggested IF-VIKOR model. The suggested model should be compared to other MCDM methods as TOPSIS, ELECTRE, TODIM, etc., under various types of fuzzy logic concepts, in order to assess its efficacy.

## References

1. Energy Atlas: Electirc consumption in Turkey (2019). https://www.enerjiatlasi.com/elektrik-tuketimi/. Accessed 26 May 2022
2. EIA (Energy Information Administration): Renewable energy explained (2021). https://www.eia.gov/energyexplained/renewable-sources/. Accessed 26 May 2022.
3. General Directorate of Energy Affairs: Wind—Rebuplic of Turkey Ministry of Energy and Natural Resources (2022). https://www.enerji.gov.tr/eigm-yenilenebilir-enerji-kaynaklar-ruzgar. Accessed 26 May 2022
4. TUREB (Turkish Wind Energy Association): Turkish Wind Energy Association Annual Report (2021). https://tureb.com.tr/lib/uploads/61d0c67b9f2adb06.pdf. Accessed 24 May 2022.
5. Opricovic, S., Tzeng, G.H.: Compromise solution by MCDM methods: a comparative analysis of VIKOR and TOPSIS. Eur. J. Oper. Res. **156**(2), 445–455 (2004). https://doi.org/10.1016/S0377-2217(03)00020-1
6. Atanassov, K.T.: Intuitionistic fuzzy sets. Fuzzy Sets Syst. **20**(1), 87–96 (1986). https://doi.org/10.1016/S0165-0114(86)80034-3
7. Abbas, S.E.: On intuitionistic fuzzy compactness. Inf. Sci. **173**, 75–91 (2005). https://doi.org/10.1016/j.ins.2004.07.004
8. Wang, Q., Li, C.: Evaluating risk propagation in renewable energy incidents using ontology-based Bayesian networks extracted from news reports. Int. J. Green Energy (2021). https://doi.org/10.1080/15435075.2021.1992411
9. Karatop, B., Taşkan, B., Adar, E., Kubat, C.: Decision analysis related to the renewable energy investments in turkey based on a fuzzy AHP-EDAS-fuzzy FMEA approach. Comput. Ind. Eng. **151**, 106958 (2021). https://doi.org/10.1016/j.cie.2020.106958
10. Hashemizadeh, A., Ju, Y., Bamakan, S.M.H., Le, H.P.: Renewable energy investment risk assessment in belt and road initiative countries under uncertainty conditions. Energy **214**, 118923 (2021). https://doi.org/10.1016/j.energy.2020.118923
11. Somi, S., Seresht, N.G., Fayek, A.R.: Framework for risk identification of renewable energy projects using fuzzy case-based reasoning. Sustainability (Switzerland) **12**(13) (2020). https://doi.org/10.3390/su12135231
12. Yatim, P., Lin, N.S., Lam, H.L., Choy, E.A.: Overview of the key risks in the pioneering stage of the Malaysian biomass industry. Clean Technol. Environ. Policy **19**(7), 1825–1839 (2017). https://doi.org/10.1007/s10098-017-1369-2
13. Abba, Z.Y.I., Balta-Ozkan, N., Hart, P.: A holistic risk management framework for renewable energy investments. Renew. Sustain. Energy Rev. **160**, 112305 (2022). https://doi.org/10.1016/j.rser.2022.112305
14. Hansen, A.C., Clarke, N., Hegnes, A.W.: Managing sustainability risks of bioenergy in four Nordic countries. Energy Sustain. Soc. **11**(1) (2021). https://doi.org/10.1186/s13705-021-00290-9
15. Ranganath, N., Sarkar, D., Patel, P., Patel, S.: Application of fuzzy TOPSIS method for risk evaluation in development and implementation of solar park in India. Int. J. Constr. Manag. (2020). https://doi.org/10.1080/15623599.2020.1826027

16. Aquila, G., Rotela Junior, P., de Oliveira Pamplona, E., de Queiroz, A.R.: Wind power feasibility analysis under uncertainty in the Brazilian electricity market. Energy Econ. **65**, 127–136 (2017). https://doi.org/10.1016/j.eneco.2017.04.027
17. Hagfors, L.I., Kamperud, H.H., Paraschiv, F., Prokopczuk, M., Sator, A., Westgaard, S.: Prediction of extreme price occurrences in the German day-ahead electricity market. Quant. Finance **16**(12), 1929–1948 (2016). https://doi.org/10.1080/14697688.2016.1211794
18. Guerrero-Liquet, G.C., Sánchez-Lozano, J.M., García-Cascales, M.S., Lamata, M.T., Verdegay, J.L.: Decision-making for risk management in sustainable renewable energy facilities: a case study in the Dominican Republic. Sustainability (Switzerland) **8**(5) (2016). https://doi.org/10.3390/su805045
19. Kul, C., Zhang, L., Solangi, Y.A.: Assessing the renewable energy investment risk factors for sustainable development in Turkey. J. Clean. Prod. **276**, 124164 (2020). https://doi.org/10.1016/j.jclepro.2020.124164
20. Apak, S., Atay, E., Tuncer, G.: Financial risk management in renewable energy sector: comparative analysis between the European Union and Turkey. Procedia. Soc. Behav. Sci. **24**, 935–945 (2011)
21. Kanamura, T., Homann, L., Prokopczuk, M.: Pricing analysis of wind power derivatives for renewable energy risk management. Appl. Energy **304**, 117827 (2021). https://doi.org/10.1016/j.apenergy.2021.117827
22. Kumar, N., Rogers, D., Burnett, T.D., Sullivan, E., Gascon, M.: Reliability, availability, maintainability (ram) for wind turbines. ASM POWER 2 (2017). https://doi.org/10.1115/POWER-ICOPE2017-3045
23. Soroudi, A., Mohammadi-Ivatloo, B., Rabiee, A.: Energy hub management with intermittent wind power. In: Hossain, J., Mahmud, A. (eds.) Large Scale Renewable Power Generation, Green Energy and Technology, pp. 413–438. Springer, Singapore (2014). https://doi.org/10.1007/978-981-4585-30-9_16
24. Macrander, A.M., Brzuzy, L., Raghukumar, K., Preziosi, D., Jones, C.: Convergence of emerging technologies: development of a risk-based paradigm for marine mammal monitoring for offshore wind energy operations. Integr. Environ. Assess. Manag. (2021). https://doi.org/10.1002/ieam.4532
25. Astiaso Garcia, D., Bruschi, D.: A risk assessment tool for improving safety standards and emergency management in Italian onshore wind farms. Sustain. Energy Technol. Assess. **18**, 48–58 (2016). https://doi.org/10.1016/j.seta.2016.09.009
26. Gatzert, N., Kosub, T.: Determinants of policy risks of renewable energy investments. Int. J. Energy Sect. Manage. **11**(1), 28–45 (2017). https://doi.org/10.1108/IJESM-11-2015-0001
27. Chou, J., Liao, P., Yeh, C.: Risk analysis and management of construction and operations in offshore wind power project. Sustainability (Switzerland) **13**(13) (2021). https://doi.org/10.3390/su13137473
28. Ioannou, A., Angus, A., Brennan, F.: Risk-based methods for sustainable energy system planning: a review. Renew. Sustain. Energy Rev. **74**, 602–615 (2017). https://doi.org/10.1016/j.rser.2017.02.082

# Chapter 14
# Some Discrete Parametric Markov–Chain System Models to Analyze Reliability

**Rakesh Gupta, Shubham Gupta, and Irfan Ali**

**Abstract** The chapter deals with the development of various reliability characteristics and the inter-relations between them when the random variable denoting the lifetime of a device follows a discrete distribution. It analyzes the reliability characteristics for an n-unit series, parallel and standby system models when the failure times of the units are discrete random variables. The results are also drawn in the case when the lifetimes of the units follow geometric distributions. Many important conclusions are drawn regarding the lower/upper bounds of the number of units in the system and the failure rate of a unit. This chapter also presents the cost–benefit analysis of two identical unit warm standby repairable system models, assuming the geometric distributions of failure and repair times. A single repairman is always considered with the system to repair a failed unit. The results may also be obtained for the following two particular cases—(i) Two identical unit cold standby systems with geometric failure and repair time distributions and (ii) Two identical unit parallel systems with geometric failure and repair time distributions. The curves for MTSF and net-expected profit per unit of time in steady state are drawn to study the system behavior in respect of different parameters, and various important conclusions from the curves of these characteristics are drawn.

**Keywords** Reliability · Availability · MTSF · Hazard rate · Series configuration · Parallel configuration · Standby system · Redundancy · Regenerative point · Transition probability · Mean sojourn time

---

R. Gupta · S. Gupta
Department of Statistics, Chaudhary Charan Singh University, Meerut, Uttar Pradesh 250005, India

I. Ali (✉)
Department of Statistics and Operations Research, Aligarh Muslim University, Aligarh, Uttar Pradesh 202002, India
e-mail: irfii.st@amu.ac.in

S. Gupta
Department of Electronics and Communication Engineering, Manipal University Jaipur, Rajasthan, India



H. Garg (ed.), *Advances in Reliability, Failure and Risk Analysis*, Industrial and Applied Mathematics, https://doi.org/10.1007/978-981-19-9909-3_14

## 14.1 Introduction

The ever-developing technology and the increasing needs of society in the scenario of present-day cause many problems concerning the effectiveness of various articles/systems sound in industries, health services and day-to-day life. A fault or interruption in the operation of a system deteriorates its quality, and the system is declared as failed either partially or totally. Thus, the failures usually happening in the ways of life, and the penalties paid by the user in terms of money, time and security are becoming more and more severe due to the increasing application of complexities and automation. A better understanding of the causes of failure, the latest manufacturing techniques, designing of new systems with proper selection as well as an optimum network of components of high quality and the consideration of appropriate methods for reliability improvement are some of the basic techniques which can be applied to minimize the degree of failure of the system.

The reliability of a system can be improved in several ways. We can use techniques to improve the system's reliability that best suits the operating conditions and cost constraints. We know that a unit/system is composed of several components or elements, and to make the system more reliable, we have to use highly reliable components. Since highly reliable components, or the cost of producing such components is very high. In that case, we can improve the system's reliability by introducing redundancies. In a redundant system, either one of the components/units is sufficient for the system's successful operation; we deliberately use more components/units to increase the probability of success. The extra components or units used in the system are known as redundancies. There are two types of redundancies—active and passive or standby. The standby redundancies may be classified as hot, warm and cold according to how they are loaded in the standby state. The other methods to improve the system's reliability are preventive maintenance (p.m.) and repair maintenance (r.m.).

The p.m. is a repair done before the unit fails—for example, servicing an automobile, oiling a machine, etc. Usually, a periodic policy for p.m. is adopted. However, it is not always possible to perform this maintenance action exactly when desired, so one would expect the time at which the p.m. is made to be a random variable (r.v.) having small dispersion about the desired time. Repair maintenance (r.m.) is concerned with increasing system reliability by implementing major changes in the failed unit/system. In order to increase reliability, the failed components of the unit/system are repaired or replaced by new ones. When a system is intended for use over a long period or when the cost of a new unit/system is considered too high, introducing a repair facility is worth considering to improve system effectiveness.

Reliability characteristics, such as the probability of survival, the system's failure rate, mean time to failure, mean operative time of the system during a finite interval and expected frequency of failures, are some measures of system effectiveness. These measures provide necessary criteria by which alternate design policies can be compared and judged to help the system planner select one that best satisfies the objectives under certain techno-economic constraints.

## 14.2 Concepts Used in Analyzing System Models

Several research articles in the field of reliability modeling have been analyzed during the past four decades by many authors on the system models of continuous parametric Markov–Chain using various real existing important concepts. A few of them are listed with related references as: Priority Unit [1–9], Partial Failure [5, 10–14], Imperfect and Slow Switching Device [2, 15–19], Administrative Delay in Repair [4, 20, 21], Preparation for Repair of a Failed Unit [22–24], Common Cause Failure [25–27], Preventive Maintenance [28–30].

Random Shocks [3, 6, 29], Repair Machine Failure [31–33], Rest to Operator/Repairman [34, 35], Random Appearance and Disappearance of Repairman [14, 36–38], Two Types of Failure and Repair [11, 39, 40], Man–Machine System /Physical Conditions [36, 41, 42], Abnormal Weather Condition [41, 43, 44], Repair and Replacement Policies [45, 46], Repair and Post-Repair Policies [47, 48], Two-Phase Repair [28, 49], Correlated Failure and Repair [10, 22, 23, 37, 48], Correlated Life Times [33, 46, 49, 50] and Correlated Working and Rest Time of Repairman/Operator [51, 52].

## 14.3 Concept of Discrete Failure and Repair Time Models

In practice, the situations exist when the failure and repair of a unit occur at discrete random epochs so that the lifetime and repair time of a unit follow discrete distributions like geometric, negative binomial, Poisson, etc. Discrete failure data arise in several common situations. For example:

(a) A device is monitored only once per period (i.e., an hour, a day, a minute), and the observation is the number of periods completed before the failure.
(b) A piece of equipment operates in cycles, and the experimenter observes the number of completed cycles before failure.

**Some examples of discrete lifetimes are as follows**

1. In a photocopy machine, the bulb is lightened whenever the machine takes paper for Xerox purposes. Thus, the lifetime of the bulb is a discrete random variable.
2. In an on/off switching device, the lifetime of the switch is a discrete random variable.
3. A spring may break down, completing a certain number of cycles of 'to and fro' movements, and then, the lifetime of the spring is a discrete random variable.
4. The bulb is lightened whenever the door is opened in a refrigerator. Thus, the lifetime of the bulb is a discrete random variable.
5. The strength in terms of the number of shocks that a product can withstand a discrete random variable.

More so, let the continuous time period $(0, \infty)$ is divided as $0, 1, 2, \ldots, n$, of equal distance on the real line, and the probability of failure of a unit during time $(i, i+1); i = 0, 1, 2 \ldots$ is $p$. Then, the probability that the unit will fail during $(t, t+1)$, i.e., after passing successfully $t$ intervals of time, is given by $p(1-p)^t; t = 0, 1, 2, \ldots$. This is the probability mass function (p.m.f.) of the geometric distribution. Similarly, if $r$ denotes a failed unit's probability of being repaired $(i, i+1); i = 0, 1, 2 \ldots$. Then, the probability that the unit will be repaired $(t, t+1)$ is given by $r(1-r)^t; t = 0, 1, 2 \ldots$.

## 14.4 Development of Some Important Results

Let the discrete r.v. $T$ be the lifetime of a device having p.m.f.

$$P(T = t) = g(t); \quad t = 0, 1, 2, \ldots \infty \tag{14.4.1}$$

Then, the reliability or survival function of the device is the probability that the system has completed for at least $t$ epochs (cycles) and is defined as

$$R(t) = P(T > t-1) = P(T \geq t) = \sum_{j=t}^{\infty} g(j) \tag{14.4.2}$$

and the failure rate function of the discrete distribution is given by

$$r(t) = \frac{g(t)}{\sum_{j=t}^{\infty} p_j} = \frac{g(t)}{R(t)} \tag{14.4.3}$$

Clearly, $R(0) = 1$ and $r(t) \leq 1$.

The unreliability of the system is equal to the C.d.f. of time to system failure and is defined as

$$F(t) = P(T < t) = \sum_{j=0}^{t-1} g(j)$$

Obviously,

$$F(t) + R(t) = \sum_{j=0}^{t-1} g(j) + \sum_{j=t}^{\infty} g(j) = 1$$

By the definition of expectation, the mean life of the device is

$$E(T) = \sum_{t=0}^{\infty} t g(t) = \sum_{t=0}^{\infty} t[R(t) - R(t+1)]$$

$$= \sum_{t=1}^{\infty} R(t) \tag{14.4.4}$$

To establish the inter-relations between failure rate, reliability function and failure time p.m.f., let us consider

$$\begin{aligned} r(t) &= \frac{g(t)}{R(t)} = \frac{R(t) - R(t+1)}{R(t)} \\ &= 1 - \frac{R(t+1)}{R(t)} \end{aligned}$$

Therefore,

$$\frac{R(t+1)}{R(t)} = 1 - r(t) \tag{14.4.5}$$

Now writing $R(t)$ as

$$\begin{aligned} R(t) &= \frac{R(t)}{R(t-1)} . \frac{R(t-1)}{R(t-2)} \cdots \frac{R(1)}{R(0)} . R(0) \\ &= [1 - r(t-1)][1 - r(t-2)] \ldots [1 - r(0)].R(0), \text{ using (4.5)} \\ &= \prod_{i=0}^{t-1} [1 - r(i)], \quad \text{as} \quad R(0) = 1 \end{aligned} \tag{14.4.6}$$

Now, using (14.4.3) and (14.4.6), we get an interesting relationship between failure rate and failure time p.m.f. as follows:

$$g(t) = r(t) \prod_{i=0}^{t-1} [1 - r(i)] \tag{14.4.7}$$

when the lifetime $T$ of the device (system) follows a geometric distribution with p.m.f.

$$g(t) = pq^t; \quad q = 1 - p, \quad t = 0, 1, 2 \ldots$$

Then, reliability function is given by

$$\begin{aligned} R(t) &= p[T > t - 1] \\ &= \sum_{j=t}^{\infty} pq^j = q^t \end{aligned} \tag{14.4.8}$$

and the hazard rate is

**Table 14.1** Inter-relations among $g(t)$, $F(t)$, $R(t)$ and $r(t)$

| Function | Expressed by | | | |
|---|---|---|---|---|
| | $g(t)$ | $F(t)$ | $R(t)$ | $r(t)$ |
| $g(t)$ | – | $F(t+1) - F(t)$ | $R(t) - R(t+1)$ | $r(t)\prod_{i=0}^{t-1}\{1 - r(i)\}$ |
| $F(t)$ | $\sum_{j=0}^{t-1} g(j)$ | – | $1 - R(t)$ | $1 - \prod_{i=0}^{t-1}\{1 - r(i)\}$ |
| $R(t)$ | $\sum_{j=t}^{\infty} g(j)$ | $1 - F(t)$ | – | $\prod_{i=0}^{t-1}\{1 - r(i)\}$ |
| $r(t)$ | $\frac{g(t)}{\sum_{j=t}^{\infty} g(j)}$ | $\frac{F(t+1)-F(t)}{1-F(t)}$ | $\frac{R(t)-R(t+1)}{R(t)}$ | – |

$$r(t) = \frac{g(t)}{R(t)} = p(\text{probability of failure of the device in each cycle}) \tag{14.4.9}$$

Also, the expected life of the device is (Table 14.1)

$$\begin{aligned} E(T) &= \sum_{t=1}^{\infty} R(t) \\ &= \sum_{t=1}^{\infty} q^t = \frac{q}{1-q} = \frac{q}{p} \end{aligned} \tag{14.4.10}$$

A large number of discrete parametric Markov–Chain models pertaining to the two-unit reparable redundant systems have been analyzed in respect of reliability and cost-benefit measures of system effectiveness during the last fifteen years, including [32, 53–57].

## 14.5 Analysis of n-Unit Series System

Let us suppose that we have $n$ independent functioning units $U_1, U_2, \ldots, U_n$ arranged in a series configuration with lifetime $T_i$ and reliability function $R_i$ of the $i$th unit. Then, the reliability of the series system is

$$\begin{aligned} R(t) &= P(T > t - 1) \\ &= P[\min.(T_1, T_2, \ldots T_n) > t - 1] \\ &= P[T_1 > t - 1, T_2 > t - 1, \ldots T_n > t - 1] \\ &= \prod_{i=1}^{n} P[T_i > t - 1] = \prod_{i=1}^{n} R_i(t) \end{aligned} \tag{14.5.1}$$

To obtain system failure rate $r(t)$ in terms of unit failure rates $r_1(t).r_2(t)\ldots r_n(t)$, we apply result (14.4.6) on both sides of (14.5.1) so that

$$\prod_{j=0}^{t-1}[1-r(j)] = \prod_{i=1}^{n}\left[\prod_{j=0}^{t-1}\{1-r_i(j)\}\right]$$

$$= \prod_{j=0}^{t-1}\prod_{i=1}^{n}\{1-r_i(j)\}$$

$$\Rightarrow 1-r(t) = \prod_{i=1}^{n}\{1-r_i(t)\}$$

$$\Rightarrow r(t) = 1-\prod_{i=1}^{n}\{1-r_i(t)\} \tag{14.5.2}$$

When the lifetime $T_i$ $(i = 1, 2, \ldots, n)$ of $i$th unit $U_i$ follows a geometric distribution with parameter $p_i$, i.e.,

$$P(T_i = t) = p_i q_i^t; \quad t = 0, 1, 2, \ldots; \quad q_i = 1 - p_i$$

Then, using (14.5.1), the system reliability is

$$R(t) = \prod_{i=1}^{n} R_i(t) = \prod_{i=1}^{n} q_i^t$$

$$= \left(\prod_{i=1}^{n} q_i\right)^t \tag{14.5.3}$$

Now using (14.5.2), the failure rate of the system is given by

$$r(t) = 1-\prod_{i=1}^{n}[1-p_i] = 1-\prod_{i=1}^{n} q_i \tag{14.5.4}$$

Therefore, by result (14.4.7), the system lifetime p.m.f. is

$$g(t) = \left[1-\prod_{i=1}^{n} q_i\right]\left[\prod_{i=1}^{n} q_i\right]^t \tag{14.5.5}$$

which is the p.m.f. of geometric distribution with parameter $\left[1-\prod_{i=1}^{n} q_i\right]$.

The expected life of the system is

$$E(T) = \sum_{t=1}^{\infty} R(t) = \sum_{t=1}^{\infty}\left[\prod_{i=1}^{n} q_i\right]^t$$

$$= \frac{\prod_{i=1}^{n} q_i}{1 - \prod_{i=1}^{n} q_i} \quad (14.5.6)$$

In the case of identical units, we consider

$$p_1 = p_2 = \cdots = p_n = p (= 1 - q)$$

Then, the results (14.5.3) to (14.5.6) become

$$R(t) = q^{nt}$$

$$r(t) = 1 - q^n$$

$$g(t) = \left(1 - q^n\right)\left(q^n\right)^t$$

and

$$E(T) = \frac{q^n}{1 - q^n} \quad (14.5.7\text{–}14.5.10)$$

The curves for $R(t)$ and $r(t)$ with respect to $n$ for different values of $q$ when $t =$ 200 h are sketched, respectively, in Figs. 14.1 and 14.2.

Similarly, the curves for $R(t)$ and $r(t)$ with respect to p for different values of $n$ when $t = 50$ h are sketched, respectively, in Figs. 14.3 and 14.4.

In case, if one is to achieve system reliability, at least $K_1$ units, i.e.,

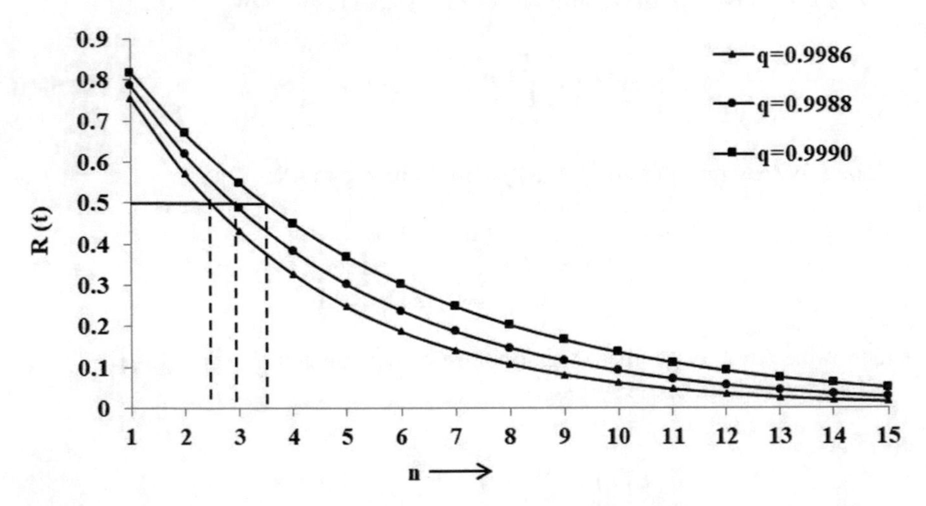


**Fig. 14.1** Reliability curves for series system with respect to '$n$' for different values of '$q$' ($t =$ 200 h)

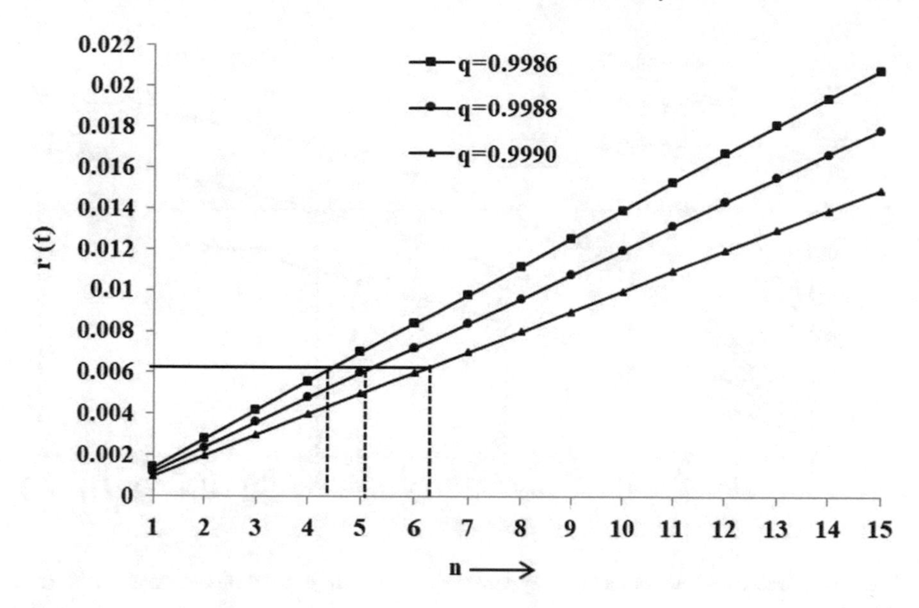


**Fig. 14.2** Failure rate curves for series system with respect to '$n$' for different values of '$q$' ($t$ = 200 h)

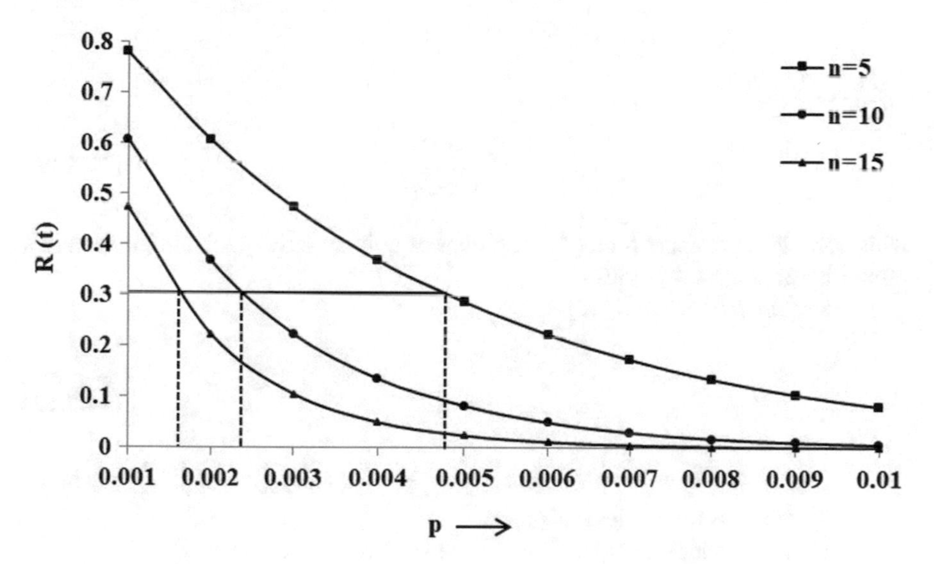


**Fig. 14.3** Reliability curves for series system with respect to '$p$' for different values of '$n$' ($t$ = 50 h)

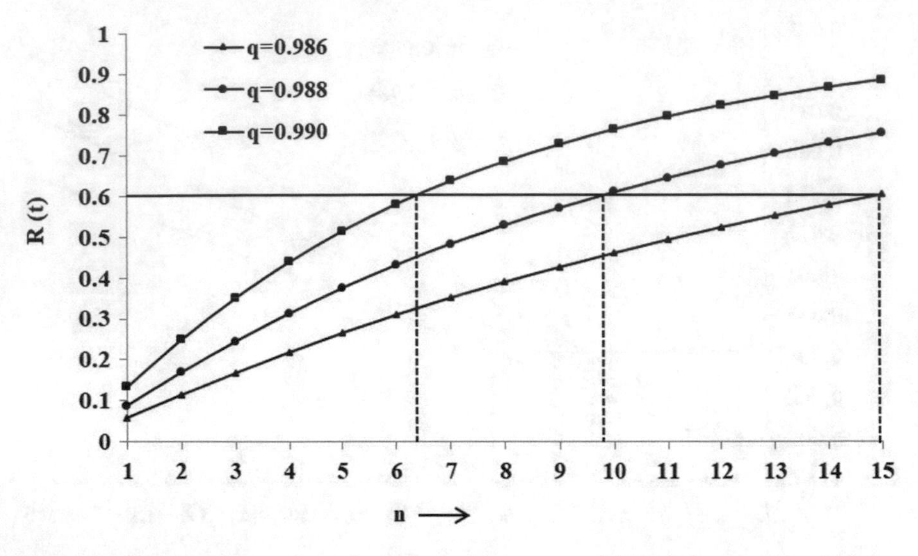


**Fig. 14.4** Failure rate curves for series system with respect to '$p$' for different values of '$n$' ($t =$ 50 h)

$$R(t) = q^{nt} \geq K_1$$

Then,

$$n \leq \frac{\log K_1}{t \log q} \tag{14.5.11}$$

which provides the upper limit of the number of units in series to obtain the system's reliability of at least $K_1$ units.

Further, for $R(t) = q^{nt} \geq K_2$

We have

$$p \leq 1 - K_2^{1/nt} \tag{14.5.12}$$

i.e., the failure rate of each unit should not exceed the value $\left(1 - K_2^{1/nt}\right)$ to achieve reliability of the system at least $K_2$ units.

Now, if one decides the upper limit of the system failure rate as

$$r(t) = \left(1 - q^n\right) \leq K_3$$

then

$$n \leq \frac{\log(1 - K_3)}{\log(q)} \tag{14.5.13}$$

which provides the upper limit of the number of units in series to obtain the system failure rate, not more than $K_3$.

Also, for

$$r(t) = \left(1 - q^n\right) \leq K_4$$

we have

$$p \leq 1 - (1 - K_4)^{1/n} \tag{14.5.14}$$

i.e., the failure rate of each unit should not exceed the value $\left\{1 - (1 - K_4)^{1/n}\right\}$.

**Concluding Remarks**

As an illustration, to achieve at least reliability 0.50 units when $t = 200$ h, the result (14.5.11) provides that the number of components in series system should not exceed 3.46 (= 3), 2.89 (= 2) and 2.47 (= 2), respectively, for $q = 0.9990$, 0.9988 and 0.9986 which are verified from Fig. 14.1. Similarly, to achieve failure rate at the most 0.006, when $t = 200$, we find from result (14.5.13) that the number of components in series system should not exceed 6.03 (= 6), 5.01 (= 5) and 4.30 (= 4), respectively, for $q = 0.990$, 0.9988 and 0.9986 which are verified from Fig. 14.2. Similarly, to achieve reliability of 0.30 units when $t = 50$, the result (14.5.12) provides that the failure rate of each unit should not exceed 0.0048, 0.0024 and 0.0016, respectively, for $n = 5$, 10 and 15. It is also verified from Fig. 14.3. Finally, to achieve the failure rate of the series system at the most 0.03 when $t = 50$, the result (14.5.14) reveals that the failure rate of each unit should not exceed the values 0.0061, 0.0030 and 0.0019, respectively, for $n = 5$, 10 and 15, which are verified from Fig. 14.4.

## 14.6 Analysis of n-Unit Parallel System

We know that in the parallel configuration of the units, $U_1, U_2, \ldots, U_n$ the system reliability is given by

$$R(t) = 1 - \prod_{i=1}^{n} [1 - R_i(t)] \tag{14.6.1}$$

In case the lifetimes of the units $U_1, U_2, \ldots, U_n$ follow geometric distributions with parameters $p_1, p_2, \ldots, p_n$, respectively.

Then,

$$R(t) = 1 - \prod_{i=1}^{n} \left[1 - q_i^t\right]$$

$$= \sum_{i=1}^{n} q_i^t - \sum_{i<j=1}^{n} (q_i q_j)^t + \sum_{i<j<k=1}^{n} (q_i q_j q_k)^t + \cdots + (-1)^{n-1} \left( \prod_{i=1}^{n} q_i \right)^t \tag{14.6.2}$$

and the system lifetime p.m.f. is

$$\begin{aligned} g(t) &= R(t) - R(t+1) \\ &= \sum_{i=1}^{n} q_i^t (1 - q_i) - \sum_{i<j=1}^{n} (q_i q_j)^t (1 - q_i q_j) \\ &\quad + \cdots + (-1)^{n-1} \left( \prod_{i=1}^{n} q_i \right)^t \left( 1 - \prod_{i=1}^{n} q_i \right) \end{aligned} \tag{14.6.3}$$

Therefore, the failure rate of the system is

$$r(t) = \frac{g(t)}{R(t)} = \frac{\prod_{i=1}^{n} (1 - q_i^{t+1}) - \prod_{i=1}^{n} (1 - q_i^t)}{1 - \prod_{i=1}^{n} (1 - q_i^t)} \tag{14.6.4}$$

Now, the mean lifetime of the system is

$$\begin{aligned} E(T) &= \sum_{t=1}^{\infty} R(t) \\ &= \sum_{i=1}^{n} \frac{q_i}{(1 - q_i)} - \sum_{i<j}^{n} \frac{q_i q_j}{1 - q_i q_j} + \cdots + (-1)^{n-1} \frac{\prod_{i=1}^{n} q_i}{1 - \prod_{i=1}^{n} q_i} \end{aligned} \tag{14.6.5}$$

When the units are identical, i.e., $p_i = p \, \forall i = 1, 2, \ldots, n$; then the result (14.6.2) to (14.6.5) become

$$R(t) = 1 - \left(1 - q^t\right)^n$$

$$g(t) = \left(1 - q^{t+1}\right)^n - \left(1 - q^t\right)^n$$

$$r(t) = \frac{\left(1 - q^{t+1}\right)^n - \left(1 - q^t\right)^n}{1 - (1 - q^t)^n}$$

and

$$E(T) = \sum_{t=1}^{\infty} \left[1 - \left(1 - q^t\right)^n\right]$$

$$= \sum_{j=1}^{n} \binom{n}{j} (-1)^{j-1} \frac{q^j}{1 - q^j} \quad (14.6.6\text{–}14.6.9)$$

The curves for $R(t)$ with respect to n for three different values of $q = 0.990, 0.998$ and 0.996 for fixed $t = 200$ h are sketched in Fig. 14.5, and the curves for $R(t)$ with respect to $p$ for $n = 5$, 10 and 15 for fixed $t = 50\,h$ are sketched in Fig. 14.6.

In a parallel system, it is evident that the system's reliability increases as we increase the number of units in the system. So, in order to get the reliability of the system, at least $K_5$ units, i.e.,

$$R(t) = 1 - \left(1 - q^t\right)^n \geq K_5$$

The minimum number of units required in the system is given by

$$n \geq \frac{\log(1 - K_5)}{\log(1 - q^t)} \quad (14.6.10)$$

Further, to achieve $R(t) = 1 - \left(1 - q^t\right)^n \geq K_6$, we get

$$p(= 1 - q) \leq 1 - \left[1 - (1 - K_6)^{1/n}\right]^{1/t} \quad (14.6.11)$$

i.e., the failure rate of each unit should not exceed the value

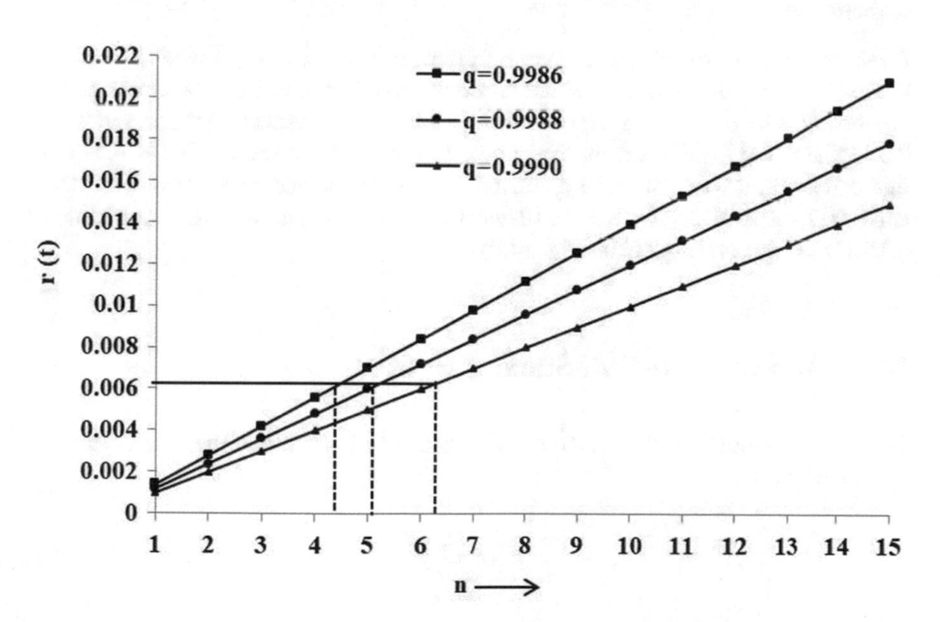


**Fig. 14.5** Reliability curves for parallel system with respect to '$n$' for different values of '$q$' ($t =$ 200 h)

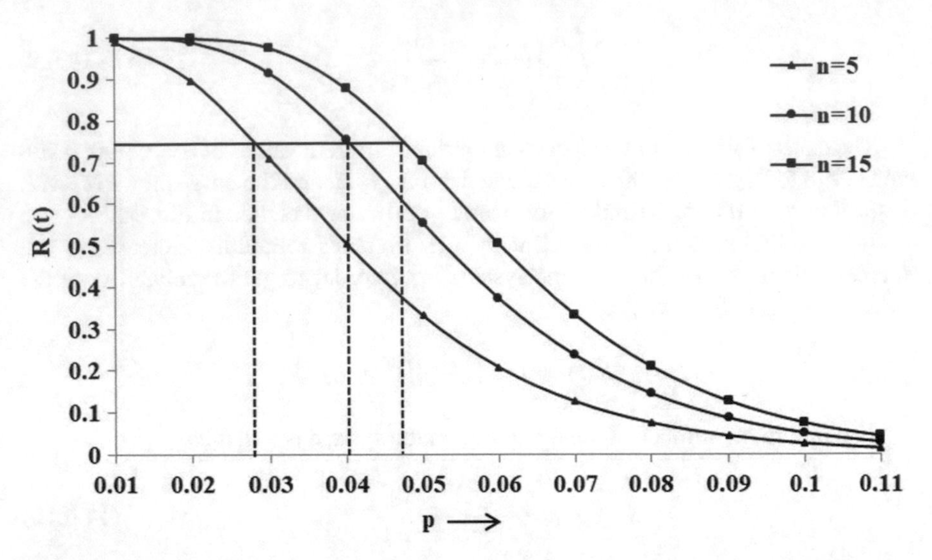


**Fig. 14.6** Reliability curves for parallel system with respect to '$p$' for different values of '$n$' ($t =$ 50 h)

$$1 - \left[1 - (1 - K_6)^{1/n}\right]^{1/t}$$

**Concluding Remark**

As an illustration, to achieve at least a system reliability of 0.60 when $t = 200$ h, the result (14.6.10) provides that the minimum number of units to be considered in parallel must be 6.4 (= 7), 9.10 (= 10) and 14.9 (= 15), respectively, for $q = 0.990$, 0.988 and 0.986 (verified from Fig. 14.5). Similarly, from the result (14.6.11), we find that the failure rate of each system unit should not exceed the values 0.028, 0.040 and 0.047, respectively, for $n = 5$, 10 and 15 to achieve at least system reliability of 0.75 at $t = 50$ h (verified from Fig. 14.6).

## 14.7 Analysis of n-Unit Standby System

In an n-unit standby system, if $T_i$ is the lifetime of $i$th unit, then the system lifetime $T$ is given by

$$T = \sum_{i=1}^{n} T_i \tag{14.7.1}$$

So, the expected lifetime of the system is

$$E(T) = \sum_{i=1}^{n} E(T_i) \tag{14.7.2}$$

If $g_i(t)$ denotes the p.m.f. of $T_i$ and $g(t)$ denotes the p.m.f. of $T$. Then $n = 2$, we get

$$g(t) = \sum_{u=0}^{t} g_1(u)\, g_2(t-u) = g_1(t) \copyright g_2(t) \tag{14.7.3}$$

When $T_i$ follows a geometric distribution with parameter $p_i$

$$\begin{aligned} g(t) &= \sum_{u=0}^{t} p_1 q_1^u p_2 q_2^{t-u} \\ &= \begin{cases} \dfrac{p_1 p_2 \left(q_2^{t+1} - q_1^{t+1}\right)}{q_2 - q_1}; & \text{if } \; p_1 \neq p_2 \\ p^2 (t+1) q^t = \dbinom{t+1}{1} p^2 q^t; & \text{if } \; p_1 = p_2 = p \end{cases} \\ &= \text{Negative binomial}(2, p) \end{aligned} \tag{14.7.4}$$

In general, for n-unit standby system

$$g(t) = g_1(t) \copyright g_2(t) \copyright \ldots \copyright g_n(t) \tag{14.7.5}$$

where

$$g_i(t) = p_i q_i^t; \quad i = 1, 2, \ldots, n$$

and for $p_1 = p_2 = \cdots = p_n = p$

$$\begin{aligned} g(t) &= \binom{t+n-1}{n-1} p^n q^t \\ &= \text{Negative binomial}(n, p) \end{aligned} \tag{14.7.6}$$

The reliability of the system $n = 2$ is given by

$$R(t) = \begin{cases} \sum\limits_{j=1}^{\infty} \frac{p_1 p_2 \left(q_2^{t+1} - q_1^{t+1}\right)}{q_2 - q_1} = \frac{p_1 q_2^{t+1} - p_2 q_1^{t+1}}{q_2 - q_1}; & \text{if } \; p_1 \neq p_2 \\ p^2 \sum\limits_{j=t}^{\infty} (j+1) q^j = (1 + pt) q^t; & \text{if } \; p_1 = p_2 = p \end{cases} \tag{14.7.7}$$

Alternatively, the reliability of a two-unit standby system can be obtained as follows:

$$
\begin{aligned}
R(t) &= P[T > t - 1] = P[T \geq t] \\
&= P[\text{Unit 1st survives up to time}(t-1)] + P[\text{Unit 1st fails at epoch } u, \\
&\quad u < t - 1;\ \text{2nd unit switched on, and it survives up to } (t - 1 - u)] \\
&= q_1^t + \sum_{u=0}^{t-1} p_1 q_1^u q_2^{t-u} \\
&= \begin{cases} \dfrac{p_1 q_2^{t+1} - p_2 q_1^{t+1}}{q_2 - q_1}; & \text{when } p_1 \neq p_2 \\ (1 + pt)q^t; & \text{when } p_1 = p_2 = p \end{cases}
\end{aligned}
\tag{14.7.8}
$$

In general, the reliability of the n-unit standby system is given by

$$
\begin{aligned}
R(t) = R_1(t) &+ g_1(t-1) © R_2(t-1) + g_1(t-1) © g_2(t-1) © R_3(t-1) + \cdots \\
&+ g_1(t-1) © g_2(t-1) © \ldots g_{n-1}(t-1) © R_n(t-1)
\end{aligned}
\tag{14.7.9}
$$

To find the expected lifetime of the n-unit standby system, we have

$$
E(T_i) = \sum_{t=0}^{\infty} t p_i q_i^t = \frac{q_i}{p_i}
$$

Therefore,

$$
E(T) = \begin{cases} \sum_{i=1}^{n} \frac{q_i}{p_i}; & \text{for non-identical units} \\ \frac{nq}{p}; & \text{for identical units} \end{cases}
\tag{14.7.10}
$$

If one wants to achieve the system's expected life at least $K_7$ with $n_0$ identical units, then the result (14.7.10) implies that he must consider each unit in the system such that the failure rate of each unit should not exceed the value $n_0/(n_0 + K_7)$.

## 14.8 A Two Identical Unit Warm Standby System Model with Geometric Failure and Repair Time Distributions

Here, we have considered two identical units. Each unit has two modes—normal (N) and total failure (F). Initially, the system starts functioning from a state $S_0$ where both the units are in N-mode, with one unit operative and the other as a warm standby. The system reaches state $S_1$ if in state $S_0$ either the operating unit fails with rate '$p$'

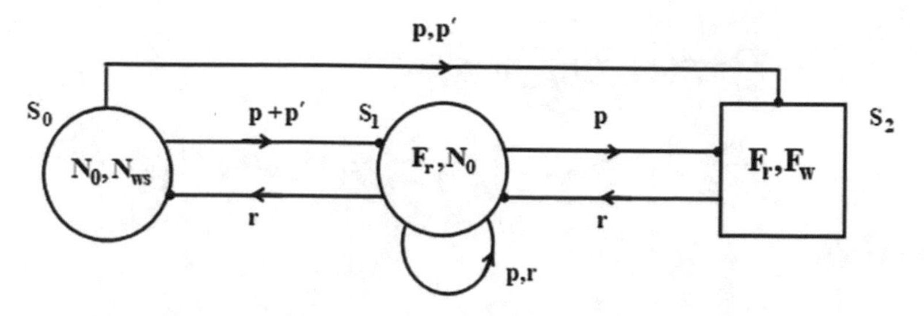


**Fig. 14.7** The transition diagram of the system model

or the warm standby unit fails with rate $p'$. The system also reaches state $S_2$ from $S_1$ if both the units (operative and warm standby) in state $S_0$ fail simultaneously at the same epoch. As soon as the system transits from state $S_0$ to $S_1$, the failed unit goes into repair, and other units operate in N-mode. Further, as the system reaches state $S_2$ from $S_0$, one of the failed units goes into repair, and the other waits for repair as we assume a single repairer with the system to repair a failed unit.

Similarly, we have three possible transitions from $S_1$ and one from $S_2$ where '$r$' represents the repair rate of a failed unit. A repaired unit is also assumed always to work as well as new. The transition diagram of the system model is shown in Fig. 14.7.

We define,

$pq^x(p+q=1)$: p.m.f. of failure time of a operative unit; $x=0,1,\ldots$

$p'q'^x(p'+q'=1)$: p.m.f. of failure time of a warm standby unit; $x-0,1,\ldots$

$rs^x(r+s=1)$: p.m.f. of repair time of a failed unit; $x=0,1,\ldots$

$q_{ij}(t); i,j=0,1,2$: p.m.f. of transition time $T_{ij}$ from state $S_i$ to $S_j$.

### 14.8.1 Transition Probabilities

Let

$$Q_{ij}(t)=P\left[\text{System transits from } S_i \text{ to } S_j \text{ at any epoch} \le t\right]$$

Therefore,

$$Q_{01}(t)=\sum_{u=0}^{t} P\left[\text{Either operative or warm standby unit fails at epoch } u\right]$$

$$= \sum_{u=0}^{t} pq^u (q')^{u+1} + \sum_{u=0}^{t} p' (q')^u q^{u+1}$$

$$= \frac{pq'}{1-qq'}\left[1-(qq')^{t+1}\right] + \frac{p'q}{1-qq'}\left[1-(qq')^{t+1}\right]$$

Similarly,

$$Q_{02}(t) = \sum_{u=0}^{t} P[\text{Operating unit as well as warm standby unit in } S_0 \text{ fail at same epoch } u]$$

$$= \sum_{u=0}^{t} pq^u p' (q')^u = \frac{pp'}{1-qq'}\left[1-(qq')^{t+1}\right]$$

$$Q_{10}(t) = \sum_{u=0}^{t} rs^u q^{u+1} = \frac{rq}{1-qs}\left[1-(qs)^{t+1}\right]$$

$$Q_{11}(t) = \sum_{u=0}^{t} pq^u rs^u = \frac{pr}{1-qs}\left[1-(qs)^{t+1}\right]$$

$$Q_{12}(t) = \sum_{u=0}^{t} pq^u s^{u+1} = \frac{ps}{1-qs}\left[1-(qs)^{t+1}\right]$$

$$Q_{21}(t) = \sum_{u=0}^{t} rs^u = \left(1-s^{t+1}\right)$$

The steady-state transition probability from state $S_i$ to $S_j$ is given by

$$p_{ij} = \lim_{t\to\infty} Q_{ij}(t)$$

So that

$$p_{01} = \frac{pq'}{1-qq'} + \frac{p'q}{1-qq'}, \quad p_{02} = \frac{pp'}{1-qq'}, \quad p_{10} = \frac{rq}{1-qs},$$
$$p_{11} = \frac{pr}{1-qs}, \quad p_{12} = \frac{ps}{1-qs}, \quad p_{21} = 1 \qquad (14.8.1.1\text{–}14.8.1.6)$$

Obviously,

$$p_{01} + p_{02} = 1, \quad p_{10} + p_{11} + p_{12} = 1 \qquad (14.8.1.7, 14.8.1.8)$$

### 14.8.2 Mean Sojourn Time in the Various States

Let $T_i$ be the sojourn time in the state $S_i$; $i = 0, 1, 2$, then the mean sojourn time in the state $S_i$ is

$$\psi_i = \sum_{t=1}^{\infty} P(T_i > t - 1) = \sum_{t=1}^{\infty} P(T_i \geq t]$$

So that

$$\psi_0 = \sum_{t=1}^{\infty} q^t \left(q'\right)^t = \frac{q\,q'}{1 - q\,q'}$$

Similarly,

$$\psi_1 = \sum_{t=1}^{\infty} q^t s^t = \frac{qs}{1 - qs}$$

$$\psi_2 = \sum_{t=1}^{\infty} s^t = \frac{s}{r} \qquad (14.8.2.1\text{–}14.8.2.3)$$

### 14.8.3 Analysis of Reliability and MTSF

We define $R_i(t)$ it as the probability that the system does not fail during the first '$t$' cycles, i.e., up to the epoch $(t - 1)$ when the system initially starts from upstate $S_i$ $(i = 0, 1)$. To determine it, we regard the failed state $S_2$ as absorbing. In particular,

$$\begin{aligned} R_0(t) = {} & P[\text{System sojourns in } S_0 \text{ up to epoch } (t-1)] \\ & + P[\text{System transits from } S_0 \text{ to } S_1 \text{ at epoch } u \leq t-1 \\ & \text{and then starting from } S_1 \text{ system remains up during} \\ & \text{the remaining epochs } (t-1-u)] \\ = {} & q^t \left(q'\right)^t + \sum_{u=0}^{t-1} q_{01}(u) R_1(t-1-u) \\ = {} & Z_0(t) + q_{01}(t-1) \copyright R_1(t-1) \end{aligned} \qquad (14.8.3.1)$$

Similarly,

$$R_1(t) = Z_1(t) + q_{10}(t-1) \copyright R_0(t-1) + q_{11}(t-1) \copyright R_1(t-1) \qquad (14.8.3.2)$$

where $Z_1(t) = (qs)^t$.

Taking geometric transforms defined as $R^*(h) = \sum_{t=0}^{\infty} h^t R(t)$, of relations (14.8.3.1) and (14.8.3.2), we get

$$R_0^*(h) = Z_0^*(h) + hq_{01}^*(h)R_1^*(h)$$
$$R_1^*(h) = Z_1^*(h) + hq_{10}^*(h)R_0^*(h) + hq_{11}^*(h)R_1^*(h)$$

Solving the above equations for $R_0^*(h)$, we get

$$R_0^*(h) = \frac{\left[1 - hq_{11}^*(h)\right]Z_0^*(h) + hq_{01}^*(h)Z_1^*(h)}{1 - hq_{11}^*(h) - h^2 q_{01}^*(h)q_{10}^*(h)} \tag{14.8.3.3}$$

The value $R_0(t)$ is obtained by collecting the coefficient of $h^t$ in (14.8.3.3).
The MTSF is

$$E(T) = \sum_{t=1}^{\infty} R_0(t) = \lim_{h\to 1} \sum_{t=1}^{\infty} h^t R_0(t)$$
$$= \lim_{h\to 1} R_0^*(h) - 1$$

Observing that

$$\lim_{h\to 1} Z_i^*(h) = \sum_{t=0}^{\infty} Z_i(t) = Z_i(0) + \sum_{t=1}^{\infty} Z_i(t) = (1 + \psi_i)$$

and

$$\lim_{h\to 1} q_{ij}^*(h) = p_{ij}$$

we get

$$E(T) = \frac{(1 - p_{11})(1 + \psi_0) + p_{01}(1 + \psi_1)}{1 - p_{11} - p_{01}p_{10}} \tag{14.8.3.4}$$

### *14.8.4 Availability Analysis*

We define

$$A_i(t) = P[\text{System is up (operative) during the } t\text{th cycle } (t-1, t)$$
$$\text{when system initially starts from } S_i; \quad i = 0, 1, 2]$$

In particular,

$$A_0(t) = q^t (q')^t + \sum_{u=0}^{t-1} q_{01}(u) A_1(t-1-u) + \sum_{u=0}^{t-1} q_{02}(u) A_2(t-1-u)$$
$$= Z_0(t) + q_{01}(t-1) \copyright A_1(t-1) + q_{02}(t-1) \copyright A_2(t-1)$$

Similarly,

$$A_1(t) = Z_1(t) + q_{10}(t-1) \copyright A_0(t-1) + q_{11}(t-1) \copyright A_1(t-1) + q_{12}(t-1) \copyright A_2(t-1)$$
$$A_2(t) = q_{21}(t-1) \copyright A_1(t-1) \quad (14.8.4.1\text{–}14.8.4.3)$$

Taking geometric transforms of the above relations and solving the resulting set of algebraic equations $A_0^*(h)$, we get

$$A_0^*(h) = \frac{N_1(h)}{D(h)} \quad (14.8.4.4)$$

where

$$N_1(h) = \left[1 - hq_{11}^*(h) - h^2 q_{12}^*(h) q_{21}^*(h)\right] Z_0^*(h) + \left[hq_{01}^*(h) + h^2 q_{02}^*(h) q_{21}^*(h)\right] Z_1^*(h)$$

and

$$D(h) = \left[1 - hq_{11}^*(h) - h^2 q_{12}^*(h) q_{21}^*(h)\right] - hq_{10}^*(h)\left[hq_{01}^*(h) + h^2 q_{02}^*(h) q_{21}^*(h)\right]$$

The steady-state availability of the system is given by

$$A_0 = \lim_{t \to \infty} A_0(t) = \lim_{h \to 1} (1-h) A_0^*(h)$$

Observing that $q_{ij}^*(1) = p_{ij}$ and $D(1) = 0$, we get

$$A_0 = \frac{p_{10}(1+\psi_0) + (1+\psi_1)}{p_{10}(1+\psi_0) + (1+\psi_1) + (p_{12} + p_{10} p_{02})(1+\psi_2)} \quad (14.8.4.5)$$

The expected uptime of the system up to the cycles $(t-1, t)$ is

$$\mu_{\text{up}}(t) = \sum_{u=0}^{t} A_0(u) \quad \text{so that} \quad \mu_{\text{up}}^*(h) = \frac{A_0^*(h)}{(1-h)} \quad (14.8.4.6)$$

### 14.8.5 Busy Period Analysis

Let $B_i(t)$ be the probability that the repairman is busy repairing a failed unit during the $t$th cycle $(t-1, t)$ when initially system starts from $S_i$; $i = 0, 1, 2$. Using the same probabilistic arguments as in the case of availability analysis, we have the following recurrence relations for $B_i(t)$

$$\begin{aligned}
B_0(t) &= q_{01}(t-1) © B_1(t-1) + q_{02}(t-1) © B_2(t-1) \\
B_1(t) &= Z_1(t) + q_{10}(t-1) © B_0(t-1) + q_{11}(t-1) © B_1(t-1) \\
&\quad + q_{12}(t-1) © B_2(t-1) \\
B_2(t) &= Z_2(t) + q_{21}(t-1) © B_1(t-1)
\end{aligned} \quad (14.8.5.1\text{–}14.8.5.3)$$

where $Z_2(t) = s^t$.

Taking geometric transforms of the above relations and solving the resulting set of equations $B_0^*(h)$, we get

$$B_0^*(h) = \frac{N_2(h)}{D(h)} \quad (14.8.5.4)$$

where

$$\begin{aligned}
N_2(h) &= \left[hq_{01}^*(h) + h^2 q_{02}^*(h) q_{21}^*(h)\right] Z_1^*(h) \\
&\quad + \left[h^2 q_{01}^*(h) q_{12}^*(h) + hq_{02}^*(h)\left\{1 - hq_{11}^*(h)\right\}\right] Z_2^*(h)
\end{aligned}$$

In the long run, the probability that the repairman will be busy is given by

$$\begin{aligned}
B_0 &= \lim_{t\to\infty} B_0(t) = \lim_{h\to 1} (1-h) B_0^*(h) \\
&= \frac{(1+\psi_1) + (p_{12} + p_{10} p_{02})(1+\psi_2)}{p_{10}(1+\psi_0) + (1+\psi_1) + (p_{12} + p_{10} p_{02})(1+\psi_2)}
\end{aligned} \quad (14.8.5.5)$$

The expected busy period of the repairman up to the $t$th cycle $(t-1, t)$ is

$$\mu_b(t) = \sum_{u=0}^{t} B_0(u) \quad \text{so that} \quad \mu_b^*(h) = \frac{B_0^*(h)}{1-h} \quad (14.8.5.6)$$

### 14.8.6 Profit Function Analysis

Let $C_0$ be the revenue per cycle by the system when it is operative and $C_1$ be the repair cost per cycle when the system is under repair. Then, the net-expected profit earned by the system when it is observed in '$t$' cycles is given by

$$\begin{aligned} P(t) &= \text{Total expected revenue up to } t\text{th cycle} - \text{Total cost of repair up to } t\text{th cycle} \\ &= C_0\mu_{\text{up}}(t) - C_1\mu_b(t) \end{aligned} \tag{14.8.6.1}$$

The expected profit per epoch in steady state is

$$P = \lim_{t\to\infty} \frac{P(t)}{t} = C_0 A_0 - C_1 B_0 \tag{14.8.6.2}$$

The results can be obtained for two identical unit cold standby and parallel system model, respectively, when $p' = 0$ and $p' = p$.

### 14.8.7 Graphical Conclusions

For a more detailed view of the behavior of system characteristics with respect to the various parameter involved, we plot curves for **MTSF** and **profit function** in Figs. 14.8 and 14.9 with respect to the failure parameter '**p**' for three different values of repair parameter '**r**' and two different values of failure parameter '**p′**'. From the curves of Fig. 14.8, we observe that **MTSF** increases uniformly as the values of '**r**' increase and '**p′**' decrease and decrease with the increase in '**p**'. Further, we also observe from the smooth curves of Fig. 14.8 that the values of '**p**' must be less than **0.167, 0.179** and **0.190** corresponding to **r = 0.20, 0.25** and **0.30** to achieve at least **15** units of **MTSF** when **p′ = 0.1**. From the dotted curves, we observe that the values of '**p**' must be less than **0.21, 0.229** and **0.249** corresponding to **r = 0.20, 0.25** and **0.30** to achieve at least **15** units of **MTSF** when **p′ = 0**.

Similarly, Fig. 14.9 reveals the variations in **profit** with respect to '**p**' for varying values of '**r**' and '**p′**', when the values of other parameters are kept fixed as $\boldsymbol{C_0 = 200}$ and $\boldsymbol{C_1 = 100}$. From this figure, it is clearly observed from the smooth curves that the system is profitable if the value of parameter '**p**' is less than **0.178, 0.278** and **0.380,** respectively, for **r = 0.10, 0.15** and **0.20** when **p′ = 0.1**. From dotted curves, we conclude that system is profitable if the value of parameter '**p**' is less than **0.2, 0.3** and **0.4,** respectively, for **r = 0.10, 0.15** and **0.20** for fixed value of **p′ = 0.**

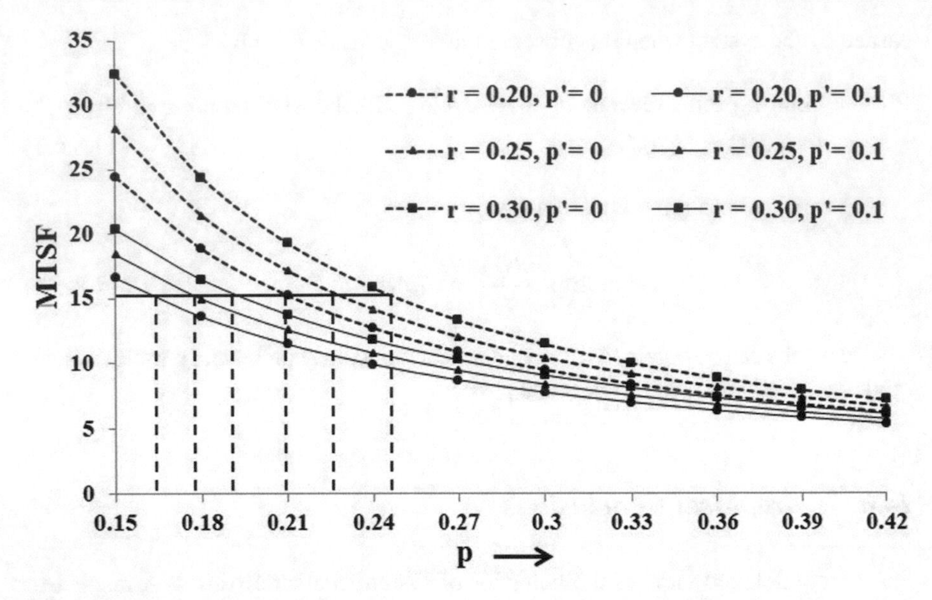


**Fig. 14.8** Behavior of MTSF with respect to '$p$', '$r$' and '$p'$'

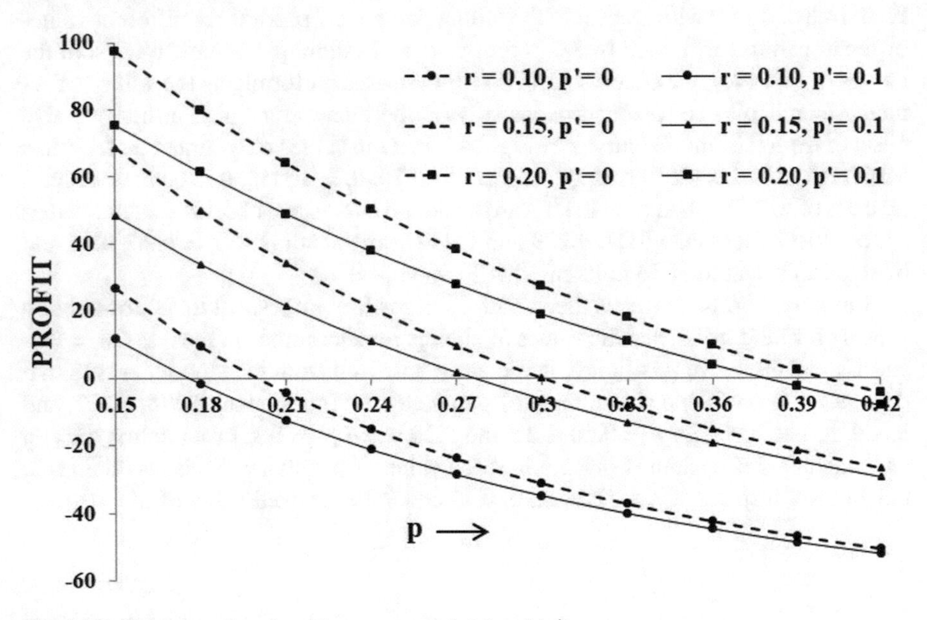


**Fig. 14.9** Behavior of profit with respect to '$p$', '$r$' and '$p'$'

## References

1. Arora, J.R.: Reliability of a two unit priority standby redundant system with finite repair capability. IEEE Trans. Reliab. **25**(3), 205–207 (1976)
2. Goel, L.R., Gupta, R., Singh, S.K.: Cost analysis of a two-unit priority standby system with imperfect switch device and arbitrary distributions. Microelectron. Reliab. **25**, 65–69 (1985)
3. Gupta, R., Chaudhary, A.: A two unit system subject to partial failure mode and Gamma repair time distribution. Microelectron. Reliab. **33**(15), 2277–2280 (1993)
4. Gupta, R., Goel, L.R.: Profit analysis of a two-unit priority standby system with administrative delay in repair. Int. J. Syst. Sci. **29**(9), 1703–1712 (1989)
5. Gupta, R., Bansal, S.: Profit analysis of a two unit priority standby system subject to degradation. Int. J. Syst. Sci. **22**(1), 61–72 (1991)
6. Gupta, R., Chaudhary, A.: Profit analysis of a two unit priority standby system subject to degradation and random shocks. Microelectron. Reliab. **33**(8), 1073–1079 (1993)
7. Nakagawa, T., Osaki, S.: Stochastic behavior of a two unit priority standby redundant system with repair. Microelectron. Reliab. **14**, 309–313 (1975)
8. Sinha, S.M., Kumar, A., Agarwal, M.: Reliability analysis of a two unit priority standby system with different operative states. J. Math. Sci. 65–79 (1978)
9. Subramanium, R., Chandran, N.R.: Studies of two unit priority standby system. IEEE Trans. Reliab. **30**, 388 (1981)
10. Goel, L.R., Shrivastava, P.: A two unit cold standby system with three modes and correlated failure and repair. Microelectron. Reliab. **31**(5), 835–840 (1991)
11. Goel, L.R., Kumar, A., Rastogi, A.K.: Analysis of a two-unit redundant system under partial failure and two types of repairs. Microelectron. Reliab. **24**(5), 873–876 (1984)
12. Gupta, S.M., Jaiswal, N.K., Goel, L.R.: Analysis of a two-unit standby redundant system under partial failure and preemptive repair priority. Int. J. Syst. Sci. **13**, 675 (1982)
13. Proctor, C.L., Singh, B.: A three state system Markov model. Microelectron. Reliab. **14**, 403 (1975)
14. Tuteja, R.K., Malik, S.C.: Reliability and profit analysis of a two single unit models with three modes and different repair policies of repairman who appears and disappears randomly. Microelectron. Reliab. **32**(3), 351–356 (1992)
15. Gopalan, M.N.: Availability and reliability of one server two unit system with imperfect switchover. IEEE Trans. Reliab. **24**, 218 (1975)
16. Gopalan, M.N., Marathe, K.Y.: Availability of one server 2 dissimilar unit system with slow switch. IEEE Trans. Reliab. **27**(3), 230–231 (1978)
17. Gopalan, M.N., Marathe, K.Y.: Availability analysis of a one server n-unit system with slow switch. IEEE Trans. Reliab. **27**(3), 231–232 (1978)
18. Nakagawa, T., Osaki, S.: Stochastic behavior of a two unit standby redundant system with imperfect switch over. IEEE Trans. Reliab. **24**, 143–146 (1975)
19. Sinha, S.M., Kapil, D.V.S.: Two unit redundant system with slow-switch. IEEE Trans. Reliab. **28**, 418 (1979)
20. Jaiswal, N.K., Krishna, J.V.: Analysis of two dissimilar units standby redundant system with administrative delay in repair. Int. J. Syst. Sci. **11**, 495–511 (1989)
21. Pandey, D.K., Gupta, S.M.: Analysis of a two unit standby redundant system with administrative delay in repair. Microelectron. Reliab. **25**(5), 917–920 (1985)
22. Goel, L.R., Gupta, R., Tyagi, P.K.: Analysis of a two-unit standby system with preparation time and correlated failures and repairs. Microelectron. Reliab. **35**(8), 1163–1165 (1995)
23. Goel, L.R., Gupta, R., Tyagi, P.K.: Analysis of a two unit standby system with preparation time and correlated failures and repairs. Microelectron. Reliab. **35**(5), 585 (1986)
24. Singh, S.K., Srinivasu, B.: Stochastic analysis of a two unit cold standby system with preparation time for repair. Microelectron. Reliab. **27**, 50–60 (1987)
25. Dhillon, B.S., Vishvanath, H.C.: Reliability analysis of a non-identical unit parallel system with common cause failure. Microelectron. Reliab. **31**, 429–441 (1991)

26. Goel, L.R., Gupta, R., Singh, S.K.: A two (multi-component) unit parallel system with standby and common cause failure. Microelectron. Reliab. **24**(3), 415–418 (1984)
27. Pandey, D., Jacob, M., Tyagi, S.K.: Stochastic modeling of a power loom plant with common cause failure, human error and overloading effect. Int. J. Syst. Sci. **27**(3), 309–313 (1996)
28. Gupta, R.: Probabilistic analysis of a two-unit cold standby system with two-phase repair and preventive maintenance. Microelectron. Reliab. **26**(1), 13–18 (1986)
29. Murari, K., Al-Ali, A.: One unit reliability system subject to random shocks and preventive maintenance. Microelectron. Reliab. **28**(3), 373–377 (1988)
30. Nakagawa, T., Osaki, S.: Stochastic behavior of two unit parallel redundant system with preventive maintenance. Microelectron. Reliab. **14**, 457–461 (1975)
31. Gupta, R., Chaudhary, A.: Analysis of stochastic models in manufacturing systems pertaining to repair machine failure. In: Optimization Methods for Manufacturing, pp. 7.1–7.50. CRC Press, Washington (2001)
32. Gupta, R., Chaudhary, A.: Cost-benefit analysis of a two unit standby system with provision of repair machine failure. Microelectron. Reliab. **34**(8), 1891–1394 (1994)
33. Mogha, A.K., Gupta, R., Gupta, A.K.: A two-unit parallel system with correlated lifetimes and repair machine failure. IAPQR **28**(1), 1–22 (2003)
34. Gupta, R., Bansal, S., Goel, L.R.: Cost-benefit analysis of a two-unit cold standby system with the provision of rest to a unit. Int. J. Syst. Sci. **21**(8), 1451–1462 (1990)
35. Gupta, R., Bansal, S., Goel, L.R.: Profit analysis of a two-unit priority standby system with rest period of the operator. Microelectron. Reliab. **30**(4), 649–654 (1990)
36. Gupta, R., Chaudhary, A.: A two unit priority standby system subject to random shocks and Raleigh failure time distribution. Microelectron. Reliab. **32**(12), 1713–1723 (1992)
37. Gupta, R., Tyagi, P.K., Kishan, R.: A two-unit system with correlated failures and repairs and random appearance and disappearance of repairman. Int. J. Syst. Sci. **27**(6), 561–566 (1996)
38. Singh, S.K.: Profit evaluation of a two unit cold standby system with random appearance and disappearance time of the service facility. Microelectron. Reliab. **29**(1), 21–24 (1989)
39. Dhillon, B.S.: Availability analysis of systems with two types of repair facilities. Microelectron. Reliab. **20**, 679–686 (1980)
40. Gupta, R., Goel, L.R., Agnihotri, R.K.: Analysis of a three unit redundant system with two types of repair and inspection. Microelectron. Reliab. **29**(5), 769–773 (1989)
41. Goel, L.R., Kumar, A., Rastogi, A.K.: Stochastic behavior of a man-machine system operating under different weather conditions. Microelectron. Reliab. **25**, 87–91 (1985)
42. Mokaddis, G.S., Taofek, M.L., Elashia, S.A.M.: Busy period analysis of a man-machine system operating subject to different physical conditions. J. Math. Stats. **1**(2), 160–164 (2005)
43. Dhillon, B.S., Natesan, J.: Stochastic analysis of outdoor power systems in fluctuating environment. Microelectron. Reliab. **23**, 867–881 (1983)
44. Gupta, R., Goel, R.: Profit analysis of a two-unit cold standby system with abnormal weather conditions. Microelectron. Reliab. **3**(1), 1–5 (1991)
45. Goel, L.R., Gupta, R.: A multi-standby multi failure mode system with repair and replacement policy. Microelectron. Reliab. **23**(5), 805–808 (1983)
46. Gupta, R., Sharma, P., Sharma, V.: Cost-benefit analysis of a two unit duplicate parallel system with repair/replacement and correlated life time of units. J. Rajasthan Acad. Phy. Sci. **9**(4), 317–330 (2010)
47. Goel, L.R., Gupta, R., Singh, S.K.: Cost-benefit analysis of a two-unit warm standby system with inspection, repair and post-repair. IEEE Trans. Reliab. **35**(1), 70 (1986)
48. Gupta, R., Tyagi, V., Tyagi, P.K.: Cost-benefit analysis of a two-unit standby system with post repair activation time and correlated failures and repairs. J. Qual. Maintenance Eng. **3**(1), 55–63 (1997)
49. Gupta, R., Mogha, A.K., Gupta, A.K.: A two unit active redundant system with two phase repair and correlated life times. Aligarh J. Stat. **24**, 63–79 (2004)
50. Gupta, R., Kishan, R., Kumar, P.: A two-non-identical unit parallel system with correlated lifetimes. Int. J. Syst. Sci. **30**(10), 1123–1129 (1999)

51. Gupta, R., Sharma, V.: A two non-identical unit standby system with correlated working and rest time of repairman. J. Comb. Info. Syst. Sci. **32**, 241–255 (2007)
52. Sharma, S.C., Gupta, R., Saxena, V.: A two unit parallel system with dependent failure rates and correlated working and rest time of repairman. Int. J. Agric. Stat. Sci. **8**(1), 145–155 (2012)
53. Gupta, R., Varshney, G.: A two identical unit parallel system with geometric failure and repair time distributions. J. Comb. Info. Sci. **32**(1–4), 127–136 (2007)
54. Gupta, R., Varshney, G.: A two non-identical unit parallel system with geometric failure and repair time distributions. IAPQR Trans. **31**(2), 127–139 (2006)
55. Balaguruswamy, E.: Reliability Engineering. Tata McGrraw Hill Publishing Company Ltd., New Delhi (1984)
56. Gnedenko, B.V., Belayer, Yu.K., Soloyar: Mathematical Methods of Reliability Theory. Academic Press, New York (1969)
57. Gupta, R., Mogha, A.K.: Stochastic analysis of series parallel and standby system models with geometric lifetime distributions. J. Ravi-Shankar Univ. **13**(1), 68–80 (2000)

**Dr. Rakesh Gupta** has been a Professor in the Department of Statistics C.C.S. University Meerut. He passed M.Sc., M.Phil. and Ph.D. from Ch. Charan Singh University, Meerut, in 1980, 1982 and 1984, respectively, got first position in the University both at M.Sc. and M.Phil. levels. He has been the recipient of J.R.F, S.R.F of CSIR at Meerut University, Meerut and R.A. of CSIR for Post-doctoral Research in the department of O.R., University of Delhi, Delhi. To date, 01-11-2022, he published more than 15 dozen research papers in the field of Reliability Modelling in various international and national journals of repute. He has supervised 70 M.Phil. and 29 Ph.D. students completing their degrees. He has contributed five chapters in different books published by International Publishers. Prof. Rakesh Gupta has presented research papers and delivered invited talks at various national and international conferences. He has reviewed many research papers in various national and international journals and evaluated more than 50 Ph.D. thesis from different Universities. The name of Prof. Gupta has been selected for the award of Shikshak Sri Samman-2020 by U.P. State Government, Lucknow.

**Shubham Gupta** graduated Engineer (M. Tech.) in Very Large Scale Integration (VLSI Design) from National Institute of Technology, Surathkal India and Under-graduation (B.Tech.) in Electronics and Communication from Indraprastha University, Delhi. After completing Masters, qualified NET JRF (UGC) in Electronic Science in the year 2019. Nowadays, working on "fabrication and integration of Silicon Nanostructure as smart antibacterial implantation material for health care applications" as a scholar pursuing PhD from Manipal University Jaipur, Rajasthan. Completed one-year internship in INTEL during my Masters's on Physical Designing on 10 nm (Nanometer) Double Data Rate Random Access Memory and Major Project during Under-graduation on Floating Voltage Controlled Inductor using Current Feedback Operational Amplifiers and its Implementation on Filter & Oscillator Designing.

**Dr. Irfan Ali** received B.Sc., M.Sc., M.Phil., and Ph.D. degrees from Aligarh Muslim University. He is currently working as an Assistant Professor with the Department of Statistics and Operations Research at Aligarh Muslim University. He received the Post Graduate Merit Scholarship Award during his M.Sc. (statistics) and the UGC-BSR Scholarship awarded during his Ph.D. (statistics) programs. His research interests include applied statistics, mathematical programming, Fuzzy optimization and multi-objective optimization. He has supervised M.Sc., projects, M.Phil., dissertations and Ph.D. thesis under his supervision in operations research and applied statistics topics. He has completed a research project UGC–Start-Up Grant Project, UGC, New Delhi, India. He has published more than 100+ research articles in reputed journals and serves as a Reviewer for several journals. He has published several edited books for Springer and Taylor's Francis, CRC press. He is a Lifetime Member of various professional societies: Operational Research Society of India, the Indian Society for Probability and Statistics, the Indian Mathematical Society, and The Indian Science Congress Association. He delivered invited talks at several universities and Institutions. He also serves as an Associate Editor for some journals.

# Chapter 15
# Distributed System Reliability Analysis with Two Coverage Factors: A Copula Approach

**Ibrahim Yusuf and Ismail Tukur**

**Abstract** Traditional performance and reliability analysis for distributed systems frequently overestimates the fundamental dependability of their components. To address this quandary, a copula approach for analyzing the performance of distributed system situated to two separate regions employing fault coverage factor in each region is proposed at first. Copula technique is a powerful tool for describing variable dependence and has received much attention in a variety of fields of study. Regions I and II consist of two client and one directory server each. In this chapter, we also introduce coverage factors for each region, such that the failure of each client and directory server for each region is detected by the coverage factor. Each area designed for this system has a constant failure rate. This is prone to two sorts of failure: lower failure and higher failure. When a lower failure occurs, the system is repaired using the general repaired rate; however, a higher failure can be rectified using the copula family of Gumbel-Hougaard distribution. The goal is to obtain explicitly expressed expressions for the dependability, availability, sensitivity, and cost function. As a result of this, in this chapter, we aim at examining the reliability strength of a distributed system in relation to availability, expected profit and reliability. For the sake of generality, we use the Laplace transforms and technique of supplementary variable to establish the partial differential equations related to transition diagram essential to this chapter. The numerical validation of explicit expressions for system availability, reliability, and profit function is performed. Computations for reliability characteristics are evaluated as a specific example by analyzing availability, reliability, and cost in order to reflect the impacts of coverage factor and both failure and repair rates on reliability characteristics. Tables and graphs highlight the computation of the result base on assume numerical values.

**Keywords** Availability · Coverage factor · Reliability · Sensitivity and expected profit

---

I. Yusuf (✉)
Department of Mathematical Sciences, Bayero University, Kano, Nigeria
e-mail: iyusuf.mth@buk.edu.ng

I. Tukur
Department of Art and Humanities, Kano State Polytechnic, Kano, Nigeria



H. Garg (ed.), *Advances in Reliability, Failure and Risk Analysis*, Industrial and Applied Mathematics, https://doi.org/10.1007/978-981-19-9909-3_15

## 15.1 Introduction

A network is made up of two or more systems of computers that communicate assets and communicate in some way. This discussion occurs over a common channel of communication. A network, on the other hand, is an accumulation of machines that have been mechanically and electronically linked in order to facilitate and share data. Each computer network has applications that run on numerous machines linked via a network that has become exceedingly sophisticated and difficult to rely on. Failures in computer networking systems can be classified as hardware, software, or both. Various techniques for improving distributed computer networking system performance have been proposed. Backup is one of the approaches used to improve and forecast system strength and effectiveness, which results in increased system safety, quality, production, and income mobilization.

**Definition 1.1**

**Reliability function**

The chance that a system/machine will be up and running throughout a period of time $t$ is defined as reliability. Thus, reliability $R(t) = P_r\{T > t\}$, where $T$ is the time when the system is down and not running with $R(t) \geq 0$, $R(t) = 1$ (For a full description, see Ebeling [10]). Thus,

$$R(t) = \int_{t}^{\infty} f(t_0)\mathrm{d}t_0 \tag{15.1}$$

and

$$R(t) = \mathrm{e}^{-\lambda t} \tag{15.2}$$

for exponentially distributed rate of failure.

**Definition 1.2**

**Availability function**

Availability is the likelihood that a system or piece of equipment will be operational at any time $t$. A system's availability is a merged proportion of its maintainability and dependability.

$$A(\infty) = \lim_{T\to 0} A(t) = \lim_{T\to 0} \frac{1}{T}\int_{0}^{T} A(t)\mathrm{d}t \tag{15.3}$$

## 15.2 Literature Review

Numerous researchers have looked into various aspects of the durability of sophisticated repairable systems, commencing with [28], who investigated the reliability of a production system using a combination of copula and coverage approaches and discovered that the blended copula coverage procedure enhances system reliability. Tyagi et al. [36] used copula, coverage, and copula plus coverage statistical approaches to study the probabilistic performance of a parallel system subjected to human, unit, and major failure. Chopra and Ram [8] presented dependability measurements for two distinct units in parallel using the Gumbel-Hougaard copula distribution. Jain and Meena [20] designed fault-tolerant system performance models that include actual aspects like server vacation, reboot, and incomplete coverage. Pourhassan et al. [26] examined the operation of a sugar factory subjected to stochastic and random shocks. Pourhassan et al. [25] created models to assess multistate system dependability in the presence of nonfatal and fatal shocks in a capacitor bank. Singh et al. [31] examined the copula repair strategy on the probabilistic evaluation of a CBT system that included clients, a load balancer, a database server, and a centralized database server. Niwas and Garg [24] examined the profitability and dependability of a cost-free warranty policy-based industrial system. Yusuf et al. [38] created models for a serial system that disclose the ideal level of profit that the system can achieve as well as the impact of the most crucial subsystem that leads to minimum profit and system failure. Kabiru et al. [16] evaluated the impact of copula repair on the availability and cost evaluation of a multi-server tree topology network. Ismail et al. [17] evaluated the impact of copula and general repair policies on dual-server computer network performance. Jain and Gupta [19] created performance models of machining systems including corrective maintenance failures employing various vacations and imperfect coverage. Jain et al. [21] created models for evaluating the effectiveness of a machining system composed of main units and warm standby units under the supervision of an unpredictable single server.

Several researchers used different approach to investigate the performance of distributed systems and have proclaimed improved performance in terms of their operations. We can cite few of them; Teslyuk et al. [35] developed models for reliability modeling of metrics of testing the performance of local area networks. Rotar et al. [29] presented mathematical approach to reliability determination of solar tracking system using fault coverage aware metrics. Bisht et al. [7] developed an algorithm for computing reliability metrics, component measures, and critical measures for communication network. Song [34] dealt with reliability optimization of communication network using genetic algorithm. Ram and Goyal [28] dealt with reliability analysis of fault-tolerant system exhibiting two types of repair. He et al. [14] discussed reliability optimization of computer communication network via genetic algorithm. Arora et al. [3] developed models for determining the reliability metrics of parallel system with fault coverage. Abdulwahab et al. [1] presented Markov availability improvement of distributed hardware and systems. Ahmed and Ramalashmi [2] investigated the performance distributed and centralize controllers through weighted round robin, random and round robin algorithm. Handoko et al. [13] analyzed the availability of load balancer, database cluster and virtual router protocol. Yusuf et al. [37] analyzed the measures of reliability of server-client system. Dhulavvagol et al. [9] analyzed the performance enhancement of distributed processing through partitioning and efficient shard selection technique. Muñoz-Esco and Juan-Marn [23] analyzed the synchrony level in dynamic distributed systems. Raghav et al. [27] presented prediction of reliability of distributed homogeneous software system. Bisht and Singh [4] analyzed some reliability metrics of complex network using universal generating function. Huang et al. [15] developed models for reliability analysis of distributed network. Bisht and Singh [4] used Markov processes to analyze the reliability measures of transmission network system enhancement. Bisht and Singh [6] analyzed the profit and reliability of transmission network using artificial neural network and Markov process.

The aforementioned researchers provided excellent work on reliability analysis of complicated repairable systems using a copula technique, claiming that their operations improved the performance of the distributed systems. Keeping the above facts in view, this chapter dealt with reliability and performance analysis of distributed network systems stationed in two different regions. Each region has two clients and a directory server with replication of another in a separate region. To the author's knowledge, no maintainability modeling and effectiveness study has focused on estimating the reliability, strength, effectiveness, and performance using Gumbel-Hougaard family copula based on distributed systems stationed in two separate regions with coverage factors. As a result, the current study was conducted to address this research gap. In this chapter, a novel technique, namely the copula repair technique, was used to analyze the performance of the distributed systems with coverage factors. According to the literature review, no research has been carried out on the distributed system reliability analysis with two coverage factors copula approach. Motivated by this fact, we are interested in the reliability analysis of two coverage factors distributed system operators in this present work. The impact of two coverage factors, in conjunction with the copula, on the distributed system reliability analysis has been captured. The objective of this work is to find out how two coverage factors will improve the reliability measure of the system under consideration.

The chapter is arranged with Sect. 15.1 serving as an introduction. Section 15.2 Literature review. Nomenclatures, assumptions, and model description are covered in Sect. 15.3. Section 15.4 is concerned with model formulation and solution, while Sect. 15.4 provides numerical simulations in specific cases. Section 15.5 highlights the finding of the results, while the chapter is concluded in Sect. 15.6 followed by references.

## 15.3 Nomenclatures and Model Description

### 15.3.1 Nomenclatures

$q$: Scale of time
$s$: Variables' Laplace transform
$S_i$: Transitional states, $0 \leq i \leq 9$
$\delta_1$: Rate of failure of clients in region I
$\delta_2$: Rate of failure of clients in region II
$\delta_3$: Rate of failure of directory server in region I
$\delta_4$: Rate of failure of directory server in region II
$\delta_5$: Rate of failure of replica server
$C1, C2$: Coverage factor of region I and coverage factor of region II
$\beta(y_1)$: Rate of repair for the higher failed state
$h(y_1)$: Rate of repair for the degraded state
$M_i(t)$: Chance of the system sojourning in $S_i$ state at instants for $0 \leq i \leq 9$
$\overline{M}_i(s)$: Laplace transformation of state transition probability $P(t)$
$M_i(y_1, q)$: Chance of the system sojourning in $S_i$ with $y_1$ variable of repair and variable time $q$
$\beta(y_1) = \exp\left[y_1^\theta + (\log h(y_1))^\theta\right]^{\frac{1}{\theta}}$, where $\beta(y_1)$ is the joint chance of repair from higher failed state to perfect state by copula

$$\overline{S}(s) = \int_0^\infty \beta(y_1) e^{-sy_1 - \int_0^\infty h(y_1) dy_1} dy_1$$

$E_p(q) E_p(t)$: Expected profit during the time interval [0, $q$)
$\upsilon_1, \upsilon_2$: Revenue and service cost per unit time, respectively.

### 15.3.2 Model Description

In this current study, we analyze a distributed system composed of two region and shared one replica server with the following specifications: Regions I and II consist of two client and one directory server each. We have also introduced coverage factors for each region in this work (c1 for region I and c2 for region II), so that the failure of each clients and directory server for each region will be detected by coverage factor (Table 15.1).

**Model Formulation and Solutions**

As in Gahlot et al. [11, 12], Lado et al. [22], Singh and Ayagi, [32, 33], the following equations are obtained via Fig. 15.1 (Fig. 15.2):

**Table 15.1** Description of states

| States | Description |
|---|---|
| $S_0$ | In initial state, the subsystems and the system are in perfect state, the system is up and running |
| $S_1$ | In this state, in region I, the first client fails, the second client is working, the system is up and running |
| $S_2$ | In this state, in region II, the first client fails, the second client is working, the system is up and running |
| $S_3$ | In region II, one client fails, the other client is working, the system is up and running |
| $S_4$ | In region I, one client fails, the other client is working, the system is up and running |
| $S_5$ | The system is down due to failure of clients in region I |
| $S_6$ | The system is down due to failure of clients in region II |
| $S_7$ | The system is down due to replica server failure |
| $S_8$ | The system is down due to directory server failure in region I |
| $S_9$ | The system is down due to directory server failure in region II |

$$\left(\frac{\partial}{\partial t} + 4c_1\delta_1 + 2c_2\delta_2 + \delta_3 + c_1\delta_4 + c_2\delta_5\right)M_0(q)$$
$$= \int_0^\infty h(y_1)M_1(y_1, q)\mathrm{d}x$$
$$+ \int_0^\infty h(y_2)M_2(y_2, q)\mathrm{d}y_2 + \int_0^\infty \beta_0(y_1)M_5(y_1, q)\mathrm{d}y_1$$
$$+ \int_0^\infty \beta_0(y_2)M_6(y_2, q)\mathrm{d}y_2 + \int_0^\infty \beta_0(x_0)M_7(x_0, q)\mathrm{d}x_0$$
$$+ \int_0^\infty \beta_0(x_1)M_1(x_1, q)\mathrm{d}x_1 + \int_0^\infty \beta_0(x_2)M_8(x_2, q)\mathrm{d}x_2 \tag{15.4}$$

$$\left(\frac{\partial}{\partial q} + \frac{\partial}{\partial y_1} + c_1\delta_1 + 2c_2\delta_2 + h(y_1)\right)M_1(y_1, q) = 0 \tag{15.5}$$

$$\left(\frac{\partial}{\partial q} + \frac{\partial}{\partial y_2} + c_2\delta_2 + 2c_1\delta_1 + h(y_2)\right)M_2(y_2, q) = 0 \tag{15.6}$$

$$\left(\frac{\partial}{\partial q} + \frac{\partial}{\partial y_2} + c_2\delta_2 + h(y_2)\right)M_3(y_2, q) = 0 \tag{15.7}$$

$$\left(\frac{\partial}{\partial q} + \frac{\partial}{\partial y_1} + c_1\delta + h(y_1)\right)M_4(y_1, q) = 0 \tag{15.8}$$

$$\left(\frac{\partial}{\partial q}+\frac{\partial}{\partial y_1}+\beta_0(y_1)\right)M_5(y_1,q)=0 \tag{15.9}$$

$$\left(\frac{\partial}{\partial q}+\frac{\partial}{\partial y_2}+\beta_0(y_2)\right)M_6(y_2,q)=0 \tag{15.10}$$

$$\left(\frac{\partial}{\partial q}+\frac{\partial}{\partial x_0}+\beta_0(x_0)\right)M_7(x_0,q)=0 \tag{15.11}$$

$$\left(\frac{\partial}{\partial q}+\frac{\partial}{\partial x_1}+\beta_0(x_1)\right)M_8(x_1,q)=0 \tag{15.12}$$

$$\left(\frac{\partial}{\partial q}+\frac{\partial}{\partial x_2}+\beta_0(x_2)\right)M_9(x_2,q)=0 \tag{15.13}$$

**Conditions of Boundary**

$$M_1(0,q)=2c_1\delta_1 M_0(q) \tag{15.14}$$

$$M_2(0,q)=2c_2\delta_2 M_0(q) \tag{15.15}$$

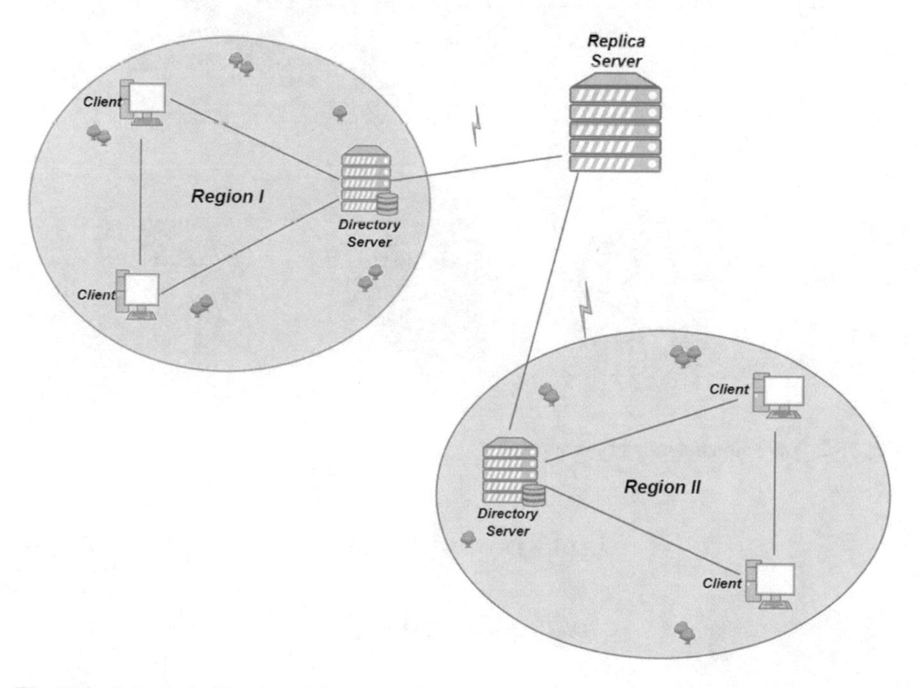


**Fig. 15.1** Schematic diagram of the network

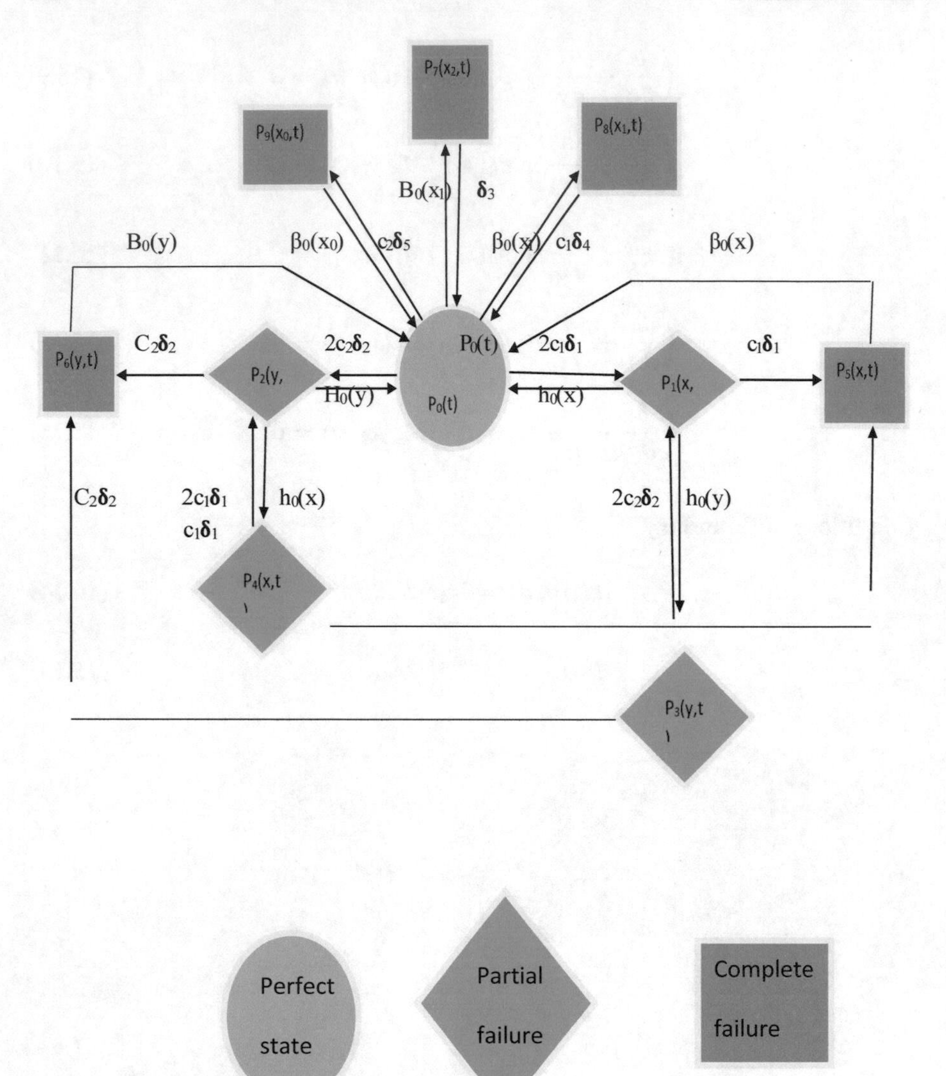


**Fig. 15.2** Transition diagram of the system

$$M_3(0, q) = 4c_1c_2\delta_1\delta_2 M_0(q) \tag{15.16}$$

$$M_4(0, q) = 4\delta_1\delta_2 M_0(q) \tag{15.17}$$

$$M_5(0, q) = 2c_1^2\delta_1^2(1 + 2c_2\delta_2) M_0(q) \tag{15.18}$$

$$M_6(0,q) = 2c_2^2\delta_2^2(1 + 2c_1\delta_1)M_0(q) \tag{15.19}$$

$$M_7(0,q) = \delta_3 M_0(q) \tag{15.20}$$

$$M_8(0,q) = c_1\delta_4 M_0(q) \tag{15.21}$$

$$M_9(0,q) = c_2\delta_5 M_0(q) \tag{15.22}$$

### 15.3.3 *Solution of the Model*

By taking the Laplace transformation of Eqs. (15.1) to (15.19), we obtain the following results

$$\begin{aligned}
&(s + 2c_1\delta_1 + 2c_2\delta_2 + \delta_3 + c_1\delta_4 + c_2\delta_5)\overline{M}_0(s) \\
&= 1 + \int_0^\infty h(y_1)\overline{M}_1(x,s)\mathrm{d}y_1 \\
&+ \int_0^\infty h(y_2)\overline{M}_2(y_2,s)\mathrm{d}y_2 + \int_0^\infty \beta_0(y_1)\overline{M}_5(y_1,s)\mathrm{d}y_1 \\
&+ \int_0^\infty \beta_0(y_2)\overline{M}_6(y_2,s)\mathrm{d}y_2 + \int_0^\infty \beta_0(x_0)\overline{M}_7(x_0,s)\mathrm{d}x_0 \\
&+ \int_0^\infty \beta_0(x_1)\overline{M}_8(x_1,s)\mathrm{d}x_1 + \int_0^\infty \beta_0(x_2)\overline{M}_9(x_2,s)\mathrm{d}x_2
\end{aligned} \tag{15.23}$$

$$\left(s + \frac{\partial}{\partial y_1} + c_1\delta_1 + 2c_2\delta_2 + h(y_1)\right)\overline{M}_1(y_1,s) = 0 \tag{15.24}$$

$$\left(s + \frac{\partial}{\partial y_2} + c_2\delta_2 + 2c_1\delta_1 + h(y_2)\right)\overline{M}_2(y_2,s) = 0 \tag{15.25}$$

$$\left(s + \frac{\partial}{\partial y_2} + c_2\delta_2 + h(y_2)\right)\overline{M}_3(y_2,s) = 0 \tag{15.26}$$

$$\left(s + \frac{\partial}{\partial y_1} + c_1\delta_1 + h(y_1)\right)\overline{M}_4(y_1,s) = 0 \tag{15.27}$$

$$\left(s + \frac{\partial}{\partial y_1} + \beta_0(y_1)\right)\overline{M}_5(y_1,s) = 0 \tag{15.28}$$

$$\left(s + \frac{\partial}{\partial y_2} + \beta_0(y_2)\right)\overline{M}_6(y_2,s) = 0 \tag{15.29}$$

$$\left(s + \frac{\partial}{\partial x_0} + \beta_0(x_0)\right)\overline{M}_7(x_0, s) = 0 \tag{15.30}$$

$$\left(s + \frac{\partial}{\partial x_1} + \beta_0(x_1)\right)\overline{M}_8(x_1, s) = 0 \tag{15.31}$$

$$\left(s + \frac{\partial}{\partial x_2} + \beta_0(x_2)\right)\overline{M}_9(x_2, s) = 0 \tag{15.32}$$

Conditions of boundary of Laplace transform are as follows:

$$\overline{M}_1(0, s) = 2c_1\delta_1\overline{M}_0(s) \tag{15.33}$$

$$\overline{M}_2(0, s) = 2c_2\delta_2\overline{M}_0(s) \tag{15.34}$$

$$\overline{M}_3(0, s) = 4c_1c_2\delta_1\delta_2\overline{M}_0(s) \tag{15.35}$$

$$\overline{M}_4(0, s) = 4c_1c_2\delta_1\delta_2\overline{M}_0(s) \tag{15.36}$$

$$\overline{M}_5(0, s) = 2c_1^2\delta_1^2(1 + 2c_2\delta_2)\overline{M}_0(s) \tag{15.37}$$

$$\overline{M}_6(0, s) = 2c_2^2\delta_2^2(1 + 2c_1\delta_1)\overline{M}_0(s) \tag{15.38}$$

$$\overline{M}_7(0, s) = \delta_3\overline{M}_0(s) \tag{15.39}$$

$$\overline{M}_8(0, s) = c_1\delta_4\overline{M}_0(s) \tag{15.40}$$

$$\overline{M}_9(0, s) = c_2\delta_5\overline{M}_0(s) \tag{15.41}$$

we get Eqs. (15.39)–(15.48) by solving Eqs. (15.20)–(15.29) with the help of boundary conditions (15.30)–(15.38).

$$\overline{M}_0(s) = \frac{1}{D(s)} \tag{15.42}$$

$$\overline{M}_1(s) = \frac{1}{D(s)}\left\{\frac{1 - \overline{S}_h(s + c_1\delta_1 + 2c_2\delta_2)}{(s + c_1\delta_1 + 2c_2\delta_2)}\right\}2c_1\delta_1 \tag{15.43}$$

$$\overline{M}_2(s) = \frac{1}{D(s)}\left\{\frac{1 - \overline{S}_h(s + c_2\delta_2 + 2c_1\delta_1)}{(s + c_2\delta_2 + 2c_1\delta_1)}\right\}2c_2\delta_2 \tag{15.44}$$

$$\overline{M}_3(s) = \frac{1}{D(s)} \left\{ \frac{1 - \overline{S}_h(s + c_2\delta_2)}{(s + c_2\delta_2)} \right\} 4c_1c_2\delta_1\delta_2 \tag{15.45}$$

$$\overline{M}_4(s) = \frac{1}{D(s)} \left\{ \frac{1 - \overline{S}_h(s + c_1\delta_1)}{(s + c_1\delta_1)} \right\} 4c_1c_2\delta_1\delta_2 \tag{15.46}$$

$$\overline{M}_5(s) = \frac{1}{D(s)} \left\{ \frac{1 - \overline{S}_{\beta_0}}{s} \right\} 2c_1^2\delta_1^2(1 + 2c_2\delta_2) \tag{15.47}$$

$$\overline{M}_6(s) = \frac{1}{D(s)} \left\{ \frac{1 - \overline{S}_{\beta_0}(s)}{s} \right\} 2c_2^2\delta_2^2(1 + 2c_1\delta_1) \tag{15.48}$$

$$\overline{M}_7(s) = \frac{1}{D(s)} \left\{ \frac{1 - \overline{S}_{\beta_0}(s)}{s} \right\} \delta_3 \tag{15.49}$$

$$\overline{M}_8(s) = \frac{1}{D(s)} \left\{ \frac{1 - \overline{S}_{\beta_0}(s)}{s} \right\} c_1\delta_4 \tag{15.50}$$

$$\overline{M}_9(s) = \frac{1}{D(s)} \left\{ \frac{1 - \overline{S}_{\beta_0}(s)}{s} \right\} c_2\delta_5 \tag{15.51}$$

where:

$$D(s) = \left\{ \begin{array}{l} (s + 2c_1\delta_1 + 2c_2\delta_2 + \delta_3 + c_1\delta_4 + c_2\delta_5) - 2c_1\delta_1\overline{S}_h(s + c_1\delta_1 + 2c_2\delta_2) \\ -\delta_3\overline{S}_{\beta_0}(s) - c_1\delta_4\overline{S}_{\beta_0}(s) - 2c_2^2\delta_2^2(1 + 2c_1\delta_1)\overline{S}_{\beta_0}(s) \\ -2c_2\delta_2\overline{S}_h(s + c_2\delta_2 + 2c_1\delta_1) - 2c_1^2\delta_1^2(1 + 2c_2\delta_2)\overline{S}_{\beta_0}(s) - c_2\delta_5\overline{S}_{\beta_0}(s) \end{array} \right\} \tag{15.52}$$

The chance that the system is up and running is,

$$\overline{M}_{\text{UP}}(s) = \overline{M}_0(s) + \overline{M}_1(s) + \overline{M}_2(s) + \overline{M}_3(s) + \overline{M}_4(s) \tag{15.53}$$

## 15.4 Analytical Analysis of the Model for Particular Cases

In order to acquire a thorough knowledge of this study, the simulation results of the models are shown in this section.

### 15.4.1 Availability Analysis

Taking the values of different parameter as: $\delta_1 = 0.01, \delta_2 = 0.02, \delta_3 = 0.03$ and $\delta_4 = 0.04, \delta_5 = 0.05, \theta(x) = 1$ and $\beta = 1$ and $h(x) = 1, c1 = 1, c2 = 1$ in Eq. (15.50) and inverting the transformation to have the following models of availability:

(a) Availability Analysis without copula and coverage factors

$$A_v(q) = \begin{Bmatrix} -0.0003143994970e^{-1.02q} + 0.08783207682e^{-1.195240242q} \\ +0.0007707542438e^{-1.047232186q} \\ +0.01079981008e^{-1.026280643q} + 0.9001982052e^{-0.001246929198q} \\ +0.00008475414236e^{-1.01q} \end{Bmatrix} \tag{15.54}$$

(b) Availability Analysis with copula only

$$A_v(q) = \begin{Bmatrix} -0.0002161941437e^{-1.02q} + 0.04381069018e^{-2.843384795q} \\ -0.006290102932e^{-1.097329896q} \\ -0.00004720059867e^{-1.046251477q} + 0.9630229281e^{-0.001333811830q} \\ -0.0002801208251e^{-1.01q} \end{Bmatrix} \tag{15.55}$$

(c) Availability with one coverage factor only (i.e., $c1 = 0.3$ and $c2 = 1$, $\beta = 1$)

$$A_v(q) = \begin{Bmatrix} -0.00007574896238e^{-1.02q} + 0.07254375879e^{-1.148259926q} \\ +0.0008324322965e^{-1.041469619q} \\ +0.006951222584e^{-1.016885473q} + 0.9197398205e^{-0.0003849812264q} \\ +0.00008514550493e^{-1.003q} \end{Bmatrix} \tag{15.56}$$

(d) Availability Analysis without copula and one coverage factor (i.e., $c1 = 1$, $c2 = 0.3$)

$$A_v(q) = \begin{Bmatrix} 0.00002091751179e^{-1.006q} + 0.06958795524e^{-1.123921410q} \\ +0.0003283496763e^{-1.024769912q} \\ +0.005438041030e^{-1.015897536q} + 0.9245787411e^{-0.0004111424846q} \\ +0.00004599536265e^{-1.01q} \end{Bmatrix} \tag{15.57}$$

(e) Availability Analysis without copula but using both coverage factors (i.e., $c1 = 0.3$, $c2 = 0.3$)

**Table 15.2** Availability with passage time for different cases

| $q$ | Availability | | | | | |
|---|---|---|---|---|---|---|
| | (a) | (b) | (c) | (d) | (e) | (f) |
| 0 | 1.0000 | 1.0000 | 1.0000 | 1.0000 | 1.0000 | 1.0000 |
| 10 | 0.8890 | 0.9503 | 0.9162 | 0.9208 | 0.9460 | 0.9788 |
| 20 | 0.8780 | 0.9377 | 0.9127 | 0.9170 | 0.9447 | 0.9775 |
| 30 | 0.8671 | 0.9252 | 0.9092 | 0.9132 | 0.9435 | 0.9762 |
| 40 | 0.8564 | 0.9130 | 0.9057 | 0.9095 | 0.9423 | 0.9749 |
| 50 | 0.8458 | 0.9009 | 0.9022 | 0.9058 | 0.9410 | 0.9735 |
| 60 | 0.8353 | 0.8890 | 0.8987 | 0.9021 | 0.9398 | 0.9722 |
| 70 | 0.8250 | 0.8772 | 0.8953 | 0.8983 | 0.9386 | 0.9709 |
| 80 | 0.8147 | 0.8656 | 0.8918 | 0.8947 | 0.9374 | 0.9696 |
| 90 | 0.8046 | 0.8541 | 0.8884 | 0.8910 | 0.9361 | 0.9683 |
| 100 | 0.7947 | 0.8428 | 0.8850 | 0.8873 | 0.9350 | 0.9670 |

$$A_v(q) = \left\{ \begin{array}{l} 0.00001216742101\mathrm{e}^{-1.006q} + 0.04984838801\mathrm{e}^{-1.078410838q} \\ \quad +0.0002860038336\mathrm{e}^{-1.014279187q} \\ \quad +0.002644243252\mathrm{e}^{-1.009179407q} + 0.9472047723\mathrm{e}^{-0.000130568120q} \\ \quad +0.000004425626607\mathrm{e}^{-1.003q} \end{array} \right\} \tag{15.58}$$

(f) Availability Analysis with present of copula and both coverage factors (i.e., $c1$ – 0.3, $c2$ – 0.3, $\beta$ – 2.7183).

$$A_v(q) = \left\{ \begin{array}{l} 0.00001952119465\mathrm{e}^{-1.006q} + 0.02076610302\mathrm{e}^{-2.775974868q} \\ \quad -0.0008463236303\mathrm{e}^{-1.030310526q} \\ \quad -0.000005587200069\mathrm{e}^{-1.013879499q} + 0.9801305761\mathrm{e}^{-0.0001351065033q} \\ \quad -0.0000252471517\mathrm{e}^{-1.003q} \end{array} \right\} \tag{15.59}$$

For $t \in [0, 100]$ in the above equation, i.e., Eqs. (15.47)–(15.51), the result of the above availability models is given in Table 15.2 and Fig. 15.3.

### 15.4.2 Reliability Analysis

Vanishing repairs to zero and considering $\delta_1 = 0.01, \delta_2 = 0.02, \delta_3 = 0.03, \delta_4 = 0.04$ and $\delta_5 = 0.05$ and applying Laplace transformation in (15.50) to have expression for reliability for system as follows:

(i) Reliability analysis without two coverage factors (i.e., $c1 = 1$, $c2 = 1$)

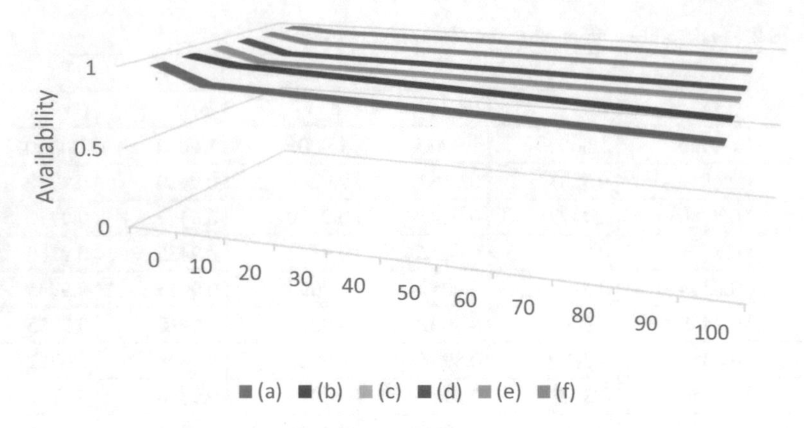


**Fig. 15.3** Availability against time for different cases

$$R(q) = \left\{ \begin{array}{l} 0.28571428557e^{-0.04q} + 0.1538461538e^{-0.05q+0.005e^{-0.02q}} \\ +0.004705882353e^{-0.01q} + 0.5507336781e^{-0.18q} \end{array} \right\} \tag{15.60}$$

(ii) Reliability analysis with one coverage factor (i.e., $c1 = 0.3$, $c2 = 1$)

$$R(q) = \left\{ \begin{array}{l} 0.3571428571e^{-0.026q} + 0.00177777778e^{-0.003q} \\ +0.002033896305e^{-0.02q} + 0.5758875720e^{-0.138q} \\ +0.06315789474e^{-0.043q} \end{array} \right\} \tag{15.61}$$

(iii) Reliability analysis with one coverage factor (i.e., $c1 = 1$, $c2 = 0.3$)

$$R(q) = \left\{ \begin{array}{l} 0.6532003995e^{-0.17q} + 0.002162162162e^{-0.006q} \\ +0.2105263158e^{-0.022q} + 0.1318681319e^{-0.26q} \\ +0.002242990654e^{-0.01q} \end{array} \right\} \tag{15.62}$$

(iv) Reliability analysis with both coverage factors (i.e., $c1 = 0.3$, $c2 = 0.3$)

$$R(q) = \left\{ \begin{array}{l} 0.1904761905e^{-0.12q} + 0.7074803313e^{-0.075q} + 0.001e^{-0.003q} \\ +0.001e^{-0.015q} + 0.001043478261e^{-0.006q} \end{array} \right\} \tag{15.63}$$

### *15.4.3 Cost Analysis*

The anticipated profit in $[0, q)$ is

$$E_p(q) = v_1 \int\limits_0^q M_{\mathrm{UP}}(q)\mathrm{d}q - v_2 q \tag{15.64}$$

Assuming $v_1 = 1$ and $v_2 = 0.6, 0.5, 0.4, 0.3, 0.2,\ v_1 = 1,\ v_2 \in [0.1, 0.6]$ and $t \in [0, 100]$ in (15.50), respectively, to have

(a) Cost Benefit Analysis without copula and both coverage factors (i.e., $c1 = 1, c2 = 1, \beta = 1$)

$$E_P(q) = v_1 \left\{ \begin{array}{l} -0.073487252\mathrm{e}^{-1,195240242q} \\ \quad -0.0007359917448\mathrm{e}^{-1.447232186q} \\ \quad -0.01052325224\mathrm{e}^{-1.026280643q} \\ \quad -721.9320926\mathrm{e}^{-0.001246929198q} \\ \quad -0.0008391499244\mathrm{e}^{-1.01q} \\ \quad -0.0003082348010\mathrm{e}^{-1.02q} \\ \quad +722.0172288 \end{array} \right\} - v_2 q \tag{15.65}$$

(b) Cost Benefit Analysis with copula and no two coverage factors (i.e., $c1 = 1, c2 = 1, \beta = 2.7813$)

$$E_P(q) = v_1 \left\{ \begin{array}{l} -0.001540793573\mathrm{e}^{-2.843384995q} \\ \quad +0.00573489522\mathrm{e}^{-1.097329896q} \\ \quad +0.00004511400825\mathrm{e}^{-1.046251497q} \\ \quad -721.0080872\mathrm{e}^{-0.001333811830q} \\ \quad +0.0002773516\mathrm{e}^{-1.01q} \\ \quad +0.0002119550428\mathrm{e}^{-1.02q} \\ \quad +722.0172288 \end{array} \right\} - v_2 q \tag{15.66}$$

Case (c) Cost Benefit Analysis with two coverage factor and no copula (i.e., $c1 = 0.3, c2 = 0.3, \beta = 1$)

$$E_P(q) = v_1 \left\{ \begin{array}{l} -0.0462239086\mathrm{e}^{-1.078410838q} \\ \quad -0.0002819774252\mathrm{e}^{-1.014279187q} \\ \quad +0.002620191448\mathrm{e}^{-1.009179407q} \\ \quad -7254.459995\mathrm{e}^{-0.0013025686120q} \\ \quad -0.0000441238943\mathrm{e}^{-1.003q} \\ \quad -0.00001209485190\mathrm{e}^{-1.006q} \\ \quad +7254.509137 \end{array} \right\} - v_2 q \tag{15.67}$$

(d) Cost Benefit Analysis with one coverage factor and no copula (i.e., $c1 = 0.3$, $c2 = 1$, $\beta = 1$)

$$E_P(q) = v_1 \left\{ \begin{array}{l} -0.00002079275526\mathrm{e}^{-1.06q} \\ -0.061915321332\mathrm{e}^{-1.0123921410q} \\ -0.0003204130727\mathrm{e}^{-1.024769912q} \\ -0.005352942435\mathrm{e}^{-0.0158975369q} \\ -2248.803701\mathrm{e}^{-0.0004111424846q} \\ -0.00004553996302\mathrm{e}^{-1.01q} \\ +2248.871356 \end{array} \right\} - v_2 q \quad (15.68)$$

Case (e) Cost Benefit Analysis with one coverage factor and copula (i.e., $c1 = 1$, $c2 = 0.3$, $\beta = 2.7183$)

$$E_P(q) = v_1 \left\{ \begin{array}{l} -0.0000829983761\mathrm{e}^{-1.006q} \\ +0.002188178547\mathrm{e}^{-1.053380418q} \\ +0.000008928235032\mathrm{e}^{-1.024374038q} \\ -0.01101737634\mathrm{e}^{-2.805113352q} \\ +0.00006896920683\mathrm{e}^{-1.01q} \\ -2248.862738\mathrm{e}^{-0.0004320238303q} \\ +2248.871406 \end{array} \right\} - v_2 q \quad (15.69)$$

Case (f) Cost Benefit Analysis with one coverage factor and copula (i.e., $c1 = 0.3$, $c2 = 1$, $\beta = 2.7183$)

$$E_P(q) = v_1 \left\{ \begin{array}{l} -0.00002956658416\mathrm{e}^{-1.02q} \\ +0.00008247606823\mathrm{e}^{-1.01003q} \\ -0.01200288992\mathrm{e}^{-2.813524595q} \\ +0.003204811886\mathrm{e}^{-1.071605575q} \\ +0.00008667258287\mathrm{e}^{-1.0397633878q} \\ -2389.113211\mathrm{e}^{-0.000405926624q} \\ +2389.121810 \end{array} \right\} - v_2 q \quad (15.70)$$

## 15.5 Results Discussion

The decision-making process for performance evaluation of the model under consideration is carried out based on Tables 15.2, 15.3, 15.4, 15.5, 15.6, 15.7, 15.8, 15.9

and Figs. 15.3, 15.4, 15.5, 15.6, 15.7, 15.8, 15.9, 15.10. First and foremost, failure rates must be determined, preferably ones with the lowest risk of error.

Table 15.2 depicts availability with passage of time for six different cases considered. From the table, it is evident that availability decreases with passage of time for each case. Nonetheless, availability is better in case (f) when compared with the other five cases. This implies that employing copula at complete failure which restored the system to the initial state and applying both coverage factors will produce better availability that the other cases.

Table 15.3 displayed the result of reliability appreciate in time for the four different cases considered in the chapter. From the table, it is clear that reliability decreases drastically with increase in time for each case. From the table, system without two coverage factors in case (a) has the least reliability than the rest. However, reliability is better when the two coverage factors are as seen in case (f). This reveals the importance of coverage factors in enhancing the performance of the system. From the table, it is clear that applying both coverage factors will produce better reliability that the other cases.

Tables 15.4, 15.5, 15.6, 15.7, 15.8 and 15.9 depict the cost with passage of time for six different cases considered for different values of service cost $K_2$. From the tables, it is evident that the cost appreciates with passage of time and service cost $K_2$ in each case. However, the cost is better in each case when the service cost $K_2$ reduces to 0.1 in each case. It can be seen from Tables 15.4, 15.5, 15.6, 15.7, 15.8 and 15.9 that the optimal cost is case (f) when employing copula and applying both coverage factors.

Figure 15.3 and Table 15.2 depict the system's availability for case (a) to (f) with passage of time. From the figure and table, it is evident that availability decreases with passage of time for each case. From the results, it is clear that case (f) in both

**Table 15.3** Reliability with passage of time for different cases

| $q$ | Reliability | | | |
|---|---|---|---|---|
| | (i) | (ii) | (iii) | (iv) |
| 0 | 1.0000 | 1.0000 | 1.0000 | 1.0000 |
| 10 | 0.3842 | 0.4647 | 0.4774 | 0.5912 |
| 20 | 0.2072 | 0.2785 | 0.2807 | 0.3836 |
| 30 | 0.1291 | 0.1930 | 0.1923 | 0.2730 |
| 40 | 0.0843 | 0.1423 | 0.1423 | 0.2096 |
| 50 | 0.0561 | 0.1075 | 0.1109 | 0.1700 |
| 60 | 0.0377 | 0.0821 | 0.0873 | 0.1428 |
| 70 | 0.0256 | 0.0630 | 0.0692 | 0.1224 |
| 80 | 0.0176 | 0.0485 | 0.0551 | 0.1062 |
| 90 | 0.0123 | 0.0374 | 0.0440 | 0.0928 |
| 100 | 0.0087 | 0.0290 | 0.0351 | 0.0814 |

**Table 15.4** Cost with passage of time for (a)

| $q$ | $E_P(q)$: $\upsilon_2 = 0.6$ | $E_P(q)$: $\upsilon_2 = 0.5$ | $E_P(q)$: $\upsilon_2 = 0.4$ | $E_P(q)$: $\upsilon_2 = 0.3$ | $E_P(q)$: $\upsilon_2 = 0.2$ | $E_P(q)$: $\upsilon_2 = 0.1$ |
|---|---|---|---|---|---|---|
| 0 | 0 | 0 | 0 | 0 | 0 | 0 |
| 10 | 3.0312 | 4.0312 | 5.0312 | 6.0312 | 7.0312 | 8.0312 |
| 20 | 5.8665 | 7.8665 | 9.8665 | 11.8665 | 13.8665 | 15.8665 |
| 30 | 8.5922 | 11.5922 | 14.5922 | 17.5922 | 20.5922 | 23.5922 |
| 40 | 11.2098 | 15.2098 | 19.2098 | 23.2098 | 27.2098 | 31.2098 |
| 50 | 13.7207 | 18.7206 | 23.7207 | 28.7207 | 33.7207 | 38.7265 |
| 60 | 16.126 | 22.126 | 28.126 | 34.126 | 40.126 | 46.126 |
| 70 | 18.4272 | 25.4272 | 32.4272 | 39.4272 | 46.4272 | 53.4272 |
| 80 | 20.6256 | 28.6256 | 36.6256 | 44.6256 | 52.6256 | 60.6256 |
| 90 | 22.7223 | 31.7223 | 40.7223 | 49.7223 | 58.7223 | 67.7223 |
| 100 | 24.7187 | 34.7187 | 44.7187 | 54.7187 | 64.7189 | 74.7187 |

**Table 15.5** Cost with passage of time for (b)

| $q$ | $E_P(q)$: $\upsilon_2 = 0.6$ | $E_P(q)$: $\upsilon_2 = 0.5$ | $E_P(q)$: $\upsilon_2 = 0.4$ | $E_P(q)$: $\upsilon_2 = 0.3$ | $E_P(q)$: $\upsilon_2 = 0.2$ | $E_P(q)$: $\upsilon_2 = 0.1$ |
|---|---|---|---|---|---|---|
| 0 | 0 | 0 | 0 | 0 | 0 | 0 |
| 10 | 3.5754 | 4.5754 | 5.5754 | 6.5754 | 7.5754 | 8.5754 |
| 20 | 7.015 | 9.015 | 11.015 | 13.015 | 15.015 | 17.015 |
| 30 | 10.3294 | 13.3294 | 16.3294 | 19.3294 | 22.3294 | 25.3294 |
| 40 | 13.5205 | 17.5205 | 21.5205 | 25.5205 | 29.5205 | 33.5205 |
| 50 | 16.5898 | 21.5898 | 26.5898 | 31.5898 | 36.5898 | 41.5898 |
| 60 | 19.5389 | 25.5389 | 31.5389 | 37.5389 | 43.5389 | 49.5389 |
| 70 | 22.3694 | 29.3694 | 36.3694 | 43.3694 | 50.3694 | 57.3694 |
| 80 | 25.083 | 33.083 | 41.083 | 49.083 | 57.083 | 65.083 |
| 90 | 27.6811 | 36.6811 | 45.6811 | 54.6811 | 63.6811 | 72.6811 |
| 100 | 30.1652 | 40.1652 | 50.1652 | 60.1652 | 70.1652 | 80.1652 |

Table 15.2 and Fig. 15.3 is the optimal having the highest availability with passage of time compare to different cases.

Thus,

$$(f) > (e) > (d) > (c) > (b) > (a)$$

Figure 15.4 and Table 15.3 depict the system's reliability for case (a) to (d) with passage of time. From the figure and table, it is observed that reliability decreases with passage of time for each case. From the results, it is clear that case (d) in both Table 15.3 and Fig. 15.4 is the optimal having the highest reliability with passage of

**Table 15.6** Cost with passage of time for (c)

| $q$ | $E_P(q)$: $v_2 = 0.6$ | $E_P(q)$: $v_2 = 0.5$ | $E_P(q)$: $v_2 = 0.4$ | $E_P(q)$: $v_2 = 0.3$ | $E_P(q)$: $v_2 = 0.2$ | $E_P(q)$: $v_2 = 0.1$ |
|---|---|---|---|---|---|---|
| 0 | 0 | 0 | 0 | 0 | 0 | 0 |
| 10 | 3.515 | 4.515 | 5.515 | 6.515 | 7.515 | 8.515 |
| 20 | 6.9685 | 8.9685 | 10.9685 | 12.9685 | 14.9685 | 16.9685 |
| 30 | 10.4097 | 13.4097 | 16.4097 | 19.4097 | 22.4097 | 25.4097 |
| 40 | 13.8386 | 17.8386 | 21.8386 | 25.8386 | 29.8386 | 33.8386 |
| 50 | 17.2551 | 22.2551 | 27.2551 | 32.2551 | 37.2551 | 42.2551 |
| 60 | 20.6594 | 26.6594 | 32.6594 | 38.6595 | 44.6594 | 50.6594 |
| 70 | 24.0514 | 31.0514 | 38.0514 | 45.0514 | 52.0514 | 59.0514 |
| 80 | 27.4311 | 35.4311 | 43.4311 | 51.4311 | 59.4311 | 67.4311 |
| 90 | 30.7986 | 39.7986 | 48.7986 | 57.7986 | 66.7986 | 75.7986 |
| 100 | 34.1539 | 44.1539 | 54.1539 | 64.1539 | 74.1539 | 84.1539 |

**Table 15.7** Cost with passage of time for (d)

| $q$ | $E_P(q)$: $v_2 = 0.6$ | $E_P(q)$: $v_2 = 0.5$ | $E_P(q)$: $v_2 = 0.4$ | $E_P(q)$: $v_2 = 0.3$ | $E_P(q)$: $v_2 = 0.2$ | $E_P(q)$: $v_2 = 0.1$ |
|---|---|---|---|---|---|---|
| 0 | 0 | 0 | 0 | 0 | 0 | 0 |
| 10 | 3.2945 | 4.2945 | 5.2945 | 6.2945 | 7.2945 | 8.2945 |
| 20 | 6.4834 | 8.4834 | 10.4834 | 12.4834 | 14.4834 | 16.4834 |
| 30 | 9.6347 | 12.6347 | 15.6347 | 18.6347 | 21.6347 | 24.6347 |
| 40 | 12.7483 | 16.7483 | 20.7483 | 25.7483 | 28.7483 | 32.7483 |
| 50 | 15.8246 | 20.8246 | 25.8246 | 30.8246 | 35.8247 | 40.8247 |
| 60 | 18.8637 | 24.8637 | 30.8637 | 36.8637 | 42.8637 | 48.8637 |
| 70 | 21.8657 | 28.8657 | 35.8657 | 42.8657 | 49.8657 | 56.8657 |
| 80 | 24.8341 | 32.8341 | 40.8341 | 48.8341 | 56.8341 | 64.8341 |
| 90 | 27.759 | 36.759 | 45.759 | 54.759 | 63.759 | 72.759 |
| 100 | 30.6506 | 40.6506 | 50.6506 | 60.6506 | 70.6506 | 80.6506 |

time compare to different cases.

$$(d) > (c) > (b) > (a)$$

This analysis illustrates the consequences of failing to restore the system. It is widely held belief that the higher maintenance and the greater the reliability.

The change in the profit with passage of time $t$ for $k_2 \in [0.1, 0.6]$ when $k_1 = 1$ is fixed at 1 for repair policy of copula is given in Tables 15.4, 15.5, 15.6, 15.7, 15.8, 15.9 and Figs. 15.5, 15.6, 15.7, 15.8, 15.9, 15.10 for case (a) to (f). From the Tables and Figures, it is clear that increasing the duration (time) and decreasing the

**Table 15.8** Cost with passage of time for (e)

| $q$ | $E_P(q)$: $v_2 = 0.6$ | $E_P(q)$: $v_2 = 0.5$ | $E_P(q)$: $v_2 = 0.4$ | $E_P(q)$: $v_2 = 0.3$ | $E_P(q)$: $v_2 = 0.2$ | $E_P(q)$: $v_2 = 0.1$ |
|---|---|---|---|---|---|---|
| 0 | 0.0000 | 0.0000 | 0.0000 | 0.0000 | 0.0000 | 0.0000 |
| 10 | 3.7033 | 4.7033 | 5.7033 | 6.7033 | 7.7033 | 8.7033 |
| 20 | 7.3562 | 9.3562 | 11.3562 | 13.3562 | 15.3562 | 17.3562 |
| 30 | 10.9675 | 13.9675 | 16.9675 | 19.9675 | 22.9675 | 25.9675 |
| 40 | 14.5373 | 18.5373 | 22.5373 | 26.5377 | 30.5373 | 34.5373 |
| 50 | 18.0659 | 23.0659 | 28.0659 | 33.0131 | 38.0659 | 43.0659 |
| 60 | 21.5534 | 27.5534 | 33.5534 | 39.5534 | 45.5534 | 51.5534 |
| 70 | 24.9999 | 31.9999 | 38.9999 | 45.9999 | 52.9999 | 59.9999 |
| 80 | 28.4058 | 36.4058 | 44.4058 | 52.4058 | 60.4058 | 68.4058 |
| 90 | 31.7712 | 40.7712 | 49.7712 | 58.7712 | 67.7712 | 76.7712 |
| 100 | 35.0961 | 45.0961 | 55.0961 | 65.0961 | 75.0961 | 85.0961 |

**Table 15.9** Cost with passage of time for (f)

| $q$ | $E_P(q)$: $v_2 = 0.6$ | $E_P(q)$: $v_2 = 0.5$ | $E_P(q)$: $v_2 = 0.4$ | $E_P(q)$: $v_2 = 0.3$ | $E_P(q)$: $v_2 = 0.2$ | $E_P(q)$: $v_2 = 0.1$ |
|---|---|---|---|---|---|---|
| 0 | 0 | 0 | 0 | 0 | 0 | 0 |
| 10 | 3.6876 | 4.6876 | 5.6876 | 6.6876 | 7.6876 | 8.6876 |
| 20 | 7.3274 | 9.3274 | 11.3274 | 13.3274 | 15.3274 | 17.3274 |
| 30 | 10.9282 | 13.9282 | 16.9282 | 19.9282 | 22.9282 | 25.9282 |
| 40 | 14.4899 | 18.4899 | 22.4899 | 26.4399 | 30.4399 | 34.4399 |
| 50 | 18.0131 | 23.0131 | 28.0131 | 33.0131 | 38.0131 | 43.0131 |
| 60 | 21.4977 | 27.4977 | 33.4977 | 39.4977 | 45.4977 | 51.4977 |
| 70 | 24.9437 | 31.9437 | 38.9437 | 45.9437 | 52.9437 | 59.9437 |
| 80 | 28.3516 | 36.3516 | 44.3516 | 52.3516 | 60.3516 | 68.3516 |
| 90 | 31.7213 | 40.7213 | 49.7213 | 58.7213 | 67.7213 | 76.7213 |
| 100 | 35.0531 | 45.0531 | 55.0531 | 65.0531 | 75.0531 | 85.0531 |

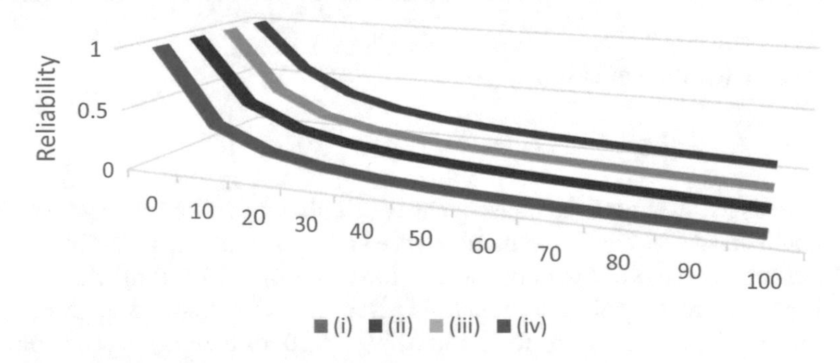


**Fig. 15.4** Reliability with passage of time for different cases

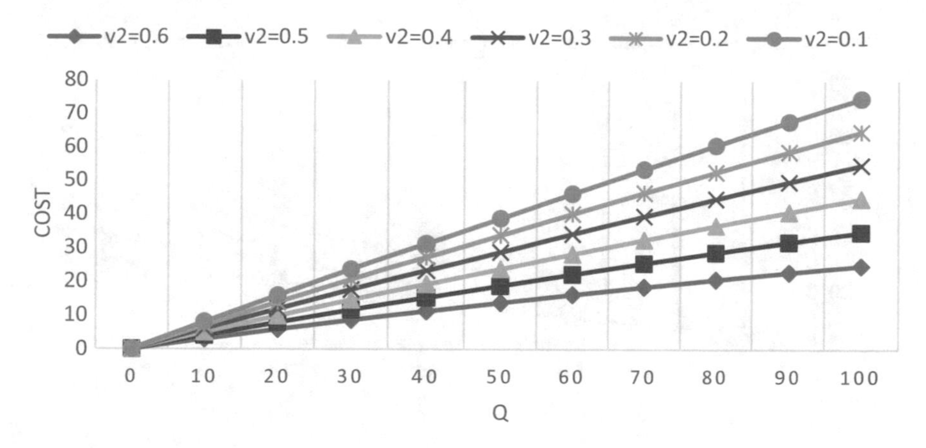


**Fig. 15.5** Cost with passage of time for different $v_2$

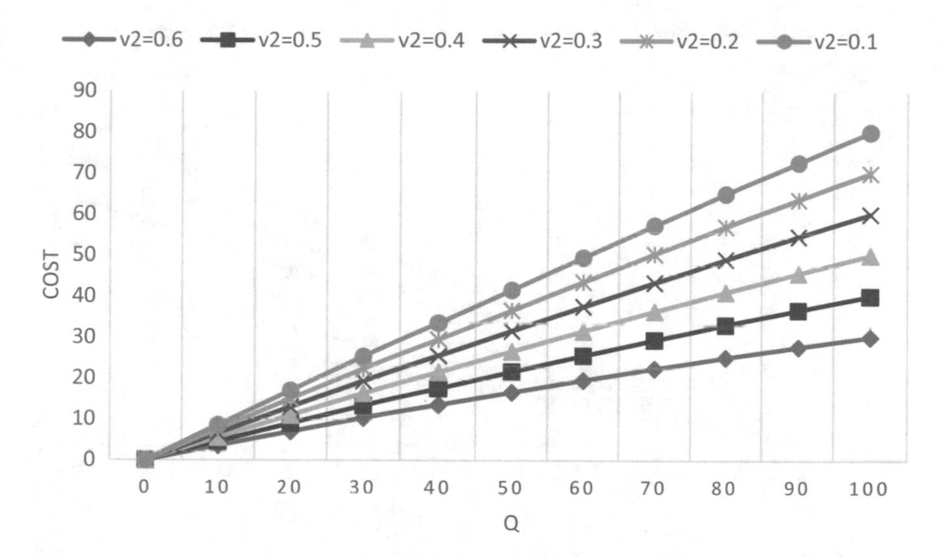


**Fig. 15.6** Cost with passage of time for different $v_2$ for Case (b)

service rate/cost will likely raise the estimated service cost for case (a) to (f). When comparing the cases, it appears that the predicted profit is larger for case (f) where the repair policy is followed by copula distribution and both coverage factors are present. In both circumstances, the predicted profit in each case is higher when the service cost is low and lowest when the service cost is highest. This analysis will assist the analyst to set the budget for the system's smooth operation in advance based on the system's usability.

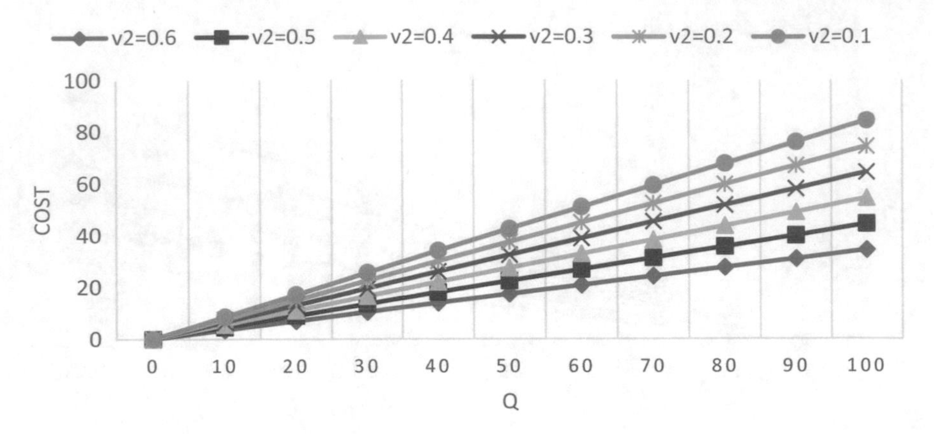


**Fig. 15.7** Cost with passage of time for different $v_2$ for (c)

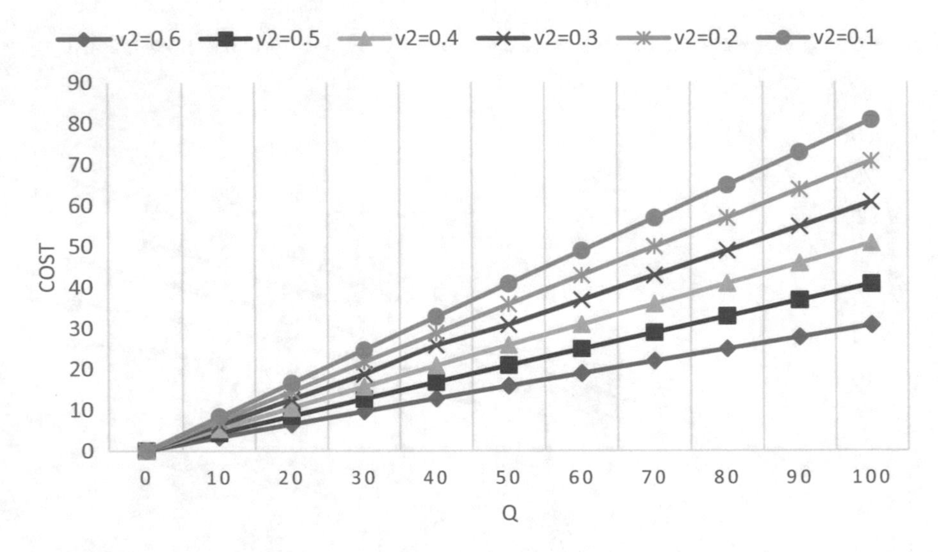


**Fig. 15.8** Cost with passage of time for different $v_2$ for Case (d)

## 15.6 Conclusion

The availability, reliability, maintenance strategy/technique, and revenue generated are some of the factors that influence the development of any process sector. So, in order to get the most out of operating production systems, they must be meticulously maintained so that the rate of failure and repair is kept to a minimum. In this manner, various maintenance policies or tactics can be planned to improve system strength and performance. As a result of the preceding, this chapter presents evaluation of performance of a distributed system through copula characteristics and coverage factors. The inclusion of copula characteristics has increased the application of the

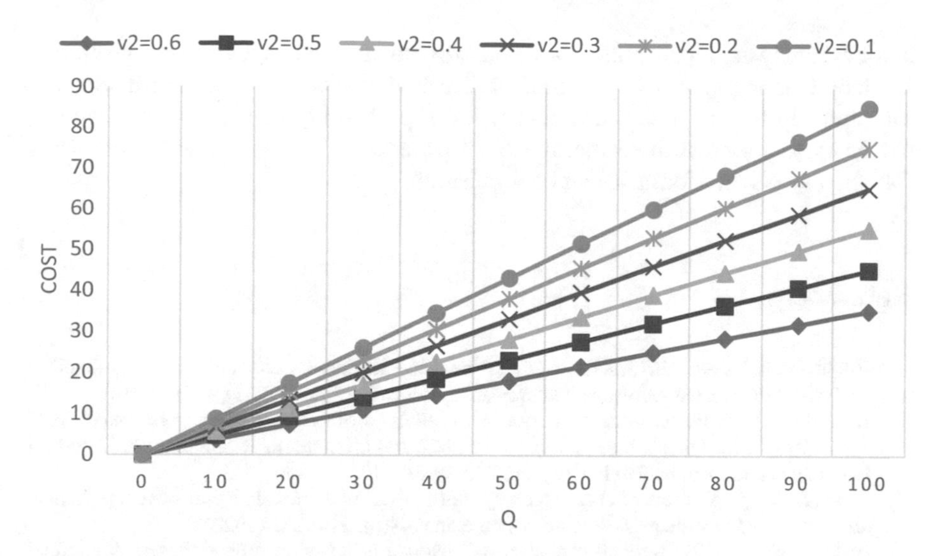


**Fig. 15.9** Cost with passage of time for different $v_2$ for (e)

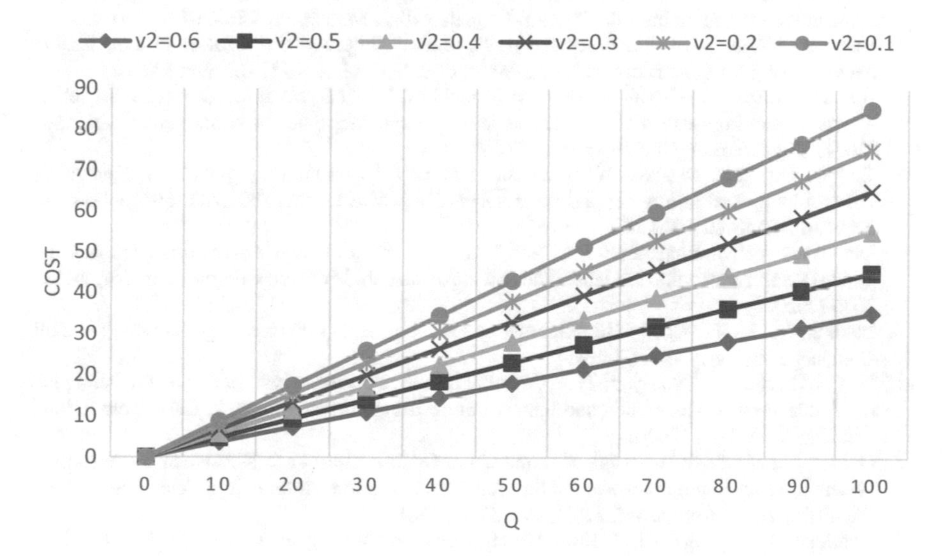


**Fig. 15.10** Cost with passage of time for different $v_2$ for (f)

developed model to a wider range of performance analysis of repairable systems operating under repair policy of copula and coverage factors. Availability, cost function, and reliability explicit expressions are established and statistically validated. Based on the availability analysis, the failure rates priority for each subsystem are identified. Through the cost function analysis, it has been discovered that higher cost of services entailed lower system profit and vice versa. On the basis of availability and cost analysis, repair policy of copula distribution enhanced system profit and availability than repair policy of general distribution. These are the contributions of this study. In future studies, different types of probability continuous distributions can be applied for distributed hardware-software to analyze the reliability measures for better system performance and optimization.

## References

1. Abdulwahab, L., Abdullahi, J.T., Yusuf, I.: Enhanced Markov-based model for the availability analysis of distributed software and hardware systems. J. Phys. **1132**(1), 1–16 (2018)
2. Ahmed, H.G., Ramalakshmi, R.: Performance analysis of centralized and distributed SDN controllers for load balancing application. In: 2018 2nd International Conference on Trends in Electronics and Informatics (ICOEI), pp. 758–764 (2018)
3. Arora, R., Tyagi, V., Ram, M.: Multi-state system analysis with imperfect fault coverage, human error and standby strategies. Rev. Inv. Operacional **41**(2), 214–231 (2020)
4. Bisht, S., Singh, S.B.: Reliability analysis of acyclic transmission network based on minimal cuts using copula in repair. Proc. Jangjeon Math. Soc. **22**, 163–173 (2019)
5. Bisht, S., Singh, S.B.: Signature reliability of binary state node in complex bridge networks using universal generating function. Int. J. Qual. Reliab. Manag. **36**, 186–201 (2019)
6. Bisht, S., Singh, S.B.: Reliability and profit function improvement of acyclic transmission network using artificial neural network. Math. Eng. Sci. Aerosp. **11**, 127–141 (2020)
7. Bisht, S., Kumar, A., Goyal, N., Ram, M., Klochkov, Y.: Analysis of network reliability characteristics and importance of components in a communication network. Mathematics **9**, 1347 (2021). https://doi.org/10.3390/math9121347
8. Chopra, G., Ram, M.: Reliability measures of two dissimilar units parallel system using Gumbel-Hougaard family copula. Int. J. Math. Eng. Manag. Sci. (2019). https://doi.org/10.33889/IJMEMS.2019.4.1-011
9. Dhulavvagol, P.M., Bhajantri, V.H., Totad, S.G.: Performance analysis of distributed processing system using shard selection techniques on elasticsearch. Procedia Comput. Sci. **167**, 1626–1635 (2020)
10. Ebeling, A.: An introduction to reliability and maintainability engineering. Tata Mcgraw Hill Company Ltd, New Delhi (2000)
11. Gahlot, G., Singh, V. V., Ayagi, H. I., Abdullahi, I.: Stochastic analysis of a two units' complex repairable system with switch and human failure using copula approach. Life Cycle Reliab. Saf. Eng. **9**(1), 1–11 (2020)
12. Gahlot, M., Singh, V.V., Ayagi, H.I. and Goel, C.K.: Performance assessment of repairable system in series configuration under different types of failure and repair policies using copula linguistics. Int. J. Reliab. Saf. **12**(4), 348–374 (2018)
13. Handoko, H., Isa, S.M., Si, S., Kom, M.: High availability analysis with database cluster, load balancer and virtual router redundancy protocol. In: 2018 3rd International Conference on Computer and Communication Systems (ICCCS), pp. 482–486 (2018)
14. He, Y., Zhang, R., Ye, N.: Genetic algorithm-based reliability of computer communication network. IETE J. Res. (2022). http://doi.org/10.1080/03772063.2021.2010610
15. Huang, C.F., Huang, D.H., Lin, Y.K.: Network reliability evaluation for a distributed network with edge computing. Comput. Ind. Eng. **147**, 106492 (2020)
16. Ibrahim, K.H., Isa, M.S., Abubakar, M.I., Yusuf, I., Tukur, I.: Availability and cost analysis of complex tree topology of computer network with multi-server using Gumbel-Hougaard family copula approach. Reliab. Theor. Appl. **16**(1(61)), 203–216 (2021)
17. Ismail, A.L., Abdullahi, S., Yusuf, I.: Performance evaluation of a hybrid series-parallel system with two human operators using Gumbel-Hougaard family copula. Int. J. Qual. Reliab. Manag. (2021). https://doi.org/10.1108/IQRM-05-2020-0137

18. Ismail, T., Kabiru, H.I., Salihu, M.I., Yusuf, I.: Availability and performance analysis of computer network with dual-server using Gumbel-Hougaard family copula distribution. Reliab. Theor. Appl. **16**(3(63)), 183–194 (2021)
19. Jain, M., Gupta, R.: Optimal replacement policy for a repairable system with multiple vacations and imperfect fault coverage. Comput. Ind. Eng. **66**, 710–719 (2013). https://doi.org/10.1016/j.cie.2013.09.011
20. Jain, M., Meena, R.K.: Fault tolerance system with imperfect coverage and reboot and server vacation. J. Ind. Eng. Int. (2016). https://doi.org/10.1007/s40092-016-0180-8
21. Jain, M., Shekhar, C., Rani, V.: N-policy for multi-component machining system with imperfect coverage, reboot, and unreliable server. Prod. Manuf. Res. **2**, 457–476 (2014)
22. Lado, A., Singh, V.V.: Cost assessment of complex repairable system consisting of two subsystems in the series configuration using Gumbel–Hougaard family copula. Int. J. Qual. Reliab. Manage. **36**(10), 1683–1698 (2019)
23. Muñoz-Esco, F.D., Juan-Marn, R.D.: On synchrony in dynamic distributed systems. Open Comput. Sci. **8**, 154–164 (2018)
24. Niwas, R., Garg, H.: An approach for analyzing the reliability and profit of an industrial system based on the cost free warranty policy. J. Braz. Soc. Mech. Sci. Eng. **40**, 265 (2018)
25. Pourhassan, M.R., Raissi, S., Apornak, A.: Modeling multi-state system reliability analysis in a power station under fatal and nonfatal shocks: a simulation approach. Int. J. Qual. Reliab. Manag. (2021). https://doi.org/10.1108/IJQRM-07-2020-0244
26. Pourhassan, M.R., Raissi, S., Fezalkotob, H.A.: A simulation approach on reliability assessment of complex system subject to stochastic degradation and random shock. Eksploatacja i Niezawodnosc Maintenance Reliab. **22**(2), 370–379 (2020). http://doi.org/10.17531/ein.2020.2.20
27. Raghav, D., Rawal, D.K., Yusuf, I., Kankarofi, R.H., Singh, V.V.: Reliability prediction of distributed system with homogeneity in software and server using joint probability distribution via copula approach. RT&A **61**(1), 217–230. http://doi.org/10.24412/1932-2321-2021-161-217-230
28. Ram, M., Goyal, N.: Bi-directional system analysis under copula-coverage approach. Commun. Stat. Simul. Comput. **47**(6), 1831–1844 (2018). https://doi.org/10.1080/03610918.2017.1327068
29. Rotar, R., Jurj, S.L., Opritoiu, F., Vladutiu, M.: Fault coverage-aware metrics for evaluating the reliability factor of solar tracking systems. Energies **14**, 1074 (2021). https://doi.org/10.3390/en14041074
30. Salihu, M.I., Abubakar, I.M., Kabiru, H.I., Yusuf, I., Ismail, T.: Performance analysis of complex series parallel computer network with transparent bridge using copula distribution. Int. J. Reliab. Risk Saf. Theor. Appl. **4**(1), 47–59 (2021). https://doi.org/10.30699/IJRRS.4.1.7
31. Singh, V.V., Lado Ismail, A.K., Yusuf, I., Abdullahi, A.H.: Probabilistic assessment of computer-based test (CBT) network system consists of four subsystems in series configuration using copula linguistic approach. J. Reliab. Stat. Stud. 401–428 (2021)
32. Singh, V. V., Ayagi, H. I.: Study of reliability measures of system consisting of two subsystems in series configuration using copula. Palest. J. Math. **6**(1), 1–10 (2017)
33. Singh, V.V., Ayagi, H.I.: Stochastic analysis of a complex system under preemptive resume repair policy using Gumbel-Hougaard family of copula. Int. J. Math. Oper. Res. **12**(2), 273–292
34. Song, B.: Reliability analysis and optimization of computer communication network based on genetic algorithm. Int. J. Commun. Syst. e4601 (2020). http://doi.org/10.1002/dac.4601
35. Teslyuk, V., Sydor, A., Karovič, V., Pavliuk, O., Kazymyra, I.: Modelling reliability characteristics of technical equipment of local area computer networks. Electronics **10**, 955 (2021). https://doi.org/10.3390/electronics10080955
36. Tyagi, V., Arora, R., Ram, M.: Copula based measures of repairable parallel system with fault. Int. J. Math. Eng. Manag. Sci. **6**(1), 322–344 (2021)
37. Yusuf, I., Ismail, A.L., Lawan, M.A., Ali, U.A., Nasir, S.: Reliability modelling and analysis of client–server system using Gumbel-Hougaard family copula. Life Cycle Reliab. Saf. Eng. (2020). https://doi.org/10.1007/s41872-020-00159-4

38. Yusuf, I., Sani, B., Yusuf, B.: Profit analysis of a serial-parallel system under partial and complete failures. J. Appl. Sci. **19**, 565–574 (2019). https://doi.org/10.3923/jas.2019.565.574

**Ibrahim Yusuf** is a Lecturer in the Department of Mathematical Sciences, Bayero University, Kano, Nigeria. He received his B.Sc. in Mathematics in 1996, M.Sc. in Mathematics in 2007, and Ph.D. in Mathematics in 2014 from Bayero University, Kano, Nigeria. He is currently an Associate Professor at the Department of Mathematical Sciences, Bayero University, Kano, Nigeria. He has reviewed papers from IJSA, *JCCE*, Life Cycle Reliability and Safety Engineering, JRSS, IJRRS, Inderscience journals, IJQRM, and Operation Research and Decision. His research includes system reliability theory, maintenance and replacement and operation research.

**Ismail Tukur** is a Lecturer in the Department of Art and Humanities, Kano State Polytechnic, Kano, Nigeria. He is currently pursuing his M.Sc. Mathematics at Department of Mathematical Sciences, Bayero University, Kano, Nigeria. His research interest includes system reliability modelling and analysis, statistics.

# Chapter 16
# Repair and Maintenance Management System of Food Processing Equipment

**Fatemeh Afsharnia and Abbas Rohani**

**Abstract** The major goal of this chapter is to present the reader with concepts and examples from several food sectors that will help to reinforce his or her understanding of food processing equipment maintenance strategies. In addition, case studies will be used to illustrate the similarities and differences between the foods processing line reliability, availability, and maintainability (RAM) analysis such as juice bottling, canned products, dairy products, and milling process. Food production lines are made up of multiple equipment and machines connected by a common transfer mechanism and control system that can fail in a variety of modes. Line reliability and production rate are negatively affected by failures. Therefore, RAM analysis is an accurate and efficient method that enables managers to minimize system downtime and costs and maximize production rate and profits.

**Keywords** Processing equipment · Perishable food · Maintenance · Availability · Reliability

## Abbreviations

| Notation | Main Acronyms |
|---|---|
| FAO | Food and Agriculture Organization |
| FMEA | Failure modes and effects analysis |
| MTBF | Mean time between failures |
| MTTF | Mean time to failure |
| OEE | Overall equipment efficiency |

F. Afsharnia (✉)
Department of Agricultural Machinery and Mechanization Engineering, Agricultural Sciences and Natural Resources University of Khuzestan, Ahvaz, Iran
e-mail: afsharniaf@yahoo.com; phd.afsharnia@asnrukh.ac.ir; f.afsharnia@ferdowsi.um.ac.ir

F. Afsharnia · A. Rohani
Department of Biosystems Engineering, Faculty of Agriculture, Ferdowsi University of Mashhad, Mashhad, Iran
e-mail: arohani@um.ac.ir



H. Garg (ed.), *Advances in Reliability, Failure and Risk Analysis*, Industrial and Applied Mathematics, https://doi.org/10.1007/978-981-19-9909-3_16

| | |
|---|---|
| RAM | Reliability, availability, and maintainability |
| RBD | Reliability block diagram |
| RCM | Reliability centered maintenance |
| TBF | Time between failures |
| TQM | Total quality management |
| TTR | Time to repair |

## 16.1 Introduction

Food processing is one of the most important parts of industries and a significant contributor to food security. Food security is one of the criteria of human development and the main goal of every country. In developing countries, population and economic growth will lead to increasing demand for food. According to the FAO, in these countries, about 43.5% of basic agricultural products for human consumption are annually lost by pests, diseases, weeds, and drought in the post-harvest stage. So, despite the increased production of crops, achieving food self-sufficiency and the export possibility of many products are prevented by factors such as population growth and non-reduction of agricultural waste, which includes 30% of total production. Conversion industries can be considered a key step that uses agricultural products as raw materials. These industries can maintain the products by changing and processing them to be consumed throughout the year. The creation and expansion of conversion industries have some economic effects, including value-added creation, job creation, foreign exchange earnings as well as more use of agricultural products and subsequently prevent the loss and waste of the crop.

Recently, technology is being initiated at a continually quickening rate, which is an all-time high in the food processing industry [1]. Industrial maintenance as a service replaces traditional reactive approaches with proactive maintenance methods [2]. The processing production lines consist of $n$ equipment in a series configuration. In a series system, if a failure occurred, the entire production line will stop. Furthermore, the raw material needed for conversion industries is severely perishable including crop, livestock, and garden products. Most crop and garden products such as apple, cabbage, pomegranates, potatoes, lemons and limes, carrots, onions, and oranges are the slowest perishable vegetables and fruits and require a dry, ventilated place with little light. In dairy filling and packing lines, raw milk as the source of most dairy products is a fresh and highly perishable food. Keeping the milk at ambient temperature ($35 \pm 2$ °C), it significantly influenced the pH and lactose content of milk that it gives soured milk and reduces its shelf life. In developing countries, raw milk has a shelf life of three days when refrigerated.

The foregoing reveals that the use of reliable equipment contributes to significantly reducing processing costs and risks. The yearly cost of unreliability borne by the facility is used to assess plant and equipment dependability. This puts reliability into a business context. Higher equipment's processing line reliability lowers the cost of equipment failure. Failure reduces output and lowers gross profitability. Because of competitive environments and overall operating production costs, Barabady and Kumar [3] reported that system reliability, availability, and maintainability (RAM) have become increasingly important in recent years, an unplanned failure can result in significantly higher repair costs than planned maintenance or repair. Moreover, the implementation of a new maintenance strategy leads to an increase in the reliability of equipment, improving the quality of products, and managing the risk regarding health, safety, and environmental impacts. Yavuz et al. [4] proposed the RCM practices in the packaging machine of the food industry and explained the effect of RCM on OEE. First, the OEE is fast improved for one month, and then, the process response phase started. Bahrudin et al. [5] compared the effect of preventive maintenance and breakdown maintenance on production achievement in the food seasoning industry. The findings indicated that production success was somewhat significantly impacted by preventative maintenance. Besides that, the attainment of output is not significantly impacted by breakdown maintenance.

Cárcel-Carrasco and Gómez-Gómez [6] used qualitative techniques to investigate the application of knowledge management techniques in the maintenance activity in the era of industry 4.0. They found that the introduction of knowledge management techniques addresses topics related to daily performance, such as the company's reliability, energy efficiency, and maintainability processes, which leads to a lower failure rate, a shorter service or availability replacement time, an improvement in the use of energy, and a lowering of the maintenance processes. Maintenance activity processes are characterized by a high human factor and a high degree of tacit knowledge.

The objective of this chapter is to present the critical points of food processing lines aimed at enhancing operational efficiency and maintenance effectiveness.

## 16.2 RAM Theory

One of the quality management approaches used to improve efficiency and productivity in food production lines is reliability, availability, and maintainability. It may be used in conjunction with other TQM methods such as failure mode and effect analysis (FMEA), Pareto analysis, statistical process control, and so on. Nonetheless, there is a paucity of research on RAM analysis in food processing lines.

### *16.2.1 Reliability*

The probability that a machine or system will perform a needed function under specified conditions over a certain amount of time $t$ is known as reliability (Eq. 16.1).

$$\text{Reliability} = 1 - \text{Probability of Failure} \tag{16.1}$$

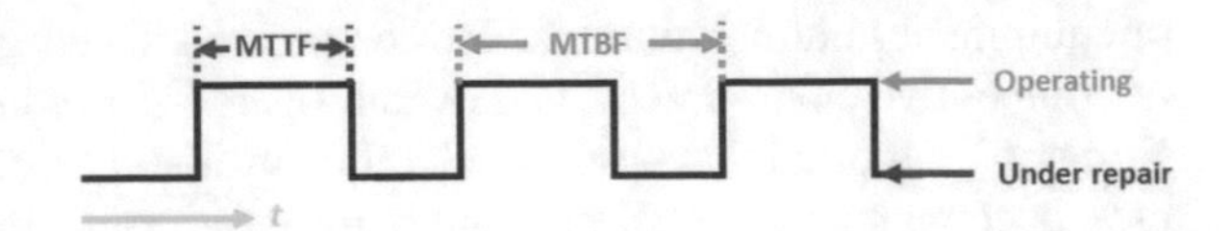


Fig. 16.1 Measurement of MTBF and MTTF

In general, we have two approaches as follows:

- More components and/or more complicated systems reduce reliability
- Simpler systems with few components increase reliability.

Reliability is calculated by Eq. 16.2 at a given time.

$$R = \mathrm{e}^{-\lambda t} \tag{16.2}$$

where $R$ is reliability that its value is between 0 and 1 that 1 indicates 100% live components and 0 indicates 0% live components. $\lambda$ is the proportional failure rate, and $t$ is the time of mission (hours).

The failure rate equals the reverse of the mean time between failures (MTBF), which can be calculated by Eq. 16.3 as follows:

$$\mathrm{MTBF} = \frac{T}{n}, \quad \lambda = \frac{1}{\mathrm{MTBF}} \tag{16.3}$$

where $n$ is the number of failures, and $T$ is defined as total time in terms of the hour. For correct measurements of $T$, the difference between MTTF and MTBF is illustrated in Fig. 16.1.

An accuracy reliability block diagram (RBD) can be constructed once a component's reliability-wise configuration has been determined. The system's reliability will also be affected by the component's or subsystem's reliability. Simple configurations can involve units arranged in parallel or series.

#### 16.2.1.1 Series Systems

With a series configuration, any failure in one component can fail the entire system (Fig. 16.2). It is usually found that at the basic subsystem level, complete systems are arranged in a series configuration, in terms of reliability. For instance, the motherboard, hard drive, power supply, and processor are the four essential subsystems that make up a personal computer. These are connected in a series system, and a failure of any of these subsystems will result in a system failure. In other words, for a series system to work, all of the units in the system must succeed. When unit 1, unit 2… and unit $\boldsymbol{n}$ succeed, and all of the other units in the system succeed, the system is said to be reliable. For the system to succeed, all $\boldsymbol{n}$ units must succeed. The system's reliability is then determined by Eq. 16.4:

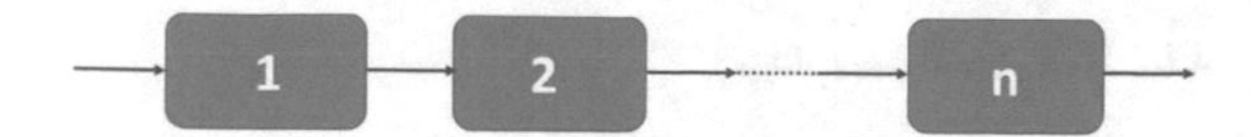

**Fig. 16.2** Series system

$$R_S = P(U_1 \cap U_2 \cap \ldots \cap U_n) P(U_1) P(U_2|U_1) P(U_3|U_1U_2) \cdots P(U_n|U_1U_2 \ldots U_{n-1}) \tag{16.4}$$

where $R_s$ is the system's reliability, $U_i$ is the event of unit ***i*** being operational, $P(U_i)$ is the probability that unit $i$ is operational.

When a component's failure impacts the failure rates of other components (i.e., when one component fails, the life distribution features of the other components change), the conditional probabilities in the equation above must be addressed. In the case of independent components, however, Eq. 16.4 becomes Eq. 16.5:

$$R_s = P(U_1)P(U_2)\ldots P(U_n) \tag{16.5}$$

Or:

$$R_s = \prod_{i=1}^{n} P(U_i) \tag{16.6}$$

Alternatively, in terms of component reliability:

$$R_s = \prod_{i=1}^{n} R_i \tag{16.7}$$

In other words, the system reliability of a pure series system is the product of the reliability of its constituent components.

***Example 1*** Four subsystems are reliability-wise in series and make up a system. Reliability of subsystem 1, 2, 3, and 4 is 95.5%, 99.1%, 98.7%, and 97.3% for a mission of 50 h, respectively. What is the overall system's reliability for a 50-h mission?

Because the subsystem reliabilities are stated for 50 h, the system's reliability for a 50-h mission is simply:

$$R_s = R_1 \cdot R_2 \cdot R_3$$
$$R_s = 0.955\,.\,0.991\,.\,0.987\,.\,0.973$$
$$R_s = 0.90888$$
$$R_s = 90.88$$

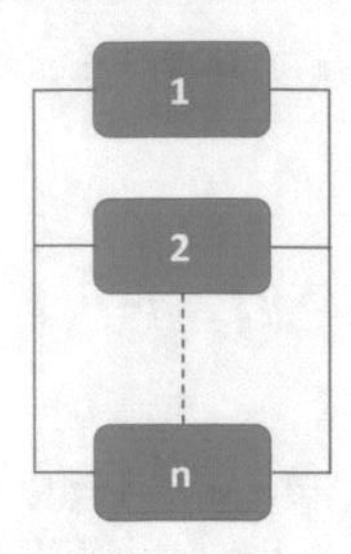


**Fig. 16.3** Parallel system

#### 16.2.1.2 Parallel Systems

As illustrated in Fig. 16.3, in a simple parallel system, the system can only succeed if at least one of the parts is successful. Units that run in parallel are often known as redundant units. Redundancy is a critical feature of system design and reliability since it is one of the strategies for increasing system reliability. It is commonly utilized in mission-critical systems in the aerospace sector. RAID computer hard disk systems, braking systems, and bridge support cables are examples of other uses.

For a system with n statistically independent parallel components, the probability of failure, or unreliability, is the chance that unit 1 fails, unit 2 fails, and all other units in the system fail. Therefore, a parallel system must collapse if all ***n*** units fail. To put it another way, if unit 1 or unit 2 or any of the remaining n units succeeds, the system succeeds. The system's unreliability is then determined by Eq. 16.8:

$$Q_s = P(U_1 \cap U_2 \cap \ldots \cap U_n) P(U_1) P(U_2|U_1) P(U_3|U_1 U_2) \ldots P(U_n|U_1 U_2 \ldots U_{n-1}) \tag{16.8}$$

where $Q_s$, $U_i$, and $P(U_i)$ are the unreliability of the system, the event of failure of unit *i*, and the probability of failure of unit *i*, respectively.

When a component's failure influences the failure rates of other components, the conditional probabilities in the equation above must be taken into account. In the case of independent components, however, the equation becomes

$$Q_s = P(U_1) P(U_2) \ldots P(U_n) \tag{16.9}$$

Or:

$$Q_s = \prod_{i=1}^{n} P(U_i) \tag{16.10}$$

Or, in terms of component unreliability:

$$Q_s = \prod_{i=1}^{n} Q_i \tag{16.11}$$

Consider the difference between the series and parallel systems: The series system's system reliability was the product of component reliabilities, but the parallel system's overall system unreliability was the result of component unreliabilities. The parallel system's dependability is then determined by:

$$\begin{aligned} R_s &= 1 - Q_s = 1 - (Q_1 \cdot Q_2 \cdot \ldots \cdot Q_n) \\ &= 1 - [(1 - R_1) \cdot (1 - R_2) \cdot \ldots \cdot (1 - R_n)] \\ &= 1 - \prod_{i=1}^{n} 1(1 - R_i) \end{aligned} \tag{16.12}$$

***Example 2*** Consider a system with three subsystems organized in parallel in terms of reliability. Reliability's subsystem 1, 2, and 3 were 95.1%, 99.7%, and 94.3% for a 50-h mission. What is the overall system's reliability for a 50-h mission?

Because the subsystem reliabilities are stated for 50 h, the system's dependability for a 50-h mission is as follows:

$$\begin{aligned} R_s &= 1 - (1 - 0.951) \cdot (1 - 0.997) \cdot (1 - 0.943) \\ &= 1 - 0.000008379 \\ &= 0.999 \\ &= 99.9\% \end{aligned}$$

### *16.2.2 Availability*

Availability is described as an item's ability to fulfill its needed function at a certain moment or over a specified length of time based on its reliability, maintainability, and maintenance support [7]. Because availability is the likelihood that a component is now in a non-failure condition, even if it has previously failed and been restored to its operational state, system availability can never be less than system reliability [8]. Consider a repairable system that is operational at time $t = 0$: When the system fails, a repair action is performed to bring the system back online. A binary variable can be used to represent the status of the system [9]:

$$X(t) = \begin{cases} 1 & \text{If the system is operating at time } t \\ 0 & \text{Otherwise} \end{cases} \tag{16.13}$$

Once the MTBF and MTTR are determined, the component's availability may be computed using Eq. 16.14:

$$A = \frac{\text{MTBF}}{\text{MTBF} + \text{MTTR}} \tag{16.14}$$

The availability in series and parallel systems is calculated by Eqs. 16.15 and 16.16, respectively.

$$A = A_x \ldots A_n \tag{16.15}$$

$$A = \begin{cases} 1 - (1 - A_x)^n & \text{If all } i = 1, \ldots, n \text{ components are identical} \\ 1 - \left(1 - A_{x1}\right).\left(1 - A_{x2}\right) \ldots \left(1 - A_{x_n}\right) & \text{If the component reliabilities differ} \end{cases} \tag{16.16}$$

Consider a system with $N$ components that are deemed operational when at least $N$-$M$ components are present (i.e., no more than $M$ components can fail). $A_{N,M}$ denotes the availability of such a system, which is expressed as follows (Eq. 16.17):

$$A_{N,M} = \sum_{i=0}^{M} \frac{N!}{i! \times (N-i)!} \times A^{(N-i)} \times (1-A)^i \tag{16.17}$$

### 16.2.3 Maintainability

The possibility that a failed machine or system will be returned to operational effectiveness within a specific time t if the repair activity is completed according to the defined protocols is known as maintainability. The possibility of performing the repair within a particular time frame is known as maintainability.

The following criteria must be addressed to achieve good maintainability [10]:

(a) The equipment or machine may fail at any time.
(b) The location of maintenance displays, checkpoints, gages, and meters, as well as the relative position of one assembly to others.
(c) The constraints imposed by the human body.
(d) The setting in which maintenance or repairs will take place.
(e) The development of testing equipment.
(f) How information is presented in the maintenance and repair handbook.

Product design, the technical level of repair staff, repair method, and repair facilities all affect maintainability. Maintainability is a probability measure of a product's ability to be maintained or restored to its original function for a given amount of time under any given set of repair conditions.

Assume $T_m$ is the repair time, and $m(t)$ is the repair density function. After that, the product's maintainability may be stated as Eq. 16.18 [11]:

$$M(t) = \begin{cases} P(0 \le T_m \le t = \int_0^t m(t)\mathrm{d}t & (t \ge 0) \\ 0 & (t < 0) \end{cases} \tag{16.18}$$

If $M(t)$ is differentiated, then:

$$m(t) = \frac{\mathrm{d}M(t)}{\mathrm{d}t} \quad (t \geq 0) \tag{16.19}$$

As a result, the repair density function $m(t)$ represents the likelihood that the defective product will be fixed to its original state in time $t$.

## 16.3 Application of RAM Analysis in the Food Processing Lines

### *16.3.1 Juice Bottling*

The automated juice bottling line is made up of several workstations linked together by a series system. There are seven workstations on the juice bottling manufacturing line (Fig. 16.4): washer, juice extractor, centrifuge, filter, pasteurizer, blender, and packer. Each workstation is made up of one or more machines, each of which is prone to distinct failure modes.

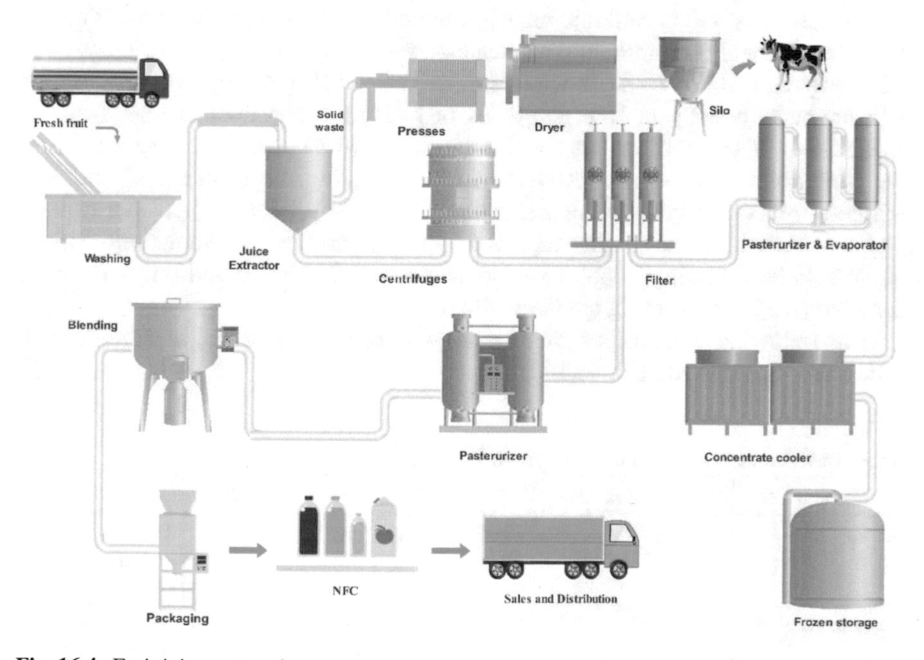


**Fig. 16.4** Fruit juice processing

Tsarouhas and Arvanitoyannis [12] carried out a reliability and maintainability analysis for the limoncello production industry to improve the operation. The production process of the limoncello production line was made up of eight separate machines including washing and peeling, peel tank, mixing, filtration and centrifugation filling, injection, and labeler and palletizer that all were connected by a single transfer mechanism and control system. Line maintenance takes place throughout the weekend. The present maintenance program includes corrective as well as preventative. When a failure occurs, the maintenance team in charge of the line's correct functioning performs the necessary breakdown maintenance to rectify the problem. Once the descriptive statistics were calculated, the best theoretical distributions were identified for TBF and TTR. They pointed out that the TBF at line level follows the Weibull distribution, whereas the TTR has a lognormal distribution. Then, the models for reliability, maintainability, hazard rate, and repair rate were developed.

Over 9 months, a reliability investigation for beer packaging was conducted Tsarouhas and Arvanitoyannis [13]. Beer production consists of sixteen stages including raw materials receipt, malting milling, mashing, lautering, boiling, clarification, cooling, fermentation, maturation, filtration, packing and sealing, bottle pasteurization, bottle inspection, bottle labeling and standardization, bottle packaging, and storage. In this study, the most common failure modes were identified, and descriptive statistics were computed at the failure and machine levels. The best match of failure data was determined using many theoretical distributions. The failure data's reliability and hazard rate models were established to predict current operation management (i.e., training, maintenance policy, spare parts) and increase line efficiency. In addition to the previous studies, Niu et al. [14] applied the AHP and EIE methods to analyze the health index of beer filling production lines systems. The results of the calculations for this production line reveal that the combined weighting approach is an effective way of calculating the health index and can correctly reflect the real production status. The health index is predicted using a support vector machine (SVM) improved by multi-parameters; simulations demonstrate that least squares support vector machine (LSSVM) based on radial basis function (RBF) has a strong prediction impact.

The following conclusions may be drawn from the features of certain juice production lines listed in Table 16.1:

**Table 16.1** Characteristics of the juice production lines

| Production line | Number of failures | TBF | TTR | Availability (%) | Refs. |
|---|---|---|---|---|---|
| Limoncello | 315 | 75,053 | 8947 | 89.34 | Tsarouhas and Arvanitoyannis [12] |
| Beer packaging | 77 | 5688.5 | 311.5 | 94.8 | Tsarouhas and Arvanitoyannis [13] |
| Juice bottling | 1261 | 23,197 | 3882 | 85.66 | Tsarouhas et al. [15] |

- The availability is between 85.66% and 94.8%, and the maximum availability is observed in the beer packaging line.
- The maximum value of TBF and TTR occurs in a limoncello production line with 75,053 and 7947 min, respectively, whereas the minimum value of TBF and TTR is observed in the beer packaging line.

### *16.3.2 Canned Products*

There are two production lines at the canned food factory. One line is dedicated to can-making (Fig. 16.5). Can production includes the process of seven stages as follows [16]:

1. **Slitting**: Producing blanks of desired dimensions from tin sheets,
2. **Welding**: Forming a cylindrical shape from two rectangular blanks welded together,
3. **Lacquering**: Applying a varnish layer to the welded blanks' inside face,
4. **Curing**: While moving to the flanging machine, the varnish is curing and drying,
5. **Flanging**: Before seaming, the can require flanging both ends,
6. **Seaming**: Cans are seamed at one end by seamer,
7. **Palletizing**: Approximately 2500 cans are placed on pallets and moved by forklift to the area where empty cans are stored.

Another production line or can filling line includes the process of sixteen stages as follows,

1. **Soaking**: Depending on the type of food, it is soaked in a hopper for 8–14 h (peas, kidney, beans, mushrooms, etc.). The plant contains five hoppers, each with a 3000-kg capacity (meat and corn do not go into the process).

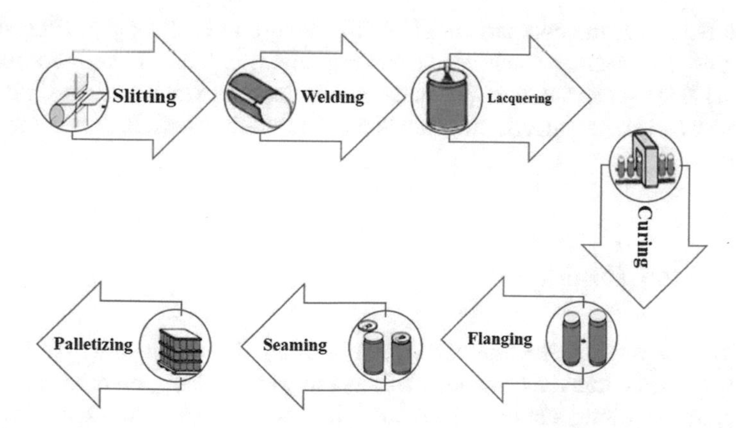


**Fig. 16.5** Can-making process

2. **Reel washing**: Showering cleans the food, and the extra water is drained. A bucket elevator transports the food to the blancher.
3. **Blanching**: To release gases and enzymes, the food is blanched for 5–30 min.
4. **De-strong**: To remove stones, the meal is transported to the de-stoner.
5. **Inspection belt**: The food is thoroughly sifted to eliminate any dark or broken bits. In the filling hopper, the food is kept.
6. **Solid filling**: The solid food is placed in empty cans.
7. **Liquid filling**: The container is filled with a liquid solution and sucked by a shower filler machine at 75–85 °C. This method extends the shelf life of canned foods while also protecting customers.
8. **Seaming**: Double seaming is used to attach the opposite lid to the can.
9. **Coding**: The coding machine prints a code on the lid of the can display the product's manufacture and expiration dates.
10. **Crate loading**: More than 500 cans are placed on a crate, and a cart transports seven layers of crates to sterilize the stage.
11. **Sterilizing**: The can in the cartons is sterilized at 121 °C. Depending on the product and the liquid used, this procedure might take anywhere from 10 to 70 min. The water is then rapidly chilled to destroy any lingering microorganisms. After that, the cans are dried.
12. **Crate unload**: The cans are removed from the container and transported to the labeler.
13. **Labeling**: The labeling machine applies labels to the cans.
14. **Label inspect**: Labels are examined to see if they have been applied appropriately.
15. **Packaging**: A tray contains 12 cans. The shrink wrapper is used to join two trays together. Two people and one forklift place every 20 boxes on a pallet.
16. **Storing**: The completed goods are kept for four days before a sample is obtained to conduct three types of testing (chemical, physical, and biological) to ensure that they fulfill specifications and are ready for distribution.

Given Fig. 16.6, the comparison of MTBF, failure rate, and repair time showed that the filler and seamer machine had the maximum failure rate and the filler and seamer, labeler, and shrink wrapper machines had the maximum repair time, while crate loader and crate unloader machines had the minimum failure rate and repair time [16].

### *16.3.3 Dairy Products*

An outline process flow diagram of the milk bottling plant is illustrated in Fig. 16.7. Because of the line imbalance in dairy filling and packing lines, extra time is needed for the upstream filling station, which has a low output rate, to work in advance to construct work-in-process. As a result, the downstream packing station can reach the target output volume by operating at greater productivity during a shift. Production

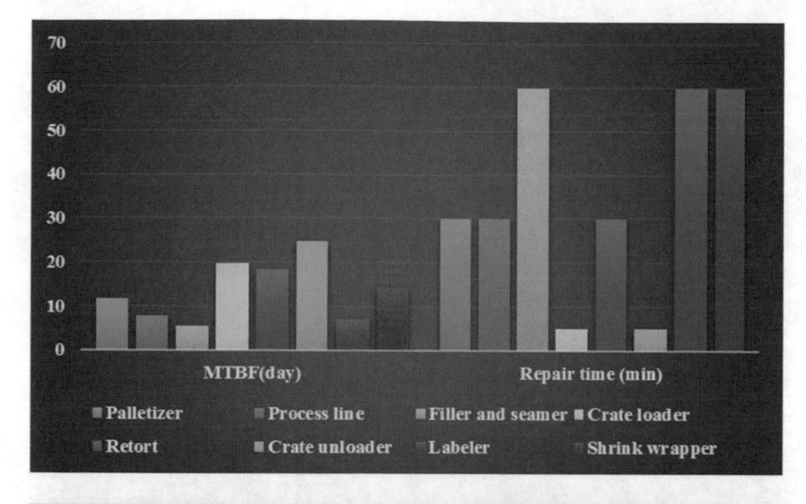


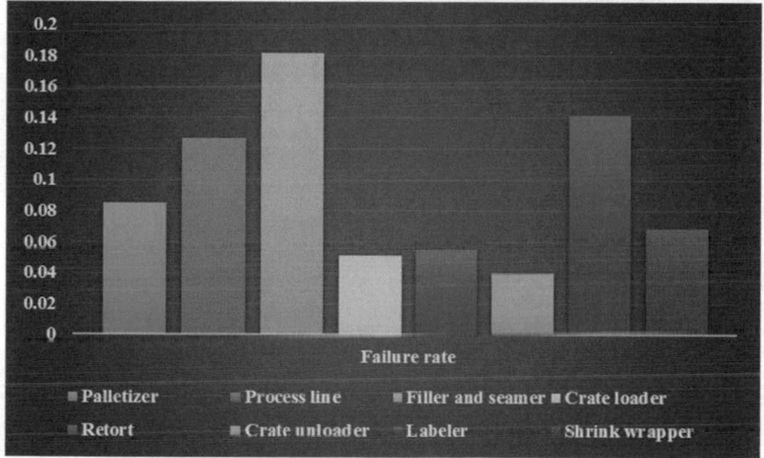


**Fig. 16.6** MTBF, failure rate, and repair time for all machines of a canned production line

lines in the food industry strive to attain a stable condition in a short amount of time. Because dairy products are perishable, they cannot be properly kept in buffers after a shift, causing systems to start the following shift with empty buffers, resulting in production loss until a steady state is attained [9].

The link between factory management and the operation of a typical Italian cheese production line is investigated in Tsarouhas's research [17]. For 26 months, failure and repair data from the line were analyzed. Descriptive statistics were obtained at the machine and line levels. The OEE's factors of availability, performance efficiency, and quality rate were also determined. The results demonstrate that the line's OEE performance is poor (76.47%) as compared to the aim of 85%. For all equipment and the full line level, a reliability analysis of an automated yogurt manufacturing line was also performed by Tsarouhas and Arvanitoyannis [18]. The assumption of

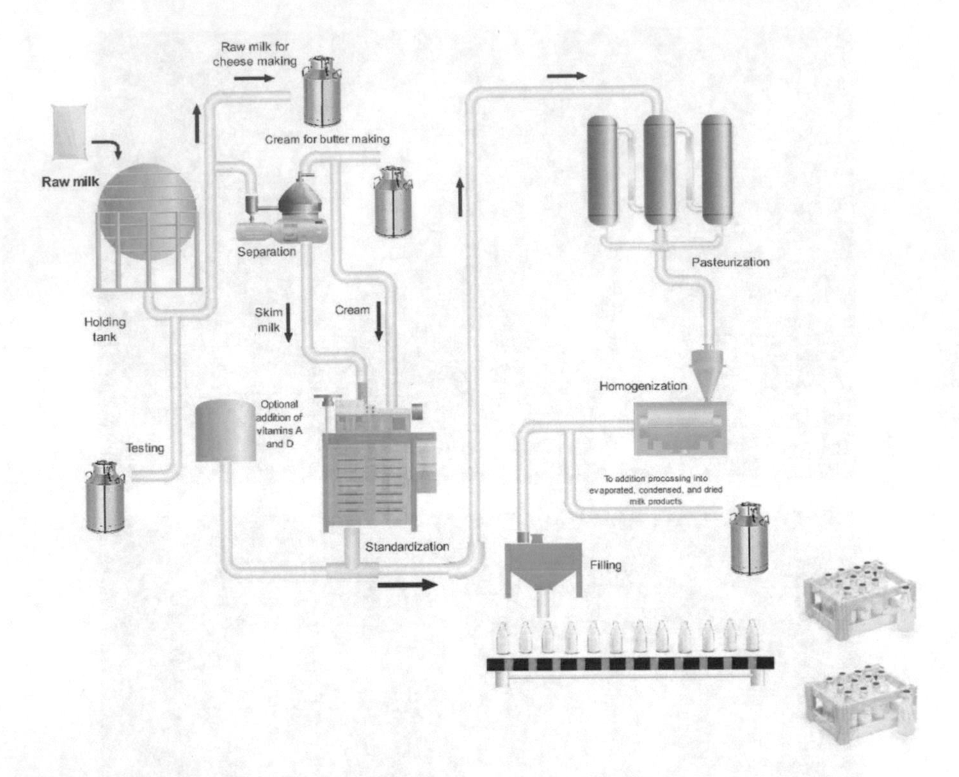


**Fig. 16.7** Milk processing flow diagram

independence was confirmed using trend and serial correlation tests, and the failure data were distributed in the same way. The best match to the failure data was found by determining the theoretical distribution parameters. In addition, models for the whole manufacturing line's dependability and failure rate were created. The models might be a beneficial tool for assessing present conditions and predicting the reliability of improving the yogurt line's operations management rules. The Weibull distribution was determined to be the best fit for the yogurt manufacturing line in terms of describing the time between failures.

Moreover, Tsarouhas [19] carried out the analysis of the RAM of a milk production line using statistical techniques of failure data. The best fitness indices were developed using descriptive statistics of the failure and repair data. Some idealized probability distributions, such as the Weibull, exponential, lognormal, and normal distributions, have their parameters determined. At various mission periods, the reliability and maintainability (R&M) of workstations and the full line have been calculated using their best-fit distribution. Both R&M proved to be valuable instruments for assessing the existing condition and predicting reliability, especially in the short term, for enhancing the milk production line's operation management.

- The availability of the milk production line was 96.13%, and the line was down or under repair for 3.9%;
- Both the heat exchanger/pasteurizer machine and the bottling machine had 57.5% of all failures on the production line; and
- Both failure and repair data had no-trend and serial correlation. Failure and repair records might be a helpful resource for milk industry manufacturers when designing new safe and reliable equipment.

In another study, the RAM analysis of ice cream analysis was created using historical data collected over 12 months. The data were subjected to a Pareto analysis, descriptive statistics, trend, and serial correlation test. For each equipment and the whole ice cream manufacturing system, failure and repair parameters were calculated. The RAM study evaluates existing operations management and makes improvements to the line's quality, productivity, and performance. It was discovered that:

- Preventative maintenance intervals for each machine and finished system were computed for various reliability periods.
- The packing machine and the freezer tunnel are the two machines with the lowest reliabilities. The importance of this equipment is critical, and it must be well-maintained to avoid quality and production losses.
- The exogenous machine, as well as the ice cream machine and the entire manufacturing system, have the poorest maintainability.

Furthermore, because the RAM indices were created to measure and improve machine performance, production managers and engineers can quickly assess the next steps and decisions they make in terms of the system's function [20].

Based on Table 16.2, the following conclusions may be derived from the features of the dairy production lines:

- The availability is between 50% and 91.2%, and the maximum availability is observed in the cheese (feta) production line.
- The maximum value of TBF and TTR occurs in the mozzarella cheese production line with 968,980 and 126,860 min, respectively, whereas the minimum value of TBF and TTR is observed in the ice cream and cheese (feta) production lines, respectively.

**Table 16.2** Characteristics of the dairy production lines

| Production line | No. of failures | TBF | TTR | Availability (%) | Refs. |
|---|---|---|---|---|---|
| Ice cream | 468 | 469 | 468 | 50 | Tsarouhas [20] |
| Traditional Italian cheese | – | 1622 | 67 | 88.41 | Tsarouhas [17] |
| Mozzarella cheese | 1889 | 968,980 | 126,860 | 88.42 | Tsarouhas [21] |
| Cheese (feta) | 292 | 748.8 | 66.15 | 91.2 | Tsarouhas et al. [15] |

### *16.3.4 Milling Process*

The economic importance of cereals and their contribution to the diets of humans and livestock cannot be disputed. Cereals are grown in most parts of the world, from near-arctic to near-equatorial latitudes. Wheat is the most important crop among the cereals by area planted and is followed in importance by corn, barley, and sorghum. The amount of wheat traded internationally exceeds that of all other grains. Furthermore, the protein and caloric content of wheat are greater than that of any other food crop. Most wheat is consumed in the form of baked goods, mainly bread; therefore, wheat grains must be milled to produce flour before consumption. Wheat is also used as an ingredient in compound feedstuffs, starch production, and as a feedstock in ethanol production. The wheat flour is produced during the mechanized process at the factory. Therefore, increasing the productivity of this mechanized process is most important. The maintenance unit is one of the best opportunities to increase the productivity of a mechanized system. For many systems, the category of human resources is considered to be the most important system capital, which has a significant impact on system performance. The manpower productivity of a factory during the ten years was evaluated taking into account the reliability of the system. The results indicated that the winnower and mill of flour factory included 44.59% and 55.41% of total failures, respectively. The results indicated that the spiral (22.7%), feeder (16.8%), and elevator (11.2%) caused the majority of recorded failures in winnower parts. Furthermore, the roller mill (21.9%), spiral (11.8%), and sieve (9.8%) parts of the mill were more prone to failure than the rest. The manpower efficiency and factory equipment failure rate showed a positive relationship as well as there is a positive relationship between manpower effectiveness and factory equipment reliability [22].

Maduekwe and Oke [2] proposed a novel Taguchi scheme-based on DEMATEL methods and DEMATEL method for the principal performance indicators of maintenance in a wheat processing plant. They calculated the direct-relationship matrix for FF, DT, MTTR, MTBF, MTTF, and availability by both Taguchi scheme–based DEMATEL and DEMATEL methods. When the DEMATEL and T-DEMATEL methodologies were applied to the issue, the results showed that downtime and availability had the greatest causal influence on other criteria. Additionally, additional variables in the selection of the key performance indicators utilizing the two methodologies have the greatest impact on the frequency of failure. To optimize and identify the causal linkages between components, the Taguchi scheme paired with the DEMATEL approach is ideal.

As illustrated in Figs. 16.8 and 16.9, grain storage drying and grain logistics contain several machines and equipment that the stoppage of each one can harm the grain supply chain. So, it is essential to investigate the failures in several sectors of this process for future research.

From the foregoing, it appears that the canned product lines had the minimum TBF and TTR as compared to the rest. Among the production lines under study, the maximum TBF and TTR occurred in dairy production lines and juice production lines, respectively. The maximum was related to the Mozzarella cheese production

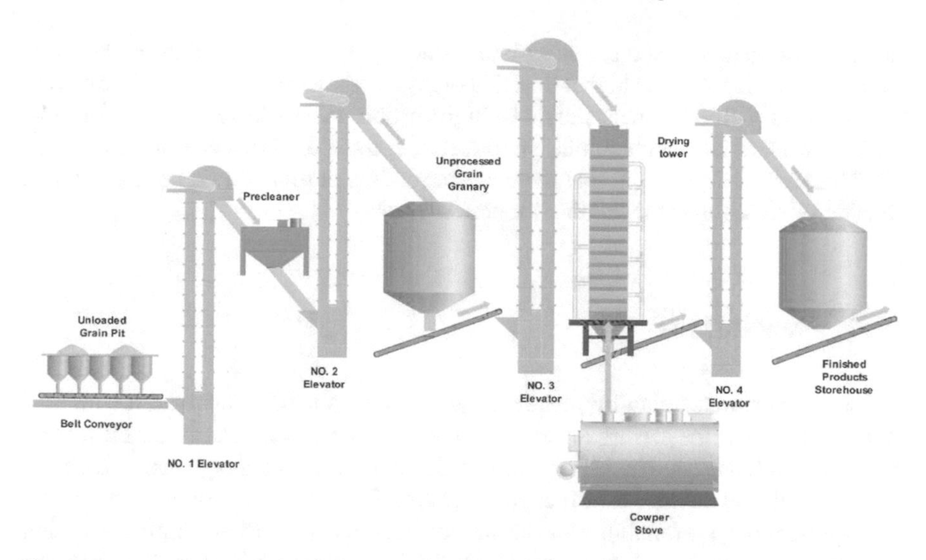


Fig. 16.8 Grain storage drying

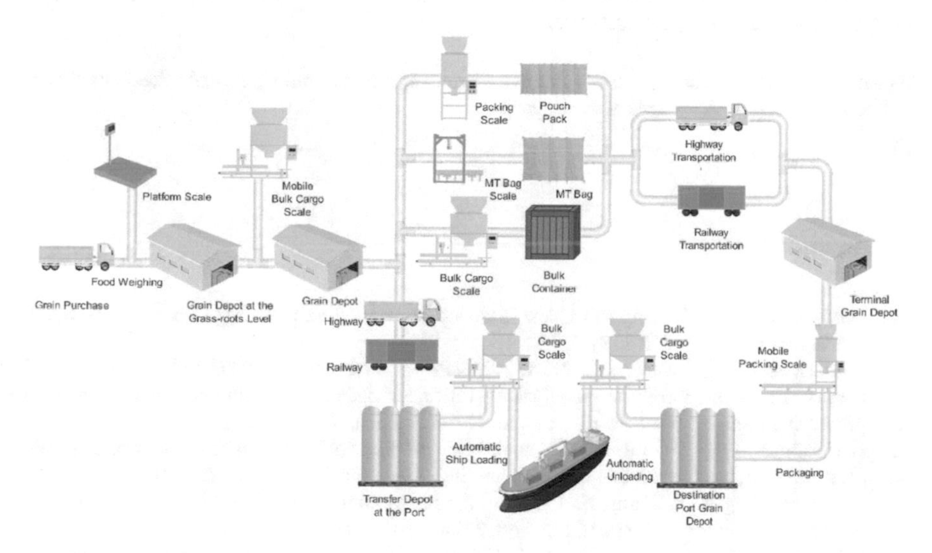


Fig. 16.9 Grain logistics

line. The maximum availability was observed in the beer packaging line, and the minimum was related to the ice cream production line. The failure rate trend of equipment is related to manufacturing defects, age of equipment, components quality, dimensioning, load rate, maintenance strategy, environment conditions, stress, etc. [23]. Moreover, the type of raw materials (such as corrosive and acidic substances) strongly affects equipment failure. Also, the compatibility of raw materials with the material of equipment is very important. For instance, acidic raw materials will cause corrosion in steel equipment and subsequently operation-stopping breakdowns. Corrosion rates of steel are higher than that of high carbon steel in all the acidic media. So, the material of equipment should be high carbon steel to decrease corrosion. The acidic raw materials include milk, yogurt, sauce, tomato, citrus, carbonated drinks, etc. Therefore, it is essential that the managers of the production line of these foods considered the equipment suitable for acidic foods.

## 16.4 Conclusions

In the food business, reliability, availability, and maintainability analysis is a method that may improve line performance and production rates, as well as assess the impact of failures and suggested modifications to present operations management. Machines with a high risk of failure can be recognized, extending the production system's useful life. Furthermore, the reliability and hazard rate models may be beneficial in assessing present circumstances and predicting reliability to improve the food production line's maintenance strategy. As a result, quantifying losses and advantages might help a corporation make better decisions.

**Acknowledgments** Authors hereby acknowledge and appreciate Iran's National Elites Foundation for financial support of postdoc research project.

## References

1. Dani, R., Rawal, Y.S.: Impact of new innovations in food and beverage service industry. J. Emerg. Technol. Innov. Res. **6**(3), 128–134 (2019)
2. Maduekwe, V.C., Oke, S.A.: Novel Taguchi scheme-based DEMATEL methods and DEMATEL method for the principal performance indicators of maintenance in a food processing industry. Int. J. Intell. Comput. Cybern. **14**(3), 363–397 (2021)
3. Barabady, J., Kumar, U.: Reliability analysis of mining equipment: a case study of crushing plan at Jajarm bauxite mine in Iran. Reliab. Eng. Syst. Saf. **93**, 647–653 (2008)
4. Yavuz, O., Doğan, E., Carus, E., Görgülü, A.: Reliability centered maintenance practices in food industry. Procedia Comput. Sci. **158**, 227–234 (2019)
5. Bahrudin, A., Setiawan, I., Arifin, M.M., Fipiana, W.I., Lusia, V.: Analysis of preventive maintenance and breakdown maintenance on production achievement in the food seasoning industry. OPSI **14**(2), 253–261 (2021)
6. Cárcel-Carrasco, J., Gómez-Gómez, C.: Qualitative analysis of the perception of company managers in knowledge management in the maintenance activity in the era of industry 4.0. Processes **9**, 121 (2021)
7. Rausand, M., Høyland, A.: System Reliability Theory: Models, Statistical Methods, and Applications, 2nd edn. Wiley, NJ, USA (2004)
8. Ebeling, C.E.: An Introduction to Reliability and Maintainability Engineering. McGraw Hill, New York, NY (1997)
9. Tsarouhas, P.: Reliability, availability and maintainability analysis in food production lines: a review. Int. J. Food Sci. Technol. **47**(11), 2243–2251 (2012)

10. Gaonkar, R.S.P., Verlekar, M.V.: Reliability and maintainability of safety instrumented system. In: Safety and Reliability Modeling and Its Applications, pp. 43–90. Elsevier, Amsterdam (2021)
11. Wang, S., Tomovic, M., Liu, H.: Comprehensive reliability design of aircraft hydraulic system. In: Commercial Aircraft Hydraulic Systems, pp. 115–169 (2016)
12. Tsarouhas, P.H., Arvanitoyannis, I.S.: Reliability and maintainability analysis to improve the operation of the limoncello production line. Int. J. Food Sci. Technol. **47**(8), 1669–1675 (2012)
13. Tsarouhas, P.H., Arvanitoyannis, I.S.: Assessment of operation management for beer packaging line based on field failure data: a case study. J. Food Eng. **98**(1), 51–59 (2010)
14. Niu, G.C., Wang, Y., Hu, Z., Zhao, Q., Hu, D.M.: Application of AHP and EIE in reliability analysis of complex production lines systems. Math. Probl. Eng. 1–10 (2019)
15. Tsarouhas, P.H., Arvanitoyannis, I.S., Varzakas, T.H.: Reliability and maintainability analysis of cheese (feta) production line in a Greek medium-size company: a case study. J. Food Eng. **94**(3–4), 233–240 (2009)
16. Savsar, M., Çiçek, H.: Effects of maintenance policies on production costs and system reliability in a canned food factory. Avrupa Bilim ve Teknoloji Dergisi **29**, 387–396 (2021)
17. Tsarouhas, P.H.: Equipment performance evaluation in a production plant of traditional Italian cheese. Int. J. Prod. Res. **51**(19), 5897–5907 (2013)
18. Tsarouhas, P.H., Arvanitoyannis, I.S.: Yogurt production line: reliability analysis. Prod. Manuf. Res. **2**(1), 11–23 (2014)
19. Tsarouhas, P.: Evaluation of reliability, availability and maintainability of a milk production line. Int. J. Ind. Syst. Eng. **31**(3), 324–342 (2019)
20. Tsarouhas, P.: Reliability, availability, and maintainability (RAM) study of an ice cream industry. Appl. Sci. **10**(12), 4265 (2020)
21. Tsarouhas, P.H.: Application of statistical approaches for analyzing the reliability and maintainability of food production lines: a case study of mozzarella cheese. In: Mathematical and Statistical Methods in Food Science and Technology, pp. 491–510 (2014)
22. Afsharnia, F., Marzban, A.: Evaluation of manpower productivity in repair shop taking into account the reliability of the system (case study: flour factory). J. Res. Mech. Agric. Mach. **8**(1), 67–76 (2019)
23. Afsharnia, F.: Failure rate analysis. In: Failure Analysis and Prevention. IntechOpen, UK (2017)

**Fatemeh Afsharnia** received her BS, MS, and Ph.D. degree in agricultural mechanization engineering form the Agricultural Sciences and Natural Resources University of Khuzestan, Ahvaz, Iran. She was also a visiting scholar at the Condition Monitoring Center at the Sharif University of Technology for six months in 2019. Dr. Fatemeh Afsharnia is a member of the Iranian Maintenance Association (IRMA) and has published over 30 publications in several journals. Moreover, she has been working as a reviewer in peer-review journals such as Measurement (Elsevier), International Journal of Quality and Reliability Management (Emerald), Journal of the Brazilian Society of Mechanical Sciences and Engineering (Springer), etc. Her main research interests are Reliability Engineering and Analysis, Maintenance Management, Maintenance Planning, Regression Modeling, Reliability Theory, Risk Assessment and Analysis, Safety Management and Engineering, Probabilistic Risk Analysis, Maintenance optimization, Condition-based Maintenance, etc.

**Abbas Rohani** born in 1980, is currently a professor at Ferdowsi University of Mashhad (FUM), Iran. He received his Ph.D. degree from University of Tabriz, Iran, in 2009. His research interests include Physical Asset Management (Reliability and Maintenance), Renewable Energy Technologies.

# Chapter 17
# Reliability, Availability, Maintainability, and Dependability of a Serial Rice Mill Plant (RMP) with the Incorporation of Coverage Factor

**Nafisatu Muhammad Usman and Ibrahim Yusuf**

**Abstract** Due to growing consumer demand, a number of businesses, including the plastics and food industries, are struggling to improve efficiency and customer satisfaction. The coverage factor "*c*" is extremely important in this direction for the manufacturing sectors' day-to-day operations. One parameter that can increase system efficiency without significantly increasing maintenance costs is the coverage factor because when a covered defect occurs, the system will automatically recover. The chance of the system recovering from the malfunction that caused it is how Arnold (1973) defined coverage factor. A serial rice mill plant with three subsystems, the cleaning and destoning section, the husking and paddy separation section, and the polishing and bagging section is the subject of this article's analysis of reliability, availability, maintainability, and dependability (RAMD). The system can be in one of three operational states while it is in use: full capacity, decreased capacity, or failed status. The Chapman-Kolmogorov differential equations are constructed using the Markov birth–death process and the transition diagrams of all subsystems that integrate the coverage factor. Each subsystem is designed to have independent failure and repair rates that follow an exponential distribution. The system's reliability, availability, maintainability, and dependability all factors crucial to system performance have been quantified for various subsystems and are shown in tables. There are also estimated values for dependability ratio, mean time to failure (MTTF), and mean time between failures (MTBF). The machine's performance has been assessed on the basis of the numerical results attained.

**Keywords** Rice mill · Probability · MTTF · Availability · Reliability

N. M. Usman (✉)
School of General Studies, Kano State Polytechnic, Kano, Nigeria
e-mail: mamankhairat2015@gmail.com

I. Yusuf
Department of Mathematical Sciences, Bayero University, Kano, Nigeria
e-mail: iyusuf.mth@buk.edu.ng



H. Garg (ed.), *Advances in Reliability, Failure and Risk Analysis*, Industrial and Applied Mathematics, https://doi.org/10.1007/978-981-19-9909-3_17

## 17.1 Introduction

A crucial step in ensuring successful operations and production is dependability, maintainability, availability, and reliability (RAMD) analysis, which also aims to pinpoint the system's weaker parts. RAMD evaluates the system using a variety of performance modeling techniques at different stages. RAMD evaluation can be used to derive the important performance measures. The following metrics are among them: MTBR, MTTR, availability, maintenance-ability, dependency ratio, reliability, and dependency minimum. Planning maintenance strategies to improve system performance frequently use these performance indicators.

A rice mill plant is made up of various parts or subsystems. The overall effectiveness of the entire system is dependent on the availability of particular components or subsystems. High levels of availability and reliability are required for RMPs. These two are frequently used to gage how well any RMPs provide their services. The features of each component or subsystem of the rice mill plant must be examined in order to identify the factors that have the greatest influence on how the quality is perceived. In order to ensure trustworthy and dependable system performance, components are designed to be highly durable in the sense that they rarely suffer from abrupt breakdowns. Even now, abrupt disruptions are impossible to completely prevent, according to system operators. As a result, this article proposes adding a coverage factor to each component or subsystem.

Because the probability of a fault tolerance system's successful reconfiguration operation is characterized as a coverage factor, according to Kumar and Kumar (2011), it offers a more realistic picture of system behavior and more support for reliability estimations. According to Ram (2013), the conditional chance of recovery after a problem has occurred is the coverage factor. Copula-based fault coverage metrics for repairable parallel systems were discussed by Tyagi et al. [1]. Incorporating the idea of coverage factor and two different types of repair facilities, they present a stochastic model for analyzing the behavior of multi-state systems made up of two non-identical units. They came to the conclusion, based on their findings, that the use of copula and coverage techniques together tends to increase system availability and reliability. A copula-coverage approach was used in Ram and Goyal's presentation of a bi-directional system analysis. In this paper, the legion stochastic model for repairable systems has been developed by assuming different types of time trends, failure modes, and repair effects. This study illustrates a fresh idea for two different types of repairable three state fault tolerant systems. The Gumbel-Hougaard family of copula method is used in this study to forecast how the coverage factor will affect the planned system's dependability properties. This demonstrates that a covered problem can cause the system to automatically recover. The RAMD or coverage techniques have been employed by many engineers to guarantee system availability and dependability as well as to enhance system features.

Numerous techniques have been employed by researchers to evaluate dependability metrics in the literature. Reliability, maintainability, and availability study on reciprocating compressors were covered by Corvaro et al. [2]. The engineering approaches, tools, and methodologies used in this written up are mean time to failure, equipment downtime, and system availability numbers, which are used to identify and quantify equipment and system failures that obstruct the accomplishment of productive goals. The study, which was carried out in collaboration with a private company we will call RC company for privacy reasons, was based on the analysis of the behavior of states defined for each individual part and component of a reciprocating compressor. It also sought to identify and evaluate the effects of RAM-type factors. The dependability, reliability, maintainability, and availability of a computer-based test (CBT) network system were examined by Sunusi et al. (2021). CBT is software that allows for the online administration of exams to local or remote candidates. The Chapman-Kolmogorov differential equations are then created using the Markov birth–death process, leading to the construction of all subsystem/component transition diagrams and the eventual acquisition of the dependability metrics. In a steam turbine power plant, the generator's dependability and maintainability were examined by Gupta et al. (2021). This study is designed to explore different generator reliability metrics used in STP using a RAMD technique at the component level. For this reason, all of the generator's subsystems' mathematical models for the Markovian birth–death process have been created and examined. Goyal et al. [3] looked into the physical processing unit of a sewage treatment plant's reliability, maintainability, and sensitivity analysis. Five components make up this system, which is configured in series. The subsystems' failure and repair rates have been assumed to be exponentially distributed. The Markovian birth–death process is used to develop Chapman-Kolmogorov differential equations, and many metrics, including dependability ratio, mean time between failures, and mean time to repair, are also obtained. The raw sewage sump was discovered to be the plant's most sensitive subsystem by numerical simulations, with a dependability of 0.382893. Reena and Basotia (2020) conducted research on the cement production patil's reliability and maintainability. The system breakdown technique described above was used to model each subsystem mathematically utilizing the Marcovian birth–death process. Copula-based measurements of repairable parallel systems with fault coverage were discussed by Tiyagi et al. [1]. Saini et al. (2019) used the RAMD technique to examine microprocessor systems. State transition diagrams have been created for each subsystem in their investigation, and differential equations have been constructed utilizing the Markov birth–death process. The performance analysis of the small and medium-sized enterprises' tire production systems was studied by Velmurugan et al. [4]. This study's primary goal is to put the novel method to use in identifying the most important subsystems in the rubber industry's tire production system. The Chapman-Kolmogorov method is used in this study to construct the equations for the subsystems and measure the impact of variation in maintenance indices, or RAMD, to determine which component of the tire production system is crucial. Additionally, each subsystem of the tire production system's dependability ratio and mean time between failures (MTBF), mean time to repair (MTTR), and other maintenance characteristics are calculated. Finally, using

MATLAB software to change the rates of failure and repair of various subsystems, RAMD analysis of tire production systems has been carried out in order to estimate the most crucial subsystem. Aggarwal et al. [5] used the RAMD technique to build a performance model of a dairy plant's system for producing skim milk powder. The goal of this research is to present a technique for computing RAMD indices to assess and enhance the performance of a dairy plant's system for producing skim milk powder under actual operating circumstances. The work being done now involves creating a performance model based on the Markov birth–death process. Six units make up the production system for skim milk powder. The mnemonic rule is used to derive the first order governing differential equations, which are then solved to calculate RAMD indices. From a maintenance standpoint, subsystem SS1, which consists of the chiller and cream separator, is the most crucial. In order to evaluate the effectiveness of the A-pan crystallization technology used by the sugar industry, Dahiya et al. [6] used the RAMD approach. In this paper, a fuzzy dependability technique has been used to attempt to develop a mathematical model of the A-pan crystallization system of a sugar factory. Four subsystems are arranged in a succession in the A-pan crystallization system. While the third and fourth subsystems are configured as a single unit, the first and second subsystems are configured as a 2-out-of-2: G with two cold standby. Consideration of the exponential distribution of failure and repair rates has led to the proposal of a mathematical model. Differential equations have been constructed by taking into account the fuzzy reliability approach and Markov birth–death model. The fuzzy availability is then obtained by solving these equations using the fourth order Runge–Kutta method in MATLAB. Tsarouhas and Arvanitoyannis [7] studied and analyzed the reliability of the yogurt manufacturing process. A reliability analysis of an automated yogurt production line was conducted at the line level and for all machines. It was determined which theoretical distribution parameters suited the failure data the best. The whole production line's reliability and failure rate models were also established. The models might prove to be a helpful tool for updating the operations management practices of the yogurt production line as well as for evaluating the current situation and predicting reliability. The following conclusions were reached: (a) the Weibull distribution best described the time between failures for the yogurt production line, (b) the failure rate of the production line increased, indicating that the current maintenance strategy is insufficient and needs to be upgraded soon; and (c) in order to prevent losses related to quality and productivity, the reliability must be improved first on the pasteurizing boiler and then on the filling machine. The TBF, TTR, and data accessibility have all been statistically examined. Aggarwal et al. [8] developed a mathematical model for assessing the effectiveness of serial processes in the sugar plant's refining system using RAMD analysis. The important component of the system is determined by computing the reliability, availability, maintainability, and dependability (RAMD) factors or indices in this study. Chapman-Kolmogorov differential equations are derived through mathematical modeling of the system using the Markov birth–death process. With the help of the mean time between failures (MTBF), mean time to repair (MTTR), and dependability ratio parameters for each component of the system, these equations are further solved, and RAMD parameters are derived. By altering the failure and repair rates of each subsystem of the system, sensitivity analysis has been done to identify the most crucial component of the system. Niwas and Garg [9] developed a method for gaging the profitability and reliability of an industrial system based on the cost-free warranty policy. In this study, the system enters a rest phase after operating for an arbitrary length of time in order to improve operational efficiency and lower the failure rate both during and after the warranty term. The mechanism resets once you have had a full night's sleep. Additionally, a negative exponential distribution is assumed for the failure and repair rates of the system's components during formulation. The many metrics for a system, including reliability, mean time to system failure, availability, and predicted profit, are generated using a mathematical model of the system based on the Markov process. A reliability and availability study of the skim industrial powder business was put up by Aggarwal et al. [10]. In this research, a numerical technique is put forth to determine the skim milk powder system's mean time between failures (MTBF) and long-term availability and reliability. It is a complicated system made up of six repairable subsystems, including the chiller, cream separator, pasteurizer, evaporator, drying chamber, and packaging subsystems. These subsystems are arranged in series or parallel configurations. The failure and repair rates of each subsystem are assumed to follow an exponential distribution in this analysis, which is based on the Markov birth–death process. The system is mathematically formulated, and Chapman-Kolmogorov differential equations are produced. The Runge–Kutta fourth order method is then utilized to numerically solve these differential equations. In order to determine the profit of an engineering system with several subsystems in a series structure, Kumar et al. [11] employed reliability and availability analysis. De Sanctis et al. [12] provided engineers with several maintenance techniques for addressing problems including excessive costs, safety, and environmental protection. They also recommended a methodology for enhancing industry performance. To do this, equipment from the oil and gas industry was used as a case study object in a RAMD analysis. A sewage treatment plant physical processing unit reliability, maintainability, and sensitivity analysis was given by Goyal et al. in [3]. The three components of a sewage treatment plant are physical processing, chemical processing, and biological process. The most crucial component is the physical process, which consists of five parts set up in a series format. The subsystems' failure and repair rates are thought to follow an exponential distribution. Several metrics, including mean time between failures, mean time to repair, and dependability ratio, are obtained from Chapman-Kolmogorov differential equations utilizing the Markovian birth–death process. The plant's dependability sensitivity analysis has also been carried out. The RAMD investigation reveals that the raw sewage sump, with a reliability of 0.382893, is the most vulnerable component of the plant. To model the performance of significant engineering systems, Sharma and Kumar [13] used the RAM technique. The use of RAM analysis in a process industry is discussed in this research. The behavior of the system is modeled using the Markovian technique. Transition diagrams are created for various subsystems to be analyzed, and differential equations related to them are created. Following the discovery of the steady-state solution, reliability and maintainability values are estimated for various mission times. The computed findings are made available to plant personnel for active review. They were able to analyze the system behavior using the data, which significantly improved the system's performance when appropriate maintenance policies and methods were adopted. Saini and Kumar [14] employed RAMD analysis to examine an evaporation system's performance in the sugar industry. The primary goal of this study is to examine how reliability, availability, maintainability, and dependability are applied in order to identify the sugar plant's most sensitive evaporation system component. All of the subsystems' transition diagrams are generated for this study, and the corresponding Chapman-Kolmogorov differential equations are derived using the Markov birth–death process. Each subsystem's failure and repair rates are all exponentially distributed. For all three subsystems, dependability ratios, mean times between failures, and mean times to repair are all computed as additional system effectiveness metrics. System reliability sensitivity analysis is also carried out. At various time instants, the maintainability and reliability are estimated. From a reliability standpoint, analysis of all the subsystems reveals that sulfite syrup is extremely sensitive. Garg et al. [15] used an artificial bee colony and fuzzy approach to analyze the performance of repairable industrial systems. In this research, an unique method for computing these parameters using readily accessible or collected data has been introduced. Its name is an artificial bee colony based Lambda-Tau. In this method, the Lambda-Tau methodology is used to generate the expression of RAM parameters, and the related membership functions are computed by creating a nonlinear programming problem. To increase system productivity, a generalized RAM-Index has been utilized to rank the system's components according to their performance. A case study of a paper industry washing unit was used to test the proposed strategy, and the calculated results are then contrasted with those of existing Lambda-Tau and evolutionary algorithm techniques. Tsarouhas et al. [16] examined the dependability, availability, and maintainability of a cheese (feta) production line at a medium-sized Greek company. A 17-month reliability, maintainability, and availability analysis of the cheese production line were conducted. The failure and repair data were fitted as closely as possible to the common theoretical distributions, and the corresponding parameters were determined. Additionally, the full production line's reliability and danger rate modes were calculated. It was discovered that (a) the cheese manufacturing line's availability was 91.20% and dropped to 87.03%, (b) the dominant four failure mechanisms accounted for 62.2% of all failures, and (c) an average failure occurs every 12.5 operation hours with a mean repair time of 66 min. In addition to providing certain maintenance priorities, Malik and Tewari [17] established a mathematical model for evaluating the performance of a water flow system. Condenser, condensate extraction pump (CEP), low pressure heater (LPH), deaerator, and boiler feed pump are the five subsystems that make up the system (BFP). With the use of a normalizing condition and the Markov technique, the Chapman-Kolmogorov equations are created based on the transition diagram and then solved recursively to produce performance modeling. With the use of various combinations of failure rates and repair rates for all subsystems, availability matrices are created. Plots representing the failure rates and repair rates of various subsystems are used to evaluate the performance of each subsystem in terms of the availability level attained. Based on the repair rate, the various WFS subsystems' maintenance priorities are set. Mehta et al. (2018) used the additional variable technique to analyze the availability of an industrial system. The development of a mathematical model for evaluating the accessibility of a butter oil producing system is the goal of this study. The heater, clarifier, filler, and granulation subsystems make up the industrial system. The Chapman-Kolmogorov differential equations have been constructed from the system's state transition diagram using the mnemonic rule, under the assumptions of constant failure rates and variable repair rates. La-grange's method was used to solve these equations, and the Runge–Kutta fourth order method was used to calculate the system's availability for various failure and repair rates. The mean time between failures has been mathematically calculated. In order to help the plant management, choose the maintenance priorities for the best use of the resources, criticality analysis has also been done to help with maintenance priority ideas. The research of reliability modeling of a parallel system with a supporting unit and two types of preventive management was presented by Yusuf [18]. Both online and offline preventive maintenance are being performed on the system. Before units or systems fail, online preventive maintenance (PM) is performed, whereas offline preventive maintenance is performed on the external supporting equipment after units or systems fail. Explicit expressions for system effectiveness that are crucial to reliability engineers, maintenance managers, system designers, etc., have been produced using the Kolmogorov forward equations method. Graphical representations are provided to highlight significant results based on presumptive numerical values provided to system parameters. To demonstrate the impact of online and offline preventative maintenance, comparisons are made. Using PSO and IFS approaches, Garg and Rani [19] suggested a novel method for examining the dependability of industrial systems. This paper's major goal is to offer a method for computing the intuitionistic fuzzy set's (IFS) membership functions using shaky, uncertain, and ambiguous data. As a result, instead of using fuzzy operations to formulate a nonlinear optimization problem, these spreads were optimized. Their membership functions have been created using particle swarm optimization (PSO). Finding the crucial system component has involved doing sensitivity and performance analyzes. The computed results are then contrasted with earlier findings. The majority of the equipment used in the mining industry is reliant on operating procedures, maintenance procedures, and working conditions, according to Barabady and Kumar [20]. And they got to the conclusion that component reliability largely determines the plant's performance. As a result, a case study-based methodology has been developed to pinpoint the parts that are less reliable. In order to examine system performance in repairable industrial systems utilizing the genetic algorithm (GA) and the Markovian technique, Komal et al. (2010) created parametric calculations and indices of RAM. This research offers a method for estimating the RAM characteristics of these systems using both available knowledge and speculative data. This is accomplished using the genetic algorithms-based Lambda-Tau (GABLT) approach. In this method, formulas for the RAM parameters of the system are produced using the conventional Lambda-Tau methodology, and these parameters are computed using a genetic algorithm using quantified data in the form of triangular fuzzy numbers. In order to rank the system's component parts according to their performance, a general RAM-Index is employed for post RAM analysis. Garg [21] examined an industrial system's performance using a hybridized approach based on soft computing. This study quantified the data uncertainties as fuzzy numbers and applied them to the various dependability characteristics of the industrial system, which represents the behavior of the system. The system's parameters' associated membership functions are calculated by creating and solving a nonlinear optimization model. The obtained results were contrasted with the traditional and existing methodologies and results, and it was discovered that there were fewer levels of uncertainty throughout the investigation. The most important system component has also been the subject of a sensitivity and performance investigation. Finally, a method has been demonstrated using a case study of a repairable industrial system, a cattle feed facility. Sharma and Sharma [22] made an effort to incorporate a framework to optimize RAM and cost choices in a processing facility. The principles of fuzzy mathematics are used in the quantitative analysis to quantify the imprecise and vague information regarding the system failure behavior in terms of fuzzy and crisp values. A resource optimization approach based on multi-stage decision making (MSDM) has also been suggested to control the system reliability for the optimal economic performance. The model uses precise output values for unit dependability in addition to pertinent system data (number of components, manpower, cost ranges). Root cause analysis (RCA) and failure mode and effects investigation are used in the qualitative analysis to conduct an in-depth analysis of the system (FMEA). Gray relation analysis (GRA) and the fuzzy decision-making system (FDMS) are used to address the ambiguities in the conventional FMEA (GRA). A case has been used to illustrate the suggested framework. In order to analyze the availability of an ice cream production unit, Kumar and Mudgil [23] developed a methodology that included three possible states for each subsystem with constant failure and repair rates. The three states of various components considered in the paper are good, reduced, and failed. Each subsystem's failure and repair rates are considered constant and statistically independent. The system's mathematical formulation uses the Markov birth–death process. State transition diagrams were used to derive the various differential equations. Following that, steady-state probabilities are calculated by combining different failure and repair rates. Decision matrices are generated based on various performance levels in terms of availability. Following an analysis of each subsystem's performance, all subsystems' maintenance plans are implemented. Using PSO and fuzzy approaches, Garg [24] assesses the industrial reliability, availability, and maintainability. This paper's goal is to propose a method for evaluating an industrial system's system performance using unknown data. In this essay, fuzzy set theory has been employed for analysis, and particle swarm optimization has been used to solve a nonlinear optimization problem to produce the relevant membership functions. A composite measure of reliability, maintainability, and availability (RAM) known as the RAM-index has been introduced to help identify the system's critical component, which has the greatest impact on the performance of the system. It influences how the failure and repair rate parameters affect the system's performance. In the analysis, time-varying failure and repair rate parameters are used rather than constant rate models. The computed findings are then finally compared to established procedures. With the aid of a case, the proposed framework has been demonstrated. Availability modeling and evaluation of repairable systems exposed to slight degradation under imperfect fixes were examined by Yusuf [25]. The modeling and availability assessment of a system subject to slight deterioration and unsatisfactory repair are the topics of this research. In this study, we developed a probabilistic explicit formulation of system availability and examined how failure, repair rate, and the number of states affected system availability. The system's maximum practicable availability level is also established. Tewari et al. [26] discussed employing genetic algorithm technique to improve the performance of a sugar plant's crystallization unit. Three primary subsystems are arranged in series in the sugar industry's crystallization unit. The mathematical formulation of the problem is done using a probabilistic approach, and differential equations are built on the basis of the Markov birth–death process. The exponential distribution for the likely failures and repairs is taken into consideration. The steady-state availability of the crystallization unit is then determined by solving these equations under normalizing conditions. Using evolutionary algorithms, the performance of each crystallization subsystem in a sugar factory has also been improved. The authors of Choudhary et al. [27] proposed a strategy for improving cement plant dependability. Over a two-year period, the system's MTBF and MTTR were determined, and RAMD indicators were examined. Reliability, availability, and maintainability (RAM) analysis of a cement plant's subsystems helps increase availability by preventing failures and cutting down on maintenance time. According to a reliable two-stage failure process that includes the defect initialization stage and the defect development stage, both of which have competing failures, Qiu and Cui [28] certified a system reliability performance. The fact that these two stages share a shock mechanism that is characterized by a non-homogeneous Poisson process illustrates the dependence between them. The two stages of the random hazard rate are what define the effect of shock damage on system failure behavior. We take into consideration two common and competitive failure mechanisms, defect-based failure, and duration-based failure, based on the actual failure behavior of industrial systems. We derive several system reliability findings and demonstrate how, with various parameter values, our model reduces to a number of traditional competing risk models. A study on the dependability analysis of a robotic system using a hybridized technique was presented by Kumar et al. (2018). The current study makes use of a hybridized methodology. With this method, uncertainties are quantified using fuzzy set theory, the system is modeled using fault trees, mathematical expressions for the system's failure and repair rates are created using the Lambda-Tau method, and the problem of nonlinear programming is solved using genetic algorithms. Different robotic system dependability metrics are estimated, and the outcomes are compared to the current method. The robotic system's parts are constant and follow an exponential distribution. Additionally, sensitivity analysis is carried out, and the impact on the mean time between failures (MTBF) of the system is addressed by changing other reliability factors. In the forming industry, Velmurugan et al. [29] presented a RAM analyzes. This study's primary objective is to analyze maintenance activities in the small and medium-sized enterprise (SME) sector and recommend the optimum maintenance management strategy for the existing working environment. Because the Markov model is a potent tool for reliability, maintainability, and safety (RMS) engineering, and because it is a straightforward modeling approach for reliability measurement with respect to value of reliability availability, Markov analysis is used in this study to predict future sequence maintenance activity models. These three functions were selected for this research analysis because they directly affect the system's maintainability (RAM). MATLAB software is used to solve all mathematical functions. The proposed model's output offers a new check sheet for planned maintenance of the specified manufacturing environment together with a new maintenance model sequence with an optimal cost structure. A brand-new method for reliability analysis with time-variant performance characteristics was put forth by Wang and Wang [30] and is known as nested extreme response surface. To create a nested response surface of time corresponding to the extreme value of the limit state function, this method makes use of the cringing model. The NERS strategy and the efficient global optimization (EGO) technique are combined to extract the limit state function's extreme time responses for any given system design. Based on the mean square error (MSE), an adaptive response prediction and model maturation (ARPMM) mechanism are created to simultaneously increase the proposed approach's accuracy and computing efficiency. The time-dependent reliability analysis can be transformed into the time-independent reliability analysis using the nested response surface of time, and existing advanced reliability analysis and design methodologies can be employed. For engineered system design with time-dependent probabilistic constraints, the NERS technique is compared to other time-dependent reliability analysis approaches and linked with RBDO. The effectiveness of the suggested NERS strategy is shown through the use of two case studies. A series–parallel industrial system's performance evaluation was recently covered by Sanusi et al. [31]. Systems of first order differential equations of the developed model were created through the transition diagram to derive the steady-state probability. Recursively, these equations were solved. Analyzes and investigations were done on the availability at steady-state. In the form of availability matrices, the consequences of failure and repair rates for each subsystem were shown. The availability tends to decrease/increase when failure/repair rates rise, according to the availability metrics.

They were able to concentrate on the many viewpoints on reliability engineering provided by a number of researchers by doing a literature survey. The numerous factors that affect system reliability have all been carefully evaluated. Studies on the accessibility of various procedures in various industries have also been conducted. However, as little to no research has been done on RAMD coverage technique, there is still a great need for further study in this area (analyzing each subsystem with the incorporation of coverage factor). According to the aforementioned literature, RAM analysis is a well-known technique for predicting a system's production availability by examining the causes, modes, and consequences of failure while also considering how these factors might affect output.

This study's objective is to give the findings of a thorough evaluation of the reliability, availability, maintainability, and dependability (RAMD) of a rice mill plant system, which includes a study of coverage factors under the assumption that a failure rate will occur. When the overall system failure rate is as low as possible, the system reliability can be as reliable as possible.

There are 7 sections in this article. An introduction and a few quick reviews that are necessary for this subject are included in the first Section. Section 17.2 discusses the resources and procedures. The system description is covered in Sect. 17.3. Section 17.4 provides an overview of the system's RAMD analysis findings. In Sect. 17.5, numerical simulation is discussed. Section 17.6 provided the discussion of the results, and Sect. 17.7 wrapped up the chapter.

## 17.2 Materials and Method

This section discusses the resources available for computing RAMD measurements for the model in question. Since all failure and repair rates are exponentially distributed and statistically independent during a steady-state period, all data utilized in this study are only accurate during that period.

### *17.2.1 Reliability Function*

The probability that a system/machine will function throughout a period of time $t$ is defined as reliability. Reliability can be expressed mathematically as $R(t) = P_r\{T > t\}$, where $T$ is a continuous random variable that denotes the time of failure of system with $R(t) \geq 0$, $R(t) = 1$. (For a full description, see Ebeling 2000). In terms of failure rate, a component's reliability can be represented as follows:

$$R(t) = \int_{t}^{\infty} f(t_0)\mathrm{d}t_0. \tag{17.1}$$

For a component with an exponentially distributed failure rate, Eq. (17.1) is reduced to:

$$R(t) = \mathrm{e}^{-\lambda t}. \tag{17.2}$$

### 17.2.2 Availability Function

According to Ebeling (2000), availability is the likelihood that a component will perform its required function at a specific moment when used under a set of operational circumstances. There are three types of availability: steady-state, interval, and point availability. It is written as follows mathematically:

$$(t) = \lim A(T) = \frac{\text{MTBF}}{\text{MTBF} + \text{MTTR}} \tag{17.3}$$

### 17.2.3 Maintainability

Ebeling (2000) defined system maintainability as the likelihood that a failing component will be repaired or returned to a specific condition within a predetermined amount of time depending on the necessary method. Mathematically, system maintainability is stated as follows:

$$M(t) = P(T \leq t) = 1 - \mathrm{e}^{\left(\frac{-t}{\text{MTTR}}\right)} = 1 - \mathrm{e}^{-\mu t}. \tag{17.4}$$

where $\mu$ is the constant system's repair rate.

### 17.2.4 Dependability

Dependability is a metric that measures how consistently a system performs, and it is practically synonymous with operational availability (View Aggarwal 2007). Dependability was first highlighted by Wohl as a design requirement in 1966. Dependability has the benefit of making it possible to compare costs, reliability, and maintainability. The following is the dependability ratio for randomly distributed variables with exponential distribution:

$$d = \frac{\mu}{\theta} = \frac{\text{MTBF}}{\text{MTTR}} \tag{17.5}$$

The high dependability ratio score reflects the significance of maintenance. The value of dependability increases when availability exceeds 0.9 and decreases when availability is less than 0.1. The minimum value of dependability is determined using the formula below:

$$D_{\text{min}} = 1 - \left(\frac{1}{d-1}\right)\left(\mathrm{e}^{-\text{Ind}/d-1} - \mathrm{e}^{d\text{Ind}/d-1}\right) \tag{17.6}$$

### 17.2.5 MTBF

The MTBF stands for the mean time between failures. Hours are typically used to express it. The system becomes more reliable as the MTBF rises. The following is the MTBF for an exponentially distributed system:

$$\text{MTBF} = \int_0^\infty R(t)\mathrm{d}t = \int_0^\infty \mathrm{e}^{-\theta t}\mathrm{d}t = \frac{1}{\theta}. \quad (17.7)$$

### 17.2.6 MTTR

The reciprocal of the system repair rate is specified as MTTR. It is mathematically expressed as follows:

$$\text{MTTR} = \frac{1}{\mu}. \quad (17.8)$$

where $\mu$ is the system's repair rate.

### 17.2.7 Exponential Distribution

A random variable $X$ is said to obey an exponential distribution with parameter $\theta > 0$, if its probability density function is given by:

$$f(x,\theta) = \begin{cases} \theta \mathrm{e}^{-\theta x}, & \text{if } x \geq 0 \\ 0, & \text{otherwise} \end{cases} \quad (17.9)$$

### 17.2.8 Constant Failure Rate

The constant hazard rate function can be written as follows:

$$f(t,\theta) = \begin{cases} \theta \mathrm{e}^{-\theta t}, & \text{if } t \geq 0 \\ 0, & \text{otherwise} \end{cases} \quad (17.10)$$

where $\theta$ is constant with probability density function, with $F(t) = 1 - \mathrm{e}^{-\theta t}$ and $R(t) = \mathrm{e}^{-\theta t}$.

### 17.2.9 Notations

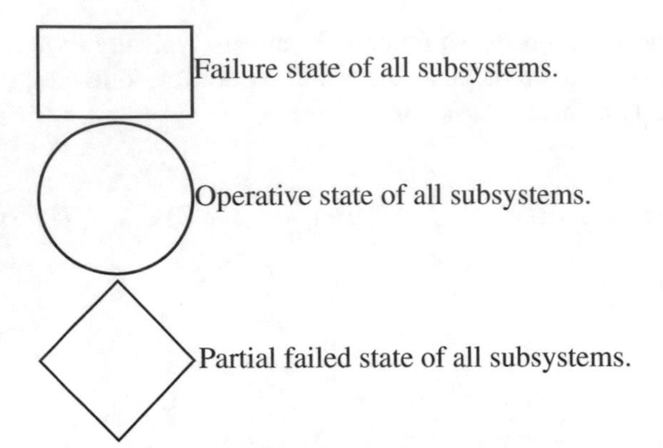


$G$, $J$, and $T$: Denote the conditions in which a subsystem is performing at its best.
$H$, $I$, $K$, $L$, $M$ and $V$: Represent the states of the subsystems that are working less.
$P$, $N$ and $r$: Represent the failure states of subsystem A, B, and C, respectively.
$c$: Coverage factor.
$\delta_i$, $i = 1, 2, 3$: Rate of failure of subsystems A, B, and C, respectively.
$\eta_i$, $i = 1, 2, 3$: Rate of repair of subsystems A, B, and C, respectively.
$P_0(t)$: Probability that the system is operating at maximum capacity when it starts up.
$P_i$; $i = 0, 1, 2, 3$: Steady-state probability that the system is in $i$th state.

## 17.3 System Description

The distributed parallel system studied in this chapter consists of three distinct subsystems:

### 17.3.1 Description

Subsystem A: comprises three active units. Two units must be operational for the system to function. The capacity of subsystem A is reduced when one of its units malfunctions.

Subsystem B: This subsystem consists of four active components. For the system to function, at least two units must be operational. Reduced system capacity results from a system unit failing.

Subsystem C: The two active servers in this subsystem are arranged in parallel. The system performs at a reduced capacity when one of the two active units in this subsystem malfunctions. While the failure of the two units' causes the system to fail completely.

### *17.3.2 Objectives*

1. To componently analyze the system reliability metrics.
2. To identify the most important subsystem.
3. To look for potential solutions for escaping dangerous situations.
4. To demonstrate the effectiveness of a covered fault system.

### *17.3.3 Assumption*

1. Moving from standby to operation is ideal.
2. The restored item functions flawlessly.
3. The distribution of failure and repair time is thought to be exponential.
4. Except in a complete failed state, a unit's failure has no impact on the system's operation (Fig. 17.1).

## 17.4 RAMD Analysis of the System

In order to mathematically model a rice mill plant system, Chapman-Kolmogorov differential equations have been constructed for each subsystem utilizing the Markov birth–death process. Using the nomenclature from Sect. 17.2.1 above, Figs. 17.2,

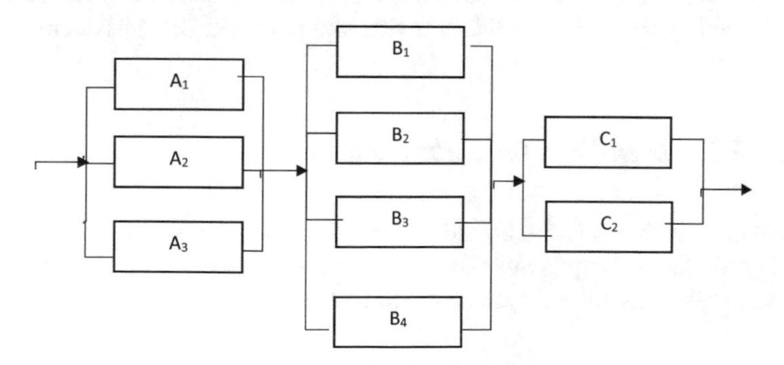


**Fig. 17.1** Distributed parallel system

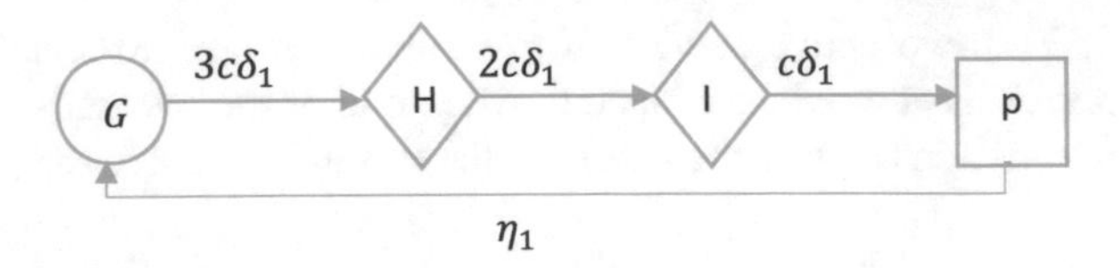


Fig. 17.2 Transition diagram of subsystem A

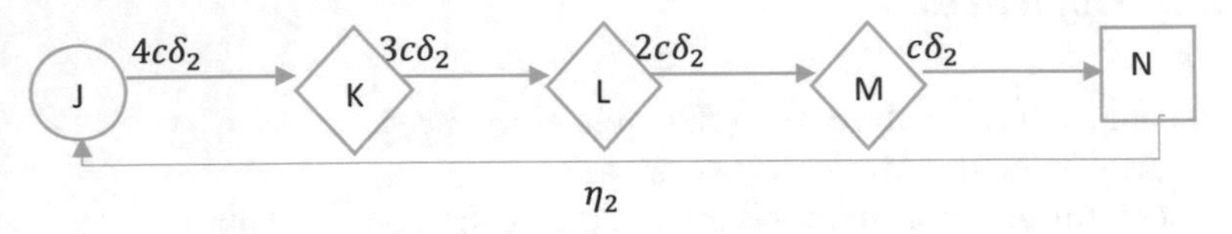


Fig. 17.3 Transition diagram of subsystem B

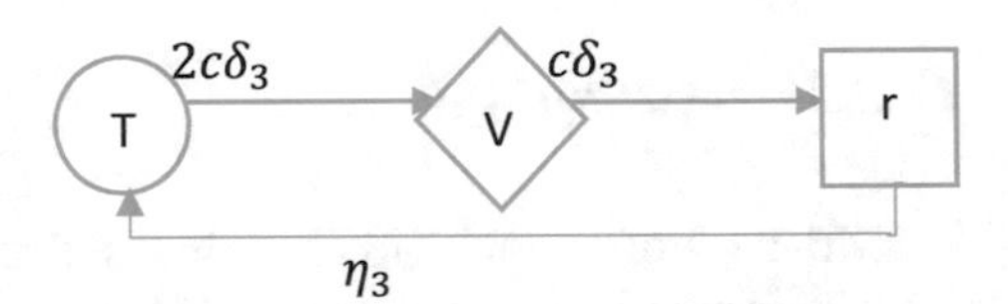


Fig. 17.4 Transition diagram of subsystem C

Table 17.1 Subsystem failure and repair rates in the system of the rice mill plant

| Subsystem | Failure rate | Repair rate |
|---|---|---|
| SSA | $\delta_1 = 0.005$ | $\eta_1 = 0.082$ |
| SSB | $\delta_2 = 0.002$ | $\eta_2 = 0.35$ |
| SSC | $\delta_3 = 0.0015$ | $\eta_3 = 0.45$ |

17.3, and 17.4 provide transition diagrams for each of the three subsystems. By solving the relevant Chapman-Kolmogorov differential equations in a steady state while simultaneously applying normalization conditions, system performance measures such as availability, reliability, maintainability, and dependability have been developed. Table 17.1 shows various subsystem maintenance and failure rates.

The following are RAMD indices for rice mill plant (RMP) subsystems:

### *17.4.1 RAMD Indices for Subsystem A*

There are three active units in this subsystem. The subsystem as a whole fails when any one of the three components fails since they all have the same failure rate. The differential equations for subsystem A are represented in Fig. 17.2 and are given below:

$$P_0^1(t) = -3c\delta_1 P_0 + \eta_1 P_3, \tag{17.11}$$

$$P_1^1(t) = -(2c\delta_1)P_1 + 3c\delta_1 P_0, \tag{17.12}$$

$$P_2^1(t) = -(c\delta_1)P_2 + 2c\delta_1 P_1, \tag{17.13}$$

$$P_3^1(t) = -\eta_1 P_3 + c\delta_1 P_2. \tag{17.14}$$

Using the initial circumstances and taking $t \to \infty$, Eqs. (17.11) through (17.14) can be simplified to the following

$$-3c\delta_1 P_0 + \eta_1 P_3 = 0, \tag{17.15}$$

$$-(2c\delta_1)P_1 + 3c\delta_1 P_0 = 0, \tag{17.16}$$

$$-(c\delta_1)P_2 + 2c\delta_1 P_1 = 0, \tag{17.17}$$

$$-\eta_1 P_3 + c\delta_1 P_2 = 0. \tag{17.18}$$

Using the normalization condition ($P_0 + P_1 + P_2 + P_3 = 1$) and recursively solving Eqs. (17.15) through (17.18), we have as follows:

$$P_0 = \frac{2\eta_1}{11\eta_1 + 6c\delta_1},\ P_1 = \frac{3\eta_1}{11\eta_1 + 6c\delta_1},\ P_2 = \frac{6\eta_1}{11\eta_1 + 6c\delta_1},\ \text{and } P_3 = \frac{6c\delta_1}{11\eta_1 + 6c\delta_1}.$$

The steady-state availability is now calculated as the product of all functioning state probability as follows:

$$\mathrm{Av}_{\mathrm{SSA}}(t) = P_0 + P_1 + P_2. \tag{17.19}$$

Thus, we have the availability of subsystem A as follows:

$$\mathrm{Av}_{\mathrm{SSA}}(t) = \frac{11\eta_1}{11\eta_1 + 6c\delta_1} = \frac{0.902}{0.902 + 0.030c} \tag{17.20}$$

1. The system's accessibility when coverage is disregarded;

$$\mathrm{Av}_{\mathrm{SSA}}(t) = \frac{11\eta_1}{11\eta_1 + 6\delta_1} = 0.9678111588$$

2. The availability of the system with coverage;

$$\mathrm{Av}_{\mathrm{SSA}}(t) = \frac{11\eta_1}{11\eta_1 + 6c\delta_1} = 0.9933920705 \text{ at } c = 0.2$$

$$\mathrm{Av}_{\mathrm{SSA}}(t) = \frac{11\eta_1}{11\eta_1 + 6c\delta_1} = 0.9868708972 \text{ at } c = 0.4$$

$$\mathrm{Av}_{\mathrm{SSA}}(t) = \frac{11\eta_1}{11\eta_1 + 6c\delta_1} = 0.9804347826 \text{ at } c = 0.6$$

Equation provides the system's reliability information (17.1). Equation (17.1) is reduced to for a component with an exponentially distributed failure rate:

$$R(t) = \mathrm{e}^{-\theta t}. \tag{17.21}$$

As a result, subsystem A's reliability is determined as follows:

$$R_{\mathrm{SSA}}(t) = \mathrm{e}^{-0.005t}. \tag{17.22}$$

The maintainability of the system is determined by Eq. (17.4).
Equation (17.23) following so presents the maintainability of subsystem A.

$$M_{\mathrm{SSA}}(t) = 1 - \mathrm{e}^{-0.082t}. \tag{17.23}$$

Other subsystem A performance indicators are listed below using Eqs. (17.4), (17.5), (17.6), (17.7), and (17.8):

$$\mathrm{MTBF} = 200\,\mathrm{h},\ \mathrm{MTTR} = 12.1951\,\mathrm{h},\ d = 16.4000,\ D_{\min(\mathrm{SSA})}(t) = 0.9395.$$

### 17.4.2 RAMD Indices for Subsystem B

This subsystem has four active components. The system will collapse as it did in subsystem A if any one of the four units—which all have the same rate of failure—fails. Figure 17.3 depicts the differential equations for subsystem B. They are as follows:

$$P_0^1(t) = -4c\delta_2 P_0 + \eta_2 P_4, \tag{17.24}$$

$$P_1^1(t) = -(3c\delta_2) P_1 + 4c\delta_2 P_0, \tag{17.25}$$

$$P_2^1(t) = -(2c\delta_2) P_2 + 3c\delta_2 P_1, \tag{17.26}$$

$$P_3^1(t) = -c\delta_2 P_3 + 2c\delta_2 P_2. \tag{17.27}$$

$$P_4^1(t) = -\eta_2 P_4 + c\delta_2 P_3. \tag{17.28}$$

Equations (17.24) through (17.27), when combined with the beginning circumstances and the value of $t \to \infty$, result in the following

$$-4c\delta_2 P_0 + \eta_2 P_4 = 0, \tag{17.29}$$

$$-(3c\delta_2)P_1 + 4c\delta_2 P_0 = 0, \tag{17.30}$$

$$-(2c\delta_2)P_2 + 3c\delta_2 P_1 = 0, \tag{17.31}$$

$$-c\delta_2 P_3 + 2c\delta_2 P_2 = 0. \tag{17.32}$$

$$-\eta_2 P_4 + c\delta_2 P_3 = 0 \tag{17.33}$$

Solving Eqs. (17.26)–(17.30) recursively and using normalizing condition (i.e., $P_0 + P_1 + P_2 + P_3 = 1$), we get:

$$P_0 = \frac{3\eta_2}{25\eta_2 + 12c\delta_2},\ P_1 = \frac{4}{3}P_0,\ P_2 = 2P_0,\ P_3 = 4P_0 \text{ and } P_4 = \frac{4c\delta_2}{\eta_2}P_0.$$

After adding up all of the working state probabilities, the steady-state availability is calculated as follows:

$$\mathrm{Av}_{\mathrm{SSB}}(t) = P_0 + P_1 + P_2 + P_3. \tag{17.34}$$

Thus, we have the availability of subsystem B as follows:

$$\mathrm{Av}_{\mathrm{SSB}}(t) = \frac{25\eta_2}{25\eta_2 + 12c\delta_2} = \frac{8.75}{8.75 + 0.0024c} \tag{17.35}$$

1. If coverage is disregarded, the system's availability is as follows:

$$\mathrm{Av}_{\mathrm{SSB}}(t) = \frac{25\eta_2}{25\eta_2 + 12\delta_2} = 0.9972646451$$

2. The system's coverage and accessibility;

$$\mathrm{Av}_{\mathrm{SSB}}(t) = \frac{25\eta_2}{25\eta_2 + 12c\delta_2} = 0.9994517295 \text{ at } c = 0.2$$

$$\mathrm{Av_{SSB}}(t) = \frac{25\eta_2}{25\eta_2 + 12c\delta_2} = 0.9989040591 \text{ at } c = 0.4$$

$$\mathrm{Av_{SSB}}(t) = \frac{25\eta_2}{25\eta_2 + 12c\delta_2} = 0.9983569899 \text{ at } c = 0.6$$

Equation provides the system's reliability information (17.1). Equation (17.1) is reduced to for a component with an exponentially distributed failure rate:

$$R(t) = \mathrm{e}^{-\theta t}. \tag{17.36}$$

The subsystem B's reliability is calculated as follows:

$$R_{\mathrm{SSB}}(t) = \mathrm{e}^{-0.002t}. \tag{17.37}$$

Equation (17.4) estimates the system's maintainability.

Therefore, Eq. (17.36) below presents the maintainability of subsystem B.

$$M_{\mathrm{SSB}}(t) = 1 - \mathrm{e}^{-0.35t}. \tag{17.38}$$

Using Eqs. (17.4), (17.5), (17.6), (17.7), and (17.8), other performance indicators of subsystem A are given below:

$$\mathrm{MTBF} = 500\,\mathrm{h},\ \mathrm{MTTR} = 2.8571\,\mathrm{h},\ d = 175.0026,\ D_{\min(B)}(t) = 0.9395.$$

### 17.4.3 RAMD Indices for Subsystem C

Two active units are connected in parallel by this subsystem's two active units. The subsystem's capability is reduced when one of its active units malfunctions. On the other hand, when the two components fail, the system as a whole fails. The differential equations for subsystem C are displayed in Fig. 17.4 and are as follows:

$$P_0^1(t) = -2c\delta_3 P_0 + \eta_3 P_2, \tag{17.39}$$

$$P_1^1(t) = -(c\delta_3)P_1 + 2c\delta_3 P_0, \tag{17.40}$$

$$P_2^1(t) = -\eta_3 P_2 + c\delta_3 P_1. \tag{17.41}$$

Using the initial circumstances and taking $t \to \infty$, Eqs. (17.37) through (17.39) can be simplified to the following

$$0 = -2c\delta_3 P_0 + \eta_3 P_2, \tag{17.42}$$

$$0 = -(c\delta_3)P_1 + 2c\delta_3 P_0, \tag{17.43}$$

$$0 = -\eta_3 P_2 + c\delta_3 P_1, \tag{17.44}$$

Using the normalization condition (i.e., $P_0 + P_1 + P_2 = 1$) and recursively solving Eqs. (17.40) through (17.42), we arrive at:

$$P_0 = \frac{\eta_3}{3\eta_3 + 2c\delta_3},\ P_1 = 2P_0,\ \text{and}\ P_2 = \frac{2c\delta_3}{\eta_3} P_0$$

Now, all of the working state probabilities are added together to determine the steady-state availability of subsystem C as follows:

$$\mathrm{Av}_{\mathrm{SSC}}(t) = P_0 + P_1. \tag{17.45}$$

As a result, we obtain subsystem C's availability as:

$$\mathrm{Av}_{\mathrm{SSC}}(t) = \frac{3\eta_3}{3\eta_3 + 2\delta_3} = \frac{1.35}{1.35 + 0.003}. \tag{17.46}$$

1. The system's availability if coverage is disregarded;

$$\mathrm{Av}_{\mathrm{SSC}}(t) = \frac{3\eta_3}{3\eta_3 + 2\delta_3} = 0.9977827051$$

2. The availability of the system with coverage;

$$\mathrm{Av}_{\mathrm{SSC}}(t) = \frac{3\eta_3}{3\eta_3 + 2c\delta_3} = 0.9995557530 \text{ at } c = 0.2$$

$$\mathrm{Av}_{\mathrm{SSC}}(t) = \frac{3\eta_3}{3\eta_3 + 2c\delta_3} = 0.9991119006 \text{ at } c = 0.4$$

$$\mathrm{Av}_{\mathrm{SSC}}(t) = \frac{3\eta_3}{3\eta_3 + 2c\delta_3} = 0.9986684420 \text{ at } c = 0.6$$

Equation provides the system's reliability information (17.1). Equation (17.1) is simplified to: for a component with an exponentially distributed failure rate.

$$R(t) = \mathrm{e}^{-\theta t}. \tag{17.47}$$

The reliability of subsystem C is thus determined as follows:

$$R_{\text{SSC}}(t) = \mathrm{e}^{-0.0015t}. \tag{17.48}$$

The maintainability of the system is determined by Eq. (17.4). Thus, Eq. (17.47) below presents the maintainability of subsystem C.

$$M_{\text{SSC}}(t) = 1 - \mathrm{e}^{-0.45t}. \tag{17.49}$$

The following list of additional subsystem C performance indicators is based on Eqs. (17.4), (17.5), (17.6), (17.7), and (17.8):

$$\text{MTBF} = 666.6666\,\text{h},\ \text{MTTR} = 2.2222\,\text{h},\ d = 300.0029,\ D_{\min(\text{SSC})}(t) = 0.9968.$$

## 17.5 Numerical Simulation

Numerical simulations of the reliability metrics are discussed in this section.

(a) **System reliability**

In a serial order, all four subsystems are linked. The system will fail completely if one or more components fails. The entire system reliability is calculated as follows

$$\begin{aligned}
R_{\text{sys}}(t) &= R_{\text{SSA}}(t) \times R_{\text{SSB}}(t) \times R_{\text{SSC}}(t),\\
R_{\text{sys}}(t) &= \mathrm{e}^{-0.005t} \times \mathrm{e}^{-0.002t} \times \mathrm{e}^{-0.0015t},\\
R_{\text{sys}}(t) &= \mathrm{e}^{-(0.005+0.002+0.0015)t},\\
R_{\text{sys}}(t) &= \mathrm{e}^{-(0.0085)t}.
\end{aligned} \tag{17.50}$$

(b) **System availability**

Each of the four subsystems is linked to the others in a sequential manner. A single failure causes the entire system to fail. The following formula gives the overall system availability:

$$\text{Av}_{\text{sys}}(t) = \text{Av}_{\text{SSA}}(t) \times \text{Av}_{\text{SSB}}(t) \times \text{Av}_{\text{SSC}}(t),$$

1. The availability of the system when coverage is ignored;

$$\text{Av}_{\text{sys}}(t) = 0.9678111584 \times 0.9972646451 \times 0.9977827051 = 0.9630237936$$

2. The availability of the system with coverage;

**Table 17.2** RAMD indices for distributed system

| RAMD indices of subsystems | Subsystem A | Subsystem B | Subsystem C |
|---|---|---|---|
| Availability | $\frac{0.164}{0.902+0.030c}$ | $\frac{8.75}{8.75+0.0024c}$ | $\frac{1.35}{1.35+0.003}$ |
| Reliability | $e^{-0.005t}$ | $e^{-0.002t}$ | $e^{-0.0015t}$ |
| Maintainability | $1-e^{-0.082t}$ | $1-e^{-0.35t}$ | $1-e^{-0.45t}$ |
| Dependability | 0.9395 | 0.9395 | 0.9968 |

$$\frac{11\eta_1}{11\eta_1+6c\delta_1}+\frac{25\eta_2}{25\eta_2+12\delta_2}+\frac{3\eta_2}{3\eta_2+2c\delta_2}=\frac{0.902}{0.902+0.030c}+\frac{8.75}{8.75+0.024c}+\frac{1.35}{1.35+0.0030c}$$

(c) **System maintainability**

All four subsystems are linked in a sequential manner. One failure causes the entire system to fail. The following formula calculates the overall system maintainability:

$$M_{\text{sys}}(t)=M_{\text{SSA}}(t)\times M_{\text{SSB}}(t)\times M_{\text{SSC}}(t),$$
$$M_{\text{sys}}(t)=\left(1-e^{-0.082t}\right)\times\left(1-e^{-0.35t}\right)\times\left(1-e^{-0.45t}\right). \quad (17.51)$$

(d) **System dependability**

Each of the four subsystems is linked to the others in a sequence. When one part of the system fails, the entire system fails. The following criteria are used to assess overall system dependability:

$$D_{\text{min(sys)}}(t)=D_{\text{min(SSA)}}(t)\times D_{\text{min(SSB)}}(t)\times D_{\text{min(SSC)}}(t),$$
$$D_{\text{min(sys)}}(t)=0.9395\times0.9395\times0.9968=0.8798. \quad (17.52)$$

The result summary of RAMD indices is presented below (Table 17.2 and Figs. 17.5, 17.6):

Table 17.7 shows the variation in each system's ability to be maintained over time.

## 17.6 Result Discussion

Tables 17.3, 17.4 and 17.5 and the figures that go with it make it abundantly evident that the system's availability reduces gradually and eventually becomes constant, that is, that it declines as the coverage variable increases. When no coverage approach is used and the repair rate follows an exponential distribution for one type of repair between two transition states of the system, the availability of the subsystems is

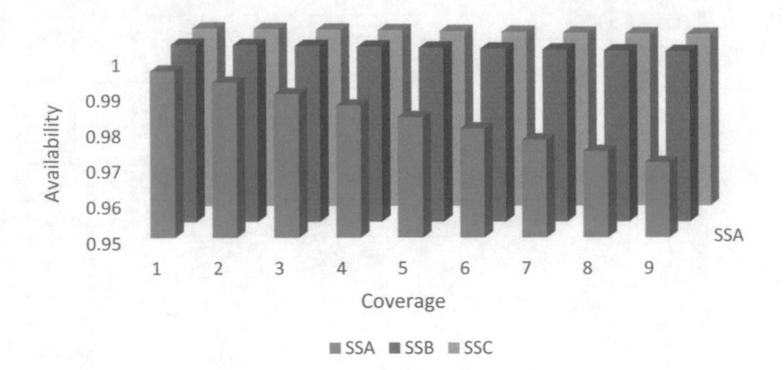


**Fig. 17.5** Subsystem's availability against coverage ($c$)

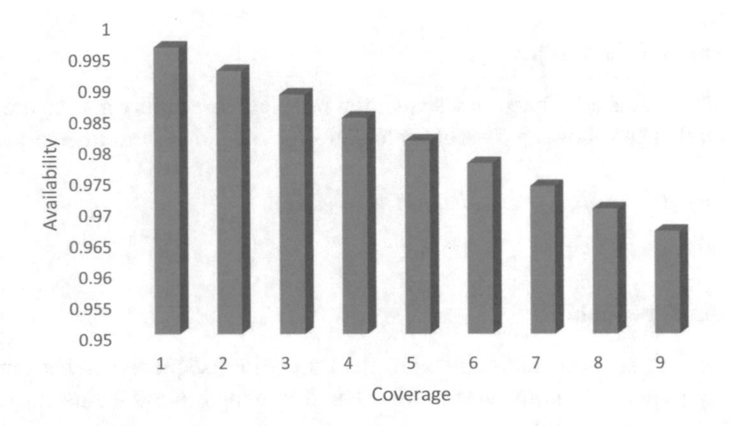


**Fig. 17.6** System availability against coverage ($c$)

0.9678111584, 0.9972646451, and 0.9977827051 for subsystems A, B, and C, respectively. While the system's availability when the subsystems are connected serially is 0.9630237936, this figure represents the system's availability when no coverage approach is used and the repair rate exhibits exponential distribution. When the repair rate follows an exponential distribution and the coverage factor varies, Table 17.4 displays the system's availability. Looking closely at Table 17.3 and the associated figure, it is evident that each subsystem's availability is maximum at a coverage value of 0.2 and lowest at a coverage value of 0.9. The availability value for each subsystem is 0.9666123428. The system's availability drops by up to 0.9630 when there is no coverage. This illustrates how beneficial a covered fault system is to system performance.

Another point is that, despite a rise in failure rates, subsystem C is the one that is most reliable. Subsystems A and B are therefore the most important and sensitive parts of the system because they have the highest failure rates. This is seen in Tables 17.6, 17.7, 17.8, 17.9, 17.10, 17.11, 17.12, and 17.13 as well as their Figs. 17.7,

**Table 17.3** Variation in subsystems availability with coverage

| $c$ | Subsystem A | Subsystem B | Subsystem C |
|---|---|---|---|
| 0.1 | 0.9966850832 | 0.9997257891 | 0.9997778272 |
| 0.2 | 0.9933920706 | 0.9994517295 | 0.9995557530 |
| 0.3 | 0.9901207466 | 0.9991778195 | 0.9993337775 |
| 0.4 | 0.9868708974 | 0.9989040591 | 0.9991119006 |
| 0.5 | 0.9836423120 | 0.9986304492 | 0.9988901222 |
| 0.6 | 0.9804347828 | 0.9983569899 | 0.9986684420 |
| 0.7 | 0.9772481043 | 0.9980836792 | 0.9984468604 |
| 0.8 | 0.9740820735 | 0.9978105182 | 0.9982253772 |
| 0.9 | 0.9709364910 | 0.9975375078 | 0.9980039920 |

**Table 17.4** Variation of systems availability with coverage

| $c$ | $Av_{sys}(t)$ |
|---|---|
| 0.1 | 0.9961904057 |
| 0.2 | 0.9924063535 |
| 0.3 | 0.9886475902 |
| 0.4 | 0.9849138663 |
| 0.5 | 0.9812049363 |
| 0.6 | 0.9775205577 |
| 0.7 | 0.9738604893 |
| 0.8 | 0.9702244950 |
| 0.9 | 0.9666123428 |

**Table 17.5** Variation of subsystem availability with no coverage

| | Av with no coverage |
|---|---|
| Subsystem A | 0.1759656652 |
| Subsystem B | 0.9972646451 |
| Subsystem C | 0.9977827051 |

17.8, 17.9, 17.10, 17.11, 17.12, 17.13, and 17.14. Table 17.6 and Fig. 17.7 make it abundantly evident that subsystem C, which has the lowest failure rate among the four and offers the highest level of system reliability. This sensitivity analysis shows that when the total system failure rate is low and the supporting units have been activated, the optimum system dependability can be attained. In order to increase the reliability of the system, efficient maintenance solutions should be developed, and redundant procedures may be employed.

According to the surface plots, tables, and figures, one of the benefits of the coverage factor "$C$" for an industrial system is that it aids in enhancing system dependability, which in turn raises the system's output capacity. Additionally, it is

**Table 17.6** Subsystem reliability variation over time

| Time ($t$) in (days) | $R_{SSA}(t)$ | $R_{SSB}(t)$ | $R_{SSC}(t)$ | $R_{sys}(t)$ |
|---|---|---|---|---|
| 0 | 1.0000 | 1.0000 | 1.0000 | 1.0000 |
| 20 | 0.9048 | 0.9608 | 0.9704 | 0.8437 |
| 30 | 0.8607 | 0.9418 | 0.9559 | 0.7749 |
| 40 | 0.8187 | 0.9231 | 0.9418 | 0.7118 |
| 50 | 0.7788 | 0.9048 | 0.9277 | 0.6537 |
| 60 | 0.7408 | 0.8869 | 0.9139 | 0.6005 |
| 70 | 0.7047 | 0.8694 | 0.9003 | 0.5516 |
| 80 | 0.6703 | 0.8521 | 0.8869 | 0.5066 |
| 90 | 0.6376 | 0.8353 | 0.8737 | 0.4653 |

**Table 17.7** Change of subsystems' maintainability has over time

| Time ($t$) in (days) | $M_{SSA}(t)$ | $M_{SSB}(t)$ | $M_{SSC}(t)$ | $M_{sys}(t)$ |
|---|---|---|---|---|
| 0 | 0.0000 | 0.0000 | 0.0000 | 0.0000 |
| 20 | 0.8060 | 0.9991 | 0.9999 | 1.0000 |
| 30 | 0.9146 | 0.9999 | 0.9999 | 1.0000 |
| 40 | 0.9624 | 0.9999 | 0.9999 | 1.0000 |
| 50 | 0.9834 | 0.9999 | 0.9999 | 1.0000 |
| 60 | 0.9927 | 0.9999 | 1.0000 | 1.0000 |
| 70 | 0.9968 | 1.0000 | 1.0000 | 1.0000 |
| 80 | 0.9986 | 1.0000 | 1.0000 | 1.0000 |
| 90 | 0.9994 | 1.0000 | 1.0000 | 1.0000 |

**Table 17.8** Variation in subsystem A's failure rate as a function of its reliability

| Time ($t$) in (days) | $\delta_1 = 0.005$ | $\delta_1 = 0.006$ | $\delta_1 = 0.007$ | $\delta_1 = 0.008$ |
|---|---|---|---|---|
| 0 | 1.0000 | 1.0000 | 1.0000 | 1.0000 |
| 20 | 0.9048 | 0.8869 | 0.8694 | 0.8521 |
| 30 | 0.8607 | 0.8353 | 0.8106 | 0.7866 |
| 40 | 0.8187 | 0.7866 | 0.7558 | 0.7261 |
| 50 | 0.7788 | 0.7408 | 0.7047 | 0.6703 |
| 60 | 0.7408 | 0.6977 | 0.6570 | 0.6188 |
| 70 | 0.7047 | 0.6570 | 0.6126 | 0.5712 |
| 80 | 0.6703 | 0.6188 | 0.5712 | 0.5273 |
| 90 | 0.6376 | 0.5827 | 0.5326 | 0.4868 |

**Table 17.9** Subsystem B's failure rate varies with its reliability

| Time ($t$) in (days) | $\delta_2 = 0.002$ | $\delta_2 = 0.003$ | $\delta_2 = 0.004$ | $\delta_2 = 0.005$ |
|---|---|---|---|---|
| 0 | 1.0000 | 1.0000 | 1.0000 | 1.0000 |
| 20 | 0.9608 | 0.9418 | 0.9231 | 0.9048 |
| 30 | 0.9418 | 0.9139 | 0.8869 | 0.8607 |
| 40 | 0.9231 | 0.8869 | 0.8521 | 0.8187 |
| 50 | 0.9048 | 0.8607 | 0.8187 | 0.7788 |
| 60 | 0.8869 | 0.8353 | 0.7866 | 0.7408 |
| 70 | 0.8694 | 0.8106 | 0.7558 | 0.7047 |
| 80 | 0.8521 | 0.7866 | 0.7261 | 0.6703 |
| 90 | 0.8353 | 0.7634 | 0.6977 | 0.6376 |

**Table 17.10** Subsystem C's Failure rate and reliability variation

| Time ($t$) in (days) | $\delta_3 = 0.0015$ | $\delta_3 = 0.0016$ | $\delta_3 = 0.0017$ | $\delta_3 = 0.0018$ |
|---|---|---|---|---|
| 0 | 1.0000 | 1.0000 | 1.0000 | 1.0000 |
| 20 | 0.9704 | 0.9685 | 0.9666 | 0.9646 |
| 30 | 0.9559 | 0.9531 | 0.9503 | 0.9474 |
| 40 | 0.9418 | 0.9380 | 0.9343 | 0.9305 |
| 50 | 0.9277 | 0.9231 | 0.9185 | 0.9139 |
| 60 | 0.9139 | 0.9085 | 0.9030 | 0.8976 |
| 70 | 0.9003 | 0.8940 | 0.8878 | 0.8816 |
| 80 | 0.8869 | 0.8799 | 0.8728 | 0.8659 |
| 90 | 0.8737 | 0.8659 | 0.8581 | 0.8504 |

**Table 17.11** Variation in system reliability owing to subsystem A failure rate variation

| Time ($t$) in (days) | $\delta_1 = 0.005$ | $\delta_1 = 0.006$ | $\delta_1 = 0.007$ | $\delta_1 = 0.008$ |
|---|---|---|---|---|
| 0 | 1.0000 | 1.0000 | 1.0000 | 1.0000 |
| 20 | 0.4449 | 0.4360 | 0.4274 | 0.4190 |
| 30 | 0.2967 | 0.2879 | 0.2794 | 0.2712 |
| 40 | 0.1979 | 0.1901 | 0.1827 | 0.1755 |
| 50 | 0.1320 | 0.1256 | 0.1194 | 0.1136 |
| 60 | 0.0880 | 0.0829 | 0.0781 | 0.0735 |
| 70 | 0.0587 | 0.0547 | 0.0510 | 0.0476 |
| 80 | 0.0392 | 0.0362 | 0.0334 | 0.0308 |
| 90 | 0.0261 | 0.0239 | 0.0218 | 0.0199 |

**Table 17.12** Variation in the systems reliability as a result of subsystem B's failure rate

| Time ($t$) in (days) | $\delta_2 = 0.002$ | $\delta_2 = 0.003$ | $\delta_2 = 0.004$ | $\delta_2 = 0.005$ |
|---|---|---|---|---|
| 0 | 1.0000 | 1.0000 | 1.0000 | 1.0000 |
| 20 | 0.4449 | 0.4360 | 0.4274 | 0.4190 |
| 30 | 0.2967 | 0.2879 | 0.2794 | 0.2712 |
| 40 | 0.1979 | 0.1901 | 0.1827 | 0.1755 |
| 50 | 0.1320 | 0.1256 | 0.1194 | 0.1136 |
| 60 | 0.0880 | 0.0829 | 0.0781 | 0.0735 |
| 70 | 0.0587 | 0.0547 | 0.0510 | 0.0476 |
| 80 | 0.0392 | 0.0362 | 0.0334 | 0.0308 |
| 90 | 0.0261 | 0.0239 | 0.0218 | 0.0199 |

**Table 17.13** Variance in system reliability owing to variation in subsystem C failure rate

| Time ($t$) in (days) | $\delta_3 = 0.0015$ | $\delta_3 = 0.0016$ | $\delta_3 = 0.0017$ | $\delta_3 = 0.0018$ |
|---|---|---|---|---|
| 0 | 1.0000 | 1.0000 | 1.0000 | 1.0000 |
| 20 | 0.4449 | 0.4440 | 0.4431 | 0.4422 |
| 30 | 0.2967 | 0.2958 | 0.2949 | 0.2941 |
| 40 | 0.1979 | 0.1971 | 0.1963 | 0.1955 |
| 50 | 0.1320 | 0.1313 | 0.1307 | 0.1300 |
| 60 | 0.0880 | 0.0875 | 0.0870 | 0.0865 |
| 70 | 0.0587 | 0.0583 | 0.0579 | 0.0575 |
| 80 | 0.0392 | 0.0389 | 0.0385 | 0.0382 |
| 90 | 0.0261 | 0.0259 | 0.0257 | 0.0254 |

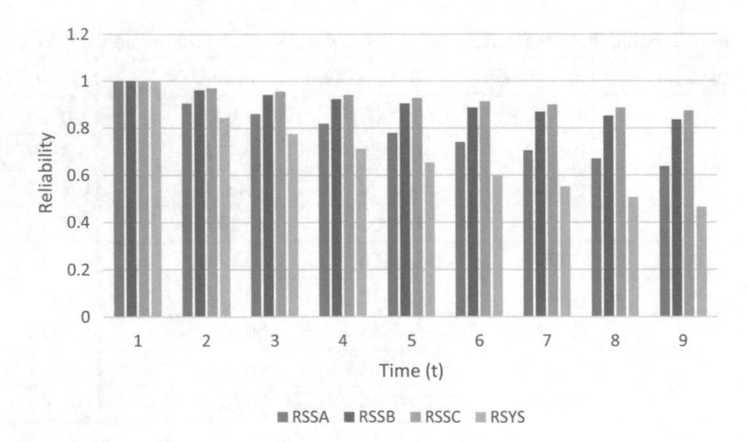


**Fig. 17.7** System reliability against time $t$

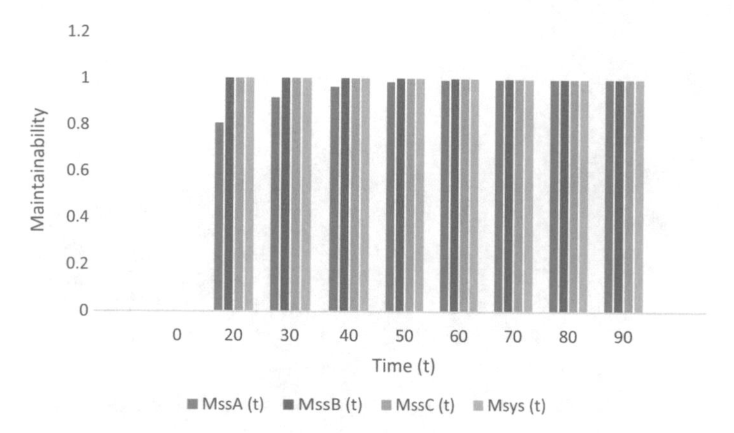


**Fig. 17.8** System maintainability against time $t$

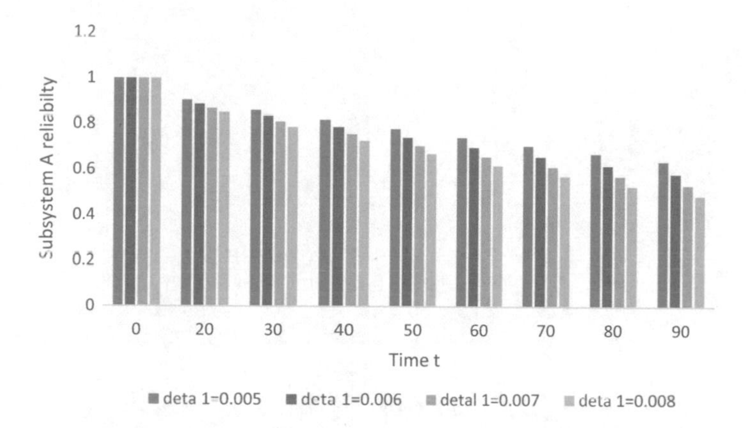


**Fig. 17.9** Effect of $\delta_1$ on the reliability of subsystem A

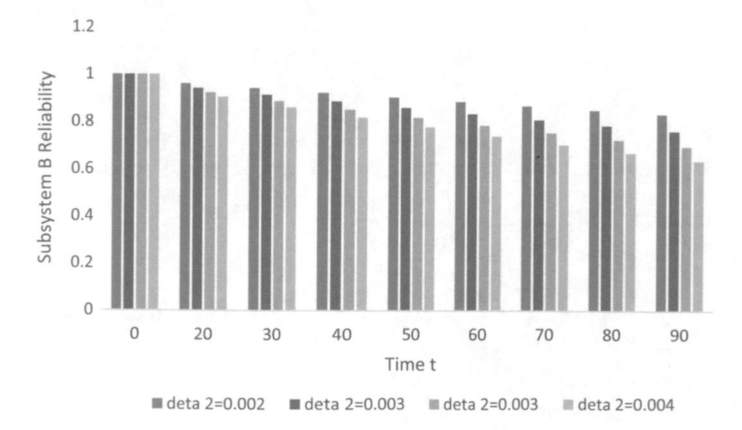


**Fig. 17.10** Effect of $\delta_2$ on subsystem B's dependability

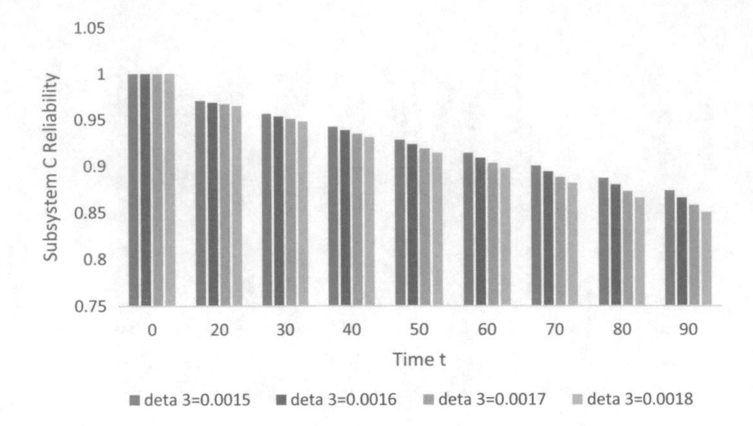


**Fig. 17.11** Effect of $\delta_3$ on the reliability of subsystem C

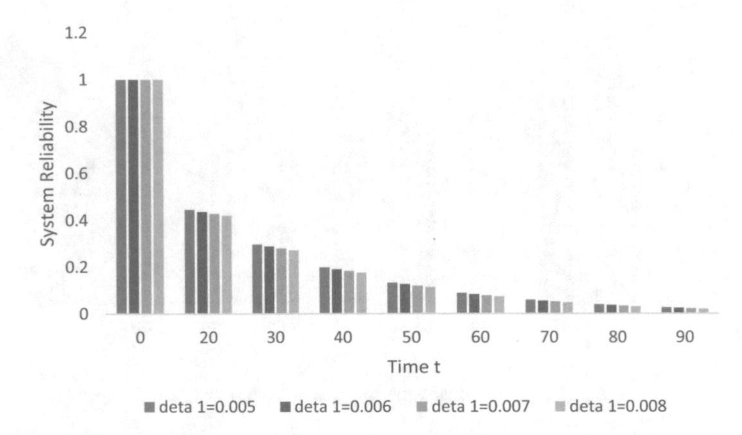


**Fig. 17.12** Impact of subsystem A's failure rate ($\delta_1$) on the system's reliability

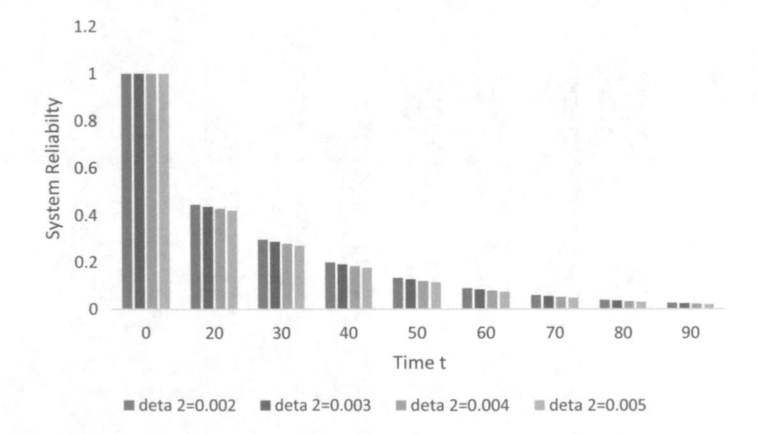


**Fig. 17.13** Impact of subsystem B's failure rate ($\delta_2$) on the system's reliability

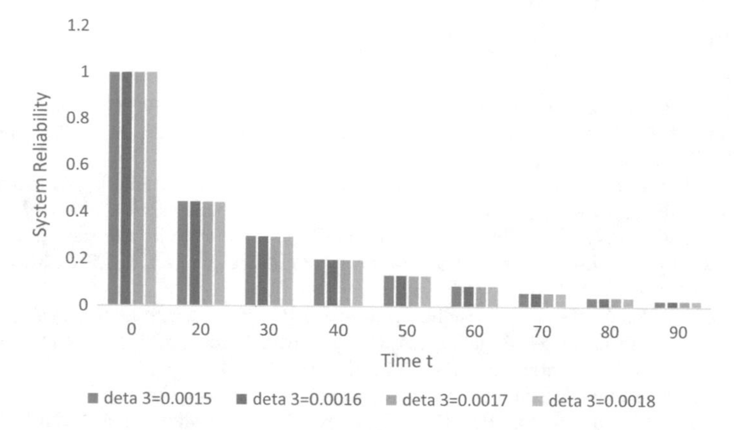


**Fig. 17.14** Impact of subsystem C's failure rate ($\delta_3$) on the system's reliability

clear that higher repair rates combined with reduced failure rates can increase availability and profit. Therefore, early unit failure repair, individual subsystem replacement, good maintenance planning to prevent catastrophic failure, and the addition of fault tolerant units/subsystems can all lead to higher system availability and income. The current work can be further expanded to include a system with several subsystems and multiple units that can be solved by human reliability analysis techniques.

## 17.7 Conclusion

In this article, the RAMD indices for each subsystem are carefully examined in order to evaluate the system's performance. Through numerical experiment, the fundamental formulations for RAMD measurements for each subsystem were discovered and validated. Table 17.1 lists the presumptive numbers for failure and repair rates for each subsystem. While Table 17.4 shows the variation of the systems' availability with coverage, Table 17.3 compiles all RAMD measurements of each subsystem's availability in relation to the coverage factor. Table 17.5 depicts each subsystem's availability without the coverage factor. The numerical results indicate that the presence of a coverage factor significantly improves the performance of the system. If updated, the models and results presented in this paper will enable management to avoid catastrophes and expensive maintenance costs that would otherwise necessitate unnecessary spending. In order to increase customer satisfaction and reduce failure rates, the adopted approach (RAMD analysis) for maintenance policies for the model under review could be proposed and put into practice. The results of the recent study are as follows.

## References


1. Tiyagi, V., Arora, R., Ram, M., Triantafyllou, I.S.: Copula based measures of repairable parallel system with fault coverage. Int. J. Math. Eng. Manag. Sci. **6**(1), 322–344 (2021). http://doi.org/10.33889/IJMEMS.2021.6.1.021
2. Corvaro, F., Giacchetta, G., Marchetti, B., Recanati, M.: Reliability, availability, maintainability (RAM) study, on reciprocating compressors API 618. Petroleum **3**, 266–272 (2017)
3. Goyal, D., Kumar, A., Saini, M., Joshi, H.: Reliability, maintainability and sensitivity analysis of physical processing unit of sewage treatment plant. SN Appl. Sci. **1**, 1507 (2019). https://doi.org/10.1007/s42452-019-1544-7
4. Velmurugan, K., Venkumar, P., Sudhakarapandian, R.: Performance analysis of tyre manufacturing system in the SMEs using RAMD approach. Math. Probl. Eng. **2021**(Article ID 6616037), 14 p (2021). http://doi.org/10.1155/2021/6616037
5. Aggarwal, A., Kumar, S., Singh, V.: Performance modeling of the skim milk powder production system of a dairy plant using RAMD analysis. Int. J. Qual. Reliab. Manag. **32**(2), 167–181 (2015)
6. Dahiya, O., Kumar, A., Saini, M.: Mathematical modeling and performance evaluation of A-Pan crystallization system in a sugar industry. SN Appl. Sci. (2019). https://doi.org/10.1007/s42452-019-0348-0
7. Tsarouhas, P.H., Arvanitoyannis, I.S.: Yogurt production line: reliability analysis. Prod. Manuf. Res. **2**(1), 11–23 (2014)
8. Aggarwal, A.K., Kumar, S., Singh, V.: Performance modeling of the serial processes in refining system of a sugar plant using RAMD analysis. Int. J. Syst. Assur. Eng. Manag. **8**(2), 1910–1922 (2017)
9. Niwas, R., Garg, H.: An approach for analyzing the reliability and profit of an industrial system based on the cost-free warranty policy. J. Braz. Soc. Mech. Sci. Eng. **40**, 1–9 (2018)
10. Aggarwal, A.K., Kumar, S., Singh, V.: Reliability and availability analysis of the serial processes in skim milk powder system of a dairy plant: a case study. Int. J. Ind. Syst. **22**(1), 36–62 (2016)
11. Kumar, A., Pant, S., Singh, S.B.: Availability and cost analysis of an engineering system involving subsystems in series configuration. Int. J. Qual. Reliab. Manag. **34**(6), 879–894 (2017)
12. De Sanctis, I., Paciarotti, C., Di Giovine, O.: Integration between RCM and RAM: a case study. Int. J. Qual. Reliab. Manag. **33**(6), 852–880 (2016)
13. Sharma, R.K., Kumar, S.: Performance modeling in critical engineering systems using RAM analysis. Reliab. Eng. Syst. Saf. **93**, 891–897 (2008)
14. Saini, M., Kumar, A.: Performance analysis of evaporation system in sugar industry using RAMD analysis. J. Braz. Soc. Mech. Sci. Eng. **41**, 4 (2019)
15. Garg, H., Rani, M., Sharma, S.P.: Performance analysis of repairable industrial systems using artificial bee colony and fuzzy methodology. Int. J. Artif. Intell. Tools **23**(5), 1450008 (23 p) (2014)
16. Tsarouhas, P., Arvanitoyannis, I., Varzakas, T.: Reliability and maintainability analysis of cheese (feta) production line in a Greek medium-size company: a case study. J. Food Eng. **94**(34), 233–240 (2009)
17. Malik, S., Tewari, P.C.: Performance modeling and maintenance priorities decision for water flow system of a coal based thermal power plant. Int. J. Qual. Reliab. Manag. **35**(4), 996–1010 (2018)
18. Yusuf, I.: Availability modeling and evaluation of repairable system subject to minor deterioration under imperfect repairs. Int. J. Math. Oper. Res. **7**, 42–51 (2015)
19. Garg, H., Rani, M.: An approach for reliability analysis of industrial systems using PSO and IFS techniques. ISA Trans. **52**(6), 701–710 (2013)
20. Barabady, J., Kumar, U.: Reliability analysis of mining equipment: a case study of crushing plant at Jajarm Bauxite Mine in Iran. Reliab. Eng. Syst. Saf. **93**(4), 647–653 (2008)
21. Garg, H.: Performance analysis of an industrial system using soft computing based hybridized technique. J. Braz. Soc. Mech. Sci. Eng. **39**(4), 1441–1451 (2017)

22. Sharma, K.R., Sharma, P.: Integrated framework to optimize RAM and cost decisions in a process plant. J. Loss Prev. Process Ind. **25**(6), 883–904 (2012)
23. Kumar, V., Mudgil, V.: Availability optimization of ice cream making unit of milk plant using genetic algorithm. Int. J. Res. Manag. Bus. Stud. **4**(3), 17–19 (2014)
24. Garg, H.: Reliability, availability and maintainability analysis of industrial system using PSO and fuzzy methodology. MAPAN J. Metrol. Soc. India **29**(2), 115–129 (2014)
25. Yusuf, I.: Reliability modeling of a parallel system with a supporting device and two types of preventive maintenance. Int. J. Oper. Res. **25**(3), 269–287 (2016)
26. Tewari, P.C., Khaduja, R., Gupta, M.: Performance enhancement for crystallization unit of a sugar plant using genetic algorithm technique. J. Ind. Eng. Int. **8**(1), 1–6 (2012)
27. Choudhary, D., Tripathi, M., Shankar, R.: Reliability, availability and maintainability analysis of a cement plant: a case study. Int. J. Qual. Reliab. Manag. (2019). https://doi.org/10.1108/IJQRM-10-2017-0215
28. Qiu, Q., Cui, L.: Reliability evaluation based on a dependent two-stage failure process with competing failures. Appl. Math. Model. **64**, 699–712 (2018)
29. Velmurugan, K., Venkumar, P., Sudhakarapandian, R.: Reliability availability maintainability analysis in forming industry. Int. J. Eng. Adv. Technol. **91S4**, 822–828 (2019)
30. Wang, Z., Wang, P.: A new approach for reliability analysis with time-variant performance characteristics. Reliab. Eng. Syst. Saf. (2013). http://doi.org/10.1016/j.ress.2013.02.017
31. Sanusi, A., Yusuf, I., Mamuda, B.Y.: Performance evaluation of an industrial configured as series-parallel system. J. Math. Comput. Sci. **10**, 692–712 (2020)

**Nafisatu Muhammad Usman** is a Lecturer in the Department of Art and Humanities, Kano State Polytechnic, Kano, Nigeria. She received her B.Sc. in Mathematics in 2010 and M.Sc. in Mathematics in 2018 from Bayero University, Kano, Nigeria. She is currently pursuing her Ph.D. in Mathematics at Department of Mathematical Sciences, Bayero University, Kano, Nigeria. She has reviewed papers from Reliability Theory and Application (RTA) and Life Cycle Reliability and Safety Engineering. Her research includes system reliability theory, maintenance and replacement and operation research.

**Ibrahim Yusuf** is a Lecturer in the Department of Mathematical Sciences, Bayero University, Kano, Nigeria. He received his B.Sc. in Mathematics in 1996, M.Sc. in Mathematics in 2007, and Ph.D. in Mathematics in 2014 from Bayero University, Kano, Nigeria. He is currently an Associate Professor at the Department of Mathematical Sciences, Bayero University, Kano, Nigeria. He has reviewed papers from IJSA, *JCCE*, Life Cycle Reliability and Safety Engineering, JRSS, IJRRS, Inderscience journals, IJQRM, and Operation Research and Decision. His research includes system reliability theory, maintenance and replacement and operation research.

Printed in the United States
by Baker & Taylor Publisher Services